AF324669

A Garden of Quanta

Essays in Honor of
Hiroshi Ezawa

A Garden of Quanta

Essays in Honor of
Hiroshi Ezawa

Editors

J Arafune
National Institute for Academic Degrees, Japan

A Arai
Hokkaido University, Japan

M Kobayashi
Accelerator Research Organization (KEK), Japan

K Nakamura
Meiji University, Japan

T Nakamura
Sundai Preparatory School, Japan

I Ojima
Kyoto University, Japan

N Sakai
Tokyo Institute of Technology, Japan

A Tonomura
Hitachi Ltd, Japan

K Watanabe
Meisei University, Japan

Published by

World Scientific Publishing Co. Pte. Ltd.

5 Toh Tuck Link, Singapore 596224

USA office: Suite 202, 1060 Main Street, River Edge, NJ 07661

UK office: 57 Shelton Street, Covent Garden, London WC2H 9HE

British Library Cataloguing-in-Publication Data
A catalogue record for this book is available from the British Library.

First published 2003
Reprinted 2004

A GARDEN OF QUANTA
Essays in Honor of Hiroshi Ezawa

ISBN 981-238-445-6

This book is printed on acid-free paper.

Printed in Singapore by Mainland Press

Professor Hiroshi Ezawa

Preface

The *Garden of Quanta* is in full bloom of quantum theory, and everyone who visits there can enjoy variety of the blossoms.

The *Garden* is in fact filled up with a collection of reviews and essays about the recent developments in the areas of quantum physics, covering a wide range of themes such as Quantum mechanics, Path integrals and stochastic processes, Quantum field theory, Statistical mechanics, Mathematical problems, and History. The authors are the experts both from east and west. The articles are written on the graduate level, but most of them are accessible, if not in fine details, to aspiring undergraduates. The book as a whole will serve as a good guide map of vast, fertile grounds of quantum physics for beginning graduate students and undergraduates who are looking for directions to go into.

The articles in *A Garden of Quanta* are dedicated to Professor Hiroshi Ezawa by his friends and students to celebrate his seventieth birthday. Born in 1932 in Tokyo, graduated from Department of Physics, the University of Tokyo in 1955, and earned Ph.D. in 1960, Ezawa took a post of research associate at the University of Tokyo. Then, after spending four years in U.S.A. and Germany doing research in mathematical aspects of quantum field theory, he moved to Gakushuin University in Tokyo in 1967, and has been engaged in reserch and teaching of physics and mathematics for about thirty-five years as an associate professor at first and then as a professor.

His research career embraces vast areas from quantum to statistical physics, particle physics, solid state physics, theory of gravity, and mathematics. The variety of subject matters presented in the *Garden* is a reflection of his wide interests. A detailed review of his activities is given in the article by Keiji Watanabe in this book. Also, a list of his papers and books can be found in the pages that follow.

His activities have not been limited to research and teaching, but have extended further to social matters. As a president of the Physical Society of Japan (1995/96) and a member of the Science Council of Japan (1997 – present), he has contributed much to developing and improving the environment of research and education of the science in Japan. He has also been strongly interacting with physics teachers in high schools, once coauthoring a textbook of high school physics.

We wish him many more years of good health and successes.

We would like to express our sincere gratitudes to all the contributors to this book for their collaborations. Many thanks are also due to Dr. Jitan Lu of the World Scientific Publishing Co. Pte. Ltd. for his collaboration and fine works in producing this book.

March, 2003

Editors

Contents

3. Quantum Field Theory

4. Statistical Mechanics

5. Mathematical Problems

6. History

List of Selected Publications by Hiroshi Ezawa

March 15, 2002

1. Papers

Quantum Statistics of Fields and Multiple Production of Mesons,
with Y. Tomozawa and H. Umezawa, Nuov. Cim. **5** (1957), 810–841.

Ionization Loss near the Origin of an Electron Pair of Very High Energy,
with I. Mito, Prog. Theor. Phys. **18** (1957), 437–447.

To Introduce the Impact Parameter into the Analysis of Multiple-Production of Mesons,
Nuov. Cim. **11** (1959), 745–759.

Impact Parameter and Perturbation Treatment of the Distant Collision of Nucleons with Pion Emission,
with O. Kamei, K. Mori, H. Shimoida and T. Yoneyama, *Proc. of Moscow Conf. on Cosmic Ray Physics (1959)*, English version, pp.274–284, Russian version, pp.276–283.

Green Functions for Elementary Particles,
with H. Umezawa, Phys. Rev. **116** (1959), 463–464.

Pion–Nucleon Interaction and the Multiple-Production Experiment,
with O. Kamei, K. Mori, H. Shimoida and T. Yoneyama, Prog. Theor. Phys. **25** (1961), 667–683.

Quantum Statistical Analogue of Ward's Identity,
with K. Watanabe and O. Kamei, Prog. Theor. Phys. **25** (1961), 735–742.

A Formulation of Bound State Problem,
with K. Kikkawa and H. Umezawa, Nuov. Cim. **25** (1962), 1141–1166.

An Approach to the Elementarity of Particles,

with T. Muta and H. Umezawa, Prog. Theor. Phys. **29** (1963), 877–892.

Some Examples of Asymptotic Fields — In the Sense of Weak Convergence, Ann. Phys. **24** (1963), 46–62.

A Note on the Van Hove–Miyatake Catastrophe, Prog. Theor. Phys. **30** (1963), 545–549.

A Perturbation Theory without the Adiabatic Hypothesis, *Proc. of Eastern U.S. Theor. Phys. Conf. (U. of North Carolina, Oct. 1963)*, II. pp.94–103.

The Representation of Canonical Variables as the Limit of Infinite Space Volume: the Case of the BCS Model, J. Math. Phys. **5** (1964), 1078–1090.

Particle-Mixture Theory and Apparent CP Violation in K-Meson Decay, with Y.S. Kim, S. Oneda and J.C. Pati, Phys. Rev. Lett. **14** (1965), 673–676.

A Criterion for Bose–Einstein Condensation and Representation of Canonical Commutation Relations, with M. Luban, Jour. Math. Phys. **8** (1967), 1285–1311.

Spontaneous Breakdown of Symmetries and Zero-Mass States, with J.A. Swieca, Commun. Math. Phys. **5** (1967), 330–336.

Remarks on the Quantum Field Theory in Lattice Space, I and II Commun. Math. Phys. **8** (1968), 261–268, **9** (1968), 38–52.

A Renormalization Scheme for the Strong-Coupling $\lambda\phi^4$ Theory, with K. Nakamura and Y. Yamamoto, Nuov. Cim. **54A** (1968), 512–515.

Regge Trajectories in the Wick–Cutkosky Model, with J. Arafune and K. Nakamura, Nuov. Cim. **2** (1969), 394–398.

Theory of the Kapitza–Dirac Effect, with H. Namaizawa, J. Phys. Soc. Jpn. **26** (1969), 458–468.

Numerical Solution of an Anharmonic Oscillator Eigenvalue Problem by Milne's Method, with K. Nakamura and Y. Yamamoto, Proc. Japan Acad. **46** (1970), 168–172.

Phonons in a Half Space,

Ann. Phys. **67** (1971), 438–460.

Electrons and "Surfons" in a Semiconductor Inversion Layer,
with T. Kuroda and K. Nakamura, Surface Sci. **24** (1971), 654–658.

Electron Mobility in a Semiconductor Inversion Layer,
with S. Kawaji, T. Kuroda and K. Nakamura, Surface Sci. **24** (1971), 659–662.

Regge Trajectories in the Wick–Cutkosky Model,
with N. Murai and K. Nakamura, Prog. Theor. Phys. **46** (1971), 909–937.

Surfons and the Electron Mobility in a Semiconductor Inversion Layer,
with S. Kawaji and K. Nakamura, Surface Sci. **27** (1971), 218–220.

Semiconductor Inversion Layers and Phonons in a Half-Space,
with S. Kawaji and K. Nakamura, Jour. Vac. Sci. and Technology **9** (1972), 762–768.

On the Divergence of Certain Integrals of the Wiener Process,
with L.A. Shepp and J.R. Klauder, Ann. Inst. Fourier, **24** (1974), 189–193.

A Path Space Picture for Feynman–Kac Averages,
with J.R. Klauder and L.A. Shepp, Ann. Phys. **88** (1974), 588–620.

Surfons and the Electron Mobility in Silicon Inversion Layers,
with S. Kawaji and K. Nakamura, Jpn. J. Appl. Phys. **13** (1974), 126–155.

Vestigial Effects of Singular Potentials in Diffusion Theory and Quantum Mechanics,
with J.R. Klauder and L.A. Shepp, J. Math. Phys. **16** (1975), 783–799.

The Effect of Large Field on the Lifetime of the Forbidden Lines in Quasars,
with M. Leventhal, J. Phys. B. Atom. Molec. Phys. **8** (1975), 1824–1830.

Inversion Layer Mobility with Intersubband Scattering,
Surface Sci. **58** (1976), 25–32.

Many-Body Effects in the Subband Structure of Si-MOS Inversion Layer,
with K. Nakamura and K. Watanabe, Surface Sci. **73** (1978), 258–265.

Einstein's Contribution to Statistical Mechanics,
Albert Einstein — His Influence on Physics, Philosophy and Politics, ed. P.C. Aichelberg and R.U. Sexl, Friedr. Vieweg (1979), pp.69–87.

Einstein's Contribution to Statistical Mechanics, Classical and Quantum, Jap. Studies in the Hist. of Sci., No. 18 (1979), 27–72.

Many-Body Effects in the Si Metal-Oxide-Semiconductor Inversion Layer: Subband Structure,
with K. Nakamura and K. Watanabe, Phys. Rev. **B 22** (1980), 1892–1904.

Many-Body Effects and the Electron Mobility in Si Inversion Layer at Room Temperature,
with K. Nakamura and K. Watanabe, Surface Sci. **98** (1980), 202–209.

Stochastic Calculus and Some Models of Irreversible Process,
with H. Hasegawa, Prog. Theor. Phys. **66** (1981), 1612–1626.

Gravitational Radiation of a Particle Falling towards a Black Hole,
with Y. Tashiro, Prog. Theor. Phys. **66** (1981), 1612–1626.

Remarks on a Stochastic Quantization of Scalar Fields,
with J.R. Klauder, Prog. Theor. Phys. **69** (1983), 664–673.

Fermions without Fermions — The Nicolai Map Revisited,
with J.R. Klauder, Prog. Theor. Phys. **74** (1985), 904–915.

From Infinite Time to Finite Temperature,
Quantum Field Theory — Proc. of the Int'l Symp. in Honor of H. Umezawa, ed. F. Mancini, North Holland (1986), pp.405–414.

Thermo Field Dynamics of Heat Conduction,
Progress in Quantum Field Theory, H. Ezawa and S. Kamefuchi ed., North Holland (1986), pp.305–324.

Reconstruction of Lagrangian from a given Hamiltonian — A Challenge in Mechanics Course,
Proc. Int'l Conf. on Physics Education, Japan Soc. for Physics Education (1986), p.297.

Towards a Quantum Theory of Heat Conduction,

Wandering in the Fields — Festschrift for Professor Kazuhiko Nishijima on the occasion of his sixtieth birthday, ed. K. Kawarabayashi and A. Ukawa, World Scientific (1987), pp.191–206.

Demonstration of Single-Electron Build Up of an Interference Pattern, with A. Tonomura, J. Endo, T. Matsuda and T. Kawasaki, Am. J. Phys. **57** (1989), 117–120.

Characterization of Thermal Coherent and Thermal Squeezed States, with A. Mann, K. Nakamura and M. Revzen, Ann. Phys. **209** (1991), 216–230.

Difference in Elastic Scattering Effects on Resonant Tunneling in One Dimention and Three Dimensions, with Y. Zohta, K. Nakamura, Solid State Commun. **80** (1991), 885–889.

Quantum Field Theory of Thermal Diffusion, with K. Watanabe and K. Nakamura, *Thermal Field Theories*, ed. H. Ezawa, T. Arimitsu and Y. Hashimoto, North Holland (1991), pp.79–94.

Thermal Coherent and Thermal Squeezed States, with A. Mann, K. Nakamura and M. Revzen, *Thermal Field Theories*, ed. H. Ezawa, T. Arimitsu and Y. Hashimoto, North Holland, (1991), pp.201–205.

Feynman Path Integral Approach to Resonant Tunneling, with Y. Zohta and K. Nakamura, *Resonant Tunneling in Semiconductors*, L.L. Chang *et al.* ed., Plenum (1991), pp.285–295.

Effect of Inelastic Scattering on Resonant Tunneling studied by the Optical Potential and Path Integrals, with Y. Zohta, J. Appl. Phys. **72** (1992), 3584–3588.

Tomonagas Studien unter Heisenberg in Leipzig, *Werner Heisenberg als Physiker und Philosoph in Leipzig*, H. Rechenberg Hrsg., Spektrum Akad. Verlag, Heidelberg (1993), pp.78–86.

Quantum Control and Measurement, *Proc. Int'l Workshop on Quantum Control and Measurement*, ed. H. Ezawa and Y. Murayama, Elsevier (1993), pp.3–8.

Quantum Mechanical Representation of Laser Beam Splitting,
Proc. of Int'l Symp. on Advanced Topics of Quantum Physics — Shanxi,
ed. J.Q. Liang, M.L. Wang and S.N. Qiao, Science Press (1993),
pp.342–353.

Particle- and Wave-Descriptions are not Equivalent,
*On Klauder's Path: A Field Trip — Essays in Honor of John R.
Klauder*, ed. G.G. Emch, G.C. Hegelfeld and L. Streit, World Scientific (1994), pp.55–61.

Impurity Scattering in Resonance Tunneling,
Proc. Int'l Conf. on Physics and Technology in 1990's, ed. C.C.
Bernido, M.V. Carpio-Bernido and D.M. Yanga, Samhang Pisika
ug Pilipinas (1994), pp.1–19.

Particle and Wave Pictures: Are they Equivalent?
*Field Theory and Collective Phenomena — In Memory of Professor
Hiroomi Umezawa, Perugia, Italy, 28–31 May 1992* , ed. S. De Lillo
and P. Sodano, F.C. Khanna, G.W. Semenoff., World Scientific
(1995), pp.456–469.

Change of Independent Variable in Variational Problems,
Jour. Mod. Phys. B **10** (1996), 1555–1561.

Quantization of Fields in a Fabri–Perot Cavity,
Physics Essays **9** (1996), 604–608.

Constructive Proof of the Existence of Bound States in One Dimension,
Foundation of Physics **27** (1997), 1495–1509.

Diversity and Unity in Physics Research in the Age of Communication,
*Proc. of the 2nd Int'l Conf. on Research and Communication in
Physics, Sept. 18–22, Tokyo*, ed. H. Toki, Phys. Soc. Jpn. (1997),
pp.3–15.

The Casimir Force from Lorentz's,
with K. Nakamura and K. Watanabe, *Frontiers in Quantum
Physics, Proc. of Int'l Conf. on Frontiers in Quantum Physics,
Kuala Lumpur, Malaysia, 9–11 July, 1997*, ed. S.C. Lim, R. Abd-
Shukor and K.H. Kwek, Springer (1998), pp.160–169.

Thermofield Dynamics of Irreversible Processes,

with K. Watanabe and K. Nakamura, *Frontiers in Quantum Physics, Proc. of Int'l Conf. on Frontiers in Quantum Physics, Kuala Lumpur, Malaysia, 9–11 July, 1997*, ed. S.C. Lim, R. Abd-Shukor and K.H. Kwek, Springer (1998), pp.219–223.

Irreversibility from a Reversible Equation,
with K. Watanabe and K. Nakamura, *Proc. of the 2nd Jagna Int'l Workshop, Mathematical Methods of Quantum Physics — Essays in Honor of Professor Hiroshi Ezawa*, ed. C.C. Bernido, K. Nakamura, M.V. Carpio-Bernido and K. Watanabe, Gordon and Breach (1999), pp.43–59.

Nonstandard Analytical Construction of a Path Space Measure for the 4-D Dirac equation,
with T. Nakamura and K. Watanabe, *Proc. of the 2nd Jagna Int'l Workshop, Mathematical Methods of Quantum Physics — Essays in Honor of Professor Hiroshi Ezawa*, ed. C.C. Bernido, K. Nakamura, M.V. Carpio-Bernido and K. Watanabe, Gordon and Breach (1999), pp.155–166.

Numerical Solutions of the Eigenvalue Problems of Sturm–Liouville Type by using Power Series Approximation about Regular–Singular Points,
with T. Yano, Y. Ezawa, and T. Wada, Prog. Theor. Phys. Suppl. No. 138, (2000), 747–749.

Ornstein–Uhlenbeck Path Integral and Its Application,
Acta Applicandae Mathematicae **63** (2000), 119–135; *Recent Developments in Infinite-Dimensional Analysis and Quantum Probability — Papers in Honour of Takeyuki Hida's 70th Birthday*, ed. L. Accardi, Hui-hsiung Kuo, N. Obata, K. Saito, Si Si and L. Streit, Kluwer Academic Publishers (2001), pp.119–135.

Obituary: Takehiko Takabayashi (1919–1999),
Historia Scientiarum **10** (2001), 286–304.

Criterion for Bose–Einstein Condensation in a Trap,
with K. Watanabe and K. Nakamura, *Proc. 7th Int'l Symposium on Foundations of Quantum Mechanics in the Light of New Technology*, ed. Y.A. Ono and K. Fujikawa, World Scientific (2002), pp.132–137.

Long-Time Average of Field Values Measured by a Brownian Wanderer,
J. Phys. Soc. Jpn. **71** (2002), 35–42.

Can Milne's Method Work for the Coulomb-like Potentials?,
> with A. Yano and Y. Ezawa, Journal of Computational and Applied Mathematics **152** (2003), 597–611.

2. Books (in Japanese)

Theory of Scattering,
> *Quantum Mechanics — Problems and Solutions*, ed. M. Kotani and H. Umezawa, Syokabo (1959), pp.274–375.

Representation of Canonical Commutation Relations,
> *Topics in the Theory of Elementary Particles*, with H. Umezawa and K. Kawarabayashi, Kyoritsu Shuppan (1963), pp.145–218.

Physics B, High School Textbook,
> with M. Nogami, I. Imai, J. Iwaoka, K. Kinoshita, M. Kojima, M. Kondo, H. Takami and J. Hayashi, Jikkyo Shuppan (1963, Revised 1967, 1969).

Structure of Quantum Mechanics,
> *Quantum Mechanics III*, Iwanami Lecture Series, "Foundation of Contemporary Physics," ed. H. Yukawa and T. Toyoda, Iwanami Shoten (1972), pp.9–180, 229–286.

Who Has Seen The Atom?
Iwanami Junior Science, Iwanami Shoten (1976).

Perspectives of Quantum Mechanics — 50 Years of Its History, I and II,
> edited with T. Tsuneto, Iwanami Shoten (1977, 1978); Translated into Russian, ed. A. S. Davidova, Naukova Dumka, Kiev (1982).

Structures of Typical States, Approximation Methods, and Quantum Mechanics and Relativity,
> *Quantum Mechanics I*, Iwanami Lecture Series, "Foundation of Contemporary Physics," ed. H. Yukawa and T. Toyoda, Iwanami Shoten (1978), pp.225–429.

Structure of Quantum Mechanics,
> *Quantum Mechanics II*, Iwanami Lecture Series, "Foundation of Contemporary Physics," ed. H. Yukawa and T. Toyoda, Iwanami Shoten (1978), pp.254–484.

Quantum Field Theory and Statistical Mechanics,
> with A. Arai, Nihon-Hyoron-sha (1988).

Modern Physics,
Society for Promoting Education by the University-of-the-Air (1989).

Method of Renormalization Group,
with K. Watanabe, M. Suzuki and H. Tasaki, Iwanami Lecture Series, "Contemporary Physics", Iwanami Shoten (1994).

Groups and Their Representations,
with K. Shima, Iwanami Lecture Series, "Applied Mathematics," ed. H. Fujita *et al.*, Iwanami Shoten (1994).

Asymptotic Analysis,
Iwanami Lecture Series, "Applied Mathematics," ed. H. Fujita *et al.*, Iwanami Shoten (1995).

Modern Physics,
Asakura Shoten (1996).

Yoshio Nishina — Atomic Physicist,
Pioneers of New Science/Technology, ed. by S. Okada *et al.*, Iwanami Lecture Series, "Science/Technology and Man", Iwanami Shoten (1999).

Quantum Mechanics, I and II,
Syokabo (2002).

Quantum Mechanics — Problems and Solutions,
Syokabo (2002).

Professor Hiroshi Ezawa

K. Watanabe

Department of Physics, Meisei University
Hino, Tokyo 191–8506, Japan
E-mail: watanabk@meisei-u.ac.jp

Professor H. Ezawa's career, scientific works and social activities are reviewed. Some of the works are picked up from large number of papers and books, which cover a wide range of field, and are briefly explained. Influences and prospects of his works are discussed.

It is a great honor to introduce Professor Hiroshi Ezawa and his scientific works on the occasion of his 70th birthday.

1. Career

Ezawa was born on June 2, 1932 in Tokyo. He got B.Sc. from Department of Physics, University of Tokyo in 1955, and his Ph.D. from the same institution in 1960 under the guidance of late Professor Mokichiro Nogami first, and subsequently of late Professor Hiroomi Umezawa, then becoming his research associate. In 1963, Ezawa went to the U.S.A. on Fulbright Travel Grant, working as a research associate at the University of Maryland, where he met Dr. H.C.S. Lam from China and Dr. J.N. Islam from Bangladesh. Here also, he had an opportunity to attend a seminar at Princeton and to meet Professor A.S. Wightman and Dr. J.R. Klauder, then a Member of Technical Staff, Bell Laboratories. In 1965 he became a research associate to Professor R. Haag at the University of Illinois, where he met Dr. J.A. Swieca from Brazil. As Professor Haag moved back to Germany, Ezawa followed him in 1966 to be a visiting scientist at the Institut für Theoretische Physik, Universität Hamburg. In 1967, he came back to Japan, to be Associate Professor at Gakushuin University in Tokyo, and then became Professor in 1970. In this year, he participated in the Les Houches summer school on Statistical Mechanics and Quantum Field Theory, where he met Klauder again. By an invitation by Klauder, he worked for 1972–1974 as a Member of Technical Staff, and in 1977 as a consultant at Bell Laboratories, Murray Hill, New Jersey, U.S.A., where he was introduced to Dr. L.A. Shepp, another Member of Technical Staff, and Professor L. Streit, a visitor from Germany, among others. He was invited to the University of Alberta, Canada by Professor Umezawa in 1983, and to Universität Bielefeld, Germany by Streit in 1984 *etc.* In international conferences and other opportunities, he met Professor S.K. Wang and Ke Wu from China, Professor M. Revzen and A. Man from Israel, and others. Many of these professors and doctors, could he invite to visit Gakushuin.

In Gakushuin University, Ezawa organized seminars, in which a number of mathematicians and physicists from universities in Tokyo area participated. He has never been satisfied with the generally accepted explanations and understanding, but sought to find out fundamental meaning and facts lying behind them and to start from the first principles. Sometimes the seminars were endless; discussions after discussions continued until late at night. This style of seminars may be a heritage from Professor Umezawa. Besides Gakushuin, he taught at many universities as visiting lecturer. Thanks to this activities and to his guidance as a teacher, grew up many excellent researchers, some of whom are the authors of this book.

Ezawa's works can be classified into two groups: One, the original scientific papers, and the other, a number of textbooks for research physicists and students, undergraduate as well as graduate, and sometimes for high-school students. He wrote some enlightenment books such as *Who Has Seen the Atoms?* (in Japanese), which was awarded Sankei Prize for Children's Science Books. He has been a popular lecturer for high-school teachers and students.

In addition to these works, social activities are also notable. Above all, he served one-year term in 1995/96 as President of the Physical Society of Japan, in which term the Society celebrated its 100th anniversary, 1987–1992 as Dean of the Faculty of Science, Gakushuin University, this length of time having been necessary to create Institute for Bio-Molecular Science at Gakushuin, 1994–2000 as a member of Philippines–Japan Science Exchange Project, through which he became acquainted with Dr. C. Bernido and Dr. M.V. Carpio-Bernido, and 1987–1992 as a visiting professor of the University of the Air.

2. Scientific works

It is almost impossible to give a complete account of Ezawa's large number of works which covers a wide range of fields. Let me pick up some of them to indicate how his works have had influences; the paper referred to may be identified in the list of his publications by the author(s) and the year of publication. For other works, the readers are reffered also to the list.

2.1. *Statistical mechanics of fields*

Ezawa started his research career in 1955 by studying the multiple production of mesons by nucleon collisions at extremely high energy. In those days, statistical theories of E. Fermi and L.D. Landau were discussed; one assumed that the collision creates a cloud of particles at extremely high temperature, for which a simple equation of state, (pressure) = (energy density)$/3$, was assumed to derive the multiplicity-collision energy relation. Ezawa, Y. Tomozawa and Umezawa (1957) wanted to find a crue in this relation as to whether the meson-nucleon interaction was renormalizable or non-renormalizable. They tried to derive the equation of state for a given interaction by using Dyson-expansion method for calculating partition function $\mathrm{Tr}\, e^{-\beta \mathcal{H}}$ T. Matsubara had proposed a couple of years earilier, which however was not covariant for it treated the inverse temperature β (which corresponds to time) as it stands without Fourier transforming unlike the 3-dimensional space coordinates. In fact, it seemed at the beginning that a covariant treatment was impossible because the range of integration over β was finite in place of being $(-\infty, \infty)$ for the time-integral in the S-matrix calculation which was essential in producing energy-conservation delta function at each vertex of the Feynman diagrams. Then, it was discovered that the propagators were periodic and anti-periodic for Bose and Fermi fields, respectively. Thanks to this discovery, they could formulate a complete 4-dimensional Feynman diagram technique for calculating the partition function for a given interaction by introducing what is nowadays called the Matsubara frequency [see the recollections by Ezawa, "From infinite time to $\cdots$," 1986]. They proposed a method of renormalization to remove divergences, which appears in the calculation of the partition function; their method is commonly used now in the thermal field theory. The (anti) periodicity property of the propagators is an important part of the KMS-condition introduced later in the algebraic formulation of statistical mechanics.

The work was extended by O. Kamei, Ezawa and myself (1961) to interacting electron system, and the analogue of the Ward-Takahashi identity was shown to hold also for statistical mechanical Green function due to its periodicity property. I studied the divergence problems, ultraviolet and infrared, for the electron system [Prog. Theor. Phys. **26** (1961) 453, **28** (1962) 951]. Importance of the field theoretical methods were not appreciated in those days by the people in statistical mechanics and solid state physics in Japan.

Ezawa's interest is now directed towards deriving irreversibility from the first principles. Ezawa, K. Nakamura and myself [Watanabe *et al.* (1998), Ezawa *et al.* (1999)] could show that the irreversibility results from time-symmetric microscopic equation of motion for the evolution for the time $t > 0$ starting from a mixture state at $t = 0$; this solution exhibits damping but is symmetric in time in that the solution for $t < 0$ is a time-reflection of the solution for $t > 0$.

2.2. *Condensed matter physics*

In 1967, Ezawa and M. Luban wanted to see if any inter-particle interactions make the Bose-Einstein condensation unstable. They considered replacement $a_0^\sharp \to \sqrt{N_0} + b_0^\sharp$ in place of Bogoliubov's $a_0^\sharp \to \sqrt{N_0}$, where N_0 and $a^\sharp$ are the average occupation number and the creation or the annihilation operators of the lowest single-particle level and $b^\sharp$'s (as well as $a^\sharp$'s) satisfy the canonical commutation relations. They tried to derive effective Hamiltonian for $b^\sharp$'s to see if it has a positive excitation energy from vacuum. Since for infinitely extended (translation-invariant) system, there is no energy gap by the Goldstone theorem, they had to consider finite-size effect, which was extremely difficult. Since in these days, the Bose-Einstein condensation is realized with trapped particle, Ezawa, Nakamura and myself have been studying the problem for trapped system of particles [Ezawa *et al* (2002)].

To account for the mobility of electrons confined in a thin layer on the surface of a semiconductor by an applied electric field F perpendicular to the surface (as in MOS device) in its relation to temperature and the strength of the field F, S. Kawaji, an experimentalist in Gakushuin, once proposed a model of electrons interacting with elastic waves confined in the layer.

Ezawa (1971) took up the task of completing the model, and carried out quantization of elastic waves in a solid occupying a half space by constructing a complete system of orthonormal modes of elastic vibration. The phonons were named surfons, and were used to calculate the electron mobility [Ezawa, Kawaji and K. Nakamura (1974)]. Then, the work was extended to take the electron-electron interaction (many-body effects) into account [K. Nakamura, Ezawa and Watanabe (1980)].

2.3. *Quantum mechanics*

A. Tonomura *et al.* used, upon Ezawa's request, the highly coherent electron beam from his field-emission tip and position-sensitive set of photomultipliers to take a time-series of pictures of single-electron build up of interference pattern [Tonomura, Endo, Matsuda, Kawasaki and Ezawa (1989)]. The electrons arrive one by one on the detection plane leaving arrival spots. While the spots, when their number is small, look randomly situated, the interference pattern appears gradually as the number of the spots increases, showing that electrons, which appeared to arrive at random positions at first, have never arrived at the minima of the interference fringes. The series of pictures appears in many textbooks of quantum mechanics world-wide.

Ezawa contributed chapters on "mathematical structure of quantum mechanics" to *Quantum Mechanics* III (1972) and (revised) II (1978) edited by H. Yukawa and T. Toyoda, which have been widely read.

2.4. *Quantum field theory*

In addition to the field-theoretical treatment of statistical mechanics and solid-state problems which are already mentioned, Ezawa wrote a number of papers, say, on: Van Hove-Miyatake catastrophe (1963), spontaneous breakdown of symmetries and zero-mass states (with J.A. Swieca, 1967), strong-coupling ϕ^4 theory (with K. Nakamura and Yamamoto, 1968), Regge trajectry of the Wick-Cutkosky model (with J. Arafune and K. Nakamura, 1969). With A. Arai, he wrote a book on mathematical treatment of quantum field theory and statistical mechanics in 1988.

2.5. *Gravitational waves*

The gravitational wave radiated by a massive particle falling towards a black hole was once studied with Y. Tashiro (1981). Ezawa has been participating in the project of constructing a gravitational wave detector at the National Astronomical Observatory. He is interested in the noise caused by the quantum fluctuation of light pressure on the mirror of the interferometer, and tried to quantize the radiation field inside a Fabri-Perot cavity

(1996) by constructing a complete orthonormal set of eigenmodes in the cavity.

2.6. *Stochastic process*

Besides many problems, Ezawa is recently interested in the long-time average of the field $V(x)$ $x \in \mathbb{R}^d$ measured by a Brownian wanderer in d-dimensional space. More precisely, the problem is to find out an appropriate factor $\nu(T)$ such that $X(T, x_0) = \nu(T)^{-1} \int_0^T V(\omega(s)) ds$ has a limit $T \to \infty$, and its probability distribution where $\omega(s)$ is the d-dimensional Brownian motion starting from x_0 at $s = 0$.

Ezawa studied the case of $d = 1$ with L. Shepp many years ago. Stimulated by a recent remark by Professor F.W. Wiegel about the relevance of the problem to physics of immune system of living bodies (see the article by Wiegel in this book), Ezawa wrote up the $d = 1$-case (2002). Then, the work by Ezawa, T. Nakamura and myself revealed that $\nu(T) = \log T$, $=$const. for 2- and 3-dimensional cases, respectively. Some more interesting facts have been revealed, and a paper is now in preparation.

I have not mentioned number of books Ezawa has written. I really wish that in the future they be translated into English so that they may be read world-wide.

Acknowledgements

I would like to take this opportunity to express my sincere gratitude to Professor Hiroshi Ezawa for the strong influence and advice he has had on my scientific work and life. In the name of all friends and the contributors to this book, we cordially wish Professor Ezawa many more years of good health and outstanding activities.

I wish to thank Dr. T. Nakamura for his valuable comments and advices.

Direct Observation of the Microscopic World by Using Phase Shifts of Electron Waves

Akira Tonomura

Advanced Research Laboratory, Hitachi, Ltd.,
Hatoyama, Saitama 350–0395, Japan
Frontier Research System, RIKEN, Wako, Saitama 351–0198, Japan
SORST, Japan Science and Technology Corporation (JST),
Tokyo 103–0027, Japan

The wave properties of electrons can now be directly observed thanks to the advent of bright and monochromatic field-emission electron beams. This paper reports several electron-interference experiments by using the bright electron beams. Examples are fundamental experiments on quantum mechanics, such as the observation of single electron build-up of an interference pattern and the conclusive experiments of the Aharonov–Bohm effect, and observations of quantum objects such as quantized vortices in superconductors. Recent results on unusual behavior of vortices in high-T_c superconductors obtained by using our newly developed brightest 1-MV electron beam are also reported.

1. Introduction

Electrons were first identified as elementary particles that constitute atoms in the 1890's.[1] Twenty years later, to explain the stability of atoms and other unexplicable data obtained up to then, electrons were considered also to have the wave properties. The experimental evidence for the wave properties of electrons was given by Davisson and Germer,[2] J. P. Thomson[3] and Kikuchi[4] *et al.* They irradiated parallel electron beams onto materials and observed the electron beams diffracted from the arrays of atoms that make up the materials; it was found that the scattered electrons from the atoms interfere with each other to form diffracted beams.

The wave nature of electrons is not limited in microscopic regions in atomic dimensions. Thirty years later, macroscopic interference patterns formed by two electron beams were observed by using an electron biprism.[5]

Furthermore, electron interferometry has progressed greatly by the development of highly collimated, monochromatic field-emission electron beams,[6] just like optical interferometry did so by the advent of lasers. Our recently developed 1-MV field-emission electron beam[7] has the highest brightness (2×10^{10} A/cm^2 sr) and narrowest monochromaticity ($\Delta E/E = 5 \times 10^{-7}$ min^{-1}) ever obtained. This paper reports on both the quantum-mechanical experiments made possible by bright electron beams and on the observation of phase objects by using the wave properties of electrons.

2. Bright and monochromatic electron beam

The experiments were carried out using electron beams field-emitted from a pointed needle at room temperatures, the accelerating voltages of which ranged from 80 kV to 1 MV.

An electron beam can behave as a wave when it is both collimated and monochromatic.[8] More quantitatively, an electron of the beam whose uncertainty in the propagation direction is 2β and whose uncertainty in the kinetic energy is ΔE is regarded as a wave packet extended in the propagation direction by $(2E/\Delta E)\lambda$ and in a plane perpendicular to the propagation direction by $\lambda/2\beta$, where E and λ are, respectively, the average kinetic energy and wavelength of the electron. Within this packet region, an electron behaves as a wave unless detected, and interference phenomena can be observed when two parts of this packet are made to overlap.

An electron beam can be collimated to get a small β, for example, by passing it through two small pinholes located at different positions along the beam direction, and it can be monochromatized to get a small ΔE by passing it through an energy filter. Doing so makes the electron wave coherent even in a macroscopic region, at least in principle. In that case, however, the beam intensity becomes low. To get an observable electron interference pattern, we thus need to use a bright electron beam.

Brightness B of an electron beam is defined as the current per unit area and per unit solid angle:

$$B = I/(dA \cdot d\Omega) = I/(\pi r^2 \cdot \pi \beta^2), \tag{1}$$

where $dA, d\Omega$, and I are, respectively, the area of the cross section and the divergence solid angle of the electron beam, and the total current passing through that area. dA and $d\Omega$ can be expressed in an axially symmetric case by $\pi\beta^2$, and πr^2, respectively, where β is the divergence angle and r is the radius of the cross section of the beam.

For a 100-kV thermal electron beam used in conventional electron microscopes, the total current of a coherent beam is 4×10^{-13} C/sec, or the average interval between coherent electrons is approximately 100 m and depends only on B and λ. When the beam is focused onto an area of 1 cm^2, the current density of the beam is two orders of magnitude lower than that needed for taking photographs on an electron-microscope film with a 1-s exposure time.

The value of the brightness is a constant and never changes, even if electron optical parts such as pinholes or lenses are inserted in the beam path. The brightness of a field-emission electron beam is more than two orders of magnitude greater than that of a thermal electron beam, and the energy spread of the field-emitted electron beam is several times narrower than that of the thermionic electron beam.

With the 80-kV field-emission beam developed in 1979,[6] it became possible to increase the number of biprism interference fringes that could be recorded on a film from 300 to 3000. And when their number was less than 50, interference fringes could be observed on the fluorescent screen with the naked eye. The development of the field-emission beam has thus made the use of electron interferometry possible in practical applications as described in the following.

3. Experiments concerning fundamentals of quantum mechanics

The relative phase of an electron wave function can now be directly observed in the form of interference patterns by using a field-emission electron beam, and possibilities to carry out even "thought experiments" on fundamentals of quantum mechanics are opened up.

3.1. *Single-electron build-up of an interference pattern*

An electron interference pattern is formed in such a way that a single electron exists at one time in the experimental apparatus. Such an experiment is referred to as "impossible, absolutely impossible to explain in any classical way, and has in it the heart of quantum mechanics." Feynman[9] continues, "This experiment has never been done in just this way, since the apparatus would have to be made on an impossibly small scale."

However, such a thought experiment has now become feasible with the progress of advanced technologies; Electron microscopy allows the observation with large magnification, and individual electrons can be observed with a photon-counting detector modified for counting electrons with almost 100% detection efficiency. The experiments Feynman described were actually carried out in a field-emission electron microscope equipped with both an electron biprism[5] and a two-dimensional position-sensitive electron-counting system.[10] As shown in Fig. 1, electrons emitted from a field-emission source are sent to the biprism. The biprism interference pattern is then enlarged by magnifying lenses, omitted in the figure, detected by the detection system, and displayed on a monitor. The detector has the same function as that of the films, and the electron intensity is accumulated over time.

When the intensity of coherent electrons is high, an interference pattern is soon formed. But when the intensity becomes extremely weak, electrons are detected one by one on the monitor [Fig. 2(a)]. The electron distribution seems quite random. However, when the number of the detected electrons increases, such an interference pattern is gradually formed as is obtained when two electron waves pass through both sides of the biprism [Fig. 2(c)–(e)]. Even when the electron arrival rate was as low as 10 electrons/s in the entire field of view, so that at most only a single electron existed at a time, the accumulation of single electrons formed the interference pattern shown

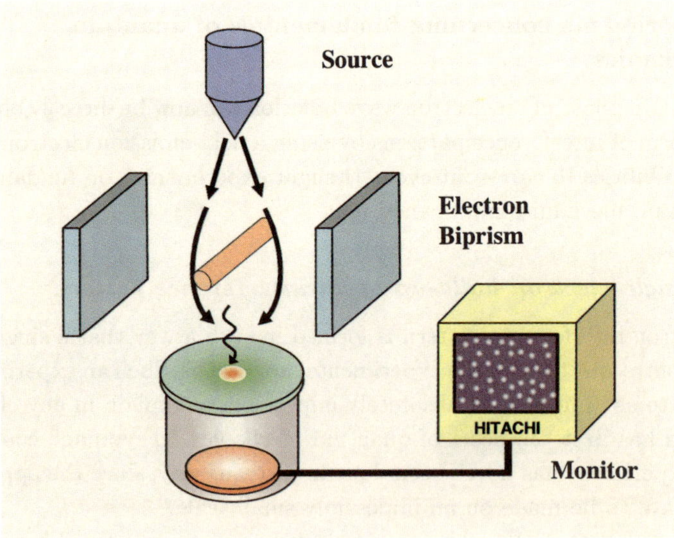

Fig. 1.   Double-slit experiment for electrons.

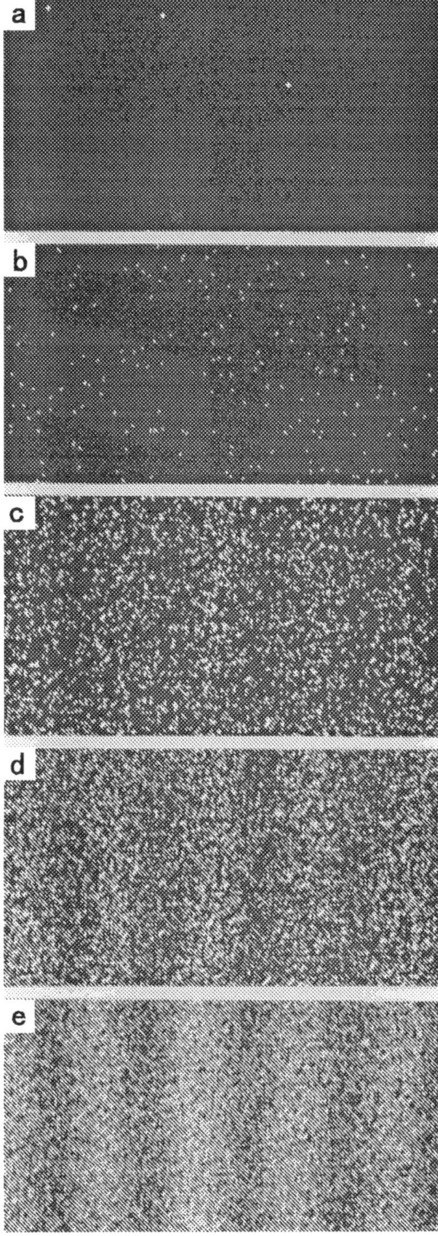

Fig. 2. Build-up process of electron interference pattern. Number of electrons: (a) 10; (b) 100; (c) 3000; (d) 20,000; (e) 70,000

12   A. Tonomura

in Fig. 2(e).

In quantum-mechanical terms, even a single electron can split into two
and they can pass though both sides of the biprism simultaneously in the
form of a wave function. Two partial electron waves overlap to interfere on
the observation plane and form a probability interference pattern. When
detected, the overlapped waves can be observed as a single electron, never
as two. It is considered that the measurement makes the extended wave
function collapse into a single point instantly.

The mystery of this collapse increases when such an experiment as
shown in Fig. 3 is carried out. In this case, two partial waves having passed
through both sides of the biprism filament are thoroughly separated by
applying a positive potential to the biprism filament [as in Fig. 3(b)] instead of a negative one [as in Fig. 3(a)]. When measured, a single electron
is detected on either side of the biprism filament. Since the two partial
wavefunctions must be located at different positions on opposite sides of
the filament, the partial wavefunction on one side must have jumped to
the other side. It should be noted here that the coherence length of the
electron-wave packet in the traveling direction is only 1 μm, while the dis-

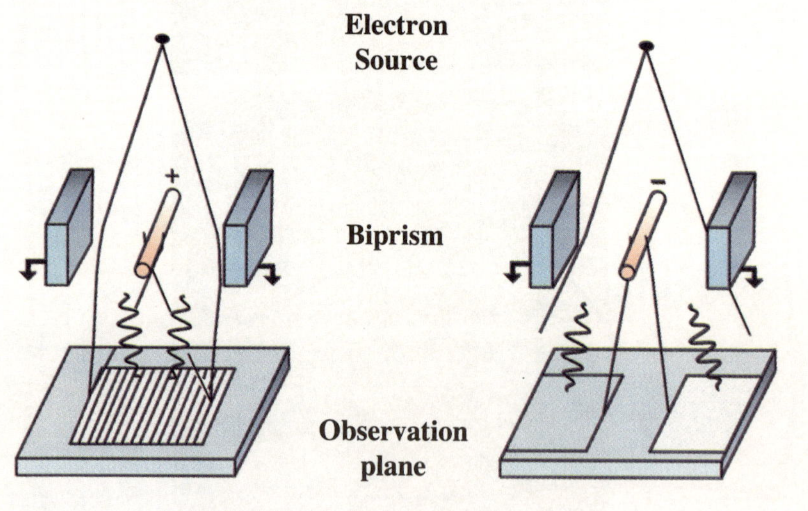

Fig. 3.   Wave packet in the electron biprism.(a) Application of a positive potential to
the central filament; (b) Application of a negative potential to the central filament.

tance between the biprism and the patrial waves at the detector plane is, say, 10 cm. That means, the two partial waves are separated by the biprism and travel a long distance in the opposite directions before the two waves collapse into a point.

Many experiments questioning quantum mechanics including the above-described double-split experiments have been tested using advanced technologies, and the predictions of quantum mechanics have been proven to be correct up to now. Our result in Fig. 2 have often been cited in physics textbooks[12] and has recently been selected in Physics World[13] as the first rank of the most beautiful experiments together with the electron interference experiment with the actual double-slit instead of the electron carried out by Jönsson in 1961.[14]

3.2. *The Aharonov–Bohm effect*

Strange behavior of electrons is not restricted to the double-slit experiment. Another example is the interaction of an electron wave with electromagnetic fields. Electric and magnetic fields are defined by forces exerting on charged particles. However, there are cases where a beam of charged particles can be physically influenced, in the form of phase shifts, even when it passes through a region free of magnetic field B outside an infinitely long solenoid; therefore, it is subjected to no forces. Aharonov and Bohm[15] attributed this effect to vector potential A. The produced phase shift ϕ is given by

$$\phi = -\frac{e}{\hbar} \int A \cdot ds. \tag{2}$$

Concerning the reality of vector potentials, there has been much debate over the past 100 years.[16] Especially in the late 1970s, when vector potentials were extended to gauge fields and regarded as a fundamental physical entity in theories unifying all fundamental forces in nature, much attention was paid to the AB effect, since the AB effect was regarded as indicating the gauge principle,[17] and even the existence of the AB effect was questioned.[18]

Accordingly, we carried out a series of experiments in order to remove any ambiguities presented in the controversy, and the last of them,[19] considered to be the most conclusive, is introduced here. We used a toroidal ferromagnet instead of a straight solenoid. An infinite solenoid is experimentally unattainable, but an ideal geometry can be achieved by a finite system consisting of a toroidal magnetic field. Furthermore, the toroidal ferromagnet was covered with a superconducting niobium layer to completely confine the magnetic field by the Meissner effect.

An electron wave was incident around a tiny toroidal sample, and the relative phase shift $\Delta\phi$ between two waves passing through inside the hole and outside the toroid was measured as an interferogram. The phase difference is given by $-e/\hbar$ times the magnetic flux enclosed by the two waves:

$$\Delta\phi = -\frac{e}{\hbar} \oint \boldsymbol{A} \cdot d\boldsymbol{s} = -\frac{e}{\hbar} \int \boldsymbol{B} \cdot d\boldsymbol{S}. \tag{3}$$

Although samples with various magnetic flux values were measured, the relative phase shift was always either 0 or π. The conclusion is now obvious. The photograph in Fig. 4 indicates that a relative phase shift of π is produced even when the magnetic fields are confined within the superconductor and shielded from the electron wave; and that the AB effect exists. An electron wave must be physically influenced by vector potentials.

The experimental evidence that the niobium layer surrounding the magnet actually became superconductive is obtained from the quantization of the relative phase shift, either 0 or π. When a superconductor completely surrounds a magnetic flux, the flux is quantized to an integral multiple of the quantized flux, $h/2e$. When an odd number of vortices are enclosed inside the superconductor, the relative phase shift becomes π (mod . 2π). For an even number of vortices, the phase shift is zero. Therefore, the oc-

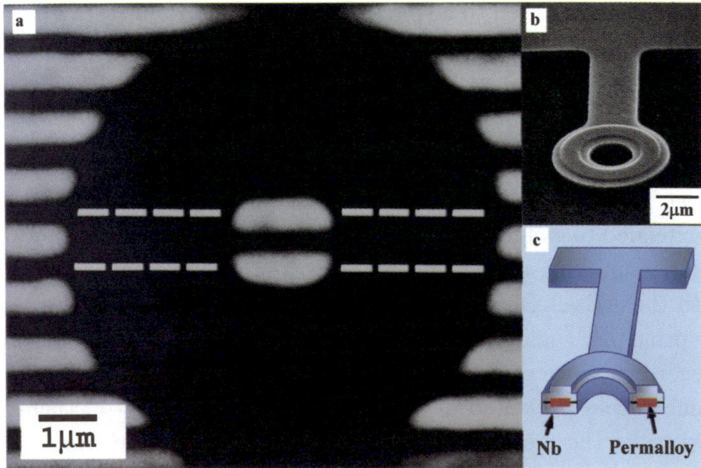

Fig. 4. Conclusive experiment of the Aharonov–Bohm effect: (a) Interference pattern; (b) Scanning electron micrograph of a toroidal feromagnet; (c) Schematic of a sample. Electron waves passing through inside and outside the toroidal magnet are phase-shifted by π by the quantized magnetic flux of $h/2e$.

currence of flux quantization is evidence that the niobium layer actually becomes superconducting, that the superconductor completely surrounds the magnetic flux, and that the Meissner effect prevents any flux from leaking out.

4. Observation of microscopic phase objects

The AB effect has been used to observe the microscopic distributions of electromagnetic fields. To be more specific, the thickness distribution of a specimen can be observed as phase contours in its interference micrograph obtained through electron holography.[8] This is because the phase of an electron wave is shifted by the inner potential of the specimen when the wave passes through it. Phase shifts can usually be detected from a normal interference pattern with a precision of $2\pi/4$, but the precision increases $2\pi/100$ by using phase-amplification technique peculiar to holography. In fact, thickness changes resulting from monatomic steps[20] and carbon nanotubes[21] have actually been detected using these techniques.

4.1. *Magnetic lines of force*

In the case of pure magnetic fields, the electron phase shift is produced by vector potentials. When the phase distribution is displayed as an interference micrograph, the micrograph can be interpreted in the following two straightforward ways.

(1) Contour fringes in the interference micrograph indicate magnetic lines of force, since there is no relative phase shift between two beams passing through two points along a magnetic line.
(2) Contour fringes show magnetic flux in units of h/e, since the phase difference between two beams enclosing a magnetic flux of h/e is 2π.

An example of the observation of magnetic lines of force inside a ferromagnetic fine particle is shown in Fig. 5. Narrow fringes parallel to the edges indicate thickness contours. The circular fringes in the inner region indicate magnetic lines of force, since the thickness is uniform there.

Interference microscopy is not the only technique that can be used to visualize phase distribution. For example, a phase object can be observed in an out-of-focus image because the phase change is transformed into an intensity change when the image is defocused. A vortex in a superconductors, which acts as a pure-weak phase object to an illuminating electron

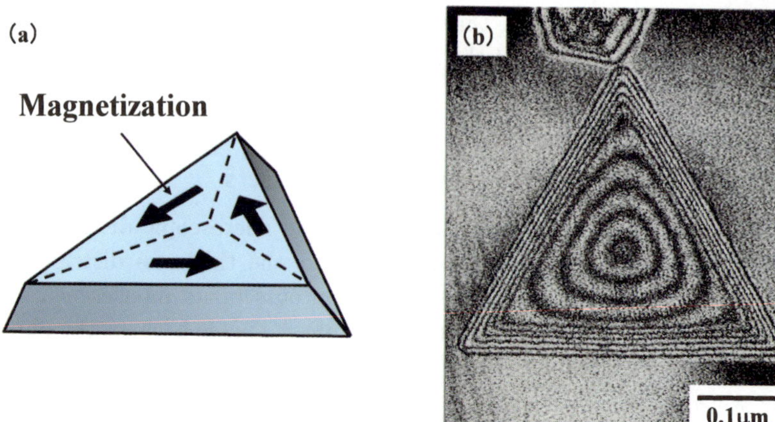

Fig. 5.   Cobalt fine particle: (a) Schematic diagram; (b) Interference micrograph.Only the triangular outline of this particle is observed by electron microscopy. Two kinds of contour fringes appear in its interference micrograph: narrow fringes parallel to the edges indicate the thickness contours, and circular fringes in the inner region indicate in-plane magnetic lines of force.

beam, has been visualized as a black-and-white spot in a defocused image, namely, a Lorentz micrograph.[22]

4.2. *Vortices in superconductors*

Vortices inside a superconducting thin film can be observed when an electron beam passes through it.[22] The experimental arrangement for observing vortices in a superconducting thin film is shown in Fig. 6. When the film is tilted and a magnetic field is applied to produce vortices, electrons passing through vortices inside the film are phase-shifted, or deflected, by the magnetic fields of the vortices. The vortices can be observed by simply defocusing the electron-microscopic image. That is, when the intensity of electrons is observed in an out-of-focus plane, the phase change is transformed into an intensity change and a vortex appears as a pair of bright and dark contrast features as shown in the simulated image (Fig. 6).

We can thus observe the dynamics of vortices in real time, by Lorentz microscopy such as the behaviors of vortices at pinning centers and surface steps, under various sample temperatures and applied magnetic fields. In fact, vortices move in an interesting manners, as if they were living creatures.

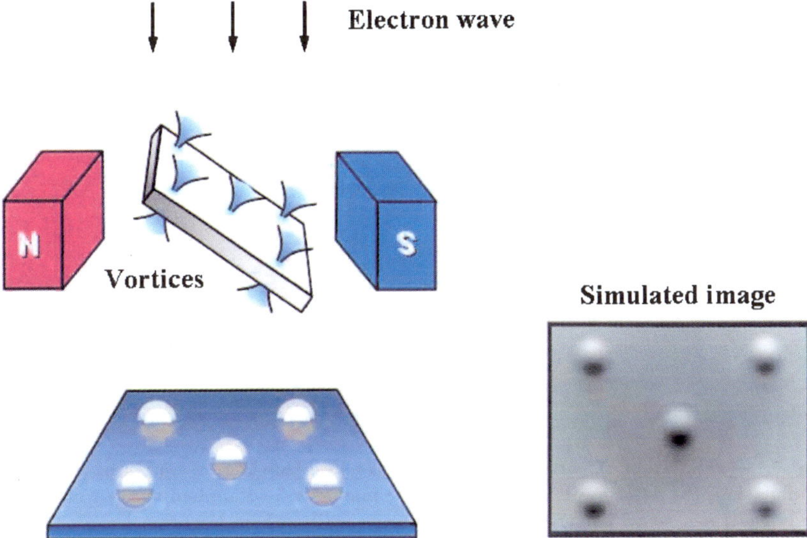

Fig. 6.   Principle behind vortex observation An incident electron wave is phase-shifted or deflected by the magnetic fields of vortices. A vortex appears as a spot consisting of black-and-white contrast by image defocusing (Lorentz microscopy).

An interesting example is shown in Fig. 7, where two kinds of the vortex images appear in a single field of view. The two images are reversed in contrast and, therefore, correspond to vortices and antivortices. They are produced in a niobium thin film when the direction of the magnetic field applied to the film is suddenly reversed. The original vortices try to leave the film, but cannot do so instantly since they are pinned at defects, while the oppositely oriented vortices begin to penetrate the film from its edges. Where two streams of vortices and antivortices collide head-on, the vortex-antivortex pairs of the heads of the two streams annihilate each other. The direct observation of such annihilation of vortices and antivortices can simulate that of particles and antiparticles.

The next example is the obervation of vortices in high-T_c superconductors.

Vortices in high-T_c superconductors behave quite differently due to both the high-temperature operation and the layered structure of the materials. However, because the magnetic radius (penetration depth) of vortices in high-T_c superconductors is ten times larger than that in a niobium film, the film has to be ten times thicker. We were unable to observe them in such

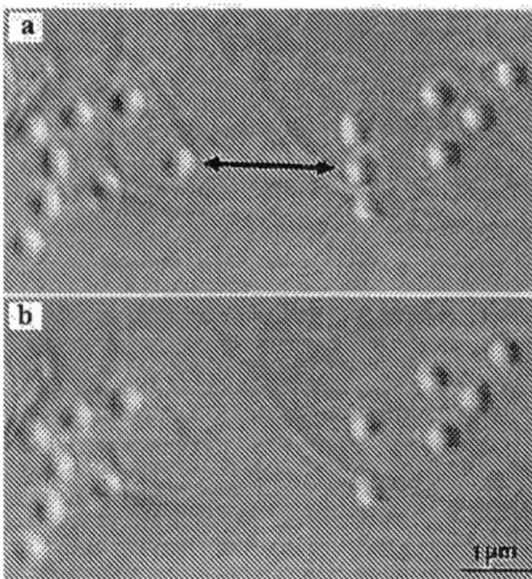

Fig. 7. Annihilation of vortices and antivortices in a thin film of niobium: (a) Before annihilation; (b) After annihilation. When the magnetic field applied to the film is suddenly reversed, some vortices remain at defects while others begin to leave the film. Antivortices begin to move in from the edges of the film. Where streams of vortices and antivortices collide head-on, the vortex-antivortex pairs of the heads of the two stream annihilate each other.

a thick film by using existing 350-kV electrons. To observe the vortices in high-T_c superconductors, we have recently developed a 1-MV field-emission electron microscope[7] (see Fig. 8).

Using this microscope, we observed the internal behavior of vortices inside high-T_c Bi-2212 thin films.[23] The columnar defects, which are produced by the irradiation of high-energy heavy ions and are considered to be optimal pinning traps for vortices in layered structure materials, are produced in Bi-2212 films in a tilted direction [see the electron micrograph shown in Fig. 9(a)]. The tilted columns can be seen as tiny black lines. When these images are defocused, they are blurred and, eventually, disappear completely by spreading out. But when they are defocused even further, vortex images appear, since they are produced by phase contrast. The resultant Lorentz micrograph is shown in Fig. 9(b).

Two different kinds of images, circular images and elongated, weak-contrast images, can be clearly seen in the micrograph. Elongated images (indicated by the arrows in the micrograph) are produced just at the locations of columnar defects and, thus, correspond to vortices trapped along the tilted columns. This conclusion was also confirmed by image simulation. The circular images are produced in regions without any defects; therefore, they correspond to vortices penetrating perpendicularly into the film. Such observation revealed under what conditions vortices are trapped by columnar defects.

Fig. 8.   1-MV field-emission transmission electron microscope.

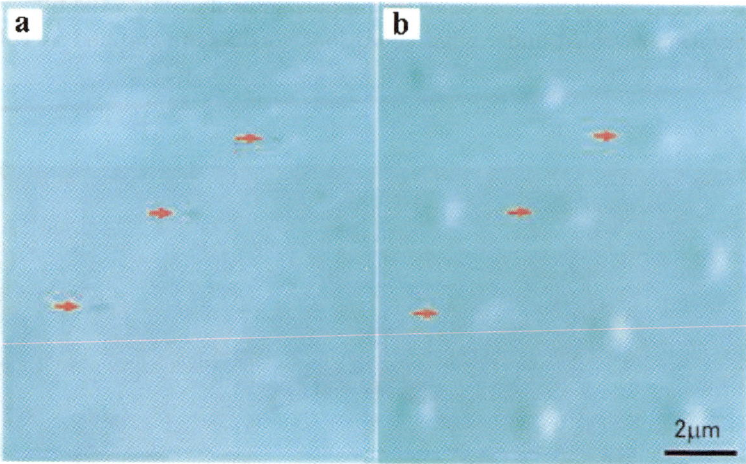

Fig. 9. Comparison of columnar-defect images and vortices in Bi-2212 thin film: (a) Electron micrograph; (b) Lorentz micrograph. Some vortices are trapped at columnar defects and others are untrapped. The images of untrapped vortex lines perpendicular to the film plane are circular spots having bright and dark regions. Vortex images located at the positions of columnar defects are elongated spots with lower contrast indicated by the arrows, since these vortex lines are trapped at columnar defects tilted at 70°.

5. Conclusions

Thanks to the recent developments of advanced technologies such as coherent electron beams, highly sensitive electron detectors, and photolithography, even experiments once regarded as thought experiments can now be carried out. In addition, the wave nature of electrons is utilized in order to observe microscopic objects hitherto unobservable. Examples are the quantitative observation of both microscopic distribution of magnetic lines of force in h/e units and dynamics of quantized vortices in superconductors. These microscopic observations are expected to play an important role in the research and development in nano-science and nano-technology.

References

1. J.J. Thompson, *Proc. Roy. Inst.* **34**, (1897) 419.
2. C. Davisson, L.H. Germer, *Nature* **119**, (1927) 558.
3. G.P. Thomson and A. Reid, *Nature* **119**, (1927) 890.
4. S. Kikuchi, *Jpn. J. of Phys.* **5**, (1928) 83.
5. G. Möllenstedt, H. Düker, *Z. Phys.* **5**, (1928) 83.
6. A. Tonomura, T. Matsuda, J. Endo, H. Todokoro, T. Komoda, *J. Electron Microsc.* **28**, (1979) 1.

7. T. Kawasaki, T. Yoshida, T. Matsuda, N. Osakabe, A. Tonomura, I. Matsui, K. Kitazawa, *Appl. Phys. Lett.* **76**, (2000) 1342.

8. A. Tonomura, *Electron Holography* 2nd Edition, Springer, Heidelberg (1999).

9. R.P. Feynman, R. B. Leighton, M. Sands, *The Feynman Lectures on Physics* Vol. III, Addison-Wesley, Reading, Mass. (1965) 1.1–1.5.

10. Y. Tsuchiya, E. Inuzuka, T. Kurono, M. Hosoda, *Adv. Electron. Electron Phys.* **64**A, (1982) 21.

11. A. Tonomura, J. Endo. T. Matsuda, T. Kawasaki, H. Ezawa, *Am. J. Phys.* **57**, (1989) 117.

12. For example, P.A. Tipler, *Physics,* Freeman Worth (1999) 304.

13. Physics World, **15**, (September 2002) 19.

14. C. Jönsson, *Zeitschrift für Physik* **161**, (1961) 454.

15. Y. Aharonov, D. Bohm, *Phys. Rev.* **115**, (1959) 485.

16. C.N. Yang, *Vector Potential, Gauge Field and Connection on a Fiber Bundle*, K. Fujikawa and Y.A. Ono eds., *Quantum Coherence and Decoherence* — Proc. of Int'l Symp. on Foundations of Quantum Mechanics in the Light of New Technology (ISQM–Tokyo'95), Hatoyama, Saitama, 1995, Elsevier, Amsterdam (1996) 307.

17. T.T. Wu, C.N. Yang, *Phys. Rev.* D**12**, (1975) 3845.

18. M. Peshkin, A. Tonomura, The Aharonov–Bohm Effect, *Lecture Notes in Physics* **340**, Springer, Heidelberg (1989).

19. A. Tonomura, N. Osakabe, T. Matsuda, T. Kawasaki, J. Endo, S. Yano, H. Yamada, *Phys. Rev. Lett.* **56**, (1986) 792.

20. A. Tonomura, T. Matsuda, T. Kawasaki, J. Endo, N. Osakabe, *Phys. Rev. Lett.* **54**, (1985) 60.

21. Q. Ru, G. Lai, K. Aoyama, J. Endo, A. Tonomura, *Ultramicroscopy* **5**, (1994) 209.

22. K. Harada, T. Matsuda, J. Bonevich, M. Igarashi, S. Kondo, G. Pozzi, U. Kawabe, A. Tonomura, *Nature* **360**, (5 November 1992) 51.

23. A. Tonomura, H. Kasai, O. Kamimura, T. Matsuda, K. Harada, Y. Nakayama, J. Shimoyama, K. Kishio, T. Hanaguri, K. Kitazawa, M. Sasase and S. Okayasu, *Nature* **412**, (9 August 2001) 620.

Electron Correlations in Atoms: Hyperspherical Approach to Multiply Excited States of Atoms

Toru Morishita

Department of Applied Physics and Chemistry,
University of Electro-Communications,
Chofu-shi, Tokyo 182–8585, Japan

C. D. Lin

Department of Physics, Kansas State University,
Manhattan, KS 66506, USA

We describe the hyperspherical coordinate method and discuss its application to multiply excited states of atoms. A new classification scheme for triply excited states is presented. It is shown that triply excited states exhibit collective modes similar to rotational and vibrational motions of XY_3 molecules. We also present a new development of the hyperspherical approach to a four-electron atomic system.

1. Introduction

The description of a many-electron atom is based mostly on the independent particle model. In this model each electron is described to be under the influence of a mean field due to the other electrons and the field from the nucleus. The many-electron atomic states are then designated by a collection of quantum numbers from each individual electrons. Such an independent particle model has been used widely to describe the properties of the ground states and singly excited states of atoms. When more than two electrons are simultaneously excited, independent particle model is not adequate any more due to the strong electron correlations. Since the experimental observation of the doubly excited states of He using synchrotron radiation in early 1960's,[1] large effort has been developed to describe electron correlations in the doubly excited states, both experimentally and theoretically. Two electron correlation patterns are now understood to be isomorphic to the stretching and rotational and vibrational modes of a floppy XY_2 molecule, with X being the nucleus and Y the electron.[2-7] Furthermore, the correlation quantum numbers, $_n(K,T)_N^A$, which is based on the molecule like modes, have been proposed to classify doubly excited states in the hyperspherical adiabatic viewpoint.[6] Here, n and N are the approximate principal quantum numbers of the inner and the outer electrons, respectively. $A = +, -$, or 0 represents the symmetry in the radial stretching motion. K describes the bending vibration of the two electrons with nucleus at the center, and T is a projection of the total orbital angular momentum onto the quantization axis of the body-fixed frame which is taken to be along the interelectronic line.[8,9] Nowadays, these correlation quantum numbers are widely used in the literature to replace the quantum numbers based on the independent particle model.

In the last few years, efforts have been made toward the understanding of triply excited states of atoms. With synchrotron radiation light sources, triply excited states of the Li atom have been investigated in many experiments.[10-17] For collisions of multiply charged ions with multielectron targets, triply excited states are also populated through multiple electron capture processes.[18] At the same time, a number of theoretical methods, such as R-matrix method,[19-22] multiconfigurational Hartree-Fock (MCHF) method,[23,24] complex-rotation method,[25-28] K-matrix method,[30,31] and configuration interaction (CI) method,[32-35] have been extended to obtain accurate energies and widths of several triply excited states.

In this work, we present our recent progress on the understanding of electron correlations in triply excited states of atoms using the hyperspherical coordinates. We also present the new development of the hyperspherical approach for four-electron atomic systems.

Atomic units $\hbar = e = m_e = 1$ are used throughout unless explicitly stated otherwise.

2. The hyperspherical method

In this section, we briefly outline the hyperspherical method. A key point of the hyperspherical formalism is to treat the mean square radius, R, as an adiabatic parameter.[36] Correlated motions of the particles are analyzed with respect to the gradual evolution of the system in R.

The Schrödinger equation for an N-election atom in the independent electron coordinates with nucleus at the center is

$$\left[\sum_{i=1}^{N}\left(-\frac{1}{2}\Delta_i^2 + \frac{Z}{r_i}\right) - \sum_{i>j}\frac{1}{|\mathbf{r}_i - \mathbf{r}_j|} - E\right]\Psi = 0,\qquad(1)$$

where Z is the nuclear charge and E is the total energy measured from the N-electron ionization threshold. Here, $E = 0$ represents the threshold energy of the complete break-up of the N electron atom. In the hyperspherical method, $3N$ dimensional configuration space of an N-electron atom is parameterized by a hyperradius

$$R = \sqrt{r_1^2 + \cdots + r_N^2},\qquad(2)$$

characterizing the size of the system, and by $(3N - 1)$ bounded hyperangles Ω describing the relative positions of the electrons. The principle of the adiabatic formalism is independent of particular choices of Ω. For atomic systems, one choice of Ω consists of the $2N$ polar angles of individual electrons $\hat{\mathbf{r}}_i = (\theta_i, \phi_i)$ and the $N - 1$ mock angles α_k defined by[37]

$$r_1 = R\sin\alpha_{N-1}\sin\alpha_{N-2}\cdots\sin\alpha_1$$
$$r_2 = R\sin\alpha_{N-1}\sin\alpha_{N-2}\cdots\cos\alpha_1$$
$$\cdots$$
$$r_{N-1} = R\sin\alpha_{N-1}\cos\alpha_{N-2}$$
$$r_N = R\cos\alpha_{N-1}.\qquad(3)$$

Coordinates in a body-fixed frame with an appropriate choice of Euler angles are also used.[38]

In hyperspherical coordinates, the Schrödinger equation for the rescaled wave function $\psi = R^{\frac{3N-1}{2}}\Psi$ can be written as

$$\left[-\frac{1}{2}\frac{\partial^2}{\partial R^2} + H_{\mathrm{ad}}(\Omega; R) - E\right]\psi = 0,\qquad(4)$$

where $H_{\mathrm{ad}}(\Omega; R)$ is the adiabatic Hamiltonian which depends parametrically on R, namely,

$$H_{\mathrm{ad}}(\Omega; R) = \frac{\Lambda_N^2(\Omega) + (3N - 1)(3N - 3)/4}{2R^2} + \frac{C(\Omega)}{R}.\qquad(5)$$

The kinetic energy scales as R^{-2} as in the first term in the right hand side of Eq. (5). The operator $\Lambda_N^2(\Omega)$ is the square of the grand angular momentum, which contains derivative terms in the variables Ω, and can be written as

$$\Lambda_N^2(\Omega) = R \sum_{i=1}^{N} \frac{l_i^2(\hat{\mathbf{r}}_i)}{r_i} + \ell_N^2(\alpha_{N-1},...,\alpha_1), \tag{6}$$

where l_i is the usual angular momentum operator for each electron, and ℓ_N is the N-dimensional angular momentum operator associated with the mock angles α_k. Details of the properties of the grand angular momentum can be seen in Ref.[39] Total interparticle Coulomb interaction in the second term in the right hand side of Eq. (5) is scaled as R^{-1}, characterized by the effective charge,

$$C(\Omega) = R \left(\sum_i \frac{Z}{r_i} - \sum_{i>j} \frac{1}{|\mathbf{r}_i - \mathbf{r}_j|} \right). \tag{7}$$

In the adiabatic picture introduced by Macek,[36] the total wave function for the state with total spin S can be expanded as

$$\psi^S = \sum_\mu F_\mu^n(R) \left(\sum_{\{S_k\}} \Phi_\mu^{S,\{S_k\}}(\Omega; R)\chi_{\{S_k\}}^S \right), \tag{8}$$

where $\Phi_\mu(\Omega; R)$ is the hyperspherical adiabatic channel function, which contains all the information about electron correlations for states within channel μ, and $\chi_{\{S_k\}}^S$ is the N-electron spin functions with a set of intermediate spins $\{S_k\}$.[40] The channel function $\Phi_\mu(\Omega; R)$ and its associated adiabatic potential $U_\mu(R)$ are obtained by solving the adiabatic eigenvalue problem at each R,

$$[H_{\mathbf{ad}}(\Omega; R) - U_\mu(R)] \Phi_\mu(\Omega; R) = 0. \tag{9}$$

The hyperradial function $F_\mu^n(R)$ is the solution of the second order coupled differential equations,

$$\left[-\frac{1}{2} \frac{\partial^2}{\partial R^2} + U_\mu(R) - E \right] F_\mu(R) = \sum_\nu W_{\mu\nu} F_\nu(R), \tag{10}$$

where

$$W_{\mu\nu} = \left\langle\!\!\left\langle \Phi_\mu \left| \frac{\partial}{\partial R} \right| \Phi_\nu \right\rangle\!\!\right\rangle \frac{d}{dR} + \frac{1}{2} \left\langle\!\!\left\langle \Phi_\mu \left| \frac{\partial^2}{\partial R^2} \right| \Phi_\nu \right\rangle\!\!\right\rangle \tag{11}$$

is the nonadiabatic coupling term. The brackets $\langle\!\langle \cdots \rangle\!\rangle$ in the above equation represent integration with respect to Ω. All the eigen states are obtained by solving the above coupled equations with appropriate boundary conditions.[41] When the couplings among different adiabatic channels are small, the single channel approximation would be adequate to determine the energy levels of the bound states and resonant states. This is mathematically equivalent to the molecular Born-Oppenheimer theory, where the nuclear distance is treated as an adiabatic parameter.

We note that the hyperspherical adiabatic approach has been applied to solve bound states and scattering states as well as to analyze particle correlations in many different fields of physics, such as general Coulomb three-body systems,[42,43] chemical reactions,[44] and recently developed atomic many-body Bose-Einstein condensations.[45]

3. Triply excited states of the three electron systems

We solved the eight-dimensional eigenvalue problem, Eq. (9), numerically for three-electron atomic systems for each $^{2S+1}L^{\pi}$ symmetry.[46] Here, L, S, and π are well-defined total orbital angular momentum, total spin, and parity, respectively. We used basis functions consisting of coupled spherical harmonics for each electron, *i.e.*, products of spherical harmonics in $(\hat{\mathbf{r}}_1, \hat{\mathbf{r}}_2, \hat{\mathbf{r}}_3)$ and direct products of discrete variable representation (DVR) functions in (α_1, α_2). In a typical calculation, the orbital angular momentum for each electron ranges from 0 to 3.

3.1. *Hyperspherical potentials*

In order to see the global features of the eigen states of a three-electron atom, we first examine the adiabatic potentials. As an example, we show the adiabatic potential curves as a function of R calculated for Li(^{2}S^e) states in Fig. 1.

At small R, each potential curve goes up as $1/R^2$ due to the dominance of the kinetic energy in R, which is controlled by the grand angular momentum. At large R, where Coulomb potential energy dominates, each curve approaches one of the two-electron Li$^+$ states as R^{-1}. In the intermediate, each curve has the minimum with a potential well where bound states or resonances can be formed.

The potentials can be classified into three groups by their asymptotic limits. The first group consists of a single curve – the lowest curve, labelled 'I' in Fig. 1. This curve approaches the Li$^+$(1s^2 ^{1}S^e) state as $R \to \infty$. The ground state and all the singly excited states which are normally designated as 1s^2ns ^{2}S^e ($n \geq 2$) in the independent particle model are obtained by solving the hyperradial equation from this curve. For instance, the binding energy of the ground state calculated from the single hyper-spherical adiabatic channel is -7.466 a.u.. which is to be compared with experimental value of -7.478 a.u. More accurate result can be obtained by including more channels. The second group, labelled 'II' in Fig. 1, consists of potential curves that approach the $1snl(n \geq 2)$ singly excited states of Li^+ at large R. These curves support doubly excited states of the types $1snl\,n'l'(n, n' \geq 2)$. The correlation patterns of these doubly excited states can be understood in terms of the motion of the two outer electrons with a core. Namely, if the motion of the innermost electron is averaged out, the correlations of two outer electrons are essentially identical with those in the doubly excited states of a two-electron atom, such as He. Thus, the same set of correlation quantum numbers, $_n(K,T)_N^A$, can be applied to classify those doubly excited states of Li.[47] The third group, labelled 'III', consists of potential curves that approach the $nln'l'(n, n' \geq 2)$ doubly excited states of Li^+ at large R. These curves support triply excited states of the types $nln'l'n''l''(n, n', n'' \geq 2)$. Clearly, the avoided crossings among the different groups are very sharp. Thus, the hyperspherical adiabatic channels can be used to separate singly, doubly, and triply excited states. We will study the wave functions of the triply excited states in the next subsection.

3.2. *Hyperspherical channel functions*

To identify the features that characterize the adiabatic channels among the triply excited states, we examined the channel functions, $\Phi(\Omega; R)$'s. Since correlation is a property of the relative motion among the electrons, the overall rotation of the system does not play a role. Thus, we analyze the wave functions in the body-fixed frame of the atom.[48] In the present analysis of the triply excited states of a three-electron atom, we use α_1 and α_2 in Eq. (3), and three relative angles among the electrons. Let us define the three relative angles θ, η, and ϕ.[49,50,51] The domain of $(r_3 \geq r_2 \geq r_1)$ is enough to be considered for their definition. We first define a σ plane formed by the three electrons. Then we define θ to be the angle between $\mathbf{r}_1$ and $\mathbf{S}_z$ perpendicular to the plane spanned by the three electrons. It takes two other angles η and ϕ to describe the shape of the triangle of the three electrons. On the σ plane, the angle between electrons 1 and 2 is defined to be 2η, chosen along the arc that includes electron 3. The angle between electron 3 and the line bisecting electrons 1 and 2 is defined to be ϕ. See Fig. 2(a). The ranges of the angles are $0 \leq \theta \leq \pi$, $0 \leq \eta \leq \pi$, and $-\eta \leq \phi \leq \eta$. These three angles, together with α_1, α_2, and R, specify a

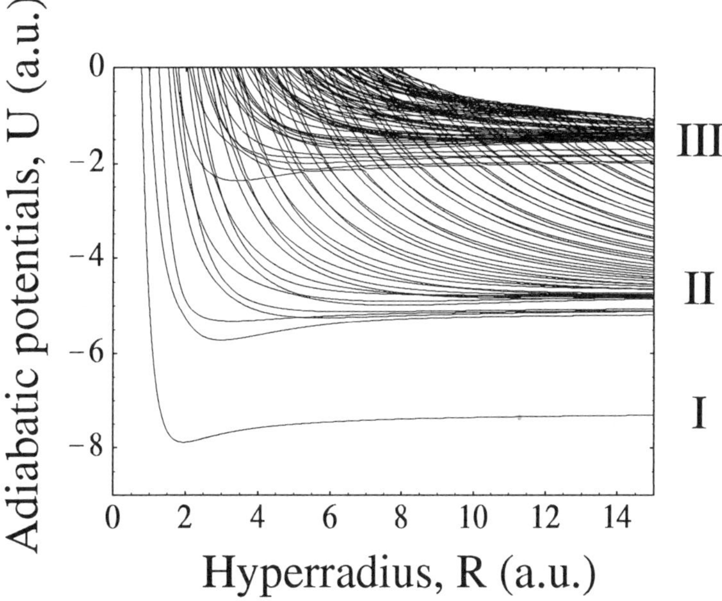

Fig. 1. Hyperspherical adiabatic potentials for the $^2\mathrm{S}^{\mathrm{e}}$ symmetry of Li.

definite shape and size of the three electrons with respect to the nucleus.

To visualize the collective motion of the three electrons, we introduce three-electron density function $\rho_\mu(\Omega_I; R)$ which is defined as the rotation-averaged density distribution for each channel function,

$$\rho_\mu(\Omega_I; R) = \sum_{S_{12}} \int |\Phi_\mu^{S,S_{12}}(\Omega; R)|^2 d\omega, \tag{12}$$

where ω is the Euler angles representing the orientation of the body-fixed frame with respect to the space fixed frame, and Ω_I is the five dimensional internal coordinates of $(\alpha_1, \alpha_2, \theta, \eta, \phi)$. S_{12} denotes an intermediate two-electron spin state. This density represents the probability for the three electrons to take specific shapes. In Fig. 2 we show the density for a triply excited state of Li as an example. The plots in Fig. 2(b) and (c) represent the contour surfaces where the density is 60% of the maximum at a given (α_1, α_2, R), taking to be where the density is maximum for the state. A contour surface of higher density would fit inside the surface. Such contour surfaces would provide information on the most probable shape of the three electrons for each triply excited state.

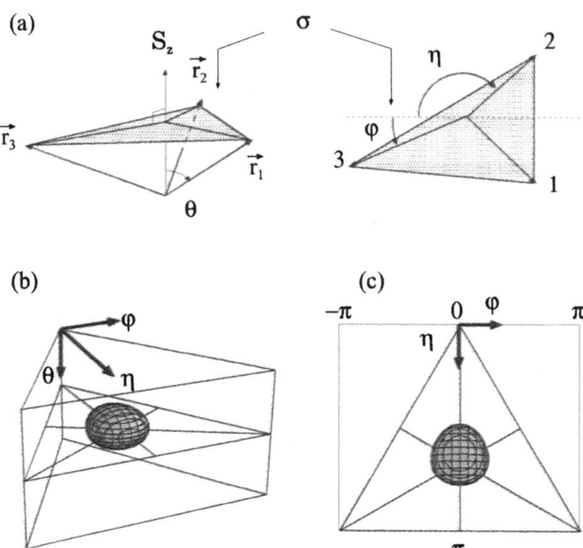

Fig. 2.   Definition of the three angles used to describe the three electrons on a sphere. The three electrons form a σ plane. On the plane (the right figure) the three electrons are confined to a circle.

Here, we analyze the $2l2l'2l''$ intrashell triply excited states, where the three electrons are at about equal distance from the nucleus. Within the independent particle model there are eight intrashell states that can be formed from the 2s and 2p orbitals. They are $2s^2 2p\ ^2P^o$, $2s2p^2$ ($^4P^e$, $^2S^e$, $^2P^e$, $^2D^e$) and $2p^3$ ($^4S^o$, $^2D^o$, $^2P^o$). In Fig. 3 we show the equidensity surfaces of each of these states at $\alpha_1 = \pi/4, \alpha_2 = \arctan\sqrt{2}$ (*i.e.*, $r_1 = r_2 = r_3$) and at $R \sim 3.5$, where the corresponding hyperradial wave functions have maxima. These surfaces can be classified into three groups. In group A we found that there are four states which have similar equidensity surfaces in Fig. 4(a). They are the $^2P^o$, $^4P^e$, $^2D^e$, and $^2D^o$ states. The similarity of the contour surfaces for these four states shows that they have nearly identical internal shape. In fact, the maximum of the density for each figure occurs at $\theta = \pi/2$, $\eta = 2\pi/3$ and $\phi = 0$; that is, the most favorable geometry is a coplanar equilateral triangle with the nucleus at the center of the triangle. This geometry is similar to that of the planar BF_3 molecule. Since these states basically have the same shape, their relative energies are expected to be similar to the rotational excitation of an oblate symmetric top.[48,49] For an oblate symmetric top, the rotation energy is

$$E(L,T) = \frac{1}{2I} \left[2L\left(L+1\right) - T^2 \right],\tag{13}$$

where I is the moment of inertia and T is the projection of L along the direction perpendicular to the plane spanned by the three electrons. To show that this is indeed the case, in Fig. 4 we display the energy levels of these eight states, arranged according to their internal density distributions. The $^2P^o$, $^4P^e$, $^2D^e$ and $^2D^o$ states have the most probable shape of a coplanar equilateral triangle and their relative energies indeed appear to be similar to that of the rotational excitation of an oblate symmetric top.

In Fig. 3(b) we show the equidensity surfaces for the $^4S^o$ and $^2P^e$ states. The $\theta = \pi/2$ plane is clearly forbidden to the three electrons, indicating that the coplanar geometry is not allowed for these two states. Above or below this plane, the three electrons still prefer to form an equilateral triangle. This geometry is similar to the antisymmetric state of the pyramidal NH_3 molecule where the wave function vanishes when the plane of the three hydrogen atoms coincides with the nitrogen atom. The existence of a nodal surface when the σ plane intersects the nucleus implies higher excitation energy since minimum electrostatic repulsion occurs when the three electrons and the nucleus are coplanar. Note that in Fig. 3(b) the energy levels of group B are higher than the levels in group A except for the highest $^2D^o$ state which has higher rotational energy.

The remaining two states belong to Group C and their densities are shown in Fig. 3(c). Both states have the maximum density at the $\theta = \pi/2$ plane, *i.e.*, the three electrons are coplanar. However, the density vanishes in the middle, indicating that the three electrons cannot form an equilateral triangle. We are not aware of any examples of low-lying vibrational states of any polyatomic molecules exhibiting such modes. The 'forbidden region' for the latter two groups B and C originates from the quantum symmetry in that a state with well-defined quantum numbers L, S, and π would incur nodal surfaces in a multidimensional wave function.[48,49,50]

We also examined the internal wave functions for the $2l2l'3l''$ intershell triply excited states of Li, where one electron is far from the other two. For these states, radial correlations as well as angular correlations play an important role in the classification. To extend the classification scheme to include $2l2l'3l''$ intershell states, additional quantum numbers describing the phase of the relative radial stretch in (α_1, α_2) has to be included, such as the A quantum number for doubly excited states. Pairs of superscripts "++" and "+−" are used for such purposes. The first superscript for each

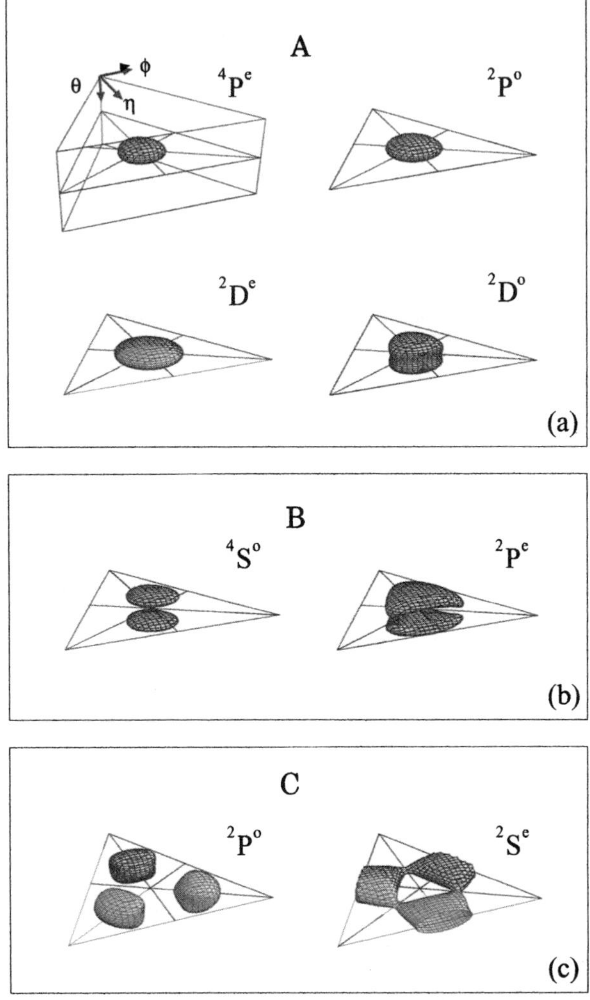

Fig. 3. The equidensity surface plots of the three-electron wave functions for the eight intrashell states at $r_1 = r_2 = r_3$. The surface represents 60% of the maximum density. Each 'slice' represents the whole range of the three angles ($0 \leq \theta \leq \pi$, $0 \leq \eta \leq \pi$, $-\eta \leq \phi \leq \eta$).

pair is to describe the phase of the stretch mode of the two inner electrons, while the second one is to describe the phase of the stretch mode of the outermost electron with respect to the two inner ones. For the "++" states, the radial stretch of all the three electrons are in phase, and their radial

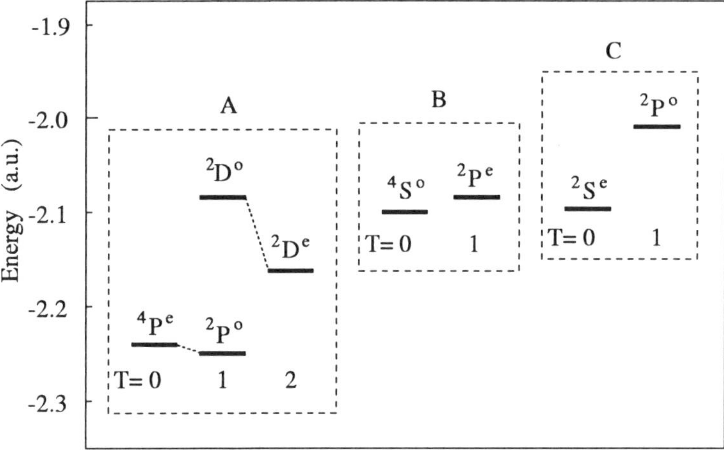

Fig. 4. Energy Levels of the eight $2l2l'2l''$ intrashell triply excited states of Li. Data are taken from B. Davis and K. T. Chung[32] for the $^4S^o$ state, and from K. T. Chung and B. Gou[25] for the other states.

motion is totally symmetric. For the $2l2l'3l''$ states, the two inner electrons form a $2l2l'$ intrashell doubly excited state core. Their radial motion is a symmetric stretch mode and thus the first superscript is always "+" for the $2l2l'3l''$ states. Angular correlation patters of intershell states are essentially the same as for those of intershell states except that the degeneracy in the states in group C disappears and separate into two groups of C_s and C_h. Fully descriptions of the classification of $2l2l'3l''$ states are given in Ref.[51]

4. Quadruply excited states of the four-electron system

The next major step is to understand the four-electron atomic systems, where one can expect much richer electron correlation effects. Theoretically, angular correlations of quadruply excited states have been studied using a model atom.[52] Experimentally, triply excited states of four-electron atoms, such as Li$^-$[53] and Be,[54] are started being investigated using synchrotron radiation. Searching for the quadruply excited states using Synchrotron radiation[54] or in ion-atom collisions[55] are being initiated.

We solved the hyperspherical adiabatic problem for Be($^{1,3,5}S^e$) within the s^4 configurations. The resulting hyperspherical adiabatic potential curves for the $^1S^e$ symmetry are shown in Fig. 5. The general appearance of the adiabatic potentials does not differ markedly from those of three-electron atoms such as Li. At large R, each curve converges to the

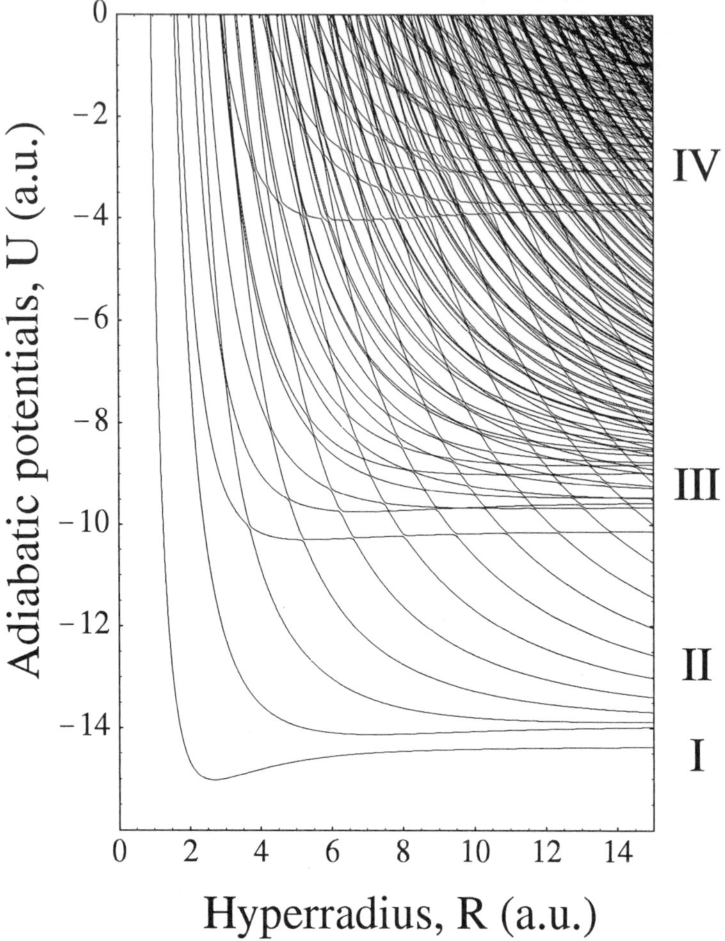

Fig. 5.   Hyperspherical adiabatic potential curves for the $^1S^e$ symmetry of Be within the s^4 configuration.

three-electron Be$^+$ states. The groups labelled 'I', 'II', and 'III' consist of potential curves that support singly, doubly, and triply excited states of Be, respectively, and they are similar to the three-electron systems as shown in Fig. 1. In addition to these three groups, we can clearly observe the fourth group, labeled 'IV'. The curves in the fourth group converge to the triply excited states of Be$^+$ and they support quadruply excited states of Be. We note that the avoided crossings among the different groups are very sharp.

This suggests that quadruply excited states of Be are rather stable against autoionization to singly, doubly, and triply ionized states of Be.

5. Conclusions

We have presented our recent progress made in the application of hyperspherical coordinate method to three- and four-electron atomic systems. By examining the hyperspherical wave functions, we have shown that electron correlations in the triply excited states are understood to be isomorphic to the stretching and rotational and vibrational modes of a floppy XY_3 molecule. We have also shown that the hyperspherical adiabatic potentials for a four-electron atom, and they have been identified with the four groups for singly, doubly, triply, and quadruply excited states. The study of hyperspherical potentials including effects from higher angular momentum states and the classification of the quadruply excited states is currently underway. The analysis of the four-electron correlations is a challenging task for the years to come.

6. Acknowledgments

T.M. would like to express his gratitude to Professor Hiroshi Ezawa for his early guidance into the wonderful world of the "Garden of Quanta". Through Professor Ezawa's instructions he learned the quantum theory of scattering and basic atomic and molecular physics in his undergraduate studies at Gakushuin University. His encouragement and advice over the years to his students is the nourishment that is essential to the continuing bloom of the garden he planted. T.M. also thanks Prof. Y. Azuma for the fruitful discussions on this subject from the experimental point of view. Thanks are also due for Professor M. Matsuzawa and Professor S. Watanabe for their interest in this topic and encouragement throughout this work. This work was supported in part by a Grant-in-Aid for Scientific Research, Ministry of Education, Science, and Culture, Japan, and in part by Chemical Sciences, Geosciences and Biosciences Division, Office of Basic Energy Sciences, Office of Science, U. S. Department of Energy.

References

1. R. P. Madden and K. Codling, Phys. Rev. Lett. **10** 516 (1963); R. P. Madden and K. Codling, Astrophys. J. **141** 363 (1965).
2. J. W. Cooper, U. Fano, and F. Prats, Phys. Rev. Lett. **10**, 518 (1963).
3. M. E. Kellman and D. R. Herrick, J. Phys. B **11**, L755 (1978).

4. M. E. Kellman and D. R. Herrick, Phys. Rev. A **22**, 1536 (1980).

5. G. S. Ezra and R. S. Berry, Phys. Rev. A **28**, 1974 (1983).

6. C. D. Lin, Phys. Rev. A **29**, 1019 (1984).

7. S. Watanabe and C. D. Lin, Phys. Rev. A **34**, 823 (1986).

8. D. R. Herrick and O. Sinanoğlu, Phys. Rev. **11**, 97 (1975).

9. O. Sinanoğlu and D. R. Herrick, J. Chem. Phys. **62**, 886 (1975).

10. L. M. Kiernan, M.-K. Lee, B. F. Sonntag, P. Sladeczek, P. Zimmermann, and E. T. Kennedy, J. Phys. B **28**, L161 (1995).

11. Y. Azuma, S. Hasegawa, F. Koike, G. Kutluk, T. Nagata, E. Shigemasa, A. Yagishita, and I. A. Sellin, Phys. Rev. Lett. **74**, 3768 (1995).

12. Y. Azuma, F. Koike, J. W. Cooper, T. Nagata, G. Kutluk, E. Shigemasa, R. Wehlitz, and I. A. Sellin, Phys. Rev. Lett. **79**, 2419 (1997).

13. L. Journel, D. Cubaynes, J.-M. Bizau, S. Al Moussalami, B. Rouvellou, F. J. Wuilleumier, L. VoKy, P. Faucher, and A. Hibbert, Phys. Rev. Lett. **76**, 30 (1996).

14. S. Diehl, D. Cubaynes, J.-M. Bizau, L. Journel, B. Rouvellou, S. Al Moussalami, F. J. Wuilleumier, E. T. Kennedy, N. Berrah, C. Blancard, T. J. Morgan, J. Bozek, A. S. Schlachter, L. VoKy, P. Faucher, and A. Hibbert, Phys. Rev. Lett. **76**, 3915 (1996).

15. S. Diehl, D. Cubaynes, K. T. Chung, F. J. Wuilleumier, E. T. Kennedy, J.-M. Bizau, L. Journel, C. Blancard, L. VoKy, P. Faucher, A. Hibbert, N. Berrah, T. J. Morgan, J. Bozek, and A. S. Schlachter, Phys. Rev. A **56**, R1071 (1997) .

16. S. Diehl, D. Cubaynes, F. J. Wuilleumier, J.-M. Bizau, L. Journel, E. T. Kennedy, C. Blancard, L. VoKy, P. Faucher, A. Hibbert, N. Berrah, T. J. Morgan, J. Bozek, and A. S. Schlachter, Phys. Rev. Lett. **79**, 1241 (1997).

17. D. Cubaynes, S. Diehl, L. Journel, B. Rouvellou, J.-M. Bizau, S. Al Moussalami, F. J. Wuilleumier, N. Berrah, L. VoKy, P. Faucher, A. Hibbert, C. Blancard, E. Kennedy, T. J. Morgan, J. Bozek, and A. S. Schlachter, Phys. Rev. Lett. **77**, 2194 (1996).

18. M. Zamkov, H. Aliabadi, E. P. Benis, P. Richard, H. Tawara, and T. J. M. Zouros, Phys. Rev. A **65**, 032705 (2002).

19. K. Berrington and S. Nakazaki, J. Phys. B **31**, 313 (1998).

20. L. Vo Ky, P. Faucher, H. L. Zhou, A. Hibbert, Y.-Z. Qu, J. M. Li, and F. Bely-Dubau, Phys. Rev. A **58**, 3688 (1998).

21. H. L. Zhou, S. T. Manson, L. Vo Ky, P. Faucher, F. Bely-Dubau, A. Hibbert, S. Diehl, D. Cubaynes, J.-M. Bizau, L. Journel, and F. J. Wuilleumier, Phys. Rev. A **59**, 462 (1999).

22. H. L. Zhou, S. T. Manson, P. Faucher, and L. Vo Ky, Phys. Rev. A **62**, 012707 (2000).

23. C. A. Nicolaides, N. A. Piangos, and Y. Komninos, Phys. Rev. A **48**, 3578 (1993).

24. C. A. Nicolaides and N. A. Piangos, J. Phys. B **34**, 99 (2001).

25. K. T. Chung and B. C. Gou, Phys. Rev. A **52**, 3669 (1995).

26. K. T. Chung and B. C. Gou, Phys. Rev. A **53**, 2189 (1996).

27. Y. Zhang and K. T. Chung, Phys. Rev. A **58**, 3336 (1998).
28. L. B. Madsen, P. Schlagheck, and P. Lambropoulos, Phys. Rev. A **62**, 062719 (2000).
29. Y. Zhang and K. T. Chung, Phys. Rev. A **58**, 1098 (1998).
30. T. K. Fang and K. T. Chung, Phys. Rev. A **63**, 020702 (2001).
31. K. T. Chung and T. K. Fang, Phys. Rev. A **63**, 062716 (2001).
32. B. Davis and K. T. Chung, Phys. Rev. A **42**, 5121 (1990).
33. G. Verbockhaven and J. E. Hansen, Phys. Rev. Lett. **84**, 2810 (2000).
34. M. J. Conneely and L. Lipsky, Phys. Rev. A **61**, 032506 (2000).
35. M. J. Conneely and L. Lipsky, At. Data Nucl. Data Tables **482**, 115(2002).
36. J. Macek, J. Phys. B**1**, 831 (1968).
37. M. Cavagnero, Phys. Rev. A **30**, 1169 (1984);*ibid* **33**, 2877 (1986);*ibid* **36**, 523 (1987).
38. X. Chapuisat,Phys. Rev. A **45**, 4277 (1992).
39. J. Avery, Hyperspherical harmonics (Kluwer Academic Publishers, 1989).
40. M. Kotani, A. Amemiya, E. Ishiguro, and T. Kimura, Table of Molecular Integrals (Maruzen Co. Ltd., Tokyo, 1955).
41. D. Kato and S. Watanabe, Phys. Rev. A **56**, 3687 (1997).
42. C. D. Lin, Phys. Rep. **257**, 1 (1995).
43. S. Watanabe, T. Morishita, D. Kato, O. I. Tolstikhin, K. Hino, and M. Matsuzawa, Nucl. Instrum. Methods B **124**, 218 (1997).
44. A. Ohsaki and H. Nakamura, Phys. Rep. **187** 1 (1990).
45. J. L. Bohn, B. D. Esry, and C. H. Greene, Phys. Rev. A **58**, 584 (1998).
46. T. Morishita and C. D. Lin, Phys. Rev. A **57**, 4268 (1998).
47. M. Le Dourneuf and. S. Watanabe, J. Phys. B **23**, 3205 (1990).
48. S. Watanabe and C. D. Lin, Phys. Rev. A **36**, 511 (1987).
49. C. G. Bao, X. Yang, and C. D. Lin, Phys. Rev. A **55**, 4168 (1997).
50. T. Morishita and C. D. Lin,Phys. Rev. A, **57**, 1835 (1999).
51. T. Morishita and C. D. Lin,Phys. Rev. A, **67**, 022511 (2003).
52. B. Cheng-guang and T.-K. Lim, J. Phys. B **25**, 3245 (1992); B. Cheng-Guang,J. Phys. B **26**, 4671 (1993); B. Cheng-guang, Phys. Rev. A **47**, 1752 (1993); B. Chengguang and D. Yiwu, Phys. Rev. A **49**, 818 (1994); B. Chengguang, Phys. Rev. A **50**, 2182 (1994).
53. N. Berrah, J. D. Bozek, A. A. Wills, G. Turri, H.-L. Zhou, S. T. Manson, G. Akerman, B. Rude, N. D. Gibson, C. W. Walter, L. VoKy, A. Hibbert, and S. M. Ferguson, Phys. Rev. Lett. **87**, 253002 (2001).
54. Y. Azuma, private communication.
55. T. J. M. Zouros, private communication.

Bifurcation of Periodic Instantons and Quantum–Classical Transition in a Biaxial Anisotropy Nano-Ferromagnet

Yi-Hang Nie

Institute of Theoretical Physics and Department of Physics, Shanxi University,
Taiyuan, Shanxi 030006, China.
Department of Physics, Yanbei Normal Institute, Datong, Shanxi 037000, China

Zhi-Jian Li

Institute of Theoretical Physics and Department of Physics, Shanxi University,
Taiyuan, Shanxi 030006, China

J.-Q. Liang

Institute of Theoretical Physics and Department of Physics, Shanxi University,
Taiyuan, Shanxi 030006, China.
Institute of Physics, Chinese Academy of Sciences, Beijing 100080, China

Q.W. Yan

Institute of Physics, Chinese Academy of Sciences, Beijing 100080, China

Crossover from classical to quantum regimes of the barrier transition rate in a biaxial ferromagnet with magnetic field applied along hard anisotropy axis is investigated as a phase transition process. We show that there exist four types of quantum–classical transition which can be classified by counting the number of bifurcation points. Besides the known type-I and -II, the model possesses also interesting type-III and -IV transitions which were predicted recently.

1. Introduction

Recently considerable attentions have been attracted to the quantum–classical crossover of the barrier transition rate in magnetic nano-particles. The escape from a metastable state or coherent transition between two degenerate states at high temperature is dominated by classical thermal activation which follows the Arrhenins law, $\Gamma_{th} \sim \exp(-\Delta U/T)$, (here ΔU denotes the barrier height) and at ground state by quantum tunneling, $\Gamma_0 \sim \exp(-S_0)$, with tunneling action S_0 independent of temperature. It has been pointed out that between the thermal activation and quantum tunneling at ground state there exists a barrier transition process called the thermally assisted quantum tunneling at excited states with energy E mediated by periodic instantons. Peculiarly the study of the periodic instantons and their stability in double-well potential and sine-Gorden theories[1,2] began only about fifteen years ago. At some critical temperature the crossover from the quantum to classical regimes occurs, which can be viewed as a kind of phase transition. The phase transition can be first order, with a discontinuous first derivative of the action-temperature diagram at the crossover temperature, or second order, with only a jump of the second derivative in the smooth temperature-action curve.[3-5] In fact, the sharp first-order transition is there shown to appear in the plot of action versus temperature and is completely analogous to the plot of free enthalpy versus pressure of a van der Waals gas whose equation of state plotted as pressure versus volume corresponds to the plot of period (of the periodic instanton) versus energy in the consideration of spin system. Of course, the phase transition from quantum to classical regimes stands on its own foot as explained in the following section 2. It was demonstrated under certain assumption on the shape of potential barrier that the phase transition is second-order.[6,7] Chudnovsky,[3] however, showed that if the Euclidean oscillation period $P(E)$ of the instanton is nonmonotonic function of the instanton energy E, the crossover can be the first-order phase transition. Soon after, the sharp first-order transitions were indeed found theoretically in spin systems which are believed to be theoretical models of molecular magnet $Mn_{12}Ac$ and Fe_8.[8-14] In particular, a criterion for the first-order transition is derived,[15] and various effective Hamiltonians of spin systems have been studied intensively.[11,15,16] It has been also shown that the action-temperature diagrams can be classified according to the numbers of bifurcation points of instanton period.[17] Smooth second-order transition is referred to the type-I in which there is no bifurcation point. The type-II is a sharp first-order transition in which there is one bifurcation point and the Euclidean action-temperature curve exhibits an abrupt change at crossover temperature from the periodic instanton to sphaleron, *i.e.*, from quantum tunneling (QT) to thermal activation (TA). Type III has two bifurcation points. The smooth second-order transition occurs between TA and thermally assisted quantum tunneling (TAQT) near the top of barrier, while with decreasing temperature the sharp first-order transition emerges from TAQT to QT (at the ground state), *i.e.* in this type the second-order transition is followed by a first-order transition.[3,18] Although the biaxial anisotropy ferromagnet with the magnetic field along hard anisotropy axis which leads to a new quantum interference phenomenon[14,19,20] has been investigated in the context of quantum–classical transition,[16,18] we in the present paper restudy the transition from quantum to classical regimes in order to explore bifurcation points of periodic instantons in the Euclidean action-temperature diagram and thus, to understand the temperature-dependence of tunneling rate in the whole range of temperature. The external field along hard axis not only increases the barrier height but also changes barrier shape. Varying barrier shape by the change of external field value the model gives rise to the three types of transition mentioned in Ref.[17] and also the interesting type-IV transition[18] with three bifurcation points in which the first-order crossover is followed by another first-order crossover. We in the following start from the effective Lagrangian of the biaxial anisotropy nano-magnet and recover all the four types of transition from quantum to classical regimes by variation of the external magnetic field magnitude.

2. Effective Lagrangian and criterion for first-order transition

Consider a biaxial single domain anisotropy magnet with XOY easy plane and the easy X axis in the XOY plane. When an external magnetic field is applied along the hard anisotropy axis (Z axis), the Hamiltonian operator of the model can be written as

$$\hat{H} = K_1 \hat{S}_z^2 + K_2 \hat{S}_y^2 - g\mu_B B \hat{S}_z, \tag{1}$$

where K_1 and K_2 are the longitudinal and transverse anisotropy constants, respectively, satisfying the condition $K_1 > K_2$. The nano-magnet possesses (in the absence of external magnetic field) two easy orientations along positive and negative X axis respectively, which may be viewed as a kind of Schrödinger cat states denoted by $|\Psi_+\rangle$, $|\Psi_-\rangle$ since the magnetization vector in solid is considered as a classical variable. The study on quantum tunneling in nano-magnets has become an attractive field in which a notable phenomenon is the coherent tunneling through the anisotropy barrier between degenerate easy orientations of the magnet. The coherent tunneling removes the degeneracy of the ground states and then results in superposition of macroscopically distinguishable states, the understanding of which is a longstanding problem in quantum mechanics. We begin with the barrier transition amplitude between the two degenerate states $|\Psi_+\rangle$, $|\Psi_-\rangle$, *i.e.*

$$\langle \Psi_+(P)|\Psi_-(-P)\rangle = \langle \Psi_+|e^{-2P\hat{H}}|\Psi_-\rangle \tag{2}$$

which has a path-integral representation where P denotes the characteristic imaginary time period for the quantum tunneling (in the natural unit $k_B = \hbar = 1$, k_B is Boltzmann's constant). With the help of spin-coherent-state path integral the transition amplitude can be reduced to

$$\langle \Psi_+|e^{-2P\hat{H}}|\Psi_-\rangle = \int [D\phi]e^{-S_E}, \qquad S_E = \int_{-P}^{P} L_E d\tau \tag{3}$$

which is a path-integral of one variable ϕ only, seen to be the azimuthal angle of the manitization vector. The effective Euclidean Lagrangian is found as

$$L_E = \frac{1}{2}m(\phi)\left(\frac{d\phi}{d\tau}\right)^2 + V(\phi), \tag{4}$$

with a position-dependent mass defined by

$$m(\phi) = \frac{1}{2K_1\left(1 - \lambda \sin^2 \phi\right)}, \tag{5}$$

and the effective potential given as

$$V(\phi) = K_2 S^2 \sin^2 \phi \left(1 - \frac{h^2}{1 - \lambda \sin^2 \phi}\right), \tag{6}$$

where S is spin, $\lambda = K_2/K_1$ and $h = B/B_c$ $(B_c = 2K_1 S/g\mu_B)$ and τ denotes the imaginary time. From the effective potential we see that the macroscopic quantum states $|\Psi_+\rangle$, $|\Psi_-\rangle$ correspond to the two local minima of the potential at $\phi = 0, \pi$. Since the minimum is zero, the ground state energy is zero in the classical limit. The path-integral Eq.(3) can be evaluated in terms of stationary phase method by which the tunneling at finite energy E is dominated by the periodic instanton which minimizes the Euclidean action $(\delta S_E = 0)$ and therefore is the solution of the following integrated equation of motion:

$$\frac{1}{2}m(\phi)\left(\frac{d\phi}{d\tau}\right)^2 - V(\phi) = -E, \tag{7}$$

where the integration constant E is nothing but the classical energy of the system. The Euclidean action evaluated along the instanton trajectory is written as

$$S_E = \int L_E d\tau = W(E) + EP(E), \tag{8}$$

where

$$W(E) = 2\sqrt{2} \int_{\phi^i(E)}^{\phi^f(E)} \sqrt{m(\phi)\left[V(\phi) - E\right]} d\phi \tag{9}$$

is the known WKB tunneling action and

$$P(E) = \sqrt{2} \int_{\phi^i(E)}^{\phi^f(E)} \sqrt{\frac{m(\phi)}{V(\phi) - E}} d\phi \tag{10}$$

is the period of instanton as a function of energy E. $\phi^i(E)$ and $\phi^f(E)$ are turning points. In the leading order approximation the tunneling rate at given energy is seen to be

$$\Gamma_E \sim e^{-S_E}, \tag{11}$$

which reduces to the tunneling rate at ground state *i.e.* $\Gamma_0 \sim \exp(-S_0)$ when energy tends to zero. On the other hand, the tunneling rate at temperature T can be evaluated by the usual thermal average of the WKB tunneling rate over all contributions from periodic instantons with various energy such that

$$\Gamma(T) \sim \int e^{-\left[W(E') + \frac{E'}{T}\right]} dE', \tag{12}$$

which is called the thermally assisted tunneling with an obvious meaning that the system is thermally activated to excited states with energy E' and tunneling takes place at excited states. The optimum path which gives rise to the maximum contribution to the tunneling rate Eq.(12) is denoted by

$$\Gamma(T) \sim e^{-S_T}, \qquad S_T = W(E) + \frac{E}{T}, \tag{13}$$

where energy E is understood as the selected energy for the optimum path at temperature T. Comparing S_T with the Eclidean action evaluated along the instanton trajectory Eq.(8), tunneling rate Γ_E may be identified with Γ_T by the condition,

$$T = \frac{1}{P(E)}. \tag{14}$$

Thus the phase transition from quantum to classical regimes can be studied directly from the plot of S_T versus temperature T and the plot of instanton period $P(E)$ versus energy which are shown in Ref.[5] for both second-order (type-I) and first-order (type-II) transitions. The thermodynamic action is given by

$$S_{th} = \frac{\Delta U}{T},\tag{15}$$

with ΔU being the barrier height. In semiclassical approximation the transition rate at temperature T is given by Ref.[3]

$$\Gamma(T) \sim e^{-S_{\min}(T)},\tag{16}$$

where $S_{\min}(T)$ is the actual action which goes along the smaller one (therefore giving higher tunneling rate) of these two actions S_{th} and S_T. We are interested in the crossover from quantum tunneling to the thermodynamic transition at some critical temperature which can be considered as a kind of phase transition seen from the plot of instanton action versus temperature. There are two criteria for the existence of first-order transition between the classical and quantum regime. The nonmonotonic behavior of the instanton period as a function of energy, *i.e.*, the existence of a minimum in the $P \sim E$ curve, is used as a condition for first-order phase transition.[3] When the condition is satisfied, the corresponding temperature dependence of the action has an abrupt change at some critical temperature $T_0^{(1)}$, *i.e.* the first-order derivative of $S_{\min}(T)$ with respect to temperature is discontinuous. The physical reason for the first-order transition is clear. Beyond the minimum of the $P \sim E$ plot the thermally assisted tunneling would be strongly suppressed by the Boltzmann's factor *i.e.* $e^{-E/T}$ in Eq.(13) by seeing from Eq.(14) that the increasing period with energy gives rise to a decreasing temperature. The tunneling from ground state at zero temperature may directly jumps to the thermal activation. The criterion can give the action-temperature diagram of first-order transition from quantum to classical regimes and also the critical temperature $T_0^{(1)}$. Another criterion for the first-order transition was derived in Ref.[13] and can be used to obtain the phase boundary.

3. Phase transition with field along hard axis

The effective potential $V(\phi)$ in Eq.(6) has three different shapes depending on the range of field value. For the sake of convenience we study the transi-

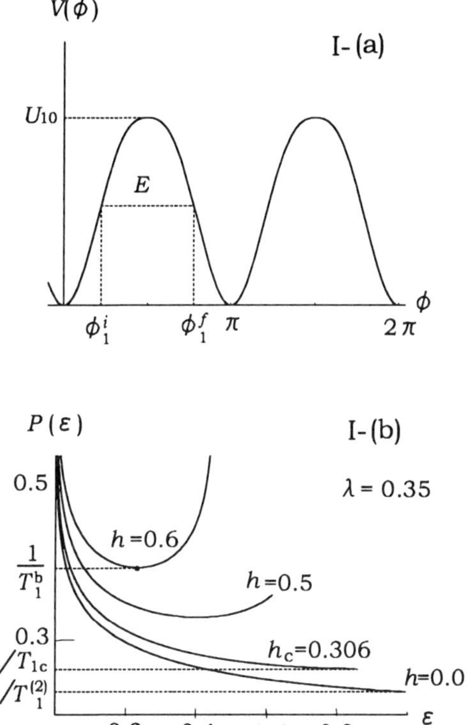

Fig. 1.   Potential, periods of periodic instantons and action-temperature diagrams for $\lambda = 0.35$ and $0 < h < 1 - \lambda$ with $S^2 = 100$ and $K_2 = 1$.   I-(a) Effective potential.   I-(b) The periods of periodic instantons.

tion of quantum to classical regime for the model in three different regions separately.

3.1. *The case of $0 < h < 1 - \lambda$*

The potential for $0 < h < 1 - \lambda$ is shown in Fig.1 I-(a) and looks like a regular barrier between $\phi = 0$ and $\phi = \pi$. The turning points corresponding to energy E are

$$\phi_1^i(\varepsilon) = \arcsin \sqrt{\frac{G_1 - Q_1}{2\lambda}}, \tag{17}$$

$$\phi_1^f(\varepsilon) = \pi - \arcsin \sqrt{\frac{G_1 - Q_1}{2\lambda}}, \tag{18}$$

where $G_1 = 1 - h^2 + \lambda\varepsilon$ ($\varepsilon = E/K_2 S^2$) and $Q_1 = \sqrt{G_1^2 - 4\lambda\varepsilon}$. From Eqs.(8), (6) and (7) we obtain the instanton period and Euclidean action as

$$P_1(\varepsilon) = \frac{g_1}{K_2 S} \mathcal{K}(k_1), \tag{19}$$

$$S_{1E}(\varepsilon) = S\left\{ \varepsilon g_1 \mathcal{K}(k_1) - 4\sqrt{\frac{Q_1}{\lambda}} \left[(1 - \alpha_1^2)\Pi(\alpha_1^2, k_1) - (1 - \bar{\alpha}_1^2)\Pi(\bar{\alpha}_1^2, k_1) \right] \right\}, \tag{20}$$

with

$$g_1 = \frac{2\sqrt{\lambda}}{\sqrt{Q_1}}, \qquad k_1^2 = \frac{1}{2}\left(1 + \frac{G_1 - 2\varepsilon}{Q_1}\right),$$

$$\alpha_1^2 = \frac{1}{2}\left(1 - \frac{G_1 - 2\lambda}{Q_1}\right), \qquad \bar{\alpha}_1^2 = \frac{2 - Q_1 - G_1}{2(1 - \lambda)}\alpha_1^2,$$

where $\mathcal{K}(k_1)$ and $\Pi(\alpha_1^2, k_1)$ are the complete elliptic integral of the first and the third kind. Corresponding thermodynamic action is

$$S_{10}(T) = \frac{U_{10}}{T}, \tag{21}$$

where $U_{10} = K_2 S^2(1 - \frac{h^2}{1-\lambda})$. The effect of the magnetic field on the behavior of period and the type of phase transition are shown in Fig.1 I-(b), Fig.2 I-(c),(d) for $S^2 = 100$, $K_2 = 1$ and $\lambda = 0.35$. The period $P_1(\varepsilon)$ decreases monotonically with energy increasing for $h < h_c$ (Fig.1 I-(b)) and corresponding action-temperature curve with smooth second-order transition at $T_1^{(2)}$ is shown in Fig.2 I-(c) where there is no bifurcation point. When $h > h_c$ the period $P_1(\varepsilon)$ has a minimum at the point of $dP_1/d\varepsilon = 0$ which corresponds to the bifurcation point (corresponding temperature is T_1^b) in the action-temperature diagram. Fig.2 I-(d) displays the behavior of the optimum path $S_{\min}$ going along the minimum action of S_T and S_{th} with temperature decreasing. S_T and S_{th} coincides at temperature $T_1^{(1)}$ but their first-order derivative with respect to T has a jump.

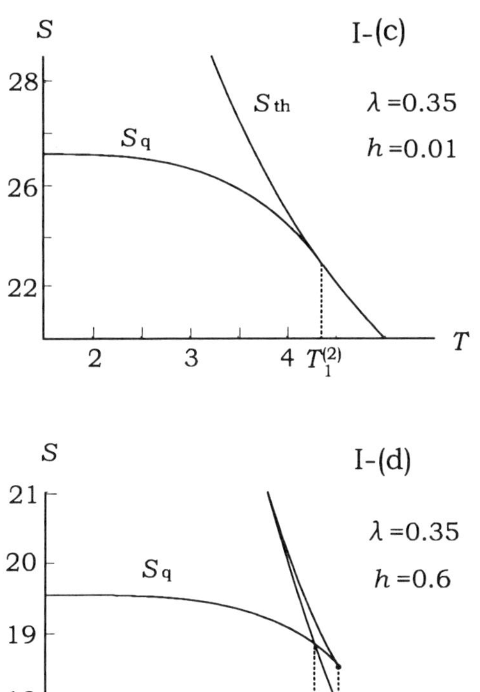

Fig. 2. Potential, periods of periodic instantons and action-temperature diagrams for $\lambda = 0.35$ and $0 < h < 1 - \lambda$ with $S^2 = 100$ and $K_2 = 1$. I-(c) Type I: second-order transition (no bifurcation point). I-(d) Type II: first-order transition (one bifurcation point).

3.2. *The case of* $1 - \lambda < h < (1 - \lambda)^{1/2}$

The effective potential for $1 - \lambda < h < \sqrt{1 - \lambda}$ is shown in Fig.3 II-(a). The minimum at $\phi = \pi/2$ is a metastable one. $\phi = 0$ and $\phi = \pi$ are two degenerate ground states. Consider a trajectory (joining ϕ_2^i and $\pi - \phi_2^i$) with the energy E such that $U_2 < E < U_{20}$ ($U_2 = K_2 S^2 (1 - \frac{h^2}{1-\lambda})$, $U_{20} = K_2 S^2 \frac{(1-h)^2}{\lambda}$). The trajectory consists of motion in imaginary time from ϕ_2^i to ϕ_2^f and from $\pi - \phi_2^f$ to $\pi - \phi_2^i$ and motion in real time from ϕ_2^f to

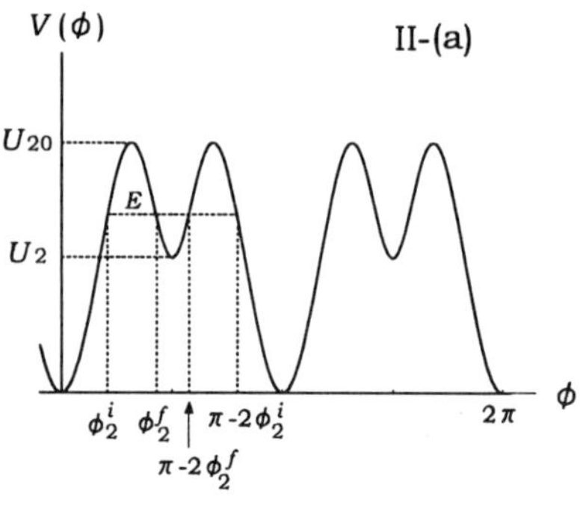

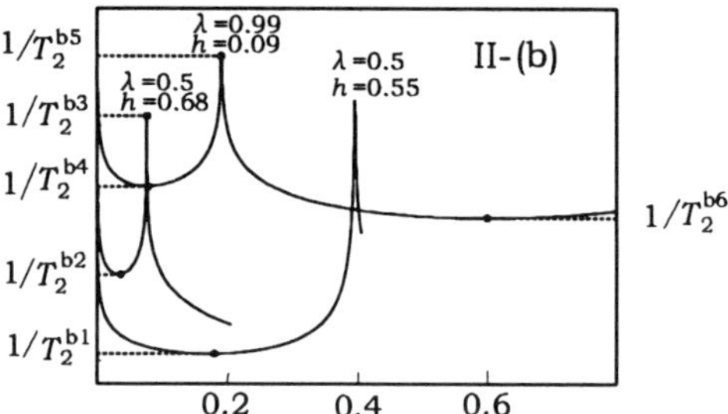

Fig. 3. Potential, periods of periodic instantons and action-temperature diagrams for $\lambda = 0.5$, 0.99 and $1 - \lambda < h < \sqrt{1 - \lambda}$ with $S^2 = 100$ and $K_2 = 1$. II-(a) Effective potential. II-(b) The periods of periodic instantons.

$\pi - \phi_2^f$. The contribution to the Euclidean action in Eq.(10) comes from two imaginary-time parts of the trajectory, while real-time part only affects the quantum phase interference. The turning points for given energy E are

$$\phi_2^i(\varepsilon) = \arcsin \sqrt{\frac{G_2 - Q_2}{2\lambda}}, \tag{22}$$

$$\phi_2^f(\varepsilon) = \arcsin \sqrt{\frac{G_2 + Q_2}{2\lambda}}, \tag{23}$$

with $G_2 = G_1$ and $Q_2 = Q_1$. From Eqs.(10), (5) and (6) the period of periodic instanton and Euclidean action for the trajectory considered above are obtained as

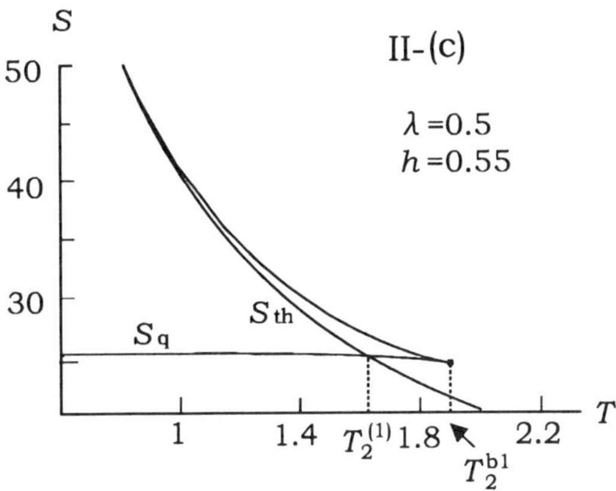

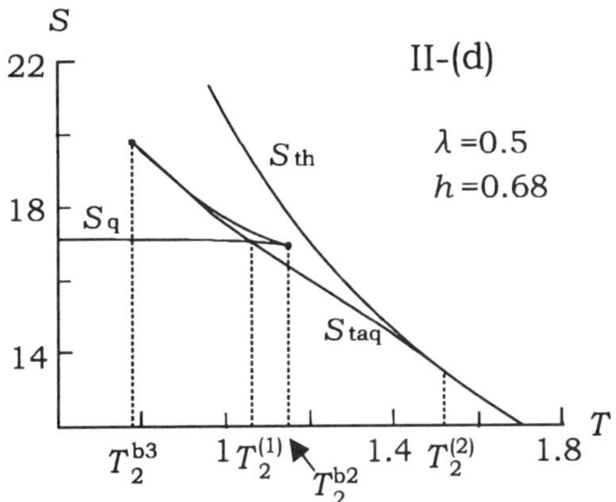

Fig. 4. Potential, periods of periodic instantons and action-temperature diagrams for $\lambda = 0.5$ and $1 - \lambda < h < \sqrt{1 - \lambda}$ with $S^2 = 100$ and $K_2 = 1$. II-(c) Type II: first-order transition (one bifurcation point). II-(d) Type III: the second-order transition followed by a first-order transition (two bifurcation points).

$$P_2(\varepsilon) = \frac{g_2}{K_2 S}\mathcal{K}(k_2), \tag{24}$$

$$S_{2E}(\varepsilon) = \varepsilon S g_2 \mathcal{K}(k_2) + S g_2 q_2 \left\{ \left(\frac{1}{\alpha_2^2} - \frac{1}{\bar{\alpha}_2^2} \right) \mathcal{K}(k_2) \right.$$
$$\left. + \left(1 - \frac{1}{\alpha_2^2} \right) \Pi\left(\alpha_2^2, k_2\right) - \left(1 - \frac{1}{\bar{\alpha}_2^2} \right) \Pi\left(\bar{\alpha}_2^2, k_2\right) \right\}, \tag{25}$$

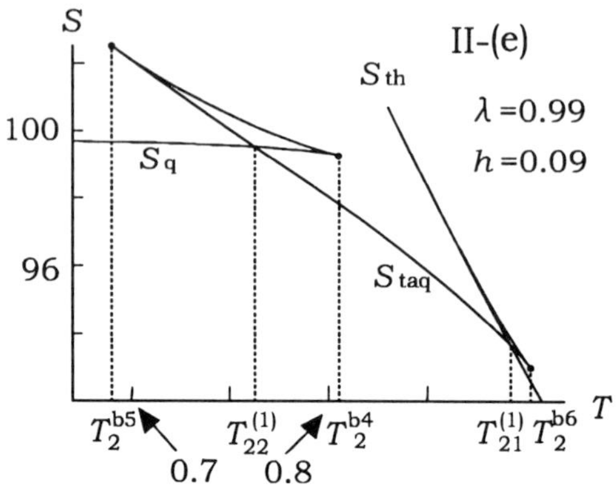

Fig. 5.   Potential, periods of periodic instantons and action-temperature diagrams for $\lambda = 0.99$ and $1 - \lambda < h < \sqrt{1 - \lambda}$ with $S^2 = 100$ and $K_2 = 1$.   II-(e) Type IV: the first-order transition by another first-order transition (three bifurcation points)

where

$$g_2 = \sqrt{\frac{8\lambda}{G_2 + Q_2 - 2\varepsilon}}, \quad k_2^2 = \frac{2Q_2}{G_2 + Q_2 - 2\varepsilon},$$

$$q_2 = \frac{2Q_2}{\lambda}, \quad \alpha_2^2 = \frac{2Q_2}{Q_2 - G_2 + 2\lambda}, \quad \bar{\alpha}_2^2 = \frac{2(1 - \lambda)}{2 - G_2 - Q_2} \alpha_2^2.$$

The corresponding thermodynamic action is

$$S_{20} = \frac{U_{20}}{T}. \tag{26}$$

For the trajectory with energy E such that $0 < E < U_2$ the period of periodic instanton and Euclidean action have the same forms as that of the case of $0 < h < 1 - \lambda$. The Fig.3 II-(b) shows the period as a function of instanton energy for $\lambda = 0.5$ and $\lambda = 0.99$. When $\lambda = 0.5$ and $h < h_c$ ($h_c = 0.618$ as seen from Fig.1 I-(b)) there is only one bifurcation point corresponding to temperature T_2^{b1} and the sharp first-order transition takes place at temperature $T_2^{(1)}$ (Fig.4 II-(c)). The periodic instanton in the range of $dP/d\varepsilon > 0$ dose not contribute to the transition rate since there is another classical solution with the same period while lower action.

It is interesting to study the behavior of quantum–classical transition for $h > h_c$. The competition between TA and TAQT near the top of barrier leads to the smooth second-order transition occurring at $T_2^{(2)}$ (Fig.4 II-(d). When $T < T_2^{(2)}$, $S_{\min}$ goes along action S_{taq} (thermally assisted quantum tunneling) and at temperature $T_2^{(1)}$ sharp first-order transition appears as the result of competion between TAQT and QT. The bifurcation point corresponding to the minimum of the period is at T_2^{b2}. The other bifurcation point appearing at T_2^{b3} is not a proper bifurcation point since it is not a proper extremum point of period $P(\varepsilon)$, but a common end of S_{taq} and the action corresponding to the periodic instanton of $dP/d\varepsilon > 0$, it may be called a bifurcation point. Thus Fig.4 II-(d) gives a behavior of type-III transition, *i.e.* the second-order transition followed by a first-order transition. For $\lambda = 0.5$, $h_c = 0.618$ is a critical field value over which the type-III of quantum–classical transition occurs. $\lambda \sim 1$ is a more interesting case in which both the type-III, and the type-IV transitions appear. For $\lambda = 0.99$ and $h = 0.09$, the period as function of instanton energy has two minima (Fig.3 II-(b)) to which corresponding bifurcation points in the action-temperature diagram are at temperature T_2^{b4} and T_2^{b6} respectively (Fig.4 II-(c)). With temperature decreasing, the third bifurcation point appears at $T = T_2^{b5}$ as the common end of S_{taq} and the action corresponding to the periodic instanton of $dP/d\varepsilon > 0$. When $T = T_{21}^{(1)}$ the first-order transition occurs. With temperature decreasing the optimum path $S_{\min}$ goes along S_{taq}, and when $T = T_{22}^{(1)}$ another first-order transition takes place. Fig.5 II-(e) displays the behavior of the type IV of transition.

3.3. *The case of* $(1 - \lambda)^{1/2} < h < 1$

When $\sqrt{1-\lambda} < h < 1$, $\phi = 0$ and $\phi = \pi$ are still two degenerate states, though the minima at $\phi = \pm\frac{\pi}{2}$ become the global minima. For the trajectory with energy E such that $0 < E < U_{30}$ ($U_{30} = U_{20}$) the period of periodic instanton and Euclidean action have the same forms as Eq.(19) and Eq.(20). In this case only type-I of transition occurs in the crossover from quantum to classical regime.

4. Conclusion

We have studied the crossover of the quantum–classical transition of the escape rate in a biaxial ferromagnet with a magnetic field applied along hard anisotropy axis. We show that the phase transition can be classified by counting the number of bifurcation points in the action-temperature diagram as four types in agreement with theoretical prediction in Refs.[17,3] The change of phase transition type is achieved by variation of the parameters λ and h which determine the shape of the effective potential. This indicates that it is important to vary the shape of potential by adjusting environmental parameters in order to observe rich quantum tunneling phenomena.

Acknowledgements

Supported by the National Natural Science Foundation of China (Grant No.10075032 and 10175039), Shanxi Natural Science Foundation (Grant No.2001007), and Fund of Chinese Academy of Sciences entitled "Nanomic Science and Technology".

References

1. N. S. Manton and T. S. Samols, Phys. Lett. B **207**, 179 (1988).
2. J.-Q. Liang, H.J.W. Müller-Kirsten, and D.H. Tchrakian, Phys. Lett. B **282**, 105 (1992).
3. E. M. Chudnovsky, Phys. Rev. A **46** , 8011 (1992).
4. D. A. Gorokhov and G. Blatter, Phys. Rev. B **56**, 3130 (1997).
5. J.-Q. Liang and H.J.W. Müller-Kirstenr, D.K. Park, and F. Zimmerschied, Phys. Rev. Lett. **81**, 216 (1998).
6. I. Affleck, Phys. Rev. Lett. **46**, 388 (1981).
7. A. I. Larkin and Yu N. Ovchinnikov, Pis'ma Zh. Eksp. Teor. Fiz. **37**, 322 (1983) [1983 JETP Lett. **37**, 382].
8. E.M. Chudnovsky and D.A. Garanin, Phys. Rev. Lett. **79**, 4469 (1997).
9. D.A. Garanin and E.M. Chudnovsky, Phys. Rev. B **56** 11102 (1997); D.A. Garanin, E.M. Martuky, Phys. Rev. B **57**, 13639 (1998).
10. D.A. Garanin and E.M. Chudnovsky, Phys. Rev. B **59**, 3671 (1999).
11. Y.B. Zhang, J.-Q. Liang , H.J.W. Müller-Kirsten, S.P. Kou, X.B. Wang, F.C. Pu, Phys. Rev. B **60**, 12886 (1999).
12. S.Y. Lee , H.J.W. Müller-Kirsten, D.K. Park, and F. Zimmerschied, Phys. Rev. B **58**, 5554 (1998).
13. W. Wernsdorfer and R. Sessoli, Science **284**, 133(1999) ; A. Garg, Phys. Rev. B **60** 6705 (1999).
14. E. M. Chudnovsky and H.X. Martcednez, Europhys. Lett. **50**, 395 (2000).
15. H.J.W. Müller-Kirsten, D.K. Park, and J.M.S. Rana, Phys. Rev. B **60**, 6662 (1999).
16. Chang-Soo Park, Sahng-Kyoon Yoo, and Dal-Ho Yoon, Phys.Rev. B **61**, 11618 (2000).
17. Hyun-Soo Min, H. Kim, D.K. Park, Soo-Young Lee, Sahng-Kyoon Yoo, Yoon

Dal-Ho Yoon, Phys. Lett. B **469**, 193 (1999); Soo-Young Lee, H. Kim, D.K. Park, and J.K. Kim, Phys. Rev. B **60**, 10086 (1999).

18. E.M. Martunovsky, J. Phys.: Condens. Matter **12**, 4243 (2000).
19. A. Garg, Europhys. Lett. **22**, 205 (1993).
20. Soo-Young Lee and Sahng-Kyoon Yoo, Phys. Rev. B **62**, 13884 (2000).
21. Y.B. Zhang, Y.H. Nie, S.P. Kou, X.B. Wang, J.-Q. Liang, and F.-C. Pu, Chin. Phys. Lett. **15**, 638 (1998).

The Feynman Path Integral:
An Historical Slice

John R. Klauder

Departments of Physics and Mathematics
University of Florida
Gainesville, FL 32611
E-mail: klauder@phys.ufl.edu

Efforts to give an improved mathematical meaning to Feynman's path integral formulation of quantum mechanics started soon after its introduction and continue to this day. In the present paper, one common thread of development is followed over many years, with contributions made by various authors. The present version of this line of development involves a continuous-time regularization for a general phase space path integral and provides, in the author's opinion at least, perhaps the optimal formulation of the path integral.

1. The Feynman path integral, 1948

Much has already been written about Feynman path integrals, and, no doubt, much more will be written in the future. A comprehensive survey after more than fifty years since their introduction would be a major undertaking, and this paper is not such a survey. Rather, it is an attempt to follow one relatively narrow development regarding a special form of regularization used in the definition of path integrals. Since we deal with several different approaches, this paper does not go too deeply into any one of them; it is intended more as a conceptual overview rather than a detailed exposition. In order to set the stage, let us start our discussion with a review of the traditional approach to path integral construction.

We begin with the Schrödinger equation

$$i\hbar \frac{\partial \psi(x,t)}{\partial t} = -\frac{\hbar^2}{2m} \frac{\partial^2 \psi(x,t)}{\partial x^2} + V(x)\psi(x,t) \tag{1}$$

appropriate to a particle of mass m moving in a potential $V(x)$, $x \in \mathbb{R}$. A solution to this equation can be written as an integral,

$$\psi(x'',t'') = \int K(x'',t'';x',t')\,\psi(x',t')\,dx' \,, \tag{2}$$

which represents the wave function $\psi(x'',t'')$ at time t'' as a linear super-position over the wave function $\psi(x',t')$ at the initial time t', $t' < t''$. The integral kernel $K(x'',t'';x',t')$ is known as the *propagator*, and according to Feynman[1] it may be given by

$$K(x'',t'';x',t') = \mathcal{N} \int e^{(i/\hbar)\int [(m/2)\dot{x}^2(t) - V(x(t))]\,dt}\,\mathcal{D}x \,, \tag{3}$$

which is a formal expression symbolizing an integral over a suitable set of paths. This integral is supposed to run over all continuous paths $x(t)$, $t' \le t \le t''$, where $x(t'') = x''$ and $x(t') = x'$ are fixed end points for all paths. Note that the integrand involves the classical Lagrangian for the system.

Unfortunately, although highly suggestive, the preceding expression for the path integral is *undefined* as it stands: For example, the normalization constant $\mathcal{N}$ diverges, and the putative translation invariant measure $\mathcal{D}x$ does not exist. To overcome these basic problems, Feynman adopted a *lattice regularization* as a procedure to yield well-defined integrals which was then followed by a limit as the lattice spacing goes to zero called the continuum limit. With $\epsilon > 0$ denoting the lattice spacing, the details regarding the lattice regularization procedure are given by

$$K(x'',t'';x',t') = \lim_{\epsilon \to 0} (m/2\pi i\hbar\epsilon)^{(N+1)/2} \int \cdots \int$$
$$\times \exp\{(i/\hbar)\Sigma_{l=0}^{N}[(m/2\epsilon)(x_{l+1} - x_l)^2 - \epsilon V(x_l)]\}\,\Pi_{l=1}^{N}\,dx_l \,, \tag{4}$$

where $x_{N+1} = x''$, $x_0 = x'$, and $\epsilon \equiv (t'' - t')/(N+1)$, $N \in \{1,2,3,\dots\}$. In this version, at least, we have an expression that has a reasonable chance of being well defined, provided, of course, that one interprets the conditionally convergent integrals involved in an appropriate manner. One common and fully acceptable interpretation adds a convergence factor to the exponent of the preceding integral in the form

$$-(\epsilon^2/2\hbar)\Sigma_{l=1}^{N} x_l^2 \,, \tag{5}$$

which is a term that formally makes no contribution to the final result in the continuum limit save for ensuring that the integrals involved are now rendered absolutely convergent.

Accepting the fact that the integrals all converge ensures that a meaningful function of ϵ emerges, but, by itself, that fact does not ensure convergence as $\epsilon \to 0$ and, even if convergence holds, it furthermore does not guarantee that the result is correct! To ensure convergence to the correct result requires that some technical condition(s) must be imposed on the potential $V(x)$. In this regard, we only observe that a correct result emerges whenever the potential has a lower bound, *i.e.*, whenever $V(x) \geq c$, for some c, $-\infty < c < \infty$, for all x.

We recall that for the free particle of mass m the potential $V(x) = 0$ for all x. In that case, Eq. (4) reads

$$
\begin{aligned}
K(x'', t''; x', t') &= \lim_{\epsilon \to 0} (m/2\pi i\hbar\epsilon)^{(N+1)/2} \int \cdots \int \\
&\quad \times \exp\{(i/\hbar)\Sigma_{l=0}^{N}[(m/2\epsilon)(x_{l+1} - x_l)^2\} \, \Pi_{l=1}^{N} \, dx_l \\
&= \sqrt{\frac{m}{2\pi i\hbar(t'' - t')}} \, \exp\left(\frac{im(x'' - x')^2}{2\hbar(t'' - t')}\right) ,
\end{aligned} \tag{6}
$$

which is the form of the quantum mechanical propagator for the free particle.

Comments

The procedure sketched above — whereby the action functional expressed as an integral over a continuous-time parameter is replaced by a natural Riemann sum approximation, which eventually is followed by a continuum limit ($\epsilon \to 0$) as the final step — provides a satisfactory procedure for a suitable and large class of potentials to define the propagator $K(x'', t''; x', t')$. However, it is important to stress that a lattice regularization followed by a continuum limit is by no means the only way to give a satisfactory definition of a path integral, nor, in the author's opinion does it even represent the most satisfactory definition, although admittedly this latter issue is in part subjective. What follows in this paper is a narrow review — an historical slice — of the development of various efforts to find a suitable *continuous-time regularization* procedure along with a subsequent limit to remove that regularization that ultimately should yield the correct propagator. While some of the work to be described did not follow directly from work that preceded it, all of this work does, with appropriate hindsight, seem to have a fairly natural progression that makes an interesting story.

2. Feynman-Kac formula, 1951

Through his own research, Mark Kac was fully aware of Wiener's theory of Brownian motion and the associated diffusion equation that describes the corresponding distribution function. Therefore, it is not surprising that he was well prepared to give a path integral expression in the sense of Feynman for an equation similar to the time-dependent Schrödinger equation save for a rotation of the time variable by $-\pi/2$ in the complex plane, namely, by the change $t \to -it$. In particular, Kac[2] considered the equation

$$\frac{\partial \rho(x,t)}{\partial t} = \frac{\nu}{2} \frac{\partial^2 \rho(x,t)}{\partial x^2} - V(x)\rho(x,t) \,. \tag{7}$$

This equation is analogous to Schrödinger's equation but of course differs from it in certain details. Besides certain constants which are different, and the change $t \to -it$, the nature of the dependent variable function $\rho(x,t)$ is quite different from the normal quantum mechanical wave function. For one thing, if the function ρ is initially real it will remain real as time proceeds. Less obvious is the fact that if $\rho(x,t) \geq 0$ for all x at some time t, then the function will continue to be nonnegative for all time t. Thus we can interpret $\rho(x,t)$ more like a probability density; in fact in the special case that $V(x) = 0$, then $\rho(x,t)$ is the probability density for a Brownian particle which underlies the Wiener measure. In this regard, ν is called the diffusion constant.

The fundamental solution of Eq. (7) with $V(x) = 0$ is readily given as

$$W(x,T;y,0) = \frac{1}{\sqrt{2\pi\nu T}} \exp\left(-\frac{(x-y)^2}{2\nu T} \right) , \tag{8}$$

which describes the solution to the diffusion equation subject to the initial condition

$$\lim_{T \to 0^+} W(x,T;y,0) = \delta(x-y) \,. \tag{9}$$

Moreover, it follows that the solution of the diffusion equation for a general initial condition is given by

$$\rho(x'',t'') = \int W(x'',t'';x',t')\rho(x',t')\,dx' \,. \tag{10}$$

Iteration of this equation N times, with $\epsilon = (t'' - t')/(N+1)$, leads to the equation

$$\rho(x'',t'') = N' \int \cdots \int e^{-(1/2\nu\epsilon)\Sigma_{l=0}^{N}(x_{l+1}-x_l)^2} \,\Pi_{l=1}^{N} dx_l \, \rho(x',t')\,dx' \,, \tag{11}$$

where $x_{N+1} \equiv x''$ and $x_0 \equiv x'$. This equation features the imaginary time propagator for a free particle of unit mass as given by

$$W(x'', t''; x', t') = N' \int \cdots \int e^{-(1/2\nu\epsilon)\Sigma_{l=0}^{N}(x_{l+1}-x_l)^2} \, \Pi_{l=1}^{N} dx_l \, . \qquad (12)$$

Since this equation holds for all $N > 0$, we may assume that it also holds in the limit $\epsilon \to 0$, *i.e.*, $N \to \infty$, which we can write formally as

$$W(x'', t''; x', t') = \mathcal{N} \int e^{-(1/2\nu)\int \dot{x}^2 \, dt} \, \mathcal{D}x \, , \qquad (13)$$

where $\mathcal{N}$ denotes a formal normalization factor. (Symbols such as $\mathcal{N}$ may stand for different factors in different expressions.)

The similarity of this expression with the Feynman path integral [for $V(x) = 0$] is clear, but there is a profound difference between these equations. In the former (Feynman) case the underlying measure is only *finitely additive*, while in the latter (Wiener) case the continuum limit actually defines a genuine measure, *i.e.*, a *countably additive measure* on paths, which is a version of the justly famous Wiener measure. In particular,

$$W(x'', t''; x', t') = \int d\mu_W^\nu(x) \, , \qquad (14)$$

where μ_W^ν denotes a measure on continuous paths $x(t)$, $t' \leq t \leq t''$, for which $x(t'') \equiv x''$ and $x(t') \equiv x'$. Such a measure is said to be a *pinned* Wiener measure, since it specifies its path values at two time points, *i.e.*, at $t = t'$ and at $t = t'' > t'$. (The traditional Wiener measure, which we shall not really deal with in this paper, specifies its values only at the initial time, and it corresponds to one additional integration over the final value x'' at the final time t''.)

We note without proof that Brownian motion paths have the property that with probability one they are concentrated on continuous paths. However, it is also true that the time derivative of a Brownian path is almost nowhere defined, which means that, with probability one, $\dot{x}(t) = \pm\infty$ for all t.

When the potential $V(x) \neq 0$ the propagator associated with (7) is formally given by

$$W(x'', t''; x', t') = \mathcal{N} \int e^{-(1/2\nu)\int \dot{x}^2 \, dt - \int V(x) dt} \, \mathcal{D}x \, , \qquad (15)$$

an expression which is well defined if $V(x) \geq c$, $-\infty < c < \infty$. A mathematically improved expression for (15) makes use of the Wiener measure and is given by

$$W(x'', t''; x', t') = \int e^{-\int V(x(t))\,dt}\, d\mu_W^\nu(x)\,. \tag{16}$$

This is an elegant relation in that it represents a solution to the differential equation (7) in the form of an integral over Brownian motion paths suitably weighted by the potential V. Incidentally, since the propagator is evidently a strictly positive function, it follows that the solution of the differential equation (7) is nonnegative for all time t provided it is nonnegative for any particular time value.

3. Gel'fand and Yaglom, 1956

In an important review article,[3] these two authors introduced the concept of a continuous-time regularization of Feynman path integrals. Although their procedure was later shown to be incorrect, their work may be said to have initiated an interesting line of development.

The idea of Gel'fand and Yaglom was relatively straightforward. Formally stated, their proposal was to define the propagator as

$$\lim_{\nu \to \infty} \mathcal{N}_\nu \int \exp\{(i/\hbar)\int [(m/2)\dot{x}^2 - V(x)]\,dt\} \exp\{-(1/2\nu)\int \dot{x}^2\,dt\}\mathcal{D}x. \tag{17}$$

Formally, therefore, their proposal consisted of introducing an auxiliary factor into the integrand that is identical to the factor which leads to Wiener measure. The purpose of such a factor is to introduce a regularization into the original formal expression. Finally, the limit $\nu \to \infty$ is taken as a last step which amounts to formally removing the auxiliary factor and leaving behind the original integrand of the Feynman path integral.

There is one glaring difficulty with this proposal. The auxiliary factor which leads to Brownian motion paths with a diffusion constant $\nu < \infty$ has the property of ensuring that the paths are continuous, but they are nowhere differentiable. For such paths, the contribution of the potential is well defined, but the contribution of the kinetic energy is divergent for all paths. Gel'fand and Yaglom hoped to get around this difficulty by combining the kinetic energy with the corresponding factor in the auxiliary term giving rise to a Wiener measure with a *complex* diffusion coefficient σ, where

$$\frac{1}{\sigma} = \frac{1}{\nu} - \frac{im}{\hbar}\,. \tag{18}$$

Thus the strategy was to take the limit of a sequence of presumably well defined integrals over Wiener measure with a complex diffusion constant σ

as the real part of σ vanished. Notice that this kind of regularization does not involve the introduction of a temporal lattice followed by a continuum limit but rather maintains a continuous time parameter throughout. It is for this reason that we refer to this kind of procedure as a *continuous-time regularization scheme*.

4. Cameron, 1960

Shortly after the appearance of the Gel'fand and Yaglom paper, Cameron[4] showed that the scheme was fatally flawed. In particular, Cameron showed that a Wiener measure defined with a complex diffusion constant σ leads to a countably additive measure only when $\sigma \equiv \mathrm{Re}\sigma > 0$, hence $\mathrm{Im}\sigma \equiv 0$. It is instructive to sketch the argument that leads to this conclusion. Consider the N-fold integral

$$\left(\frac{\lambda}{2\pi\epsilon}\right)^{(N+1)/2} \int \cdots \int \exp[-(\lambda/2\epsilon)\Sigma_{l=0}^{N}(x_{l+1}-x_l)^2]\, \Pi_{l=1}^{N}\, dx_l$$
$$= \left(\frac{\lambda}{2\pi N\epsilon}\right)^{1/2} \exp[-(\lambda/2N\epsilon)(x_{N+1}-x_0)^2]\,, \qquad (19)$$

which represents one of the primary formulas involved in discussing Wiener measures. This formula holds for any complex λ so long as $\mathrm{Re}\lambda > 0$. If such a formula exists in the limit that $\epsilon \to 0$ and $N \to \infty$ such that $N\epsilon$ remains constant, then the desired path measure will also exist. Although that condition would appear to be evident, it is not true for general λ.

Integrals exist when they converge absolutely. Absolute convergence certainly holds for (19) for all $N < \infty$, but what about when $N \to \infty$? To test this issue, we consider the absolute integral associated with (19) given by

$$\left(\frac{|\lambda|}{2\pi\epsilon}\right)^{(N+1)/2} \int \cdots \int \exp[-(\mathrm{Re}\lambda/2\epsilon)\Sigma_{l=0}^{N}(x_{l+1}-x_l)^2]\, \Pi_{l=1}^{N}\, dx_l$$
$$= \left(\frac{|\lambda|}{\mathrm{Re}\lambda}\right)^{N/2}\left(\frac{|\lambda|}{2\pi N\epsilon}\right)^{1/2} \exp[-(\mathrm{Re}\lambda/2N\epsilon)(x_{N+1}-x_0)^2] \quad (20)$$

Evidently, the final result exists as $N \to \infty$ only provided $|\lambda| \equiv \mathrm{Re}\lambda > 0$, *i.e.*, provided $\mathrm{Im}\lambda \equiv 0$, as claimed. Consequently, the proposal of Gel'fand and Yaglom fails on the ground that whenever the diffusion constant for the Wiener measure is truly complex ($\mathrm{Im}\sigma \neq 0$), then the so-defined Wiener measure is only *finitely additive and not countably additive*. In this case, finite additivity means that finite answers for general integrals involving this measure arise by delicate cancellation of both positive and negative divergent contributions. A regularization scheme that results in a finitely additive measure on paths is not qualitatively different from the original formal Feynman path integral, and therefore it does not constitute an acceptable continuous time regularization.

5. Itô, 1962

Soon after Cameron pointed out the lack of a countably additive measure for the Gel'fand-Yaglom procedure, Itô[5] proposed another version of a continuous-time regularization that resolved some of the troublesome issues. In essence, the proposal of Itô takes the form given by

$$\lim_{\nu \to \infty} \mathcal{N}_\nu \int \exp\{(i/\hbar)\int[\tfrac{1}{2}m\dot{x}^2 - V(x)]\,dt\} \exp\{-(1/2\nu)\int[\ddot{x}^2 + \dot{x}^2]\,dt\}\, \mathcal{D}x \,. \tag{21}$$

Note well the alternative form of the auxiliary factor introduced as a regulator. The additional term $\ddot{x}^2$, the square of the second derivative of x, acts to smooth out the paths sufficiently well so that in the case of (21) both $x(t)$ and $\dot{x}(t)$ are continuous functions, leaving $\dddot{x}(t)$ as the term which does not exist. However, since only x and $\dot{x}$ appear in the rest of the integrand, the indicated path integral can be well defined; this is already a positive contribution all by itself!

To proceed further with (21) we need to decide on what values to fix at the initial and final times. As was the case with the procedure proposed by Gel'fand and Yaglom, let us first suppose that a composition law of the expression in (21) holds *before* the limit $\nu \to \infty$ is taken. This requires that one fix not only $x(t'') = x''$ and $x(t') = x'$ but, in addition, that we fix $\dot{x}(t'') = \dot{x}''$ and $\dot{x}(t') = \dot{x}'$. Viewed as a solution of the Schrödinger equation, however, we are at a loss to understand the meaning of the values of x and $\dot{x}$ to be held at any time. Thus, while this interpretation may yield a limiting function it is hard to see that such a result could be a solution to Schrödinger's equation.

In point of fact, Itô did not choose the previous set of data to be fixed but he chose another set. Itô interpreted (21) as an integral in which $x(t'') = x''$ and $x(t') = x'$ are the only values held fixed. He therefore did not require any composition law on the expression (21) at finite ν, but only expected a composition law to hold for the expression that arises after $\nu \to \infty$.

With such an interpretation in mind we proceed to evaluate in a fairly heuristic manner the functional integral (21). For simplicity we shall discuss only the simple case where $V(x) = 0$; moreover, we set $t' = 0$, $t'' = T$, and choose $x(0) = 0$, $x(T) = x$. Thus, prior to taking the limit $\nu \to \infty$, let us focus on the formal integral

$$\mathcal{N}_\nu \int \exp\{(i/\hbar)\int \dot{x}\,g(t)\,dt\}\,\exp\{-(1/2\nu)\int[\ddot{x}^2 + a^2\dot{x}^2]\,dt\}\delta(x(0))\,\mathcal{D}x\,, \tag{22}$$

where $a^2 \equiv 1 - im\nu/\hbar$, $g(t)$ will be chosen below, and for the foregoing expression we assume there is no other pinning; the required second pinning at $t'' = T$ will be introduced in a moment. Let us set

$$x(t) \equiv \int_0^t \xi(u)\,du\,, \tag{23}$$

for all t, so that the former integral can be replaced by

$$\mathcal{N}_\nu \int \exp\{(i/\hbar)\int \xi\,g(t)\,dt\}\,\exp\{-(1/2\nu)\int[\dot{\xi}^2 + a^2\xi^2]\,dt\}\,\mathcal{D}\xi\,. \tag{24}$$

In this form, the integral involves an Ornstein-Uhlenbeck measure,[6] and the answer is readily given by

$$\exp[-(\nu/4a\hbar^2)\int g(t)\,g(u)\,e^{-a|t-u|}\,dt\,du]\,, \tag{25}$$

where we have chosen $\mathcal{N}_\nu$ so that the answer is unity if $g(t) = 0$ for all t. As a next step we let $g(t) = \lambda$ for $0 \le t \le T$, and $g(t) = 0$ otherwise. Thus (25) becomes

$$\exp[-(\nu\lambda^2/4a\hbar^2)\int_0^T\int_0^T e^{-a|t-u|}\,dt\,du]\,. \tag{26}$$

Since

$$\int e^{(i/\hbar)\lambda\int_0^T \xi(t)\,dt}\,e^{-(i/\hbar)\lambda x}\,d\lambda/(2\pi\hbar)$$
$$= \delta(\int_0^T \xi(t)\,dt - x)$$
$$= \delta(x(T) - x)\,, \tag{27}$$

it follows that

$$\mathcal{N}_\nu \int \delta(x(T) - x)\,e^{-(1/2\nu)\int[\dot{\xi}^2 + a^2\xi^2]\,dt}\,\mathcal{D}\xi$$
$$= \int e^{-(i/\hbar)\lambda x}\,e^{-\nu\lambda^2 F/4a\hbar^2}\,d\lambda/(2\pi\hbar)$$
$$= \sqrt{\frac{a}{\nu F\pi}}\,\exp\left(-\frac{ax^2}{\nu F}\right)\,, \tag{28}$$

where

$$F \equiv \int_0^T\int_0^T e^{-a|t-u|}\,dt\,du$$
$$= \frac{2T}{a} - \frac{2}{a^2}(1 - e^{-aT})\,. \tag{29}$$

As ν becomes large, we may set $a^2 \simeq -im\nu/\hbar$, in which case $a/\nu F$ may be replaced by $-im/2T\hbar$. Hence, we are led to the final result

$$\sqrt{\frac{m}{2\pi i T\hbar}}\, \exp\left(\frac{imx^2}{2\hbar T}\right), \tag{30}$$

which is recognized as the quantum mechanical propagator $K(x,T;0,0)$ for a free particle of mass m as given in (6).

Itô[5] made this story rigorous for constant, linear, and quadratic potentials, as well as those potentials of the form

$$V(x) = \int e^{ixs}\, w(s)\, ds \tag{31}$$

provided that $\int |w(s)|\, ds < \infty$.

In summary, by introducing a regularization involving a higher derivative $(\ddot{x}^2)$, Itô was able to soften the paths sufficiently to allow the kinetic energy to be well defined. In so doing he was able to give a satisfactory continuous-time regularization for (21) for a certain class of potentials V.

6. Feynman, 1951

It is necessary to retrace history at this point to recall the introduction of the *phase space path integral* by Feynman.[7] In Appendix B to this article, Feynman introduced a formal expression for the configuration or q-space propagator given by

$$K(q'',t'';q',t') = \mathcal{M} \int \exp\{(i/\hbar)\int [p\dot{q} - H(p,q)]\, dt\}\, \mathcal{D}p\, \mathcal{D}q. \tag{32}$$

In this equation one is instructed to integrate over all paths $q(t)$, $t' \le t \le t''$, with $q(t'') \equiv q''$ and $q(t') \equiv q'$ held fixed, as well as to integrate over all paths $p(t)$, $t' \le t \le t''$, without restriction. As customary, this is a formal statement and in practice it needs to be given a precise meaning. The lattice prescription in which the continuous time parameter is replaced by a finite set of discrete points is the procedure that is typically followed. For completeness, we illustrate a common lattice space version of the formal phase space path integral expression, as given by

$$K(q'',t'';q',t') = \lim_{\epsilon \to 0} \int \cdots \int \exp\{(i/\hbar)\Sigma_{l=0}^{N}[\tfrac{1}{2}p_{l+1/2}(q_{l+1} - q_l)$$
$$-\epsilon H(p_{l+1/2}, \tfrac{1}{2}(q_{l+1} + q_l))]\}\, \Pi_{l=0}^{N} dp_{l+1/2}/(2\pi\hbar)\, \Pi_{l=1}^{N} dq_l. \tag{33}$$

In this expression, all p and q variables are integrated over except for $q_{N+1} \equiv q''$ and $q_0 \equiv q'$, and, just as before, $\epsilon = (t'' - t')/(N+1)$. Since q_l implies a sharp q value at time $t' + l\epsilon$, we have chosen to name the conjugate variable $p_{l+1/2}$ to emphasize that a sharp p value must occur at a different time, here at $t' + (l+1/2)\epsilon$, since it is not possible to have sharp p and q values at the same time. Note that there is one more p integration than q integration in this formulation. This discrepancy becomes clear when one imposes the composition law which requires that

$$K(q''', t'''; q', t') = \int K(q''', t'''; q'', t'') K(q'', t''; q', t') \, dq'' , \qquad (34)$$

a relation which implies, just on dimensional grounds, that there must be one more p integration than q integration in the definition of each K expression.

Observe that (32) exhibits an apparent covariance under canonical coordinate transformations, but this is quite illusory. In fact, the lattice form (33) is correct only when the canonical variables are described by Cartesian coordinates, and this kind of limitation on straightforward lattice-space regularizations is quite general. In a later section of this paper we shall return to this point and explain why this limitation is necessary.

It is also instructive to consider an alternative phase space path integral for the momentum or p-space propagator formally given by

$$K(p'', t''; p', t') = \mathcal{M} \int \exp\{(i/\hbar) \textstyle\int [-q\dot{p} - H(p, q)] \, dt\} \, \mathcal{D}p \, \mathcal{D}q , \quad (35)$$

which is obtained from (32) by Fourier transformation on both end variables, q'' and q'. In this case a lattice space definition can be given by

$$K(p'', t''; p', t') = \lim_{\epsilon \to 0} \int \cdots \int \exp\{(i/\hbar) \Sigma_{l=0}^{N} [-\tfrac{1}{2} q_{l+1/2}(p_{l+1} - p_l)$$
$$- \epsilon H(\tfrac{1}{2}(p_{l+1} + p_l), q_{l+1/2})]\} \, \Pi_{l=1}^{N} dp_l \, \Pi_{l=0}^{N} dq_{l+1/2}/(2\pi\hbar) , \quad (36)$$

and we see in this expression that there is one more q integration than p integration. Again this makes sense when one considers the composition law

$$K(p''', t'''; p', t') = \int K(p''', t'''; p'', t'') K(p'', t''; p', t') \, dp'' . \qquad (37)$$

On the other hand, and following similar reasoning, one would be hard pressed to give a satisfactory lattice space formulation of the putative formal phase space path integral given by

$$\mathcal{M} \int \exp\{(i/\hbar) \textstyle\int [\tfrac{1}{2}(p\dot{q} - q\dot{p}) - H(p, q)] \, dt\} \, \mathcal{D}p \, \mathcal{D}q . \qquad (38)$$

(In the light of the previous discussion, even the physical meaning of such an expression seems uncertain, at least to the present author. On the other hand, such an expression is easily understood in the coherent state representations that follow.)

It is widely appreciated that the phase space path integral is more generally applicable than the original, Lagrangian, version of the path integral. For instance, the original configuration space path integral is satisfactory for Lagrangians of the general form

$$L(x) = \tfrac{1}{2}m\dot{x}^2 + A(x)\dot{x} - V(x) \, , \tag{39}$$

but it is unsuitable, for example, for the case of a relativistic particle with the Lagrangian

$$L(x) = -m\sqrt{1 - \dot{x}^2} \tag{40}$$

expressed in units where the speed of light is unity. For such a system — as well as many more general expressions — the phase space form of the path integral is to be preferred. In particular, for the relativistic free particle, the phase space path integral

$$\mathcal{M} \int \exp\{(i/\hbar)\textstyle\int[p\dot{q} - \sqrt{p^2 + m^2}]\,dt\} \, \mathcal{D}p\,\mathcal{D}q \, , \tag{41}$$

interpreted in the sense of (33), is readily evaluated and yields the correct propagator.

Issues of proper coordinate choice also arise for the configuration space path integral as well, and expressions such as (3) and (4) implicitly refer to Cartesian coordinates.

7. Coherent state representations, 1960

As a prelude to the following section it is pedagogically useful to recall some basic properties of coherent states and the Hilbert space representations they generate.[8] We focus on coherent states generated by Heisenberg canonical operators P and Q which obey the basic commutation relation $[Q, P] = i\hbar I$, where I denotes the unit operator. More specifically, we shall assume the Weyl form of the commutation relation holds for self-adjoint momentum and position operators in the form

$$e^{-iqP/\hbar}e^{ipQ/\hbar} = e^{-ipq/2\hbar}e^{i(pQ-qP)/\hbar} \equiv U[p, q] \, , \tag{42}$$

where p and q are arbitrary real variables. In an abstract Hilbert space, let us introduce a normalized fiducial vector $|\eta\rangle$, which is otherwise arbitrary, and define vectors of the form

$$|p, q\rangle \equiv U[p, q]\,|\eta\rangle \tag{43}$$

for all $(p, q) \in \mathbb{R}^2$. These states are the *canonical coherent states*, and they have a number of interesting properties. Since $U[p, q]$ is a unitary operator, it follows that each vector is normalized, *i.e.*,

$$\||p, q\rangle\| \equiv \langle p, q|p, q\rangle^{1/2} = 1 \ . \tag{44}$$

Next, it should be noted that the coherent states are continuously parameterized; specifically, if $(p, q) \to (p', q')$ in the sense that $|p - p'|^2 + |q - q'|^2 \to 0$, then it follows for any choice of $|\eta\rangle$, that $\||p, q\rangle - |p', q'\rangle\| \to 0$, or as one says, the vectors are strongly continuous in the parameters p and q. Finally, we want to stress the important property that these vectors admit a rather conventional looking resolution of unity as a superposition over one-dimensional projection operators onto the coherent states, and specifically that

$$\int |p, q\rangle\langle p, q|\, dp\, dq/(2\pi\hbar) = I \tag{45}$$

holds, where, as before, I is the unit operator, and this relation holds for any choice of $|\eta\rangle$. It is easy to show that the resolution of unity holds weakly in the sense that, for arbitrary vectors $|\psi\rangle$ and $|\phi\rangle$,

$$\int \langle\phi|p, q\rangle\langle p, q|\psi\rangle\, dp\, dq/(2\pi\hbar) = \langle\phi|\psi\rangle \ . \tag{46}$$

(It is less easy to show, but nevertheless true, that the resolution of unity also holds as a strong operator identity.)

The resolution of unity formula permits us to introduce novel representations of Hilbert space rather different from those customarily used, particularly in quantum mechanics. Let us introduce the functional representatives

$$\psi(p, q) \equiv \langle p, q|\psi\rangle \ , \tag{47}$$

which are defined for all $|\psi\rangle$ in the Hilbert space. As we have already done for the coherent states themselves, we suppress any dependence of the functions $\psi(p, q)$ on the state $|\eta\rangle$. Indeed, for any choice of $|\eta\rangle$, each such function is a bounded, continuous function of the variables p and q. For each pair of such functions, such as $\psi(p, q) = \langle p, q|\psi\rangle$ and $\phi(p, q) = \langle p, q|\phi\rangle$, we assign the inner product given by

$$(\phi, \psi) \equiv \int \phi(p, q)^* \, \psi(p, q)\, dp\, dq/(2\pi\hbar) = \langle\phi|\psi\rangle \ . \tag{48}$$

The last part of this relation shows that the functional inner product equals the inner product in the abstract Hilbert space. The so-defined functional representation is complete simply because every element of the abstract Hilbert space is imaged as a function of the form $\psi(p,q)$. It should be noticed that the space of functions defined in this way is a complete and proper *subspace* of the space of square integrable functions over two variables. The space of functions $\{\psi(p,q) \equiv \langle p,q|\psi\rangle\}$ defined for all $|\psi\rangle$ and all (p,q) constitutes the *coherent state functional representation* of a one-particle Hilbert space. In this representation, for example, the basic Heisenberg operators are given by

$$P \to -i\hbar\frac{\partial}{\partial q} , \qquad\qquad Q \to q + i\hbar\frac{\partial}{\partial p} , \qquad (49)$$

and thus the coherent state representation of the Schrödinger equation is given by

$$i\hbar\,\partial\psi(p,q,t)/\partial t = \mathcal{H}(-i\hbar\partial/\partial q, q + i\hbar\partial/\partial p)\psi(p,q,t) . \qquad (50)$$

What makes this a one-particle problem (rather than a special kind of two-particle problem) is the proper choice of acceptable initial conditions for this equation. In particular, it suffices to take as an initial condition the coherent state overlap function $\langle p,q|p',q'\rangle$, for arbitrary p' and q'.

The solution to this form of Schrödinger's equation can be expressed in the form

$$\psi(p'',q'',t'') = \int K(p'',q'',t'';p',q',t')\psi(p',q',t')\,dp'\,dq'/(2\pi\hbar) , \quad (51)$$

where $K(p'',q'',t'';p',q',t')$ denotes the propagator in the coherent state representation.

The coherent state representation differs from the usual Schrödinger representation given in (1), and it also has a different physical interpretation as well. For one thing, the meaning of $|\psi(x)|^2$ is the probability density to find the particle at position x. The meaning of $|\psi(p,q)|^2$, however, is rather different; this quantity is simply the probability that the state $|\psi\rangle$ can be found in the state $|p,q\rangle$. The physical meaning of the variable x in the Schrödinger representation is that of a sharp position. Correspondingly, we may ask what is the physical meaning of the parameters p and q? To answer that question it is convenient to restrict the choice of the fiducial vector very slightly by imposing the conditions that $\langle\eta|P|\eta\rangle = \langle\eta|Q|\eta\rangle = 0$. In that case, it follows that

$$\langle p,q|P|p,q\rangle = p , \qquad\qquad \langle p,q|Q|p,q\rangle = q , \qquad (52)$$

implying that unlike x in the Schrödinger representation, the variables p and q represent *mean values* in the coherent states. It is for this reason that we can specify both values simultaneously at the same time, something that could not be done if instead they both had represented sharp eigenvalues.

The propagator for the coherent state representation of Schrödinger's equation can also be given a formal phase space path integral form, namely

$$K(p'',q'',t'';p',q',t') = \mathcal{M} \int \exp\{(i/\hbar)\int[p\dot{q} - H(p,q)]\,dt\}\,\mathcal{D}p\,\mathcal{D}q \, , \quad (53)$$

which superficially is just the same expression as (32)! What makes these expressions different is what values are pinned and more explicitly what are the respective lattice space formulations. In the coherent state case, we have

$$K(p'',q'',t'';p',q',t') = \lim_{\epsilon \to 0} \int \cdots \int \exp\{(i/\hbar)\Sigma_{l=0}^{N}[\tfrac{1}{2}(p_{l+1}+p_l)(q_{l+1}-q_l)$$
$$-\epsilon H(\tfrac{1}{2}(p_{l+1}+p_l)+i\tfrac{1}{2}(q_{l+1}-q_l), \tfrac{1}{2}(q_{l+1}+q_l)-i\tfrac{1}{2}(p_{l+1}-p_l))]\}$$
$$\times \exp\{-(1/4\hbar)\Sigma_{l=0}^{N}[(p_{l+1}-p_l)^2 + (q_{l+1}-q_l)^2]\}\,\Pi_{l=1}^{N}\,dp_l\,dq_l/(2\pi\hbar) \, .$$
$$(54)$$

Observe that there are the same number of p and q integrations in this expression. Such a conclusion is fully in accord with the combination law as expressed in the coherent state representation, namely

$$K(p''',q''',t''';p',q',t')$$
$$= \int K(p''',q''',t''';p'',q'',t'')\,K(p'',q'',t'';p',q',t')\,dp''\,dq''/(2\pi\hbar) \, . (55)$$

8. Daubechies and Klauder, 1985

By any measure, phase space is the natural arena for both classical and quantum mechanics. In classical physics, dynamics evolves in phase space in a highly symmetric fashion for general Hamiltonians, while in quantum physics both P and Q appear in remarkably similar roles within the general formalism. Moreover, as stressed previously, the phase space path integral is more widely applicable than the original configuration space path integral.

In 1985, Daubechies and Klauder[9] examined the phase space path integral in a new light. In particular, they were led to consider the expression

$$\lim_{\nu \to \infty} \mathcal{M}_\nu \int \exp\{\tfrac{i}{\hbar}\int[p\dot{q} - H(p,q)]\,dt\}\,\exp\{-\tfrac{1}{2\nu}\int[\dot{p}^2 + \dot{q}^2]\,dt\}\,\mathcal{D}p\,\mathcal{D}q \, , (56)$$

which extends the continuous-time regularization procedure to phase space path integrals. Let us initially look at this expression from a general viewpoint. Evidently we can interpret the regularization as two independent Brownian motion regularizations in the limit as the diffusion constant diverges, which is rather like the original Gel'fand-Yaglom procedure. Alternatively, and since from a classical point of view p is often rather like $\dot{q}$ (depending on H, of course), this procedure has some features superficially in common with that of Itô. However, there are also important differences with those previous procedures as well. The chosen regularization ensures that both p and q are continuous functions but that $\dot{p}$ and $\dot{q}$ are almost nowhere defined. The derivatives enter the integrand only in the form $\int p\dot{q}\,dt$. However, this term can be interpreted as $\int p\,dq$, which for Brownian motion paths is a standard stochastic integral that has been well studied in both its Itô and Stratonovich versions; for reasons to be given below, when it becomes important to choose a rule we shall choose the Stratonovich rule. We interpret the integration in (56) to be pinned for both p and q at both end points: specifically, $(p(t''), q(t'')) = (p'', q'')$ and $(p(t'), q(t')) = (p', q')$. In this case it is clear that the result of (56) yields a function of the form

$$K(p'', q'', t''; p', q', t') \,. \tag{57}$$

By the time of 1985, representations of propagators for one-particle quantum systems in the form of coherent state representations, such as those described in an earlier section, were well known to many researchers. If, indeed, (56) turned out to yield coherent state representations for solutions to the Schrödinger equation — as opposed, say, to either the q-space or p-space representations — there would be no reason for concern. In fact, that is exactly what happens!

In Ref.[9] it was rigorously established that

$$\lim_{\nu \to \infty} \mathcal{M}_\nu \int e^{(i/\hbar)\int [p\dot{q} - H(p,q)]\,dt}\, e^{-(1/2\nu)\int [\dot{p}^2 + \dot{q}^2]\,dt}\, \mathcal{D}p\,\mathcal{D}q$$

$$= \lim_{\nu \to \infty} 2\pi\hbar\, e^{\nu T/2\hbar} \int e^{(i/\hbar)\int [p\,dq - H(p,q)\,dt]}\, d\mu_W^\nu(p, q)$$

$$\equiv \langle p'', q'' | e^{-i(t'' - t')\mathcal{H}/\hbar} | p', q' \rangle$$

$$\equiv K(p'', q'', t''; p', q', t') \,, \tag{58}$$

where the second line of (58) is a mathematically rigorous formulation of the heuristic and formal first line, and μ_W^ν denotes pinned Wiener measure. In fact, the connection indicated here is far reaching in that the choice of regularization also determines that

$$|p, q\rangle \equiv e^{-iqP/\hbar} e^{ipQ/\hbar} |0\rangle \,, \tag{59}$$

$$[Q, P] = i\hbar I \,, \qquad (Q + iP)|0\rangle = 0 \,, \tag{60}$$

$$\mathcal{H} = \int H(p, q)|p, q\rangle\langle p, q|\, dp\, dq/(2\pi\hbar) \,. \tag{61}$$

A sufficient set of technical assumptions ensuring the validity of this representation is given by

$$a) \qquad \int H(p, q)^2\, e^{-\alpha(p^2 + q^2)}\, dp\, dq < \infty, \text{ for all } \alpha > 0 \,, \tag{62}$$

$$b) \qquad \int H(p, q)^4 e^{-\beta(p^2 + q^2)}\, dp\, dq < \infty, \text{ for some } \beta < 1/2\hbar \,, \tag{63}$$

$$c) \qquad \mathcal{H} \text{ is essentially self adjoint on the span of finitely many}$$
$$\text{number eigenstates} \,. \tag{64}$$

A wide class of Hamiltonians — but by no means all acceptable such operators — consists of all semibounded, symmetric (Hermitian) polynomials of P and Q. We also note that the connection between the operator $\mathcal{H}$ and its symbol $H(p, q)$ given in (61) is that of anti-normal ordering. In particular, this means that

$$e^{-(r-is)(Q+iP)/2\hbar} e^{(r+is)(Q-iP)/2\hbar} = \int e^{i(sq-rp)/\hbar}|p, q\rangle\langle p, q|\, dp\, dq/(2\pi\hbar) \,,$$
$$\tag{65}$$

and any other connection follows by suitable linear combinations. As an alternative association we also note that

$$H(p, q) = e^{-(\hbar/4)(\partial_p^2 + \partial_q^2)} H_W(p, q) \,, \tag{66}$$

where $H_W(p, q)$ denotes the well-known Weyl symbol of the operator $\mathcal{H}$. [In most of the author's earlier work, the particular symbol $H(p, q)$ used here has been denoted by the lower case symbol $h(p, q)$.]

Observe that the auxiliary factor in (56) that provides the regularization involves a *metric*, specifically, $d\sigma^2 = d\sigma(p, q)^2 = dp^2 + dq^2$, on classical phase space. Such a metric characterizes a flat, two-dimensional phase space expressed in Cartesian coordinates.

It is interesting to consider the transformation of (58) under a canonical change of phase space coordinates. To this end we introduce $\bar{p} = \bar{p}(p, q)$ and $\bar{q} = \bar{q}(p, q)$ as two new classical canonical coordinates that are related to the original canonical coordinates by the relation

$$\bar{p}\,d\bar{q} = p\,dq + dF(\bar{q}, q) \, , \tag{67}$$

where F is known as the generator of the transformation. This is a standard form for this kind of relation in which $\bar{q}$ and q are regarded as the independent variables; another version that we will find more useful is given by

$$\bar{p}\,d\bar{q} + d\overline{G}(\bar{p}, \bar{q}) = p\,dq \, . \tag{68}$$

Under such a canonical coordinate transformation, the metric $d\sigma^2 = dp^2 + dq^2$ assumes the form

$$d\sigma^2 = d\sigma(\bar{p}, \bar{q})^2 = A(\bar{p}, \bar{q})\,d\bar{p}^2 + 2B(\bar{p}, \bar{q})\,d\bar{p}\,d\bar{q} + C(\bar{p}, \bar{q})\,d\bar{q}^2 \, , \tag{69}$$

for suitable A, B, and C, but, nevertheless, the metric still characterizes the same flat, two-dimensional phase space even though it is now expressed, generally speaking, in curvilinear canonical coordinates.

In the original Cartesian coordinates the stochastic integral $\int p\,dq$ evaluated by the Stratonovich rule is identical to its evaluation using the Itô rule. Under coordinate transformations, we adopt the Stratonovich definition [mid-point rule, cf.(54)] since for this rule the differential and integral laws of ordinary calculus still apply. Consequently, the differential rules (67) and (68) which apply for classical functions (say C^2), apply equally well to Brownian paths (which are only C^0). Recognizing that the Hamiltonian function transforms as a scalar, we see that

$$\overline{H}(\bar{p}, \bar{q}) \equiv H(p(\bar{p}, \bar{q}), q(\bar{p}, \bar{q})) = H(p, q) \, . \tag{70}$$

Assembling the separate parts, we learn that under a canonical coordinate transformation, (58) becomes

$$\lim_{\nu \to \infty} \mathcal{M}_\nu \int e^{(i/\hbar)\int [\bar{p}\dot{\bar{q}} + \dot{\overline{G}}(\bar{p}, \bar{q}) - \overline{H}(\bar{p}, \bar{q})]\,dt}\, e^{-(1/2\nu)\int [d\sigma(\bar{p}, \bar{q})^2/dt^2]\,dt}\,\mathcal{D}\bar{p}\,\mathcal{D}\bar{q}$$

$$= \lim_{\nu \to \infty} 2\pi\hbar\, e^{\nu T/2\hbar} \int e^{(i/\hbar)\int [\bar{p}\,d\bar{q} + d\overline{G}(\bar{p}, \bar{q}) - \overline{H}(\bar{p}, \bar{q})\,dt]}\, d\bar{\mu}_W^\nu(\bar{p}, \bar{q})$$

$$\equiv \langle \bar{p}'', \bar{q}''| e^{-i(t''-t')\mathcal{H}/\hbar} |\bar{p}', \bar{q}'\rangle$$

$$\equiv \overline{K}(\bar{p}'', \bar{q}'', t''; \bar{p}', \bar{q}', t') \, , \tag{71}$$

where $\bar{\mu}_W^\nu$ denotes pinned Wiener measure on a two-dimensional flat phase space expressed in curvilinear coordinates. The term

$$\int d\overline{G}(\bar{p}, \bar{q}) = \overline{G}(\bar{p}'', \bar{q}'') - \overline{G}(\bar{p}', \bar{q}') \, , \tag{72}$$

and thus its presence amounts to no more than a phase change of the corresponding coherent states. Specifically,

$$|\overline{p},\overline{q}\rangle \equiv e^{-i\overline{G}(\overline{p},\overline{q})/\hbar}\, e^{-ip(\overline{p},\overline{q})Q/\hbar}\, e^{iq(\overline{p},\overline{q})P/\hbar}\,|0\rangle = e^{-i\overline{G}(\overline{p},\overline{q})/\hbar}\,|p,q\rangle\;, \quad (73)$$

$$[Q,P] = i\hbar I\;, \qquad\quad (Q+iP)|0\rangle = 0\;, \quad (74)$$

$$\mathcal{H} = \int \overline{H}(\overline{p},\overline{q})\,|\overline{p},\overline{q}\rangle\langle\overline{p},\overline{q}|\,d\overline{p}\,d\overline{q}/(2\pi\hbar)\;. \quad (75)$$

Under a canonical coordinate transformation, therefore, and apart from a possible phase change, observe that the coherent state vectors are *unchanged*; it is only their *labels* that have changed. Observe further that the Hamiltonian operator $\mathcal{H}$ is completely unchanged; only the details of its representation are different in the new coordinate system.

In summary, we see that a continuous-time, Brownian motion regularization of the phase space path integral can be rigorously established. It applies to a wide class of Hamiltonians, and the formulation is fully covariant under general canonical coordinate transformations. The only additional ingredient to attain this covariance is the introduction of a *metric, $d\sigma^2$*, on classical phase space which is ultimately used to support the Wiener measure regularization. As a general rule, it is noteworthy that quantization schemes that agree with orthodox quantum mechanics — and therefore agree with a vast number of experiments — do *not* exhibit manifest covariance under canonical coordinate transformations. Hence, the fact that the present procedure exhibits covariance under canonical coordinate transformations is especially to be welcomed.

As a final comment, we remind the reader of the general requirement that canonical quantization, *i.e.*, the choice of certain phase space variables to be "promoted" to operators, should be carried out in Cartesian coordinates.[10] This rule has caused much concern over the years and its elimination has been, partially at least, a motivating factor in certain alternative quantization schemes. Instead of seeking to eliminate the Cartesian coordinates, we fully accept them, and, indeed, we understand the need for them as follows[11]: With only a symplectic form on phase space — and in the absence of any phase space metric — there is no self-referential mechanism to associate the correct *physical meaning* of any given *mathematical expression*, say, for the Hamiltonian $H(p,q)$. In particular, if $H(p,q) = p^2/2$, and in the absence of any further information, how is one to determine that this particular expression is the Hamiltonian for a free particle, or for an harmonic oscillator, or for some particular anharmonic oscillator, *etc.*, any one of which can be brought to that mathematical form by a suitable canonical coordinate transformation. It is with the addition of a flat-space phase-space metric, and with due regard to its coordinatized expression, that the physical significance of a given mathematical expression can be determined, and with that determination, the quantization procedure can lead to the correct energy spectrum for the physical system under consideration. The phase space continuous-time regularization scheme discussed in this section offers the great advantage that the introduction of the metric into the formal phase space path integral simultaneously gives mathematical and physical meaning to the path integral by assuring that the proper interpretation of the Hamiltonian is maintained throughout the quantization procedure. In point of fact, one may say that the procedure we have illustrated offers a true *geometric quantization procedure*, and consequently, the entire procedure can therefore be expressed in a *coordinate free form*. That is indeed the case, and the interested reader is referred to Ref.[12] for further discussion of this point.

Final remarks

A number of further developments and applications regarding continuous-time regularizations have appeared in recent years. We do not enter into any details of these works here, but rather list just a few additional papers for further study: Refs.[13,14,15]

Dedication

It is a distinct pleasure to dedicate this paper to my long time friend and colleague, Hiroshi Ezawa. I have known Hiroshi for approximately forty years and I have profited greatly from the many times we have been together and during which we have often worked closely on a variety of topics of mutual interest. I wish him many more years of good health and productive research.

Acknowledgments

Thanks are expressed to Lorenz Hartmann and Dae-Yup Song for their comments on the manuscript.

References

1. R.P. Feynman, Space-time Approach to Nonrelativistic Quantum Mechanics, *Rev. Modern Phys.* **20**, 367–387 (1948).

2. M. Kac, On Some Connection between Probability Theory and Differential and Integral Equations, *Proc. 2nd Berkeley Sympos. Math. Stat. and Prob.*, 189–215 (1951). Reprinted in *Mark Kac: Probability, Number Theory, and Statistical Physics — Selected Papers*, Eds. K. Baclawski and M.D. Donsker, (The MIT Press, Cambridge, MA, 1979).

3. I.M. Gel'fand and A.M. Yaglom, Integration in Function Spaces and its Applications in Quantum Physics, *Uspekhi Mat. Nauk* vol. II, 77–114 (1956) (in russian); *J. Math. Phys.* **1**, 48–69 (1960) (English translation).

4. R.H. Cameron, A Family of Integrals Serving to Connect the Wiener and Feynman Integrals, *Journal of Math. and Phys.* **39**, 126–140 (1960); The Ilstow and Feynman Integrals, *J. Analyse Math.* **10**, 287–361 (1962/63).

5. K. Itô, Wiener Integral and Feynman Integral, *Proc. Fourth Berkeley Symp. on Math., Stat. and Prob.* **2**, 227–238 (1960); Generalized Uniform Complex Measures in the Hilbertian Metric Space with their Application to the Feynman Integral, *Proc. Fifth Berkeley Symp. on Math., Stat. and Prob.* **2**, 145–161 (1965). Both articles are reprinted in *Kiyosi Itô, Selected Papers*, Eds. D.W. Stroock and S.R.S. Varadhan, (Springer-Verlag, New York, 1987).

6. See, *e.g.*, G. Roepstorff, *Path Integral Approach to Quantum Physics*, (Springer-Verlag, Berlin, 1996).

7. R.P. Feynman, An Operator Calculus having Applications in Quantum Electrodynamics, *Phys. Rev.* **84**, 108–128 (1951).

8. See, *e.g.*, J.R. Klauder, The Action Option and the Feynman Quantization of Spinor Fields in Terms of Ordinary c-Numbers, *Ann. Phys.* **11**, 123–168 (1960); V. Bargmann, On a Hilbert Space of Analytic Functions and an Associated Integral Transform. Part. I. *Commun. Pure Appl. Math.* **14**, 187–214 (1961); I.E. Segal, Mathematical Characterization of the Physical Vacuum for a Linear Bose–Einstein Field, *Illinois J. Math.* **6**, 500–523 (1962); J.R. Klauder, Continuous-Representation Theory I. Postulates of Continuous Representation Theory, *J. Math. Phys.* **4**, 1055–1058 (1963); J. McKenna and J.R. Klauder, Continuous-Representation Theory IV. Structure of a Class of Function Spaces Arising from Quantum Mechanics, *J. Math. Phys.* **5**, 878–896 (1964); *Coherent States: Applications to Physics and Mathematical Physics*, Eds. J.R. Klauder and B.-S. Skagerstam (World Scientific, Singapore, 1985).

9. I. Daubechies and J.R. Klauder, Quantum Mechanical Path Integrals with Wiener Measures for All Polynomial Hamiltonians. II, *J. Math. Phys.* **26**, 2239–2256 (1985).

10. P.A.M. Dirac, *The Principles of Quantum Mechanics* (Oxford University Press, Oxford, Third Edition, 1947), p.114.

11. J.R. Klauder, Understanding Quantization, *Found. Phys.* **27**, 1467–1483 (1997).

12. J.R. Klauder, Quantization *Is* Geometry, After All, *Ann. Phys.* **188**, 120–141 (1988).

13. I. Daubechies, J.R. Klauder, and T. Paul, Wiener Measures for Path Integrals with Affine Kinematic Variables, *J. Math. Phys.* **28**, 85–102 (1987).

14. P. Maraner, Landau Ground State on Riemannian Surfaces, *Mod. Phys. Lett.*

A **7**, 2555–2558 (1992).

15. J.R. Klauder, Noncanonical Quantization of Gravity. I. Foundations of Affine Quantum Gravity, *J. Math. Phys.* **40**, 5860–5882 (1999); Noncanonical Quantization of Gravity. II. Constraints and the Physical Hilbert Space, *J. Math. Phys.* **42**, 4440–4464 (2001); The Affine Quantum Gravity Programme, *Class. Quant. Grav.* **19**, 817–826 (2002).

Time-Sliced Approximation to Path Integral and Lie–Trotter–Kato Product Formula

Takashi Ichinose

Department of Mathematics, Faculty of Science, Kanazawa University
Kanazawa, 920-1192, Japan

Dedicated to Professor Hiroshi Ezawa on the occasion
of his seventieth birthday with my hearty appreciation

In this note we discuss the time-sliced approximation of Feynman path integral with the recent results on the selfadjoint Lie–Trotter–Kato product formula and also the unitary Trotter product formula in operator norm.

1. Introduction

In his Princeton thesis,[5] 1942, Feynman invented the idea of *path integral* to give an alternative formulation of non-relativistic quantum mechanics. The paper on the subject was published 1948 in Ref.[6] (cf. Ref.[7]) It is a very striking idea which enables us to write down in one stroke the probability amplitude $K(x,t;y,0)$ that a particle starting from position y at time 0 can be found at position x at time t as an "integral"

$$K(x,t;y,0) = \int_{X(0)=y}^{X(t)=x} \mathcal{D}[X] \exp[iS(X)]. \tag{1}$$

Here $\mathcal{D}[X]$ is a uniform "measure" on the space of all possible paths X connecting the two points $(y,0)$ and (x,t) in space-time $\mathbb{R}^d \times \mathbb{R}$, which may be thought to be someting like "constant $\times \prod_{0<\tau<t} dX(\tau)$" with $dX(\tau)$ being the Lebesgue measure on the space $\mathbb{R}^d$ for *each* fixed τ. $S(X)$ stands for the classical action for each of these paths X, which is the time integral of the Lagrangian $L(X(t),\dot{X}(t))$ depending on the trajectory $X(t)$ and the velocity $\dot{X}(t) = dX(t)/dt$ of the particle concerned: $S(X) = \int_0^t L(X(s),\dot{X}(s))ds$.

For a particle of mass 1 moving in a potential $V(x)$, we have $L(X(t), \dot{X}(t)) = \frac{1}{2}\dot{X}(t)^2 - V(X(t))$, and this probability amplitude $K(x, t; y, 0)$ is nothing but the propagator, *i.e.* the fundamental solution $e^{-itH}(x, y)$ of the Schrödinger equation

$$i\frac{\partial}{\partial t}\psi(x, t) = H\psi(x, t) = [-\frac{1}{2}\Delta + V(x)]\psi(x, t), \quad t \in \mathbb{R}, \ x \in \mathbb{R}^d, \qquad (2)$$

for the nonrelativistic Hamitonian $H = -\frac{1}{2}\Delta + V(x)$.

Although (1) is not a mathematically rigorous integral, it can give a correct answer, if suitably interpreted. It has now become, because of its universal expression, a very powerful technique applying to all kinds of problems from quantum mechanics to quantum field theory and statistical mechanics.

There are mainly two methods of imparting a mathematically rigorous meaning to the path integral (1). One is a time-sliced approximation of (1):

$$K(x, t; y, 0) = \lim_{\varepsilon \to 0} A(\varepsilon)^{-N} \prod_{k=1}^{N-1} \int_{\mathbb{R}^d} dx_k \exp\Big[i \sum_{j=1}^{N} \varepsilon L(x_{j-1}, \frac{x_j - x_{j-1}}{\varepsilon})\Big],$$

$$(3)$$

with $x_0 = y$, $x_N = x$, and $t_k = k\varepsilon$, $(k = 0, 1, 2, ..., N - 1)$, $\varepsilon = t/N$, and hence $t_N = t$. In (3), $A(\varepsilon)$ is some normalization factor dependent on ε. For instance, in the above case concerned, $A(\varepsilon)$ is equal to $(2\pi i\varepsilon)^{d/2}$, and the fundamental solution $e^{-itH}(x, y)$ for (2) becomes

$$K(x, t; y, 0)$$

$$= \lim_{\varepsilon \to 0} \frac{1}{(2\pi i\varepsilon)^{Nd/2}} \prod_{k=1}^{N-1} \int_{\mathbb{R}^d} dx_k \exp\Big[i \sum_{j=1}^{N} \varepsilon \Big(\frac{1}{2}\Big(\frac{x_j - x_{j-1}}{\varepsilon}\Big)^2 - V(x_{j-1})\Big)\Big].$$

$$(4)$$

The other is of course to construct a mathematically rigorous countably additive measure on some path space. Unfortunately, there exists no countably additive measure on the path space with which to represent the fundamental solution of the real-time Schrödinger equation (2). However, if we go to imaginary time, the Schrödinger equation (2) is changed to the imaginary-time non-relativistic Schrödinger equation, *i.e.* heat equation

$$\frac{\partial}{\partial t}u(x, t) = Hu(x, t) = [-\frac{1}{2}\Delta + V(x)]u(x, t), \quad t > 0, \ x \in \mathbb{R}^d. \qquad (5)$$

Then there is a well-known formula called the Feynman–Kac formula representing the fundamental solution of (5) with the Wiener measure on the space of the continuous paths $X : [0, \infty) \to \mathbb{R}^d$, and even in the presence of vector potential $a(x)$, the Feynman–Kac–Itô formula for the Hamiltonian $H(a) = \frac{1}{2}(-i\nabla - a(x))^2 + V(x)$ (*e.g.* Ref.[36]). In passing, there is also an analogous formula for the imaginary-time relativistic Schrödinger equation for the Weyl-quantized relativistic Hamiltonian corresponding to the classical symbol $\sqrt{(p - a(x))^2 + 1} - 1 + V(x)$ (Refs.[28,15]).

As the imaginary-time path integral, the fundamental solution $e^{-tH}(x, y)$ of the heat equation (5) should be:

$$e^{-tH}(x, y) = \int_{X(0)=y}^{X(t)=x} \mathcal{D}[X] \exp\left[-\int_0^t \left(\frac{1}{2}\dot{X}(t)^2 + V(X(t))\right) dt\right]$$

$$= \lim_{\varepsilon \to 0} \frac{1}{(2\pi\varepsilon)^{Nd/2}} \prod_{k=1}^{N-1} \int_{\mathbb{R}^d} dx_k \exp\left[-\sum_{j=1}^{N} \varepsilon\left(\frac{1}{2}\left(\frac{x_j - x_{j-1}}{\varepsilon}\right)^2 + V(x_{j-1})\right)\right].$$

$$(6)$$

As mentioned above, the "integral" in the second member of (6) has made sense as the Feynman–Kac formula. The third member is the corresponding time-sliced approximation. We may think that the time-sliced approximation in (4)/(6) is the approximation of paths $X(\cdot)$ by broken lines.

Feynman[7] himself used a time-sliced approximation to compute out the right propagators for various potential functions $V(x)$.

In 1964, Nelson[33] noted that one can use the unitary and selfadjoint Trotter product formulas for selfadjoint operators A, B and their sum H in a Hilbert space,

$$e^{-itH} = s - \lim_{N \to \infty} [e^{-itA/N} e^{-itB/N}]^N, \tag{7}$$

$$e^{-tH} = s - \lim_{N \to \infty} [e^{-tA/N} e^{-tB/N}]^N, \tag{8}$$

to justify a convergence of the time-sliced approximations (4) and (6), respectively, for the (real-time) Schrödinger equation (2) and for the imaginary-time Schrödinger equation, *i.e.* heat equation (5). Namely, he proved that, with $A = -\frac{1}{2}\Delta$ and $B = V(x)\times$, the operators on the Hilbert space $L^2(\mathbb{R}^d)$ having these (4) and (6) as their integral kernels are strongly convergent. It is because the convergence of the Trotter or Lie–Trotter–Kato product formulas was in *strong operator topology*. In the above formulas, (7) holds if A and B are selfadjoint operators such that their operator sum $H = A + B$ is essentially selfadjoint on the domain $D[H] = D[A] \cap D[B]$ and the closure of H is again denoted by the same H, while (8) holds if A and B are selfadjoint operators bounded below such that their form sum $H = A \dot{+} B$ is densely-defined, *i.e.* selfadjoint on the domain $D[H]$ (Refs.[40,29]; Ref.[34]).

In the handbook[12], Grosche and Steiner have collected, adopting the time-sliced approximation (3) as the definition of "path integral", a pretty great number of examples of propagators which can be exactly computed by the method of path integral.

The aim of this note is to point out that in *imaginary time* the convergence of the time-sliced approximation to the path integral is taking place in a very strong sense, namely, in operator norm, concluded from our recent results[26,27] on *selfadjoint* Trotter or Lie–Trotter–Kato product formula as in the following theorem. We shall also see that in *real time* it is not in general, but in some special cases it is, by giving the unitary Trotter product formula in operator norm.

Theorem 1: *Let A and B be selfadjoint operators bounded below with domains $D[A]$ and $D[B]$ such that their operator sum $H := A+B$ is selfadjoint on the domain $D[H] = D[A] \cap D[B]$. Then it holds in operator norm that*

$$\| [e^{-tB/2N}e^{-tA/N}e^{-tB/2N}]^N - e^{-tH} \| = O(N^{-1}),$$
$$\| [e^{-tA/N}e^{-tB/N}]^N - e^{-tH} \| = O(N^{-1}), \quad N \to \infty, \tag{9}$$

uniformly on each compact t-interval in $[0, \infty)$, and further, if H is strictly positive, uniformly on $[0, \infty)$. The error bound $O(N^{-1})$ is optimal and ultimate.

One of the typical examples is the Schrödinger operator

$$H = A + B := \frac{1}{2}(-i\nabla - ea(x))^2 + z|x|^{-1} + \frac{w}{2}|x|^2 + \frac{a}{6}|x|^{2002} \tag{10}$$

in $L^2(\mathbb{R}^3)$, where e is a real constant, z, w, a nonnegative constants, and $a(x)$ is a vector potential with magnetic fields $\nabla \times a(x)$ bounded. Of course, this H is selfadjoint on the domain $D[H] = D[A] \cap D[B]$.

The theorem also holds with the exponential function e^{-s} replaced by real-valued, Borel measurable functions f on $[0, \infty)$ satisfying that

$$0 \leq f(s) \leq 1, \ f(0) = 1, \ f'(0) = -1, \tag{11}$$

that for every small $\varepsilon > 0$ there exists a positive constant $\delta = \delta(\varepsilon) < 1$ such that

$$f(s) \leq 1 - \delta(\varepsilon), \quad s \geq \varepsilon, \tag{12}$$

and that

$$[f]_\kappa := \sup_{s>0} s^{-2}|f(s) - 1 + s| < \infty. \tag{13}$$

Of course, the functions $f(s) = e^{-s}$ and $f(s) = (1 + k^{-1}s)^{-k}$ with $k > 0$ are examples of functions having these properties.

Remark 1.1. Such a norm convergence of the Lie–Trotter–Kato product formula as (9) was first proved by Rogava,[35] under the additional condition that B is A-bounded, with error bound $O(N^{-1/2} \log N)$. The next was by Helffer[14] for the Schrödinger operator $H = H_0 + V \equiv -\frac{1}{2}\Delta + V(x)$ with C^∞ nonnegative potentials $V(x)$, roughly speaking, growing at most of order $O(|x|^2)$ for large $|x|$ with error bound $O(N^{-1})$. Each of these two results is independent of the other.

Then under some stronger or more general conditions, several further results are obtained. A better error bound $O(N^{-1} \log N)$ than Rogava's is obtained by Ichinose–Tamura[25] (cf. Ref.[23]) when B is A^α-bounded for some $0 < \alpha < 1$, even though the $B = B(t)$ may be t-dependent, and by Neidhardt–Zagrebnov[30,31] (cf. Ref.[32]) when B is A-bounded with relative bound less than 1. As for the Schrödinger operators, more general results were proved for continuous nonnegative potentials $V(x)$, roughly speaking, growing of order $O(|x|^\rho)$ for large $|x|$ with $\rho > 0$, together with error bounds dependent on the power ρ (for instance, of order $O(N^{-2/\rho})$, if $\rho \geq 2$), by Ichinose–Takanobu[20,21] (cf. Ref.[22]), Doumeki–Ichinose–Tamura[4], Ichinose–Tamura[24], Descombes–Dia[2] (cf. Dia–Schatzman[3]) and others. It should be noted (see Refs.[13,38]) that in all these cases of the Schrödinger operators the sum $H = H_0 + V$ is selfadjoint on the domain $D[H] = D[H_0] \cap D[V]$.

Thus the present theorem not only extends Rogava's result, but also properly contains all the results mentioned above with ultimate optimal error bounds $O(N^{-1})$.

Remark 1.2. If both A and B are bounded operators, then we have, in the non-symmetric product case, $\|[e^{-tA/N}e^{-tB/N}]^N - e^{-tH}\| = O(N^{-1})$, while, in the symmetric product case, $\|[e^{-tB/2N}e^{-tA/N}e^{-tB/2N}]^N - e^{-tH}\| = O(N^{-2})$. But the error bound $O(N^{-1})$ in Theorem 1 is optimal. In fact, also in the symmetric product case, we can give an example of two unbounded nonnegative selfadjoint operators A and B whose operator sum $H = A + B$ is selfadjoint on $D[A] \cap D[B]$ such that $\|[e^{-tB/2N}e^{-tA/N}e^{-tB/2N}]^N - e^{-tH}\| \geq L(t)N^{-1}$, with a positive continuous function $L(t)$ of $t > 0$ independent of N (see Ref.[27]).

Remark 1.3. Unless the sum $A + B$ is selfadjoint on $D[A] \cap D[B]$, the norm convergence of the Lie–Trotter–Kato product formula does not always hold, even though the sum is essentially selfadjoint there and B is A-form-bounded with relative bound less than 1. For a counterexample we refer to Ref.[39]

Remark 1.4. In a recent paper[19], the case of the form sum $H = A \dotplus B$ of nonnegative selfadjoint operators A and B has been treated where B is A-form bounded with some further auxiliary conditions among A, B and H. Here note that the operator sum $A + B$ may not be selfadjoint nor even essentially selfadjoint on $D[A] \cap D[B]$. The result is extending that in Ref.[31]

Section 2 deals with an outline of proof of Theorem 1. In Section 3, we discuss the time-sliced approximation to the imaginary-time path integral for the heat equation, and will also make some general speculative conjecture. In Section 4 we shall observe that the unitary product formula in operator norm does not in general hold, and in some special cases it holds.

2. Outline of proof

For the proof of the theorem, it is crucial that we have been able to establish the following operator-norm version of Chernoff's theorem with error bounds (see Ref.[26]).

Lemma 1: *Let C be a nonnegative selfadjoint operator in a Hilbert space $\mathcal{H}$ and let $\{F(t)\}_{t \geq 0}$ be a family of selfadjoint operators with $0 \leq F(t) \leq 1$. Define $S_t = t^{-1}(1 - F(t))$.*

Then, for $0 < \alpha \leq 1$, if it holds that

$$\|(1 + S_t)^{-1} - (1 + C)^{-1}\| = O(t^\alpha), \quad t \downarrow 0, \tag{14}$$

one has: for any $\delta > 0$ with $0 < \delta \leq 1$,

$$\|F(t/N)^N - e^{-tC}\| = \delta^{-2} t^{-1+\alpha} e^{\delta t} O(N^{-\alpha}), \quad N \to \infty, \tag{15}$$

for all $t > 0$.

Therefore, for $0 < \alpha < 1$ (resp. $\alpha = 1$), the convergence in (15) is uniform on each compact t–interval in the open half line $(0, \infty)$ (resp. in the closed half line $[0, \infty)$).

Moreover, if C is strictly positive, i.e. $C \geq \eta$ for some constant $\eta > 0$, the error bound on the right-hand side of (15) can also be replaced by $(1 + 2/\eta)^2 t^{-1+\alpha} O(N^{-\alpha})$, so that, for $0 < \alpha < 1$ (resp. $\alpha = 1$), the convergence *in (15) is uniform on the closed half line $[T, \infty)$ for every fixed $T > 0$ (resp. on the whole closed half line $[0, \infty)$).*

Sketch of Proof of Lemma 1

Put

$$F(t/N)^N - e^{-tC} = (F(t/N)^N - e^{-tS_{t/N}}) + (e^{-tS_{t/N}} - e^{-tC}). \qquad (16)$$

For the first term on the right-hand side of (16) we have by the spectral theorem

$$\|F(t/N)^N - e^{-tS_{t/N}}\| = \|F(t/N)^N - e^{-N(1-F(t/N))}\| \le e^{-1}N^{-1}, \qquad (17)$$

because

$$0 \le e^{-N(1-\lambda)} - \lambda^N \le e^{-1}/N, \quad \text{for } 0 \le \lambda \le 1.$$

For the second term, we use

$$(1 + S_\varepsilon)^{-1}[e^{-t(\delta+S_\varepsilon)} - e^{-t(\delta+C)}](1 + C)^{-1}$$
$$= \int_0^t e^{-(t-s)(\delta+S_\varepsilon)}[(1 + S_\varepsilon)^{-1} - (1 + C)^{-1}]e^{-s(\delta+C)}ds \qquad (18)$$
$$= \int_0^{t/2} + \int_{t/2}^t,$$

with $0 < \delta \le 1$ and $\varepsilon > 0$. Multiplying the integrands of these two integrals in the last member of (18) by $(1 + S_\varepsilon)$ from left and by $(1 + C)$ from right, we can bound them by $(\delta^2 t)^{-1}e^{\delta t}O(\varepsilon^\alpha)$. Taking $\varepsilon = t/N$, we have

$$\|e^{-tS_{t/N}} - e^{-tC}\| \le (\delta^2 t)^{-1}e^{\delta t}O((t/N)^\alpha) = \delta^{-2}t^{-1+\alpha}e^{\delta t}O(N^{-\alpha}).$$

Sketch of Proof of Theorem 1

We may assume without loss of generality that both A and B are nonnegative, so that H is also nonnegative.

First note that since $H = A + B$ is itself selfadjoint and so a closed operator, by the closed graph theorem there exists a constant b such that

$$\|(1 + A)u\| + \|(1 + B)u\| \le b\|(1 + H)u\|, \quad u \in D[H] = D[A] \cap D[B]. \qquad (19)$$

The proof will be done, dividing into two cases, (a) the symmetric product case $F(t) = e^{-tB/2}e^{-tA}e^{-B/2}$, and (b) the non-symmetric product case $G(t) = e^{-tA}e^{-tB}$.

(a) In the symmetric product case, put

$$S_t = t^{-1}(1 - F(t)) = t^{-1}(1 - e^{-tB/2}e^{-tA}e^{-tB/2}). \tag{20}$$

By Lemma 1 we have only to show that

$$\|(1 + S_t)^{-1} - (1 + H)^{-1}\| = O(t), \quad t \downarrow 0. \tag{21}$$

To do so, we show first that

$$\|(1 + S_t)^{-1} - (1 + H)^{-1}\| = O(t^{1/2}), \quad t \downarrow 0. \tag{22}$$

Put

$$A_t = t^{-1}(1 - e^{-tA}), \quad B_t = t^{-1}(1 - e^{-tB}), \quad H_t = t^{-1}(1 - e^{-tH}). \tag{23}$$

We have

$$\begin{aligned}
1 + S_t &= 1 + A_t + B_{t/2} - \tfrac{t}{4}B_{t/2}^2 + \tfrac{t^2}{4}B_{t/2}A_tB_{t/2} - \tfrac{t}{2}(A_tB_{t/2} + B_{t/2}A_t) \\
&= K_t^{1/2}(1 + Q_t)K_t^{1/2},
\end{aligned} \tag{24}$$

where

$$\begin{aligned}
K_t &= 1 + A_t + B_{t/2} - \tfrac{t}{4}B_{t/2}^2 \geq 1, \\
Q_t &= \tfrac{t^2}{4}K_t^{-1/2}B_{t/2}A_tB_{t/2}K_t^{-1/2} - \tfrac{t}{2}K_t^{-1/2}(A_tB_{t/2} + B_{t/2}A_t)K_t^{-1/2}.
\end{aligned} \tag{25}$$

Then we can show

$$\|(1 + Q_t)^{-1}\| \leq 2/(3 - \sqrt{5}), \tag{26}$$

$$\|(1 + S_t)^{-1}K_t^{1/2}\| = \|K_t^{-1/2}(1 + Q_t)^{-1}\| \leq 2/(3 - \sqrt{5}). \tag{27}$$

Then we have through the resolvent equation

$$\begin{aligned}
(1 + S_t)^{-1} &- (1 + H)^{-1} \\
&= (1 + S_t)^{-1}(H - S_t)(1 + H)^{-1} \\
&= (1 + S_t)^{-1}[A + B - (A_t + B_{t/2} - \tfrac{t}{4}B_{t/2}(1 - tA_t)B_{t/2} \\
&\qquad\qquad - \tfrac{t}{2}(A_tB_{t/2} + B_{t/2}A_t))](1 + H)^{-1} \\
&= (1 + S_t)^{-1}(A - A_t)(1 + H)^{-1} + (1 + S_t)^{-1}(B - B_{t/2})(1 + H)^{-1} \\
&\quad + (1 + S_t)^{-1}[\tfrac{t}{4}B_{t/2}(1 - tA_t)B_{t/2} + \tfrac{t}{2}(A_tB_{t/2} + B_{t/2}A_t)](1 + H)^{-1} \\
&\equiv R_1(t) + R_2(t) + R_3(t).
\end{aligned} \tag{28}$$

We can show the bounds

$$\|R_i(t)\| \leq ct^{1/2}, \quad i = 1, 2, 3, \tag{29}$$

with some constant $c > 0$, which yields (22). For instance, to find the bound for $R_1(t)$, rewrite it as

$$R_1(t) = [(1 + S_t)^{-1} K_t^{1/2}][K_t^{-1/2}(1 + A_t)^{1/2}]$$
$$\times [(1 + A_t)^{-1/2} - (1 + A_t)^{1/2}(1 + A)^{-1}](1 + A)(1 + H)^{-1}.$$

Then we get by (19), (27) and the spectral theorem

$$\|R_1(t)\| \le \frac{2}{3 - \sqrt{5}} b \|(1 + A_t)^{-1/2} - (1 + A_t)^{1/2}(1 + A)^{-1}\| \le c t^{1/2}.$$

Now we come to show (21). The idea of proof is simply to iterate the resolvent equation, the first identity in (28), with help of its adjoint form to get

$$
\begin{aligned}
(1 + &S_t)^{-1} - (1 + H)^{-1} \\
&= ((1 + H)^{-1} + [(1 + S_t)^{-1} - (1 + H)^{-1}])(H - S_t)(1 + H)^{-1} \\
&= (1 + H)^{-1}(H - S_t)(1 + H)^{-1} \\
&\qquad + [(H - S_t)(1 + H)^{-1}]^*(1 + S_t)^{-1}(H - S_t)(1 + H)^{-1} \\
\equiv\ & R_1'(t) + R_2'(t).
\end{aligned}
\tag{30}
$$

Then by the same arguments together with (19), (27) and the case (22) which we have proved above, we can show the bounds

$$\|R_i'(t)\| = O(t), \quad i = 1, 2. \tag{31}$$

(b) The non-symmetric product case will follow from the symmetric product case. We use the commutator argument to observe that

$$
\begin{aligned}
\|G(t/N)^N &- F(t/N)^N\| \\
&= \|[e^{-tA/N}e^{-tB/N}]^N - [e^{-tB/2N}e^{-tA/N}e^{-tB/2N}]^N\| \\
&= O(N^{-1}).
\end{aligned}
\tag{32}
$$

This proves Theorem 1.

3. Speculative thinking *via* the imaginary-time path integral

In the first part of this section, we discuss the imaginary-time path integral for the heat equation, and in the second want to make a speculative conjecture. The content of the first part was already described in a little more detail with a sketch of proof in Ref.[18], based on the results on the Kac transfer operator $K(t; H_0, V) := e^{-tV/2}e^{-tH_0}e^{-tV/2}$, where

$H = H_0 + V \equiv -\frac{1}{2}\Delta + V(x)$ is a Schrödinger operator, and was also studied by more thorough probabilistic methods in Ref.[37]

We assume $V(x)$ is a continuous function uniformly bounded from below, and for simplicity, $V(x) \geq 0$. In Refs.[14,3,20,21,4,24,2], assuming some further conditions on $V(x)$, they studied the Kac transfer operator $K(t; H_0, V)$ to estimate the operator norm of its difference from the Schrödinger semigroup e^{-tH} by the order $O(t^{1+a})$ of small $t > 0$ with a constant $a > 0$ depending on the regularity of $V(x)$ (For a background of these results and their application, see *e.g.* Ref.[17]). The Lie–Trotter–Kato product formula in operator norm was obtained as a by-product.

In the papers[20,21], a probabilistic method was used with the Feynman–Kac formula, considering as potentials $V(x)$ some general class of nonnegative continuous functions in $\mathbb{R}^d$. To briefly recapitulate it here, let us consider the following a little more restricted class of functions $V(x)$. Suppose that, for some $0 < \delta \leq 1$, $V(x)$ is a C^∞-function satisfying

$$|\partial^\alpha V(x)| \leq C_\alpha (V(x) + 1)^{(1-|\alpha|\delta)+} \tag{33}$$

for every multi-index α with constant $C_\alpha > 0$, where for a real number a, we mean $a_+ = \max\{a, 0\}$. In this case we have necessarily $2\delta \leq 1$. Then in Theorem 2.1 in Refs.[20,21], it was proved that $\|K(t; H_0, V) - e^{-tH}\| = O(t^{1+2\delta})$.

By a careful check of its proof, it turns out that we can also see, in fact, the following pointwise estimate:

$$K(t; H_0, V)(x, y) - e^{-tH}(x, y) = O(t^{1+2\delta})(2\pi t)^{-d/2} e^{-(x-y)^2/4t}, \tag{34}$$

with $O(t^{1+2\delta})$ being independent of x and y. Here $K(t; H_0, V)(x, y)$ denotes the integral kernel of $K(t; H_0, V)$ and $e^{-tH}(x, y)$ the integral kernel of e^{-tH} or the fundamental solution of the heat equation (5). Hence, since we have $K(t/N; H_0, V)^N = [e^{-tV/2N} e^{-tH_0/N} e^{-tV/2N}]^N$, whose integral kernel is denoted by $[e^{-tV/2N} e^{-tH_0/N} e^{-tV/2N}]^N(x, y)$, we conclude

$$[e^{-tV/2N} e^{-tH_0/N} e^{-tV/2N}]^N(x, y) - e^{-tH}(x, y)$$
$$= O(N^{-2\delta}) t^{1+2\delta} (2\pi t)^{-d/2} e^{-(x-y)^2/4t}, \tag{35}$$

pointwise, as $N \to \infty$, with $O(N^{-2\delta})$ independent of x, y and t.

On the other hand, for the Schrödinger equation (2), Fujiwara[8,9,10] (cf. Ref.[11]) constructed, based on the idea of path integral, the fundamental solution $\equiv e^{-itH}(x, y)$, *i.e.* the integral kernel of e^{-itH}. He obtained the following result, assuming that $V(x)$ is smooth and satisfies

$|\partial^\alpha V(x)| \leq C_\alpha (1+x^2)^{(2-|\alpha|)_+/2}$ for every multi-index α with constant C_α, though $V(x)$ need not be bounded below. Put

$$(E(t)\varphi)(x) = (2\pi it)^{-d/2} \int_{\mathbb{R}^d} e^{iS(t,x,y)} \varphi(y) dy \qquad (36)$$

for $\varphi \in C_0^\infty(\mathbb{R}^d)$, with action $S(t,x,y) = \int_0^t [\frac{1}{2}(d\widetilde{X}(s)/ds)^2 - V(\widetilde{X}(s))]ds$, where $\widetilde{X}(s)$ is the classical trajectory starting at $\widetilde{X}(0) = y$ and ending at $\widetilde{X}(t) = x$. Then for sufficiently small $t > 0$, one has

$$E(t/N)^N(x,y) - e^{-itH}(x,y) = O(N^{-1})t^2(2\pi t)^{-d/2}, \qquad (37)$$

as $N \to \infty$, uniformly in x,y, together with all the x,y-derivatives of the left-hand side, where $O(N^{-1})$ is independent of x, y and t.

The potential $V(x)$ in his result satisfies condition (33) with $\delta = 1/2$ or $2\delta = 1$, so that in this case our (35) becomes

$$[e^{-tV/2N}e^{-tH_0/N}e^{-tV/2N}]^N(x,y) - e^{-tH}(x,y)$$
$$= O(N^{-1})t^2(2\pi t)^{-d/2}e^{-(x-y)^2/4t}. \qquad (38)$$

Thus, in imaginary time, the integral kernel of the time-sliced, broken-line-approximation $[e^{-tV/2N}e^{-tH_0/N}e^{-tV/2N}]^N$ to the fundamental solution of the heat equation (5) can just bring a pointwise approximation analogous to Fujiwara's result for the Schrödinger equation (2) as above. Notice here that Fujiwara's time-sliced approximation is not the broken-line-approximation, but a more elaborate one.

Now, in this stage, we would expect the relation like (38) to hold for *any* nonnegative potential $V(x)$ for which $H = H_0 + V$ is selfadjoint on $D[H] = D[H_0] \cap D[V]$.

When the physicists accept the path integral formula (4) or (6) as a time-sliced approximation, they appear to be thinking the $(N-1)$-ple integrals on the right-hand side of (4)/ (6) are pointwise convergent to the integral kernel on the left-hand side. But what we can know from the usual Lie–Trotter–Kato product formula in the strong operator topology seems to have been merely a convergence of these integrals in a weak sense such as in distribution sense, but likely not pointwise.

However, the selfadjoint Lie–Trotter–Kato product formula established in Theorem 1 converges in the operator norm topology, in a topology far stronger than in the usual strong operator topology. So this might suggest us not only in the case of the Schrödinger operator, but also in such a case of general operator $H = A + B$ as in Theorem 1 that if the semigroups e^{-tA}, e^{-tB} and e^{-tH} possess their integral kernels, then the integral kernels of the product $[e^{-tB/2N}e^{-tA/N}e^{-tB/2N}]^N$ or $[e^{-tA/N}e^{-tB/N}]^N$ are 'nearly' convergent to the integral kernel of e^{-tH}. And we are expecting the operator norm convergence might imply the convergence of their integral kernels.

4. The unitary product formula in operator norm

In this final section, just to make sure, we shall observe that the unitary Trotter product formula in operator norm does not in general hold, and also in some special cases it holds.

Let $A = -i\frac{d}{dx}$ and $B = V(x)\times$ be the selfadjoint operators in $L^2(\mathbb{R})$ defined by the differentiation with respect to x and multiplication by a real-valued C^∞-function $V(x)$, where $D[A] = H^1(\mathbb{R})$ and $D[B] = \{f \in L^2(\mathbb{R}); Vf \in L^2(\mathbb{R})\}$. Then it is known[1] that the operator sum $A + B = -i\frac{d}{dx}+V(x)$ is essentially selfadjoint on $C_0^\infty(\mathbb{R})$ and hence on $D[A]\cap D[B] = \{f \in H^1(\mathbb{R}); Vf \in L^2(\mathbb{R})\}$, so that with H being the closure of $A + B$, the unitary Trotter product formula (7) in strong operator topology holds. But, in operator norm, $[e^{-itA/N}e^{-itB/N}]^N$ does not in general converge to e^{-itH}.

In the following case (a), we shall see this issue for $V(x) = x^k$ with k being a positive integer. Nevertheless, there are some nontrivial cases where the unitary Trotter product formula in operator norm holds; we shall see them at the end of (a) with $V(x) = x$ and in the case (b) where $V(x)$ is a real-valued C^1-function with bounded derivative.

(a) The case $V(x) = x^k$. In this case, $A + B$ is essentially selfadjoint on $C_0^\infty(\mathbb{R})$ and hence on $D[A] \cap D[B] = \{f \in H^1(\mathbb{R}); x^k f \in L^2(\mathbb{R})\}$.

But it is not selfadjoint on $D[A] \cap D[B]$. Indeed, put $u(x) = \varphi(x)\exp[-i\frac{x^{k+1}}{k+1}]$ with $\varphi(x) = (1 + x^2)^{-1/2}$. u belongs to $L^2(\mathbb{R})$. Then $\frac{d}{dx}u(x) = (\varphi'(x) - ix^k\varphi(x))\exp[-i\frac{x^{k+1}}{k+1}]$, so that $[-i\frac{d}{dx} + x^k]u(x) = -i\varphi'(x)\exp[-i\frac{x^{k+1}}{k+1}]$, which belongs to $L^2(\mathbb{R})$. But u does not belong to $H^1(\mathbb{R})$.

Now we have for $f \in C_0^\infty(\mathbb{R})$,

$$\left(e^{-itA}f\right)(x) = \left(e^{-t\frac{d}{dx}}f\right)(x) = f(x - t), \tag{39}$$

so that

$$\left([e^{-itA}e^{-itB}]f\right)(x) = e^{-t\frac{d}{dx}}\left(e^{-itx^k}f\right)(x) = e^{-it(x-t)^k}f(x - t),$$

$$\left([e^{-itA}e^{-itB}]^2 f\right)(x) = e^{-t\frac{d}{dx}}\left(e^{-it[x^k+(x-t)^k]}f\right)(x)$$

$$= e^{-it[(x-t)^k+(x-2t)^k]}f(x - 2t),$$

$$\left([e^{-itA}e^{-itB}]^N f\right)(x) = e^{-it\sum_{j=1}^{N}(x-jt)^k}f(x - Nt).$$

Then we find

$$\left([e^{-itA/N}e^{-itB/N}]^N f\right)(x) = e^{-i(t/N)\sum_{j=1}^N (x-jt/N)^k} f(x-t), \tag{40}$$

and, by letting N tend to infinity,

$$\left(e^{-itH}f\right)(x) = \exp\left[-i\int_0^t (x-s)^k ds\right] f(x-t). \tag{41}$$

Therefore we have

$$\left\|\left([e^{-itA/N}e^{-itB/N}]^N - e^{-itH}\right)f\right\|^2$$

$$= \sup_{\|f\|=1} \int \left|\exp\left[-i\frac{t}{N}\sum_{j=1}^N (x-\frac{j}{N}t)^k\right] - \exp\left[-i\int_0^t (x-s)^k ds\right]\right|^2$$

$$\times |f(x-t)|^2 dx$$

$$= \sup_{\|f\|=1} \int 2\left(1 - \cos\left[\frac{t}{N}\sum_{j=1}^N (x-\frac{j}{N}t)^k - \int_0^t (x-s)^k ds\right]\right) |f(x-t)|^2 dx$$

$$= \sup_{\|f\|=1} 4\int \sin^2 \frac{1}{2}\left[\frac{t}{N}\sum_{j=1}^N (x-\frac{j}{N}t)^k - \int_0^t (x-s)^k ds\right] |f(x-t)|^2 dx$$

$$= \sup_{\|f\|=1} 4\int \sin^2 \frac{1}{2}\left[\frac{t}{N}\sum_{j=1}^N (x+t-\frac{j}{N}t)^k - \int_0^t (x+t-s)^k ds\right] |f(x)|^2 dx.$$

It follows that

$$\left\|[e^{-itA/N}e^{-itB/N}]^N - e^{-itH}\right\|$$

$$= 2\sup_{x\in\mathbb{R}}\left|\sin\frac{1}{2}\left[\frac{t}{N}\sum_{j=1}^N (x+t-\frac{j}{N}t)^k - \int_0^t (x+t-s)^k ds\right]\right|$$

$$= 2\sup_{x\in\mathbb{R}}\left|\sin\frac{1}{2}\left[\frac{t}{N}\sum_{j=1}^N (x+t-\frac{j}{N}t)^k - \frac{1}{k+1}[(x+t)^{k+1} - x^{k+1}]\right]\right| = 2$$

for $k \geq 2$, which implies the operator norm convergence does not hold.

However, for $k = 1$, we have

$$\left\|[e^{-itA/N}e^{-itB/N}]^N - e^{-itH}\right\| = 2\left|\sin\frac{t^2}{4N}\right| = O(N^{-1}),$$

uniformly on the bounded interval on $[-T, T]$ for each $T > 0$, which yields operator norm convergence as $N \to \infty$. So the unitary Trotter product formula in operator norm holds with error bound $O(N^{-1})$.

(b) The case $V(x)$ is a real-valued C^1-function with bounded derivative, *i.e.* $\|V'\|_\infty = \sup_{x \in \mathbb{R}} |V'(x)| < \infty$. In this case, $A + B$ is essentially selfadjoint on $D[A] \cap D[B] = \{f \in H^1(\mathbb{R}); Vf \in L^2(\mathbb{R})\}$. In addition, if $V(x)$ itself is a bounded function, $B = V(x)\times$ is a bounded operator, so that $A + B$ is selfadjoint on the domain $D[A] \cap D[B] = D[A] = H^1(\mathbb{R})$.

In the same way as in the case (a), we find for $f \in C_0^\infty(\mathbb{R})$,

$$\left([e^{-itA/N}e^{-itB/N}]^N f\right)(x) = \exp[-i(t/N)\sum_{j=1}^{N} V(x - jt/N)]f(x - t), \quad (42)$$

and, by letting N tend to infinity,

$$\left(e^{-itH}f\right)(x) = \exp\left[-i\int_0^t V(x - s)ds\right]f(x - t). \quad (43)$$

Then we have

$$\||[e^{-itA/N}e^{-itB/N}]^N - e^{-itH}\|$$

$$= 2\sup_{x \in \mathbb{R}}\left|\sin\frac{1}{2}\left[\frac{t}{N}\sum_{j=1}^{N}V(x + t - \frac{j}{N}t) - \int_0^t V(x + t - s)ds\right]\right|$$

$$= 2\sup_{x \in \mathbb{R}}\left|\sin\frac{1}{2}\left[\int_0^t \left(V(x + t - [Ns/t]t/N) - V(x + t - s)\right)ds\right]\right|.$$

Here $[Ns/t]$ denotes the largest integer not exceeding Ns/t, so that $[Ns/t] \le Ns/t < [Ns/t] + 1$, and we understand $[Ns/t]t/N = 0$ for $t = 0$.

Then, since $0 \le s - [Ns/t]t/N < t/N$, we have by the mean value theorem

$$\||[e^{-itA/N}e^{-itB/N}]^N - e^{-itH}\|$$

$$= 2\sup_{x \in \mathbb{R}}\left|\sin\frac{1}{2}\left[\int_0^t ds(s - \frac{[Ns/t]t}{N})\int_0^1 V'(x + t - s + \theta(s - \frac{[Ns/t]t}{N}))d\theta\right]\right|$$

$$\le 2\left|\sin\left(\frac{\|V'\|_\infty t^2}{4N}\right)\right| = O(N^{-1}),$$

$$(44)$$

uniformly on $[-T, T]$ for each $T > 0$. So the unitary Trotter product formula in operator norm holds with error bound $O(N^{-1})$ in this case. A more general problem for such a unitary Trotter product formula in operator norm will be discussed elsewhere.

The present note has been written mainly based on our very recent results,[26,27] entirely improving our earlier versions[16,18] of this subject.

Acknowledgement

I feel especially fortunate to have been acquainted with Professor Hiroshi Ezawa these more than twenty years since I began to be interested in the Feynman path integral. I have appreciated his great knowledge not only in physics but also in mathematics, his clear explanation of problems enabling me to understand their points. As a mature researcher, I also admire the ways in which he has encouraged independent thinking to let the people around him pursue their research with integrity and warmth. I am very much grateful to him for his always kind and helpful suggestions.

Thanks are due to Professor Osanobu Yamada for a comment on the domain of the differential operator treated in Section 4 (a). This work was partially supported by the Grant-in-Aid for Scientific Research (B) No. 13440044, Japan Society of the Promotion of Science.

References

1. Paul R. Chernoff, *Schrödinger and Dirac operators with singular potentials and hyperbolic equations*, Pacific J. Math., **72**, 361–382 (1977).
2. Stéphane Descombes and Boun O. Dia, *An operator-theoretic proof of an estimate on the transfer operator*, J. Functional Analysis, **165**, 240–257 (1999).
3. Boun O. Dia and Michelle Schatzman, *An estimate on the Kac transfer operator*, J. Functional Analysis, **145**, 108–135 (1997).
4. Atsushi Doumeki, Takashi Ichinose and Hideo Tamura, *Error bound on exponential product formulas for Schrödinger operators*, J. Math. Soc. Japan, **50**, 359–377 (1998).
5. Richard P. Feynman, *The Principle of Least Action in Quantum Mechanics*, Dissertation presented to the Faculty of Princeton University in Candidacy for the Degree of Doctor of Philosophy, May 1942, 79 pages: University Microfilms, Pub. No. 2948, A XEROX Company, Ann Arbor, Michigan.
6. Richard P. Feynman, *Space-time approach to non-relativistic quantum mechanics*, Rev. Mod. Phys., **20**, 367–387 (1948).
7. Richard P. Feynman and Albert R. Hibbs, *Quantum Mechanics and Path Integrals*, McGraw–Hill, New York, 1965.
8. Daisuke Fujiwara, *A construction of the fundamental solution for the Schrödinger equations*, J. Analyse Math., **35**, 41–96 (1979).
9. Daisuke Fujiwara, *Remarks on convergence of the Feynman path integrals*, Duke Math. J., **47**, 559–600 (1980).
10. Daisuke Fujiwara, *Some Feynman path integrals as oscillatory integrals over a Sobolev manifold*, in Functional Analysis and Related Topics, Springer Lect. Notes in Math., No. **1540**, 39–53 (1993).
11. Daisuke Fujiwara and Tetsuo Tsuchida, *The time slicing approximation of the fundamental solution for the Schrödinger equation with electromagnetic fields*, J. Math. Soc. Japan, **49**, 299–327 (1997).

12. Christian Grosche and Frank Steiner, *Handbook of Feynman Path Integrals*, Springer Tracts in Modern Physics, **145**, Berlin Heidelberg 1998.

13. Denis Guibourg, *Inégalités maximales pour l'opérateur de Schrödinger*, C. R. Acad. Sci. Paris, **316**, Série I Math., 249–252 (1993).

14. Bernard Helffer, *Around the transfer operator and the Trotter–Kato formula*, Operator Theory: Advances and Appl., **78**, 161–174 (1995).

15. Takashi Ichinose, *Some results on the relativistic Hamiltonian: Selfadjointness and imaginary-time path integral*, in "Differential Equations and Mathematical Physics, Proc. of the International Conference, Univ. of Alabama at Birmingham, March 13–17, 1994", pp. 102–116, International Press, Boston 1996.

16. Takashi Ichinose, *Norm convergence of the Trotter product formula for Schrödinger operators via the Feynman–Kac formula*, Proc. of Dubna Joint Meeting of International Seminar "Path Integrals: Theory & Applications" and 5th Internatioal Conference "Path Integrals from meV to MeV", edited by V. S. Yarunin and M. K. Smondyrev, pp. 341–346, Dubna 1996.

17. Takashi Ichinose, *Norm estimate for Kac's transfer operator with applications to the Lie–Trotter product formula*, in Mathematical Methods of Quantum Physics, Essays in honor of Professor Hiroshi Ezawa edited by C. Bernido, V. Bernido, K. Nakamura and K. Watanabe (Proceedings of the 2nd Jagna International Workshop, January 4–8, 1998), pp. 145–154, Gordon and Breach Pub. 1999.

18. Takashi Ichinose, *Norm Convergence of the Lie–Trotter–Kato Product Formula and Imaginary-time Path Integral*, J. Korean Math. Soc., **38**, 337–348 (2001).

19. Takashi Ichinose, Hagen Neidhardt and Valentin A. Zagrebnov, *Trotter–Kato product formula and fractional powers of self-adjoint generators*, to appear.

20. Takashi Ichinose and Satoshi Takanobu, *Estimate of the difference between the Kac operator and the Schrödinger semigroup*, Commun. Math. Phys., **186**, 167–197 (1997).

21. Takashi Ichinose and Satoshi Takanobu, *The norm estimate of the difference between the Kac operator and the Schrödinger semigroup: A unified approach to the nonrelativistic and relativistic cases*, Nagoya Math. J., **149**, 53–81 (1998).

22. Takashi Ichinose and Satoshi Takanobu, *The norm estimate of the difference between the Kac operator and the Schrödinger semigroup II: The general case including the relativistic case*, Electronic Journal of Probability, **5**, Paper no. 5, pages 1–47 (2000).

23. Takashi Ichinose and Hideo Tamura, *Error estimates in operator norm for Trotter–Kato product formula*, Integral Equations Operator Theory, **27**, 195–207 (1997).

24. Takashi Ichinose and Hideo Tamura, *Error bound in trace norm for Trotter–Kato product formula of Gibbs semigroups*, Asymptotic Analysis, **17**, 239–266 (1998).

25. Takashi Ichinose and Hideo Tamura, *Error estimates in operator norm of exponential product formulas for propagators of parabolic evolution equations,*

Osaka J. Math., **35**, 751–770 (1998).

26. Takashi Ichinose and Hideo Tamura, *The norm convergence of the Trotter–Kato product formula with error bound*, Commun. Math. Phys., **217**, 489–502 (2001).

27. Takashi Ichinose, Hideo Tamura, Hiroshi Tamura and Valentin A. Zagrebnov, *Note on the paper "The norm convergence of the Trotter–Kato product formula with error bound" by Ichinose and Tamura*, Commun. Math. Phys., **221**, 499–510 (2001).

28. Takashi Ichinose and Hiroshi Tamura, *Imaginary-time path integral for a relativistic spinless particle in an electromagnetic field*, Commun. Math. Phys., **105**, 239–257 (1986).

29. Tosio Kato, *Trotter's product formula for an arbitrary pair of self-adjoint contraction semigroups*, in Topics in Functional Analysis (I. Gohberg and Mark Kac, eds.), Academic Press, New York, 1978, pp. 185–195.

30. Hagen Neidhardt and Valentin A. Zagrebnov, *On error estimates for the Trotter–Kato product formula*, Lett. Math. Phys., **44**, 169–186 (1998).

31. Hagen Neidhardt and Valentin A. Zagrebnov, *Fractional powers of selfadjoint operators and Trotter–Kato product formula*, Integral Equations Operator Theory, **35**, 209–231 (1999).

32. Hagen Neidhardt and Valentin A. Zagrebnov, *Trotter–Kato product formula and operator-norm convergence*, Commun. Math. Phys., **205**, 129–159 (1999).

33. Edward Nelson, *Feynman integrals and the Schrödinger equation*, J. Math. Phys., **5**, 332–343 (1964).

34. Michael Reed and Barry Simon, *Methods of Modern Mathematical Physics I: Functional Analysis*, Revised and enlarged ed., Academic Press, New York, London 1980.

35. Dzh. L. Rogava, *Error bounds for Trotter-type formulas for self-adjoint operators*, Functional Analysis and Its Applications, **27**, 217–219 (1993).

36. Barry Simon, *Functional Integration and Quantum Physics*, Academic Press, London 1979.

37. Satoshi Takanobu, *On the error estimate of the integral kernel for the Trotter product formula for Schrödinger operators*, Ann. Probab., **25**, 1895–1952 (1997).

38. Z. Shen, L^p *estimates for Schrödinger operators with certain potentials*, Ann. Inst. Fourier, Grenoble, **45**, 513–546 (1995); *Estimates in* L^p *for magnetic Schrödinger operators*, Indiana Univ. Math. J., **45**, 817–841 (1996).

39. Hiroshi Tamura, *A remark on operator-norm convergence of Trotter–Kato product formula*, Integral Equations Operator Theory, **37**, 350–356 (2000).

40. Hale F. Trotter, *On the product of semi-groups of operators*, Proc. Amer. Math. Soc., **10**, 545–551 (1959).

Innovation Approach to Some Problems in Quantum Dynamics

Takeyuki Hida

Faculty of Science and Technology
Meijo University

Si Si

Faculty of Information Science and Technology
Aichi Prefectural University

Dedicated to Professor Hiroshi Ezawa

We shall consider random complex systems, mathematical expressions of which are functionals of Gaussian and/or Poisson noise and will discuss their analysis. The idea behind the analysis is the *reductionism*, namely we first construct the innovation from the given random system. There Gaussian and Poisson noises are the systems consisting of basic elemental random variables and are members of the innovation. Then, the functionals of the innovation will be analyzed. Actual problems in quantum dynamics often give us profound problems in this direction.

1. Introduction

We are interested in the analysis of random complex systems, in particular those systems that come from quantum dynamics. Since the given system has uasually complex probabilistic structure, it would be fine, if the given system is expressed as a function (or a functional) of *elemental random variables* which are mutually independent, identically distributed and enjoy some other properties simpler than the given system. Such an idea may be considered as the *reductionism* following the famous philosopher R. Descartes.

From the view point of stochastic analysis, such a method of investigation for random systems may be called an *innovation approach*. There have been various attempts in this direction in physics, communication theory and others, where the idea may be similar to the reductionism. We shall not go into a general theory, but we specifically focus our attention on stochastic processes and random fields, and discuss analysis of them being influenced by the reductionism.

In order to concretize such an approach, we shall first clarify the notion of the innovation for random complex systems in question. What we have in mind as the term called innovation may be far from that in general cases, however we can explicitly speak of a realization of the innovation for our specified random systems. It seems quite reasonable to take a Gaussian noise (usually, simply called a white noise) and a Poisson noise, including a compound Poisson case. This will be illustrated in the next section.

It is necessary to tell our motivations in this section; actually there are two. One has come naturally from the white noise theory. Gaussian and Poisson noises (also compound Poisson noise) are considered as basic systems of idealized elemental random variables, so that stochastic analysis deals with functionals of those noises. As is well known, for the Gaussian case the theory has been well developed these years, but the Poisson case seems to be expected more development. Of course, we do not claim that Gaussian case and Poisson case should be equally investigated, but we are sure that it is now the time when more attention is paid to the Poisson case.

The other motivation has come to us from quantum dynamics, in particular quantum optics, where we are suggested to consider a characteristic property of Poisson noise; namely the optimality in a sense of information theory. This will be discussed in Section 2.

Having investigated some profound properties of Poisson noise, we come to random fields expressed as linear functionals of the noise, and discuss their innovations. There, we are recommended to remind P. Lévy's idea (see Ref.[5]) for linear processes.

It is our hope that this approach would contribute to the development of quantum computation and various problems in quantum dynamics.

2. Innovations

2.1. *White noise and Poisson noise as innovation*

To fix the idea, we first take a stochastic process $X(t)$ depending on a continuous parameter t. If one is allowed to take a relaxed view of mathematical rigor, the innovationn can be understood in the following manner. The process $X(t)$ gains, at each infinitesimal time interval $[t, t+dt)$, a new information (or randomness) which is independent of the past values $X(s), s \leq t$, if it is not degenerated in the neighbourhood of t. The information may be exressed in terms of an infinitesimal random variable, denoted by $Y(t)$ indexed by t. The $Y(t)$ is a generalized stochastic process with independent values at every t, although there are necessary additional assumptions to have the $Y(t)$ defined rigorously. Note that the $Y(t)$ is not trivial, if the process is assumed to express an evolutional random phenomenon that is actually changing as the parameter t varies.

2.1.1. *Innovation comes from a Lévy process*

Accumulation of infinitesimal independent random variables makes an additive process. Under mild assumptions. it is viewed as a Lévy process, denoted by $Z(t)$. We further assume, to fix the idea, that $Y(t)$ is stationary. We can then appeal to the Lévy decomposition of a Lévy process with stationary independent increments, together with some assumtions:

$$Z(t) = m \cdot t + X_0(t) + X_1(t),$$

where m is a constant, which shall be put $m = 0$, $X_0(t)$ is a Brownian motion up to a constant, so that we may set $X_0(t) = cB(t)$, c being a constant, and $X_1(t)$ is a compound Poisson process.

Concerning the Brownian motion or (Gaussian) white noise, very details and profound results have been obtained, and we do not go into details. We therefore come to Poisson process and Poisson noise, which is elemental.

As is well known, $\dot{P}(t)$ is a *stationary* generalized stochastic process with independent values at every instance t.

Let $P(t), t \geq 0$, be a Poisson process. Poisson noise is its derivative $\dot{P}(t)$. The time parameter is assumed to be extendend to the entire $\mathbb{R}$. The characteristic functional $C_P(\xi), \xi \in E$, of Poisson noise is given by

$$C_P(\xi) = \exp[\lambda \int (e^{i\xi(t)} - 1)dt] \quad \xi \in E,$$

where E is a nuclear space which is dense in $L^2(\mathbb{R})$.
The following property is well known.

Proposition 1: *The mesure μ_P is supported by a Sobolev space $H^{-2}(\mathbb{R})$. The sample function is a sum of delta functions, almost surely.*

As is imagined, the integral of Poisson noise recovers the original Poisson process $P(t)$.

Although various significant properties of Poisson noise are known, we should like to focus our cramming on a special side, from which a sort of optimality is discovered by using conditional probabilities.

2.2. *Some conditional probabilities*

For the detailed observation we refer to the literature.[8]

Consider the event $A(n)$ on which $P(1) = n$, where n is any non-negative integer. The collection $\{A(n), n \geq 0\}$ is a partition of the entire ω-set Ω. Namely, up to measure 0, the following relations hold:

$$A(n) \bigcap A(m) = \phi, \ n \neq m; \ \bigcup A(n) = \Omega.$$

Denote the j-th jump point by τ_j, and set $X_j = \tau_j - \tau_{j-1}$, $j = 0, 1, 2, ...(\tau_0 = 0)$.

Proposition 2: *Under the assumption that $P(1) = n$, the probability distribution of $(X_1, X_2, ..., X_{n+1})$ with $\tau_{n+1} = 1$, is uniform on the simplex $\sum_{j=1}^{n+1} x_j = 1, x_j \geq 0$.*

Corollary 1: *The probability distribution of each X_j is $1 - (1 - x)^n$.*

2.2.1. *Observation*

Let Y_k, $1 \leq k \leq n$, be a sequence on independent identically distributed random variables which are distributed uniformly on $[0, 1]$. The order statistics are such that $Y_0 \leq Y_{\pi(1)} \leq Y_{\pi(2)} \leq \cdots \leq Y_{\pi(n)} \leq Y_{\pi(n+1)}$, where $Y_0 = 0$ and $Y_{\pi(n+1)} = 1$. Then, the probability distribution of $Y_{n(k+1)} - Y_{n(k)}$ has density function $1 - (1 - x)^n$, for every $0 \leq k \leq n$.

We can say as follows.

Theorem 1: *Fix a unit time interval. Assume that the number of jumps of a Poisson process is n. Under this assumption, the conditional Poisson process enjoys the property that the sub-intervals determined by those n jump points are independent, identically distributed over the given unit interval. The distribution of n-tuples is the uniform distribution on the simplex :*

$$\sum_{1}^{n+1} x_j = 1, x_j \geq 0.$$

Because of this property we may consider that a Poisson process enjoys a certain *optimality* from the viewpoint of information theory. Incidentally, an optimality property for a Brownian motion can be seen in the paper,[4] where almost all trajectories try to occupy as much volume as possible during any time interval. Such an optimality would be a favorable property for those noises are taken to be innovations.

It should be noticed that the optimality mentioned above can be stated in terms of the sample paths, not by the notions in Hilbert space. Moreover the assertion immediately extends to a higher dimensional parameter case, and it suggests the method of projection down to a lower dimensional parameter case.

3. Effective determination of d-parameter Poisson noise

It is now clear how to define a Poisson noise with higher dimensional parameter space as a generalization of the one-dimensional parameter case. By refering the characteristic properties, in particular optimality, we can give a definition as a generalization of the one-dimensional parameter case. Namely, a sample function should be a collection of delta-functions which are randomly scattering. Having limited the parameter set, say a unit d-dimensional cube $[0,1]^d$, and fixing the event $A(n)$ on which there are delta functions exist just as many as n.

With these assumptions, consider the case where n delta functions are distributed independently and uniformly on $D = [0,1]^n$. This assumtion may be said that n delta functions are arranged on D so as to have *maximum entropy*. If a system $\{V(t), t \in D\}$ satisfies the requirements that have been stated so far is an $\mathbb{R}^d$-*parameter Poisson noise*. The last assumption eventually becomes one of the most important characteristic properties of Poisson noise. To determine the probability distribution the conditional characteristic functional is given. Let $A(n)$ be the event, as in the case $d = 1$, where n delta functions are sitting in D.

Theorem 2: *Given the event $A(n)$. The conditional characteristic functional $C_{P,n}(\xi) = E(e^{i\langle V,\xi\rangle}/A(n))$ of a Poisson noise with D-parameter is of the form*

$$C_{P,n}(\xi) = (\int_D e^{i\xi(t)} dt^d)^n.$$

We therefore have the characteristic functional of $V(t)$:

$$C_P(\xi) = \sum_{n=0}^{\infty} C_{P,n} P(A(n))$$

$$= \exp[\lambda \int_{\mathbb{R}^d} (e^{i\xi(t)} - 1) dt]$$

The expression is quite similar to the one-dimensional case, so that we shall be satisfied by our construction of Poisson noise with parameter set $[0,1]^d$. Extension to the $\mathbb{R}^d$-paramer is obvious.

4. Restriction of parameter

Restriction of the parameter of Poisson noise is sometimes necessary, in particular, in the computation of variations of random fields. Consider now a d-dimensinal parameter Poisson noise. Let n be fixed, that is, the event $A(n)$ be given. For almost every $\omega \in A(n)$ there are d-dimensional delta functions as many as n, and the support of those delta functions are d-dimensional vectors $Y_1^d, Y_2^d, \cdots, Y_n^d$. Take $d' < d$. By the projection $\pi(d, d') : [0,1]^d \to [0,1]^{d'}$, we are given n d'-dimensional vectors

$$Y_k^{d'} = \pi(d, d')Y_k^d, \ 1 \le k \le n.$$

Note that $Y_k^{d'}$'s are different almost surely, since Y_k^d's are independent and unformly distributed.

With the help of the $Y_k^{d'}$'s we can define d'-dimensional parameter Poisson noise $V(t), t \in [0,1]^{d'}$:

$$V(t) = \sum_k Y_k^{d'}.$$

It is interesting to consider the inverse of the projection $\pi(d, d')$. Given a conditional Poisson noise where the event $A(n)$ is given. The following theorem may be considered as a characteristic property of Poisson noise expressed in terms of the entropy.

Theorem 3: *Under the condition $A(n)$, suppose d'-dimensional Poisson noise is given. Then, the d-dimensional Poisson noise with $d > d'$ such that the second which is linked by the projection $\pi(d, d')$ is determined so as to have maximum entropy.*

Proof. The tail part of the coordinates of delta functions of d-dimensional parameter Poisson noise consists of independent uniformly distributed random variables. Thus the conditional entropy of d-dimensional Poisson noise should have maximum information.

The restriction of the parameter $t \in [0.1]^d$ to an ovaloid C can be done more simply by noting that C is convex in addition to the smoothness.

5. Linear random fields

Following the paper by Lévy's terminology "fonction aléatoires à corrélation linéaire" we shall call the random function simply a *linear process*. Without giving a definition and any results, we just give an example, in fact typical, of a linear process.

Example. Let a linear process $X(t)$ be given by

$$X(t) = \int_0^t F(t,u)\dot{B}(u)dt + \int_0^t G(t,u)\dot{P}(u)du,$$

where the integrals are defined not by random measures, but by the sample function-wise integrals.

There is Win Win Htay's approach,[9] which says

Theorem 4: *Observing the sample function of $X(t)$ given above, the two integrals can be discriminated, if $F(t,u)$ is a canonical kernel and $G(t,u)$ is smooth in $(t,u), u \le t$, and $G(t,t) \ne 0$. Then, the innovation of $X(t)$ is given by*

$$\dot{B}(t) + \dot{P}(t).$$

Remark 1. For the proof we use the sample function properties.
Remark 2. It is possible to rephrase this result by using the characteristic functional. Such an approach will be discussed in a separate paper.

We can generalized this result. Take a Gaussian random field

$$X_1(C) = \int_{(C)} F(C,u)W(u)du,$$

where (C) is the domain enclosed by an ovaloid C. Now, the above representation is assumed to be canonical. The notion of the *canonical representation* of such a Gaussian random fields has been discussed in Ref.[7] While, for the Poisson case, a Poisson field $X_2(C)$ is given by

$$X_2(C) = \int_{(C)} G(C,u)V(u)du.$$

The assumtions for this integral are somewhat different, sine it is important to note the integral is defined sample function-wise. Namely, the kernel $G(C,u)$ is smooth in (C,u), and we assume that $G(C,s) \ne 0$, for the purpose to have well definded variations of the $X_2(C)$.

Theorem 5: *Let a linear random field $Y(C)$ be given by*

$$Y(C) = X_1(C) + X_2(C),$$

with the assumptions stated above for the kernels $F(C, u)$ and $G(C, u)$. Then, we have

(i) *the two terms in the above formula for $Y(C)$ are discriminated by observing its sample functions.*

(ii) *the innovation is the sum*

$$W(s) + V(s), \quad s \in C.$$

Proof is given as a slight generalization of that for the case of the ordinary linear process (Theorem 4), with a remark on the restriction of the parameter to C. We can choose ovaloids in such a way that tangent spaces determine a single point.

We are interested in an effective determination of Poisson noise parameterized by a manifold, in particular by a sphere S^d.

Let d be fixed, and let Y_k, $1 \le k \le n$, be a sequence of independent random variables taking valued in S^d. Following the idea of determination of a Poisson noise with optimality in mind, we ssume that Y_k's are distributed uniformly on S^d. Denote by σ^d the uniform probability distribution on S^d. Then, the characteristic functional is

$$C_{S^d,n}(\xi) = \int_{S^d} e^{i \sum_1^n \xi(t_k)} d\sigma^d(t_1) \cdots d\sigma^d(t_n) = \left(\int_{S^d} e^{i\xi(t)} d\sigma^d(t) \right)^n.$$

This is thought of as the conditional characteristic functional for n given. The unconditional characteristic functional is to be

$$C_{S^d}(\xi) = \sum_0^\infty C_{S^d,n} \frac{\lambda^n}{n!} e^{-\lambda}$$

$$= \exp[\lambda \int (e^{i\xi(t)} - 1) d\sigma(t).$$

This expression will be used when the variation of a random field $X(C)$ is dicussed, where the $X(C)$ is a functional of Poisson noise $V(u)$, where $u \in (C)$.

Concluding remarks

When the variational problem for a linear random field based on a compound Poisson noise is given, there arises the meaurement or the computability problem. It is certainly a very profound problem, so at present we do not go into this question.

The *invariance* of Poisson noise measure under certain transformation group and the *reversibility* of linear random fields in the variable C can be discussed within this setup to some extent.

References

1. L. Accardi *et al.* eds, Selected papers of Takeyuki Hida. World Scientific Pub. Co. 2001.
2. T. Hida, Stationary stochastic processes. Princeton Mathematical Notes, Princeton Univ. Press. 1970.
3. John R. Klauder and E.C.G. Sudarshan, Fundamentals of quantum optics. Math. Phys. Monograph Series. W.A. Benjamin Inc. 1968.
4. P. Lévy, La mesure de Hausdorff de la courbe du mouvement brownien. Giornale dell'Istituto Italiano degli Attuari. 16 (1953), 1–37.
5. P. Lévy, Fonction aléatoires à corrélation linéaire. Illinois Journal of Math. 1 (1957), 217–258.
6. M.A. Man'ko, V.I. Man'ko, and R. Vilela Mendes, Quantum computation by quantumlike syatems. Physics Letters A. 288 (2001), 132–138.
7. Si Si, Gaussian processes and Gaussian random fields. Quantum Information II. T. Hida and K. Saito eds., World Scientific Pub. Co. 2000, 195-204.
8. Si Si, Effective determination of Poisson noise. to appear in Infinite Dimensional Analysis, Quantum Probability and Related Topics.
9. Win Win Htay, Note on linear processes. Paper presented at IIAS Workshop organized by M. Ohya, Jan. 2003.

Feynman Paths, Sticky Walls, White Noise

Ludwig Streit

BiBoS – University of Bielefeld, D-33615 Bielefeld
CCM – Universidade da Madeira, P-9000-390 Funchal
E-mail: streit@physik.uni-bielefeld.de

1. Thanks to Hiroshi Ezawa ...

The nice thing about contributing to a Festschrift is of course the opportunity to express gratitude, it is also the pleasure of reviving pleasant memories; at another occasion[20] I have had the good fortune to review the Hiroshi Ezawa impact on my work and life over almost a quarter of a century —so many fond memories, so many reasons for gratitude.

I have often said that Nagoya is my home town in Japan. But as we all know, the hustle and bustle of the big city is not all. From time to time one yearns for the village retreat where you have the peace and quiet to reflect and to discuss things calmly with a good friend. In this sense my village in Japan over so many years has always been Tokyo, or more precisely, the Gakushuin campus, and for all this and much more, my thanks and best wishes go out to-day to my good friend Hiroshi Ezawa and to his family .

On various occasions my work at Hiroshi's institute was focused on Feynman integrals in terms of white noise analysis. Let us use Brownian paths, that was the challenge I would bring from Nagoya. But how?

- The Wiener measure for Brownian paths is Gaussian, while Feynman's ansatz is based on a fictitious "flat", *i.e.* Euclidean invariant, measure.
- Endpoints of paths need to be pinned: we need to invoke something like Donsker's δ-function within the white noise framework.

By now these two prerequisites are well under control, see Refs.[10,16] and the review below; I mention them here because in both cases I under-

                              L. Streit

stood the 'how to' thanks to periods of reflection and discussion at Hiroshi
Ezawa's institute.

2. White noise analysis

Gaussian white noise ω appears naturally wherever one tries to model random events occurring independently at different points in time (and/or
space). Of course this strict independence, expressed in the expectation

$$E\left(\omega(s)\omega(t)\right) = \delta\left(s - t\right)$$

is an idealization, but it is one of those which physics invokes frequently
to facilitate a simple and transparent analysis, an analysis that in many
instances would become much more cumbersome for models with "colored
noise". In particular white noise appears as the velocity of Brownian motion
modelled by the Wiener process $B(t)$

$$\omega(t) = \frac{d}{dt}B(t). \tag{1}$$

With its independence and invariance properties Gaussian white noise
ω is ideally suited as a universal "coordinate system" for stochastic and
infinite dimensional analysis. Hence we shall present white noise analysis
in what follows, noting however that most results have a straightforward
extension to more general Gaussian and even non-Gaussian settings.

A certain amount of care is needed if we want to profit from the advan-
tages of white noise modelling in a nonlinear setting; a well-known example
is stochastic integration where *e.g.* the distinction between the integrals
in the sense of Ito and of Stratonovich essentially results from carefully
defining them as nonlinear functionals of white noise .

From the fact that we consider white noise as a Gaussian δ-correlated
mean zero process we are immediately led to its characteristic function

$$C(f) = E\left(e^{i\int \omega(t)f(t)dt}\right) = \int e^{i\int \omega(t)f(t)dt}d\mu(\omega) = \exp\left(-\frac{1}{2}\int f^2(t)dt\right).$$

The L^2 space of square integrable white noise functions is

$$L^2\left(d\mu\right) \equiv \left(L^2\right).$$

For nonlinear functionals

$$\varphi \in \left(L^2\right)$$

we have expansions

$$\varphi(\omega) = \sum_{n=0}^{\infty} \int d^n t \, F_n(t_1, \dots, t_n) : \omega(t_1) \dots \omega(t_n) : . \qquad (2)$$

Here

$$: \omega^{\otimes n}(t) :=: \omega(t_1) \dots \omega(t_n) :$$

is the well-known "normal" or "Wick" or "Hermite product". It amounts to an orthogonalization

$$E(: \omega^{\otimes n}(s) :: \omega^{\otimes m}(t) :) = \delta_{m,n} n! \prod_{i=1}^{n} \delta(s_i - t_i)$$

(we consider the n-tuple s and m-tuple t as ordered here), and can *e.g.* be defined recursively

$$: \omega(t_1) \dots \omega(t_n) :=: \omega(t_1) \dots \omega(t_{n-1}) : \omega(t_n) - \sum_{i=1}^{n-1} \delta(t_n - t_i) : \prod_{k \neq i}^{n-1} \omega(t_k) : .$$

In this fashion all the terms in the "chaos expansion" (2) are orthogonal to each other, and we find immediately the L^2 norm of φ :

$$\|\varphi\|_{(L^2)}^2 = E(\varphi^* \varphi) = \sum_{n=0}^{\infty} n! \int d^n t \, |F_n(t_1, \dots, t_n)|^2 .$$

On the rhs we recognize a Fock space norm:

$$\varphi \in L^2(d\mu) \sim \{F_n\} \in H_{Fock} = \bigoplus_n Sym L^2(\mathbb{R}^n, n! d^n t)$$

is the Gelfand–Ito–Segal isomorphism between square integrable white noise functionals and Boson Fock space vectors.

Smooth and generalized functionals are constructed in analogy to the usual test functions and distributions; we recall that test functions in Schwartz space are characterized by a sequence of norms

$$f \in S(\mathbb{R}^n) \text{ iff } |f|_p < \infty \text{ for all } p.$$

Likewise we introduce for white noise functionals as in (2) the norms

$$\|\varphi\|_{p,q}^2 = \sum_{n=0}^{\infty} (n!)^{1+\beta} 2^{nq} |F_n|_p^2 \qquad (3)$$

for $p, q > 0$. Putting $\beta = 0$ for the moment we define the space (S) of smooth functionals by

$$(S) = \left\{ \varphi : \|\varphi\|_{p,q} < \infty \text{ for all } p, q \right\}$$

and the space $(S)^*$ of generalized functionals ("Hida distributions") as its dual:

$$(S) \subset L^2(d\mu) \subset (S)^*.$$

Similarly, for $\beta = 1$, we have the "Kondratiev spaces" $(S)^{\pm 1}$, with

$$(S)^1 \subset (S) \subset L^2(d\mu) \subset (S)^* \subset (S)^{-1}.$$

How does one identify generalized functionals $\Phi \in (S)^*$? In many physics applications Φ will be given in terms of a "source functional", such as

$$T\Phi(f) = E\left(\Phi(\omega)e^{i \int \omega(t)f(t)dt}\right)$$

or

$$S\Phi(f) = E\left(\Phi(\omega) : e^{\int \omega(t)f(t)dt} :\right)$$

Fortunately, any of these expressions provides a complete characterization of Φ.

Theorem 1: [10] *A functional $G(f)$, $f \in S(\mathbb{R})$, is the T-transform of a unique generalized white noise functional $\Phi \in (S)^*$ iff for all $f_i \in S(\mathbb{R})$, $G(zf_1 + f_2)$ is analytic in the whole complex plane and of second order exponential growth*

$$|G(zf)| < ae^{b|zf|_p^2}.$$

Remarks:

(1) The same is true for the S-transform.
(2) Generalized functionals in the Kondratiev space are characterized by local analyticity and local boundedness of the source functionals. In our present context of Feynman integrals they arise for potentials of *e.g.* exponential growth.[14]

3. Feynman integrals

3.1. *The white noise method*

Heuristically we should deal with expressions such as

$$\langle F \rangle = \int d^\infty x(\tau) e^{\frac{i}{\hbar} \int \dot{x}^2 d\tau} e^{-\frac{i}{\hbar} \int V(x(\tau)) d\tau} F[x] \tag{4}$$

We intend to

- *consider the Feynman integral as the action of a white noise distribution I on a test functional F, and at the same time*
- *maintain the constructive approach of Feynman.*

The first point opens the door to a truly mathematical understanding and reliable manipulation of Feynman integrals, the latter point is essential if we want to explicitly construct specific quantum models.

We shall thus try to find a suitable white noise distribution I

$$\int d^\infty x(t) e^{\frac{i}{\hbar} S} F[x] = \langle I, F \rangle, \ I \in (S)^{-1} \tag{5}$$

which represents an appropriate sum over histories.

The basic ideas for this construction are:

$$x(t_0 + \tau) = x + \left(\frac{\hbar}{m} \right)^{1/2} B(\tau)$$

so that, recalling (1), the kinetic energy term in S is quadratic in white noise, and

$$e^{\frac{i}{\hbar} S} = N \exp \left(\frac{i}{2} \int_{\mathbb{R}} \omega^2 (\tau) \, d\tau \right).$$

To mimic a "flat" integration measure as in the lhs of (5) we cancel the formal Gaussian density $\exp \left(-\frac{1}{2} \int_{\mathbb{R}} \omega^2 (\tau) \, d\tau \right)$ of white noise by introducing a factor $\exp \left(\frac{1}{2} \int_{\mathbb{R}} \omega^2 (\tau) \, d\tau \right)$. Combining the two we arrive at

$$I_0 (x, t \mid x_0, t_0) = N \exp \left(\frac{i+1}{2} \int_{\mathbb{R}} \omega^2 (\tau) \, d\tau \right) \delta \left(x(t) - x_0 \right).$$

This expression admits the explicit calculation of a T-transform:

$$T I_0 (f) = \frac{1}{(2\pi i |t - t_0|)^{\frac{d}{2}}} \exp \left[-\frac{i}{2} \int_{\mathbb{R}} f^2 (\tau) \, d\tau \right.$$

$$\left. - \frac{1}{2i |t - t_0|} \left(\int_{t_0}^{t} f(\tau) \, d\tau + x - x_0 \right)^2 \right].$$

Evidently this expression satisfies the characterization theorem 1, hence I_0 is well defined mathematically as a generalized functional of white noise, but also

$$TI_0\left(f\right) \equiv K_0^{(f)}\left(x, t | x_0, t_0\right)$$

has a physical meaning: it obeys the Schrödinger equation

$$\left(i\partial_t + \frac{1}{2}\triangle_x - \dot{f}(t)\cdot x\right) K_0^{(f)}\left(x, t | x_0, t_0\right) = 0 \qquad (6)$$

with the initial condition

$$\lim_{t\searrow t_0} K_0^{(f)}\left(x, t | x_0, t_0\right) = \delta\left(x - x_0\right).$$

Note: K is the kernel of the unitary operator

$$U = e^{iH(t-t_0)}$$

where

$$H = -\frac{1}{2}\triangle_x + \dot{f}(t)\cdot x.$$

Of course this is nice, but not much as far as physical models are concerned; a good definition should allow for the inclusion of interactions as in (4). Which potentials V are admissible?

3.2. *The interactions*

Various classes of potentials have been studied, such as *e.g.* Ref.[14] Laplace transforms of complex measures

$$V\left(x\right) = \int_{\mathbb{R}^d} e^{\alpha\cdot x}\mathrm{d}m(\alpha)$$

including such examples as the Morse potential, anharmonic oscillators such as $V(x)=g \cosh x$, *etc.* For our present purposes we are particularly interested in the class given by

Example 1: [13], see also Ref.[10] Theorem 12.4.
    For d=1,

$$V\left(x\right) = \int_{\mathbb{R}} \delta\left(x - \alpha\right)\mathrm{d}m(\alpha), \qquad (7)$$

where m is a finite signed measure of compact support.

Results such as these were reported at the Ezawa–Fest in 1998.[20] So what is new?

4. Boundaries

In the examples discussed above the configuration space was always taken to be $\mathbb{R}^d$. Interesting extensions — Chern–Simons fields, M-theory — are possible if one performs some rather obvious generalizations from white noise analysis to the analogous constructions on Gaussian spaces.[9,17] In particular one is interested in quantum systems on a half-line[1,3,6,7], such as those governed by the radial (s-wave) Schrödinger operator $H_0 = -\frac{1}{2}\Delta$ on $\mathbb{R}_+$.The point of interest for us is that this operator is not essentially self-adjoint on domains such as $C_0^2\left(\mathbb{R}_+\right)$; our goal is to construct path integrals in terms of white noise not only for all the self-adjoint extensions but also to include further interactions on the half-line.[2,4].

Self-adjoint extensions are characterized by boundary conditions:

$$D\left(H_0^{\beta/\alpha}\right) = \left\{\psi \in C^2\left(\mathbb{R}_+\right) \cap L^2 \,|\, (\alpha\partial_x\psi - \beta\psi)_{x_0=0} = 0\right\},$$

where α and β are arbitrary real numbers. The ratio $\beta/\alpha > 0$ gives continuous spectra, negative values lead to a bound state ("sticky wall").

For propagators, we must have correspondingly

$$\left[\alpha\frac{\partial}{\partial x}K_0^{\beta/\alpha}(x, x_0, t) - \beta\, K_0^{\beta/\alpha}(x, x_0, t)\right]_{x=0} = 0\,.$$

Special cases:

- $\beta = 0$.

  Set

$$K_0^0(x, x_0, t) = K_0(x, x_0, t) + K_0(x, -x_0, t).$$

Of course this also solves the Schrödinger equation.

Note our restriction to $x, x_0 > 0$. Hence, on $\mathbb{R}_+$

$$\lim_{t\searrow t_0} K_0^0\left(x, t|x_0, t_0\right) = \delta\left(x - x_0\right).$$

And, because of the symmetry of K_0^0,

$$\frac{\partial}{\partial x}K_0^0(x, x_0, t)_{x=0} = 0\,.$$

- Likewise, for $\alpha = 0$, we use the antisymmetric combination

$$K_0(x, x_0, t) - K_0(x, -x_0, t),$$

and obtain the propagator for the case where the wave function itself vanishes at the origin[18,19]:

$$(K_0(x, x_0, t) - K_0(x, -x_0, t))_{x=0} = 0.$$

Now for the general case. Consider the Schrödinger equation

$$\left(i\partial_t + \frac{1}{2}\triangle_x - \gamma\delta\left(x\right)\right)\psi(x) = 0.$$

with $\psi(x) = \psi(|x|)$. Integration for a small interval around $x = 0$ gives

$$\int_{-\varepsilon}^{\varepsilon} dx \left(i\partial_t + \frac{1}{2}\triangle_x - \gamma\delta\left(x\right)\right)\psi(x) = 0$$

and in the limit

$$\psi'(+0) - \gamma\psi(0) = 0.$$

Theorem 2: [2] *For potentials of the form*

$$V\left(x\right) = \int_{(0,\infty)} \delta\left(x - a\right) dm(a)$$

and boundary condition

$$\gamma = \beta/\alpha$$

the Feynman integrand in terms of white noise is given by

$$N \exp\left(\frac{i+1}{2} \int_{\mathbb{R}} \omega^2\left(\tau\right) d\tau \right) \times$$
$$\left(\delta\left(x\left(t\right) - x_0\right) + \delta\left(x\left(t\right) + x_0\right)\right) e^{-\frac{i}{\hbar}\int \widehat{V}(x(\tau))d\tau}$$

where $\widehat{V}(x) \equiv V\left(|x|\right) + \gamma\delta(x).$

We have thus indicated how to incorporate boundary conditions into the white noise framework for Feynman integrals.

It is proper to underline that this result was obtained in collaboration with Victoria and Chris Bernido from the Philippines who would like to express here their thanks to Hiroshi Ezawa for almost twenty years of friendly encouragement and valuable support.

References

1. C. C. Bernido, M. V. Carpio-Bernido, "Path integrals for boundaries and topological constraints: a white noise functional approach," *J. Math. Phys.* **43**, 1728–1736 (2002).
2. C. C. Bernido, M. V. Carpio-Bernido, L. Streit (in preparation).
3. T. E. Clark, R. Menikoff, D. H. Sharp, "Quantum mechanics on the half-line using path integrals," *Phys. Rev.* D**22**, 3012–3016 (1980).
4. M. de Faria, M.J. Oliveira, L. Streit (in preparation).

5. M. de Faria, J. Potthoff, L. Streit, "The Feynman Integrand as a Hida Distribution" J. Math. Phys. **32**, 2123 (1991).

6. E. Farhi, S. Gutmann, "The functional integral on the half-line," *Int. J. Mod. Phys.* **5**, 3029–3051, (1990).

7. C. Grosche, F. Steiner, *Handbook of Feynman Path Integrals.* Springer, Berlin, 1998.

8. M. Grothaus, D. C. Khandekar, J. L. Silva, L. Streit, "The Feynman Integral for time-dependent anharmonic oscillators" *J. Math. Phys.* **38**(6), 3278–3299 (1997).

9. M. Grothaus, L. Streit, I. V. Volovich, "Knots, Feynman Diagrams and Matrix Models" *IDAQP* **2**, 359–380 (1999).

10. T. Hida, H. H. Kuo, J. Potthoff, L. Streit, "White Noise. An infinite dimensional calculus." Kluwer Academic, 1993.

11. T. Hida, L. Streit, "Generalized Brownian Functionals", VI. Conf. Math. Phys. Berlin 1981, Springer Lecture Notes in Physics, no. 153, 285–287, 1982.

12. T. Hida, L. Streit, "Generalized Brownian Functionals and the Feynman Integral". *Stoch. Proc. Appl.* **16**, 55 (1983).

13. D. C. Khandekar, L. Streit, "Constructing the Feynman Integrand" *Annalen d. Physik* **1**, 49 (1992).

14. T. Kuna. L. Streit, W. Westerkamp, "Feynman Integrals for a Class of Exponentially Growing Potentials" *Journ. Math. Phys.* **39**, 4476–4491 (1998).

15. A. Lascheck, P. Leukert, L. Streit, W. Westerkamp, "Quantum Mechanical Propagators in Terms of Hida Distributions", *Rep. Math. Phys.* **33**, 221–232 (1993).

16. A. Lascheck, P. Leukert, L. Streit, W. Westerkamp, "More about Donsker's delta function" , *Soochow Math. J.* **20**, 401–418 (1994).

17. P. Leukert, J. Schäfer, "A Rigorous Construction of Abelian Chern Simons Path Integrals using White Noise Analysis." *Reviews in Mathematical Physics*, Vol. 8, 445–456 (1996).

18. G. Roepstorff, "Path integral approach to quantum physics. An introduction." Texts and Monographs in Physics. Springer, Berlin, 1994.

19. L. Schulman, "Techniques and applications of path integration." Wiley-Interscience. New York, 1981.

20. L. Streit, "Some Observations on stochastic vs. quantum dynamics — following Ezawa, Klauder, and Shepp" in "Mathematical Methods of Quantum Physics". Proc. 2nd Jagna Intl. Workshop, (C. Bernido *et al.*, eds.) p. 9–18, Gordon+Breach, 1999.

White Noise Path Integrals:
Applications in Polymer Entanglement

Christopher C. Bernido

Research Center for Theoretical Physics,
Central Visayan Institute Foundation
Jagna, Bohol 6308, Philippines

M. Victoria Carpio-Bernido

Research Center for Theoretical Physics,
Central Visayan Institute Foundation
Jagna, Bohol 6308, Philippines

The white noise functional approach is used to evaluate the entanglement probabilities for a highly flexible polymer to wind n-times around a straight polymer. The entanglement probabilities are obtained for the following cases: (a) an entangled polymer confined to a narrow strip of radius R with a straight polymer at its center, and (b) a magnetic flux confined along the straight polymer. In the absence of the magnetic flux, the entanglement probability reduces to a result obtained by Wiegel.[17]

1. Introduction

One of the early memorable encounters we had with Professor Hiroshi Ezawa was the series of lectures he delivered from 31 January to 7 February 1984 at the University of the Philippines where the authors were then both members of the physics faculty. The friendship initiated during these lectures further blossomed when Professor Ezawa became our scientific host during our visits to Japan in 1990, 1995, and 1998. These visits were recriprocated by Professor Ezawa's several more visits to the Philippines. Since 1992, Professor Ezawa has been a member of the Scientific Advisory Board of the Research Center for Theoretical Physics, Central Visayan Institute, which is a small research institute that we founded that same year in the island province of Bohol, Philippines. It is therefore with great gratitude and pleasure that we contribute this work to the 70th birthday celebration in honor of Professor Ezawa on a topic with the same general theme as Professor Ezawa's memorable 1984 lectures on stochastic processes and their application to physics.

In stochastic processes, an interesting random variable that has been well studied is the Gaussian white noise variable $\omega(t)$ obtained as the time derivative of a Brownian motion $B(t)$. The white noise $\omega(t)$, therefore, may be viewed as the "velocity" of a Brownian motion. In 1975, T. Hida introduced white noise calculus[1] as a novel approach to infinite dimensional analysis. This approach has found widespread applications in the literature[2,3,4] which includes, among others, the evaluation of a quantum propagator where the Feynman path integral is expressed in terms of the variable $\omega(t)$.[5−13]

In this paper, we apply the white noise functional integral approach to evaluate entangled polymer systems. Entangled polymer problems were studied by S.F. Edwards[14] and, independently, S. Prager and H.L. Frisch,[15] for two chainlike macromolecules in the absence of intermolecular forces. This problem consists of a polymer on a plane whose motion is constrained by a straight polymer orthogonal to the plane, since the macromolecules cannot cross each other. The polymer on the plane has fixed endpoints and can be viewed as a random walk with paths that can entangle n-times around the straight polymer which intersects the origin of the plane. Interesting quantities based on this entangled polymer were also investigated by N. Saito and Y. Chen.[16] Moreover, F.W. Wiegel[17] extended this entanglement problem to include an intermolecular force where the repeating units of the entangled polymer interact with the straight polymer. Wiegel obtained the entanglement probabilities for a harmonically bound polymer, and for low temperatures, he noted that the configurations of the polymer are confined to a narrow strip in the immediate vicinity of a circle around the origin with radius R .

In section 2, we review some essential features of white noise analysis needed in our treatment. Then, we discuss in section 3 the case studied by Wiegel where the entangled polymer is confined to a narrow strip on a plane in the immediate vicinity of a circle whose center contains a straight polymer perpendicular to the plane. We then apply this white noise approach[1−5] to a case where the straight polymer perpendicular to the plane possesses a magnetic flux confined along its length. This appears to be a classical analogue of the Aharonov–Bohm effect[18] since the entanglement probability is affected by the confined magnetic flux, even if the entangled polymer lies in a region with zero magnetic field.

2. White noise analysis

A stochastic process such as Brownian motion is usually described by the random variable X satisfying the stochastic differential equation,

$$dX = a(t, X)\ dt + b(t, X)\ dB(t), \tag{1}$$

where $a(t, X)$ and $[b(t, X)]^2$ are the drift and diffusion coefficients, respectively, and $B(t)$ is the Wiener process. The familiar Langevin equation follows as,

$$\dot{X} = a(t, X) + b(t, X)\ \omega(t), \tag{2}$$

where $\dot{X} = dX/dt$ and,

$$\omega(t) = dB(t)\ /\ dt, \tag{3}$$

is called the Gaussian white noise. Alternatively, Wiener's Brownian motion $B(t)$ is represented by,

$$B(t) = \int_0^t \omega(\tau)\ d\tau = \langle \omega, 1_{[0,t)} \rangle, \tag{4}$$

where we define,

$$\langle \omega, \xi \rangle \equiv \int_0^t \omega(\tau)\xi(\tau)\ d\tau. \tag{5}$$

The basic idea in white noise analysis is to take the collection of infinitely many independent random variables, $\{\omega(t);\ t \in \mathbb{R}\}$, and treat this as the coordinate system of an infinite dimensional space. One then proceeds to investigate a generalized white noise functional, $\Phi(\omega(t);\ t \in \mathbb{R})$, instead of a functional of Brownian motion, $f(B(t);\ t \in \mathbb{R})$, and the Gaussian measure $d\mu(\omega)$ takes over the role of the Lebesgue measure dx in $\mathbb{R}^n$. In particular, the Gaussian white noise measure $d\mu(\omega)$ is characterized by its Fourier transform,

$$\int_{S^*} \exp\big(i\,\langle \omega, \xi \rangle\big)d\mu(\omega) = \exp\left(-\frac{1}{2}\int \xi^2 d\tau\right) = C(\xi), \tag{6}$$

where $C(\xi)$ is referred to as the characteristic functional.

A description of white noise functionals can also be obtained through their T- and S- transforms, which will later be shown useful in evaluating functional integrals. For instance, the T-transform for a generalized white noise functional $\Phi(\omega)$ is defined by,

$$T\Phi(\xi) = \int_{S^*} \exp(i\langle\omega,\xi\rangle)\Phi(\omega)\,d\mu(\omega), \tag{7}$$

which is like an (infinite-dimensional) Gauss–Fourier transform where the Gaussian measure $d\mu(\omega)$ has the form,

$$d\mu(\omega) = N_\omega\,\exp\left(-\frac{1}{2}\int\omega(\tau)^2\,d\tau\right)d^\infty\omega. \tag{8}$$

The exponential in $d\mu(\omega)$ is responsible for the Gaussian fall-off, and N_ω is a normalization factor. On the other hand, an S-transform is related to the T-transform as,

$$S\Phi(\xi) = C(\xi)\,T\Phi(-i\xi), \tag{9}$$

and,

$$T\Phi(\xi) = C(\xi)\,S\Phi(i\xi), \tag{10}$$

where $C(\xi)$ is given by Eq. (6).

Let us consider two examples of the S- and T-transforms of a white noise functional $\Phi(\omega)$ which are useful in treating the polymer entanglement problem.

Example 1. *The Gauss kernel,* $\Phi(\omega) = N\exp\left(-\frac{1}{2}\langle\omega,k\omega\rangle\right).$

Let us first take the S-transform of $\Phi(\omega)$ using Eqs. (6), (7) and (9), *i.e.*,

$$S\Phi(\xi) = \exp\left(-\frac{1}{2}\int\xi^2 d\tau\right)N\int\exp(\langle\omega,\xi\rangle)\exp\left(-\frac{1}{2}\langle\omega,k\omega\rangle\right)d\mu(\omega). \tag{11}$$

Using the expression for $d\mu(\omega)$ given by Eq. (8), we have,

$$S\Phi(\xi) = N\int\exp\left(-\frac{1}{2}\int\left[\xi^2 - 2\omega\xi + (k+1)\omega^2\right]d\tau\right)N_\omega\,d^\infty\omega$$

$$= N\int\exp\left[-\frac{k+1}{2}\int\left(\frac{\xi}{k+1} - \omega\right)^2 d\tau\right]$$

$$\times\exp\left(-\frac{k}{2(k+1)}\int\xi^2\,d\tau\right)N_\omega\,d^\infty\omega. \tag{12}$$

Shifting, $\omega \to \omega + (\xi/k + 1)$, we get,

$$S\Phi(\xi) = N \int \exp\left(-\frac{k}{2(k+1)} \int \xi^2 d\tau\right) \exp\left[-\frac{k+1}{2} \int \omega^2 d\tau\right] N_\omega \, d^\infty \omega$$

$$= \exp\left(-\frac{k}{2(k+1)} \int \xi^2 d\tau\right) N \int \exp\left[-\frac{k}{2} \int \omega^2 d\tau\right] d\mu(\omega). \quad (13)$$

From Eq. (13), we now take the S-transform of $\Phi(\omega) = N \exp\left(-\frac{1}{2}\langle\omega, k\omega\rangle\right)$ as,

$$S\Phi(\xi) = \exp\left(-\frac{k}{2(k+1)} \int \xi^2 d\tau\right), \quad (14)$$

by taking the normalization N to be,

$$N^{-1} = \int \exp\left[-\frac{k}{2} \int \omega^2 d\tau\right] d\mu(\omega). \quad (15)$$

Using Eqs. (10) and (14), the T-transform of $\Phi(\omega)$ can easily be shown to be,

$$T\Phi(\xi) = \exp\left[-\frac{1}{2}\langle\xi, (k+1)^{-1}\xi\rangle\right]. \quad (16)$$

Example 2. *The Donsker delta function,* $\Phi = \delta(B(t) - a)$.

From Eq. (4), we may write, $\delta(B(t) - a) = \delta(\langle\omega, \mathbf{1}_{[0,t)}\rangle - a)$, $a \in \mathbb{R}$, and express the delta function in terms of its Fourier representation as,

$$\Phi = \delta\left(\langle\omega, \mathbf{1}_{[0,t)}\rangle - a\right) = \frac{1}{2\pi} \int \exp\left[i\lambda\left(\langle\omega, \mathbf{1}_{[0,t)}\rangle - a\right)\right] d\lambda. \quad (17)$$

Taking the S-transform using Eqs. (6), (7) and (9) we have,

$$S\Phi(\xi) = \frac{1}{2\pi} \int \int C(\xi) \exp\left(i\langle\omega, -i\xi\rangle\right)$$

$$\times \exp\left[i\lambda(\langle\omega, \mathbf{1}_{[0,t)}\rangle - a)\right] d\lambda \, d\mu(\omega)$$

$$= \frac{1}{2\pi} \exp\left(-\frac{1}{2}\int \xi^2 d\tau\right) \int \int \exp\left(i\langle\omega, -i\xi + \lambda\mathbf{1}_{[0,t)}\rangle\right)$$

$$\times \exp\left(-i\lambda a\right) d\lambda \, d\mu(\omega). \quad (18)$$

Again referring to Eq. (6), we can write this as,

$$
\begin{aligned}
S\Phi(\xi) &= \frac{1}{2\pi} \exp\left(-\frac{1}{2}\int \xi^2 d\tau\right) \int C\left(-i\xi + \lambda \mathbf{1}_{[0,t)}\right) \exp\left(-i\lambda a\right) d\lambda \\
&= \frac{1}{2\pi} \exp\left(-\frac{1}{2}\int \xi^2 d\tau\right) \int \exp\left[-\frac{1}{2}\int_0^t \left(-i\xi(\tau) + \lambda \mathbf{1}_{[0,t)}\right)^2 d\tau\right] \\
&\quad \times \exp\left(-i\lambda a\right) d\lambda \\
&= \frac{1}{2\pi}\int \exp\left[-\frac{1}{2}\lambda^2 t + i\lambda\left(\int_0^t \xi(\tau)d\tau - a\right)\right] d\lambda.
\end{aligned}
\tag{19}
$$

This is just a Gaussian integral which yields the result,

$$
S\Phi(\xi) = \frac{1}{\sqrt{2\pi t}} \exp\left[-\frac{1}{2t}\left(\int_0^t \xi(\tau)d\tau - a\right)^2\right],
\tag{20}
$$

for the Donsker delta function, $\Phi = \delta\big(B(t) - a\big)$. It is a simple step to get from Eqs. (10) and (20), the corresponding T-transform given by,

$$
T\Phi(\xi) = \frac{1}{\sqrt{2\pi t}} \exp\left[-\frac{1}{2}\int \xi^2 d\tau - \frac{1}{2t}\left(\int_0^t i\xi(\tau)d\tau - a\right)^2\right].
\tag{21}
$$

The Donsker delta function is an example of a Hida distribution.[2]

3. Entangled polymers

Here we apply a white noise functional approach in evaluating the problem of two polymers where one entangles around the other. The problem consists of a polymer which lies on a plane with endpoints at r_0 and r_1, and a second polymer represented by a straight line perpendicular to the plane and intersecting the origin. The various ways in which the polymer on the plane entangles around the second polymer, whether clockwise or counterclockwise, give rise to an interesting problem with topological constraints.[14] We can view the different possible configurations of the polymer on the plane as the paths of a random walk starting at r_0 and ending at r_1 in the presence of a singularity at the origin where the second polymer is located. For this problem, S.F. Edwards[14] used the Wiener representation of the random walk in which the probability is represented by,

$$
P(r_1, r_0) = \int \exp\left[-\frac{1}{l}\int_0^L (d\mathbf{r}/ds)^2 \, ds\right] \mathcal{D}^2[\mathbf{r}],
\tag{22}
$$

where the integral is taken over all paths $r(s)$ such that $r(0) = r_0$ and $\mathbf{r}(L) = r_1$. Here, the polymer is represented as consisting of N freely hinged individual molecules, each of length l such that $L = Nl$. In view of the point singularity, a set of topologically equivalent configurations can be characterized by a winding number n, where $n = 0, \pm 1, \pm 2, \ldots$, indicating the number of times the polymer turns around the singular point at the origin ($n \geq 0$, signifies n turns counterclockwise, and $n \leq -1$ means $|n + 1|$ turns clockwise). Since we are interested in the number of possible windings around the origin that the polymer on the plane undergoes, we can simplify the calculation using polar coordinates $\boldsymbol{r} = (r, \vartheta)$ where the radial variable is fixed to $r = R$, *i.e.*, $\boldsymbol{r} = (R, \vartheta)$. The variable ϑ keeps track of the number of turns, clockwise or counterclockwise, around the origin. We note that a fixed radial part describes the entanglement scenario in the low temperature limit[17] for any polymer interaction potential $V(r)$ which has a minimum at some value $r = R$. In this case, Eq. (22) reduces to,

$$P(\vartheta_1, \vartheta_0) = \int \exp\left[-\frac{1}{l}\int_0^L R^2 \left(\frac{d\vartheta}{ds}\right)^2 ds\right] \mathcal{D}[R\,d\vartheta], \qquad (23)$$

where, $\vartheta_1 = \vartheta(L)$ and $\vartheta_0 = \vartheta(0)$.

Let us now write the integrand as a white noise functional by parametrizing the variable ϑ as,

$$\vartheta(L) = \vartheta_0 + \left(\sqrt{l}/R\right)\,B(L)$$

$$= \vartheta_0 + \left(\sqrt{l}/R\right)\int_0^L \omega(s)\,ds, \qquad (24)$$

where $\omega = dB/ds$ and $\omega(s)$ is a Gaussian random white noise variable with $B(s)$ a Brownian motion parametrized by s. Noting that, $d\vartheta/ds = (\sqrt{l}/R)\,\omega(s)$, the integrand in Eq. (23) becomes,

$$\exp\left[-\frac{1}{l}\int R^2\left(\frac{d\vartheta}{ds}\right)^2 ds\right] = \exp\left[-\int \omega(s)^2\,ds\right]. \qquad (25)$$

With the integrand expressed as a white noise functional, the integration over $\mathcal{D}[R\,d\vartheta]$ becomes an integral over the Gaussian white noise measure $d\mu(\omega)$ given by,

$$d\mu(\omega) = N_\omega \exp\left[-\frac{1}{2}\int_0^L \omega(s)^2\,ds\right] d^\infty\omega, \qquad (26)$$

where N_ω is a normalization factor. The factor, $\exp\left[-\frac{1}{2}\int \omega(s)^2 ds\right]$, is responsible for the Gaussian fall-off which may be removed by multiplying both sides of Eq. (26) by, $\exp\left[\frac{1}{2}\int \omega(s)^2\,ds\right]$. More appropriately, $\mathcal{D}[R\,d\vartheta]$ is replaced by, $\exp\left[\frac{1}{2}\int \omega(s)^2\,ds\right] d\mu(\omega) = N_\omega\,d^\infty\omega$, as was the case for the nonrelativistic quantum propagator where the path integral is expressed as a white noise functional.[5-13] The exponential multiplying $d\mu(\omega)$, can be combined with Eq. (25) which leads us to consider the white noise functional,

$$I_0 = N \exp\left(-\int_0^L \omega(s)^2 \, ds\right) \exp\left(\frac{1}{2}\int_0^L \omega(s)^2 \, ds\right)$$

$$= N \exp\left(-\frac{1}{2}\int_0^L \omega(s)^2 \, ds\right), \tag{27}$$

where N is an appropriate normalization factor.

Note that the initial point ϑ_0 has been fixed by the parametrization Eq. (24). To incorporate a fixed endpoint ϑ_1 the Donsker delta function $\delta\left(\vartheta(L) - \vartheta_1\right)$ is used, *i.e.*, we have the form, $I_0 \, \delta\left(\vartheta(L) - \vartheta_1\right)$, where I_0 is given by Eq. (27) and $\vartheta(L)$ by Eq. (24). However, note that the polymer can wind n times clockwise, or counterclockwise, around the origin, and to reflect these possibilities we should use instead, $I_0 \, \delta(\vartheta(L) - \vartheta_1 + 2\pi n)$, where $n = 0, \pm 1, \pm 2, \ldots$. Hence, summing over all possible winding number n we have the white noise functional,

$$I = \sum_{n=-\infty}^{+\infty} I_0 \, \delta\left(\vartheta(L) - \vartheta_1 + 2\pi n\right)$$

$$= \sum_{n=-\infty}^{+\infty} N \exp\left(-\frac{1}{2}\int_0^L \omega(s)^2 \, ds\right)$$

$$\times \delta\left(\vartheta_0 + \frac{\sqrt{l}}{R}\int_0^L \omega(s) \, ds - \vartheta_1 + 2\pi n\right). \tag{28}$$

Integrating over the white noise measure $d\mu(\omega)$, the probability function Eq. (23), can now be written as,

$$P(\vartheta_1, \vartheta_0) = \int \sum_{n=-\infty}^{+\infty} I_0 \, \delta\left(\vartheta(L) - \vartheta_1 + 2\pi n\right) d\mu(\omega). \tag{29}$$

To evaluate this expression, we first write the Fourier representation of the δ-function, *i.e.*,

$$P(\vartheta_1, \vartheta_0) = \frac{1}{2\pi} \sum_{n=-\infty}^{+\infty} \int \exp\left[i\lambda\left(\vartheta_0 - \vartheta_1 + 2\pi n\right)\right]$$

$$\times \int \exp\left(i\lambda\frac{\sqrt{l}}{R}\int_0^L \omega(s) \, ds\right) I_0 \, d\mu(\omega) \, d\lambda \tag{30}$$

The integration over $d\mu(\omega)$ can be done by noticing that if we set, $k = 1$, in Example 1 of the previous section (see, Eq. (16)), the T-transform of I_0, Eq. (27), is given by,

$$TI_0(\xi) = \int \exp\left(i\int \omega\xi\, ds\right) I_0(\omega)\, d\mu(\omega)$$

$$= \exp\left(-\frac{1}{4}\int_0^L \xi^2\, ds\right), \tag{31}$$

where the normalization is taken as,

$$N^{-1} = \int \exp\left[-\frac{1}{2}\int \omega(s)^2\, ds\right] d\mu(\omega)$$

(see Eq. (15)). If we let, $\xi = \lambda\sqrt{l}/R$, in Eq. (30), then the integration over $d\mu(\omega)$ is just $TI_0(\xi)$, Eq. (31). We obtain,

$$P(\vartheta_1, \vartheta_0) = \frac{1}{2\pi} \sum_{n=-\infty}^{+\infty} \int \exp\left[i\lambda\left(\vartheta_0 - \vartheta_1 + 2\pi n\right)\right]$$

$$\times \exp\left(-\lambda^2 lL/4R^2\right)\, d\lambda$$

$$= \sum_{n=-\infty}^{+\infty} P_n\,. \tag{32}$$

The P_n is the corresponding probability function for polymer configurations which entangle n - times around the origin. The sum of all the P_n's gives the total probability function. The remaining integral in P_n is a Gaussian integral over λ, *i.e.*,

$$P_n = \frac{1}{2\pi} \int \exp\left[i\lambda\left(\vartheta_0 - \vartheta_1 + 2\pi n\right) - \lambda^2(lL/4R^2)\right]\, d\lambda$$

$$= \sqrt{R^2/lL\pi}\, \exp\left[-\frac{R^2}{lL}\left(\vartheta_0 - \vartheta_1 + 2\pi n\right)^2\right]\,. \tag{33}$$

From Eq. (32) we may also use Poisson's sum formula,[20]

$$\frac{1}{2\pi} \sum_{n=-\infty}^{+\infty} \exp\left(in\phi\right) = \sum_{m=-\infty}^{+\infty} \delta\left(\phi + 2\pi m\right), \tag{34}$$

to obtain,

$$P(\vartheta_1, \vartheta_0) = \frac{1}{2\pi} \sum_{m=-\infty}^{+\infty} \int \delta\left(\lambda + m\right) \exp\left[i\lambda\left(\vartheta_0 - \vartheta_1\right) - \lambda^2(lL/4R^2)\right]\, d\lambda$$

$$= \frac{1}{2\pi} \sum_{m=-\infty}^{+\infty} \exp\left[-im\left(\vartheta_0 - \vartheta_1\right) - m^2(lL/4R^2)\right]. \tag{35}$$

For an entangled polymer with an arbitrary initial starting point we may set, $\vartheta_0 = \vartheta_1$. From Eqs. (33) and (35), the probability that the polymer winds n-times can then be calculated as,

$$
\begin{aligned}
W(n) &= P_n/P(L) \\
&= \frac{\sqrt{R^2/lL\pi}\,\exp\left[-\left(2\pi nR\right)^2/lL\right]}{\dfrac{1}{2\pi}\displaystyle\sum_{m=-\infty}^{+\infty}\exp\left[-m^2(lL/4R^2)\right]}\,.
\end{aligned}
\tag{36}
$$

For a very long polymer, $L = Nl \gg 1$, the dominant term in the denominator is that for $m = 0$, and hence,

$$
W(n) \approx (R/l)\sqrt{4\pi/N}\,\exp\left(-4\pi^2 n^2 R^2/Nl^2\right).
\tag{37}
$$

This result agrees with that obtained by Wiegel[17] for entangled polymers where an interaction potential has a minimum at some value, $r = R$.

4. Polymer with magnetic flux

Let us next consider the case where the straight polymer intersecting the origin of the plane possesses a magnetic flux confined along its length. We represent this situation by adding a potential, $V = q\mathbf{A}\cdot\dot{\mathbf{r}}$, to the probability function, Eq. (22), which describes the various configurations of the entangled polymer lying on the plane. Here, q is the net charge of each repeating unit of the polymer, and $\mathbf{A}$ is the vector potential. Incorporating this in Eq. (22) we have, for $\dot{\mathbf{r}} = d\mathbf{r}/ds$,

$$
P(\mathbf{r}_1,\mathbf{r}_0) = \int \exp\left[-\frac{1}{l}\int_0^L \left[(d\mathbf{r}/ds)^2 + lq\mathbf{A}\cdot\dot{\mathbf{r}}\right]ds\right]\mathcal{D}^2[\mathbf{r}]\,.
\tag{38}
$$

In particular, the vector potential $\mathbf{A}$ can be written as,

$$
\mathbf{A} = (\Phi_0/2\pi)\,\nabla\vartheta \quad , \quad (r > R_0)
\tag{39}
$$

where R_0 is the cross-sectional radius of the straight polymer at the origin and, $\Phi_0 = \pi R_0{}^2 B$, is the non-vanishing magnetic flux along its length. Note that outside the straight polymer, $r > R_0$, the magnetic field is, $B = \nabla \times \mathbf{A} = 0$, which is a situation analogous to the Aharonov–Bohm set-up.[18–19] With Eq. (39), the potential becomes, $q\mathbf{A}\cdot\dot{\mathbf{r}} = \Phi\dot{\vartheta}$, where $\Phi = q\Phi_0/2\pi$, and Eq. (38) can be written as,

$$
P(\mathbf{r}_1,\mathbf{r}_0) = \int \exp\left[-\frac{1}{l}\int_0^L \left[(d\mathbf{r}/ds)^2 + l\Phi\,\dot{\vartheta}\right]ds\right]\mathcal{D}^2[\mathbf{r}]\,.
\tag{40}
$$

As in the previous section, if we simplify the calculation by constraining the radial part to $r = R > R_0$, Eq. (40) becomes,

$$P(\vartheta_1, \vartheta_0) = \int \exp\left[-\frac{1}{l} \int_0^L \left[R^2 \left(\frac{d\vartheta}{ds}\right)^2 + l\Phi\,\dot\vartheta\right] ds\right] \mathcal{D}[R\,d\vartheta]. \qquad (41)$$

Parametrizing the variable ϑ as in Eq. (24), the integrand of Eq. (41) can be written as a white noise functional and the probability function acquires a form similar to Eq. (29) modified by the potential, $\Phi\,\dot\vartheta = \Phi(\sqrt{l}/R)\,\omega$, i.e.,

$$P(\vartheta_1, \vartheta_0) = \int \sum_{n=-\infty}^{+\infty} I_0 \, \exp\left(-\Phi(\sqrt{l}/R) \int_0^L \omega(s)\,ds\right)$$

$$\times \delta\left(\vartheta_0 + \frac{\sqrt{l}}{R} \int_0^L \omega(s)\,ds - \vartheta_1 + 2\pi n\right) d\mu(\omega). \qquad (42)$$

To facilitate the integration over $d\mu(\omega)$, we use the Fourier representation of the delta function and write Eq. (42) as,

$$P(\vartheta_1, \vartheta_0) = \frac{1}{2\pi} \sum_{n=-\infty}^{+\infty} \int_{-\infty}^{+\infty} \exp\left[i\lambda\,(\vartheta_0 - \vartheta_1 + 2\pi n)\right]$$

$$\times \int \exp\left[i\frac{\sqrt{l}}{R}(\lambda + i\Phi) \int_0^L \omega(s)ds\right] I_0 \, d\mu(\omega)\, d\lambda. \qquad (43)$$

If we let, $\xi = (\sqrt{l}/R)(\lambda + i\Phi)$, the integration over $d\mu(\omega)$ is just the T-transform of I_0, Eq. (31), and the probability function becomes,

$$P(\vartheta_1, \vartheta_0) = \sum_{n=-\infty}^{+\infty} \frac{1}{2\pi} \int_{-\infty}^{+\infty} \exp\left[i\lambda\,(\vartheta_0 - \vartheta_1 + 2\pi n)\right]$$

$$\times \exp\left[-(lL/4R^2)(\lambda + i\Phi)^2\right] d\lambda. \qquad (44)$$

The remaining integral is a Gaussian integral over λ and we may write Eq. (44) as,

$$P(\vartheta_1, \vartheta_0) = \sum_{n=-\infty}^{+\infty} P_n, \qquad (45)$$

where the probability for winding n-times is given by,

$$P_n = \sqrt{R^2/\pi lL}\,\exp\left\{-\left(R^2/lL\right)\left[\vartheta_0 - \vartheta_1 + 2\pi n - (lL\Phi/2R^2)\right]^2\right\}$$

$$\times \exp\left(lL\Phi^2/4R^2\right). \qquad (46)$$

Alternatively, from Eq. (44), we may apply the Poisson sum formula, Eq. (34), such that

$$P(\vartheta_1, \vartheta_0) = \sum_{m=-\infty}^{+\infty} \frac{1}{2\pi} \int_{-\infty}^{+\infty} \delta(\lambda + m) \exp\left[i\lambda\left(\vartheta_0 - \vartheta_1\right)\right]$$
$$\times \exp\left[-(lL/4R^2)(\lambda + i\Phi)^2\right] d\lambda, \tag{47}$$

and integrate λ with the help of the delta function to obtain,

$$P(\vartheta_1, \vartheta_0) = \frac{1}{2\pi} \sum_{m=-\infty}^{+\infty} \exp\left[-im\left(\vartheta_0 - \vartheta_1 - (\Phi lL/2R^2)\right)\right]$$
$$\times \exp\left[-(m^2 lL/4R^2) + (lL\Phi^2/4R^2)\right]. \tag{48}$$

For the case when the flux $\Phi = 0$, Eq. (48) reduces to Eq. (35) of the previous section.

The probability for the polymer to entangle n-times is given by,

$$W(n) = P_n/P(\vartheta_1, \vartheta_0)$$
$$= \frac{\sqrt{R^2/lL\pi} \exp\left\{-(R^2/lL)\left[\vartheta_0 - \vartheta_1 + 2\pi n - (Ll\Phi/2R^2)\right]^2\right\}}{\dfrac{1}{2\pi} \displaystyle\sum_{m=-\infty}^{+\infty} \exp\left\{-im\left[\vartheta_0 - \vartheta_1 - (\Phi lL/2R^2)\right] - m^2(lL/4R^2)\right\}}.$$
$$\tag{49}$$

For an arbitrary initial point we may set, $\vartheta_0 = \vartheta_1$, and Eq. (49) may be written as,

$$W(n) = \frac{R}{l}\sqrt{\frac{4\pi}{N}} \frac{\exp\left(-4\pi^2 n^2 R^2/Nl^2\right) \exp\left[2\pi n\Phi - (lL/4R^2)\Phi^2\right]}{\theta_3\left(\Phi Nl^2/4R^2\right)}, \tag{50}$$

where $\theta_3(u)$ is the theta function,[21]

$$\theta_3(u) = \sum_{m=-\infty}^{+\infty} q^{m^2} \exp(2mui)$$
$$= 1 + 2\sum_{m=1}^{+\infty} q^{m^2} \cos(2mu), \tag{51}$$

with $u = \Phi Nl^2/4R^2$, and $q = \exp\left(-Nl^2/4R^2\right)$. For a very long polymer, $L = Nl \gg 1$, the $\theta_3(u)$ in the denominator is approximately one, and we obtain from Eq. (50),

$$W(n) \approx (R/l)\sqrt{4\pi/N} \exp\left[-(4R^2/Nl^2)(n\pi - \Phi)^2\right]. \tag{52}$$

Curiously, even if the magnetic flux Φ is confined along the straight polymer such that $B = 0$ in the region where the entangled polymer lies on the plane, the entanglement probability is still affected by the confined magnetic field. This is reminiscent of the Aharonov–Bohm effect in quantum mechanics[18–19] where the interference pattern of the electron in a two-slit experiment is influenced by a magnetic field it never comes in contact with. Here, of course, we are dealing with "classical" entanglement probabilities of a polymer which lies in a region where the force due to the confined magnetic field B is zero, and yet, is still influenced by it. When the magnetic flux vanishes, $\Phi = 0$, the entanglement probability Eq. (52) reduces to Eq. (37), or that obtained by Wiegel.[17]

5. Conclusion

Using a white noise functional integral approach we considered two cases for the polymer entanglement problem.[22] In section 3, we took into account the observations of Wiegel[17] that, in the low temperature limit, the configurations of an entangled polymer are confined to a narrow strip in the immediate vicinity of a circle of radius R . The entanglement probability that we obtained for this case agrees with the result of Wiegel. We then extended the study of the two chainlike macromolecules by considering the second case where there is a magnetic flux confined along the straight polymer perpendicular to the plane. The probability $W(n)$ for the polymer on the plane to entangle n-times around the straight polymer is calculated, and in the limit where the magnetic flux goes to zero, the result reduces to the $W(n)$ derived by Wiegel.[17]

As noted in section 4, the case involving a magnetic flux confined along the straight polymer appears to be a classical analogue of the Aharonov–Bohm effect in quantum mechanics. The entanglement probability $W(n)$ is influenced by the magnetic flux, even if the entangled polymer lies in a region where $B = 0$, and feels no magnetic force. It would be interesting to see whether an experimental observation of this effect could be realized with the development of insulated molecular wires made from a conducting linear polymer covered with a molecular nanotube that acts as an insulator.[23]

Acknowledgement

This work is part of a joint project between the RCTP, Central Visayan Institute, and BiBoS, University of Bielefeld, supported by the Federal Ministry for Education, Science, Research and Technology (BMBF) of Germany through the Alexander von Humboldt-Stiftung under its Programme on Research Cooperation with Asia. Some local support was also given by the National Research Council of the Philippines.

References

1. T. Hida, *Analysis of Brownian Functionals*, Carleton Math. Lecture Notes No. 13 (Carleton, 1975).
2. T. Hida, H. H. Kuo, J. Potthoff and L. Streit, *White Noise. An Infinite Dimensional Calculus* (Kluwer, Dordrecht, 1993).
3. N. Obata, *White Noise Calculus and Fock Space*, Lecture Notes in Mathematics, Vol. 1577 (Springer, Berlin, 1994).
4. H.H. Kuo, *White Noise Distribution Theory* (CRC, Boca Raton, FL, 1996).
5. L. Streit and T. Hida, "Generalized brownian functionals and the Feynman integral," *Stoch. Proc. Appl.* **16** (1983) 55–69.
6. M. de Faria, J. Potthoff and L. Streit, "The Feynman integrand as a Hida distribution," *Jour. Math. Phys.* **32** (1991) 2123–2127.
7. D. C. Khandekar and L. Streit, "Constructing the Feynman Integrand," *Ann. Phys.* (Leipzig) **1** (1992) 46–55.
8. A. Lascheck, P. Leukert, L. Streit, and W. Westerkamp, "Quantum Mechanical Propagators in Terms of Hida Distributions," *Rep. Math. Phys.* **33** (1993) 221–232.
9. M. Grothaus, D. C. Khandekar, J. L. Silva and L. Streit, "The Feynman Integral for Time-Dependent Anharmonic Oscillators," *J. Math. Phys.* **38** (1997) 3278–3299.
10. T. Kuna, L. Streit and W. Westerkamp, "Feynman Integrals for a Class of Exponentially Growing Potentials," *J. Math. Phys.* **39** (1998) 4476–4491.
11. M. Cunha, C. Drumond, P. Leukert, J. L. Silva and W. Westerkamp, "The Feynman Integrand for the Perturbed Harmonic Oscillator as a Hida Distribution," *Ann. Phys.* **4** (1995) 53–67.
12. A. Lascheck, P. Leukert, L. Streit and W. Westerkamp, "More about Donsker's delta function," *Soochow Jour. Math.* **20** (1994) 401–418.
13. C.C. Bernido and M.V. Carpio-Bernido, "Path integrals for boundaries and topological constraints: a white noise functional approach," *J. Math. Phys.* **43** (2002) 1728–1736.
14. S.F. Edwards, "Statistical mechanics with topological constraints: I," *Proc. Phys. Soc. London* **91** (1967) 513–519.
15. S. Prager and H.L. Frisch, *J. Chem. Phys.* **46** (1967) 1475.
16. N. Saito and Y. Chen, *J. Chem. Phys.* **59** (1973) 3701.
17. F.W. Wiegel, "Entanglement probabilities for a harmonically bound macromolecule," *J. Chem. Phys.* **67** (1977) 469–472
18. M. Peshkin and A. Tonomura, *The Aharonov–Bohm Effect* (Springer, Berlin, 1989).
19. See, *e.g.*, C.C. Bernido and A. Inomata, "Path integrals with a periodic constraint: The Aharonov–Bohm effect," *J. Math. Phys.* **22** (1981) 715.

20. R. Bellman, *A Brief Introduction to Theta Functions* (Holt, Rinehart, & Winston, New York, 1961).

21. I.S. Gradshteyn and I.M. Ryzhik, *Table of Integrals, Series, and Products*, 5th ed. (Academic Press, San Diego, 1994) 927.

22. C.C. Bernido and M.V. Carpio-Bernido, "Entanglement probabilities of polymers: a white noise functional approach," *J. Phys. A* (2003) to appear.

23. See, *e.g.*, T. Shimomura, T. Akai, T. Abe, and K. Ito, "Atomic force microscopy observation of insulated molecular wire formed by conducting polymer and molecular nanotube," *J. Chem. Phys.* **116** (2002) 1753–1756.

Double Strata of Time for Construction of Path-Space Measure for Stochastic Differential Equations

Toru Nakamura

Department of Mathematics, Sundai Preparatory School,
Kanda–Surugadai, Chiyoda-ku, Tokyo 101–0062, Japan
E-mail: tnakamur@tsuda.ac.jp

While the nonstandard analysis has been applied to construction of a path-space measure appropriate for a stochastic differential equation $dX = a(X)dt + dW$ with bounded $a(X)$ so far, we show that the case $a(X) = -\beta X$, and further the case where $a(X)$ has growth of degree higher than one can be handled by introducing two strata of time lattices.

1. Introduction

We first consider the Ornstein–Uhlenbeck process (OU-proces), a Brownian motion in a viscous medium; the particle momentum X changes according to the stochastic differential equation,

$$dX(t,\omega) = a\big(X(t,\omega)\big)dt + dW(t,\omega) \quad (a(x) = -\beta x,\ \beta > 0). \qquad (1)$$

The first term on the right-hand side represents a resistance, the second term a random force due to a Brownian motion $W(t,\omega)$ satisfying

$$\Big\langle W(t,\omega),\, W(t',\omega) \Big\rangle = 2D\beta^2 \min\{t, t'\}.$$

We wish to find a path-space measure for the process (1). For this purpose, we have the Girsanov formula,[1] but we present in this paper a different approach, nonstandard analysis approach, applied to the forward Fokker-Planck equation,

$$\frac{\partial}{\partial t}U(t,p) = \left(\beta\frac{\partial}{\partial p}p + D\beta^2\frac{\partial^2}{\partial p^2}\right)U(t,p) \qquad (2)$$

derived from (1), which is the equation for the probability density $U(t,p)$ of the particle momentum p at time t.

Nonstandard analysis[2,3,4] has been applied to a stochastic differential equation (1) with bounded coefficient $a(X)$ so far.[5,6] We shall show in Sec. 2 that the case $a(X) = -\beta X$ $(\beta > 0)$ can also be handled by introducing two different infinitesimals for time-lattice spacings. Indeed, we observe that a time interval, short enough for the change in the drift to be small, can be long enough for a particle to undergo a large number of steps of Brownian random walks (see S. Chandrasekhar[7]). Nonstandard analysis enables us to realize these two discretizations of different orders with mathematical rigor.[8]

We first discretize the space-time and define a set of $*$-finitely many $*$-paths. Then, the differential equation (2), if rewritten into a difference equation, provides a transition probability thereby defining a $*$-measure μ for each $*$-path separately. The solution to (2) is obtained as the standard part of μ-weighted sum of the contributions from $*$-finitely many $*$-paths. A standard measure appropriate for the standard path integral can be extracted from μ by using the Loeb measure theory.[9] Let us comment that the same method was successfully applied to the Dirac equation, the fundamental equation for the relativistic quantum mechanics.[10,11,12]

An extension is discussed in Sec. 3 to a wider class of stochastic differential equations (1) with $a(x)$ having growth of degree higher than one. For such equations, some of the paths inevitably explode in finite time, so that no path integral approach has been proposed to the best of the author's knowledge.

We use the following notations from nonstandard analysis:

(i) $^*A$ is the nonstandard extension of A.

(ii) $\text{st}(B)$ is the standard part of B.

(iii) $A = o(1)$ if A is an infinitesimal number.

(iv) $A = \mathcal{O}(B)$ if A/B is finite, and $A = o(B)$ if A/B is infinitesimal, where A, B are both infinite or both infinitesimal.

(v) We say specifically that A is strictly of the order of B, or in short $A = \mathcal{O}_{\text{str}}(B)$, if $A = \mathcal{O}(B)$ and $A \neq o(B)$.

2. Path-space measure for OU-process

We start with a difference equation corresponding to the differential equation (2). In standard analysis, we first discretize the space-time with finite spacings and then take the limit of the spacings to zero. In nonstandard analysis, in contrast, we can grasp the process at a stroke by introducing infinitesimals into the number system, making the construction of the completely additive measure rigorous and elementary. Moreover, the two infinitesimal spacings of different orders we are going to use to discretize the time can be handled without difficulty.

Let us rewrite the Fokker–Planck equation (2) into a difference equation. Since we consider the initial value problem toward future, it is natural to replace t-derivative by the forward difference quotient:

$$\frac{\partial U(t,p)}{\partial t} \Rightarrow \frac{1}{\Delta t}\Big\{U(t+\Delta t,p) - U(t,p)\Big\}. \tag{3}$$

As for the p-derivative, we choose the central difference quotients

$$\frac{\partial U(t,p)}{\partial p} \Rightarrow \frac{1}{2\Delta p}\Big\{U(t,p+\Delta p) - U(t,p-\Delta p)\Big\}, \tag{4}$$

$$\frac{\partial^2 U(t,p)}{\partial p^2} \Rightarrow \frac{1}{(\Delta p)^2}\Big\{U(t,p+\Delta p) + U(t,p-\Delta p) - 2U(t,p)\Big\} \tag{5}$$

which are convenient and indeed natural for the equation having reflection symmetry in p. By the above replacements, (2) turns out to give

$$U(t+\Delta t,p)$$
$$= \frac{\beta\Delta t}{2\Delta p}\Big\{(p+\Delta p)U(t,p+\Delta p) - (p-\Delta p)U(t,p-\Delta p)\Big\}$$
$$+ \frac{D\beta^2\Delta t}{(\Delta p)^2}\Big\{U(t,p+\Delta p) + U(t,p-\Delta p)\Big\} + \Big\{1 - \frac{2D\beta^2\Delta t}{(\Delta p)^2}\Big\}U(t,p).$$

Choose $\Delta p = (2D\beta^2\Delta t)^{1/2}$ so that the coefficient of $U(t,p)$ vanishes. Then,

$$U(t+\Delta t,p) = \sum_{\pm} \frac{1}{2}\Big\{1 \pm (p\pm\Delta p)\Big(\frac{\Delta t}{2D}\Big)^{1/2}\Big\}U(t,p\pm\Delta p). \tag{6}$$

This formula suggests that the transition probability

$$\frac{1}{2}\Big\{1 \pm (p\pm\Delta p)\Big(\frac{\Delta t}{2D}\Big)^{1/2}\Big\} \tag{7}$$

should be assigned to each line segment connecting $(t,p\pm\Delta p)$ and $(t+\Delta t,p)$.

Now, we should note that the coefficient p on the right-hand side of (2) causes a problem by enlarging the error of the replacement, if it is very large. Indeed, it manifests itself as negative transition probability (7). To show that it is not actually the case, we take advantage of the two time scales of different orders the OU-process has, one for the Brownian motion and the other for the change in the drift term in (2), and introduce two infinitesimal spacings, ε and τ, of different orders. ε is the finer and stands for the time spacing of the Brownian random walk, and τ is long enough to cover many steps of the Brownian random walks, yet short enough for the change in the drift term small. More precisely, we slice the time interval $[0, T]$ with an infinite $*$-natural number N such that $\tau = T/N$:

$$0 = t_0 < t_1 < \cdots < t_N = T, \quad t_{k+1} - t_k = \tau.$$

Then, choose another infinite $*$-natural number ν and set $\varepsilon = \tau/\nu$, which we take to be the time step of the Brownian random walks (Fig. 1).

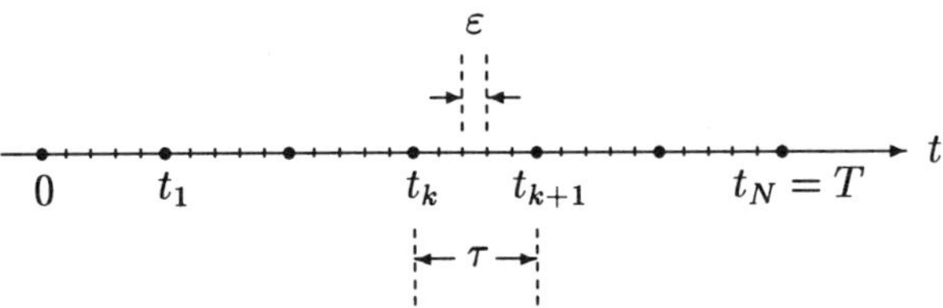

Fig. 1   The time lattice with two infinitesimal spacings τ and ε.

We first construct a $*$-path-space for the infinitesimal interval $[t_k, t_{k+1}]$ with fine-grained spacing ε. The corresponding spacing for the momentum is taken to be

$$\delta = (2D\beta^2\varepsilon)^{1/2}. \tag{8}$$

In addition, we shall find it convenient in the proof of the Lemma 1 below to discard the paths of momentum increment exceeding α,

    (C1) $\alpha = \mathcal{O}_{\text{str}}(\tau^{-2}).$

We also set

    (C2) $\nu = \mathcal{O}(e^{1/\beta\tau}),$

which makes $\varepsilon = \mathcal{O}(\tau e^{-1/\beta\tau}).$

Definition 1: (i) *Define a momentum lattice* M *by* $\mathsf{M} = \{\, n\delta \mid n \in {}^*\mathbb{Z} \,\}.$

    (ii) *Let* Ω *be the set of internal mappings from* $\{0, 1, \ldots, \nu - 1\}$ *to* $\{1, -1\}$. *For each* $\omega \in \Omega$, *define a* $*$-*path* $X_{k,\omega}(j\varepsilon)$ *starting from momentum* $p \in \mathsf{M}$ *at time* t_k *by*

$$X_{k,\omega}(j\varepsilon) = p + \sum_{l=0}^{j-1} \omega(l)\,\delta \quad (j = 1, 2, \ldots, \nu).$$

However, not all the ∗-paths are taken; we discard those of large momentum increment. Namely, we define a ∗-path-space $\mathscr{P}_k^\alpha$ for the interval $[t_k, t_{k+1}]$ by

$$\mathscr{P}_k^\alpha = \left\{ X_{k,\omega} \mid \left| X_{k,\omega}(\nu\varepsilon) - X_{k,\omega}(0) \right| < \alpha \right\}. \tag{9}$$

(iii) For each $X_{k,\omega} \in \mathscr{P}_k^\alpha$, define a ∗-measure μ_k by

$$\mu_k(X_{k,\omega}) = \prod_{j=0}^{\nu-1} \frac{1}{2} \left\{ 1 - \omega(j)p \, \frac{\varepsilon^{1/2}}{(2D)^{1/2}} \right\}, \tag{10}$$

where $p = X_{k,\omega}(0)$, the momentum at the initial time t_k.

We shall exhibit in Theorem 1 that the cutoff (9) does not cause any substantial loss of transition probability which should be carried by the ∗-path sum defined by (12) below. In fact, we shall see eventually that the α-dependence of the path-space measure we construct will be so minute as to leave no traces in its standard part (see Theorem 2).

Definition 2: *Define a set $\mathscr{P}^\alpha_{(t_{k+1},p_{k+1}|t_k,p_k)}$ of ∗-paths $X_{k,\omega}$ with end points (t_k, p_k) and (t_{k+1}, p_{k+1}) by*

$$\mathscr{P}^\alpha_{(t_{k+1},p_{k+1}|t_k,p_k)} = \left\{ X_{k,\omega} \in \mathscr{P}_k^\alpha \middle| X_{k,\omega}(0) = p_k, \ X_{k,\omega}(\nu\varepsilon) = p_{k+1} \right\}, \tag{11}$$

and a ∗-path sum $\mathcal{G}_\alpha(t_{k+1}, p_{k+1} \mid t_k, p_k)$ by

$$\mathcal{G}_\alpha(t_{k+1}, p_{k+1} \mid t_k, p_k) = \frac{1}{2\delta} \sum_{X_{k,\omega}} \mu_k(X_{k,\omega}), \tag{12}$$

where the sum is taken over the set $\mathscr{P}^\alpha_{(t_{k+1},p_{k+1}|t_k,p_k)}$.

The factor 2δ in (12) accounts for dp_k in the expression of the standard Green function

$$U(t_{k+1}, p_{k+1}) = \int G(t_{k+1}, p_{k+1} \mid t_k, p_k) U(t_k, p_k) dp_k.$$

It should be 2δ rather than δ because only the ∗-paths that start from $p_k = p_{k+1} - m\delta$ with $\nu - m$ being an even number can reach the point (t_{k+1}, p_{k+1}).

Let us calculate the ∗-path sum $\mathcal{G}_\alpha$ for the interval $[t_k, t_{k+1}]$ making use of the advantage of the double strata of time that the spacing ε is much smaller than the other spacing $\tau = \nu\varepsilon$ in the order of infinitesimals.

Lemma 1: *If* $p_k = \mathcal{O}(k\alpha)$ *and* $|p_{k+1} - p_k| < \alpha$, *then*

$$\left| \mathcal{G}_\alpha(t_{k+1}, p_{k+1} \,|\, t_k, p_k) - G_d(t_{k+1}, p_{k+1} \,|\, t_k, p_k) \right| \le \kappa_1 G_d(t_{k+1}, p_{k+1} \,|\, t_k, p_k) \tag{13}$$

holds for some infinitesimal constant $\kappa_1 = \mathcal{O}(\tau^{-9}\varepsilon^{1/2})$, *where*

$$G_d(t_{k+1}, p_{k+1} \,|\, t_k, p_k) = \frac{1}{(4\pi D\beta^2\tau)^{1/2}} \exp\left[-\frac{\{p_{k+1} - (1 - \beta\tau)p_k\}^2}{4D\beta^2\tau} \right], \tag{14}$$

an approximation of the Green function for the Fokker–Planck equation (2).

Proof: We write

$$J_k = (p_{k+1} - \alpha, \; p_{k+1} + \alpha). \tag{15}$$

Notice that $p_k \in J_k$ is necessary for a $*$-path in (9) connecting (t_k, p_k) and (t_{k+1}, p_{k+1}) to exist. Put

$$m = \frac{1}{\delta}(p_{k+1} - p_k) \quad \text{and} \quad r = \frac{1}{2}(\nu + m),$$

meaning that r is the number of j's such that $\omega(j) = 1$. By (12),

$$\mathcal{G}_\alpha(t_{k+1}, p_{k+1} \,|\, t_k, p_k)$$

$$= \frac{1}{2\delta} \frac{1}{2^\nu} \binom{\nu}{r} \left\{ 1 - p_k \frac{\varepsilon^{1/2}}{(2D)^{1/2}} \right\}^r \left\{ 1 + p_k \frac{\varepsilon^{1/2}}{(2D)^{1/2}} \right\}^{\nu - r}. \tag{16}$$

Since the order of α is given in (C1) in terms of the coarse-grained spacing τ, r and $\nu - r$ are $\mathcal{O}_{\text{str}}(\nu)$. Hence, we can apply the Stirling formula, to get

$$\log\left[\frac{1}{2^\nu} \binom{\nu}{r} \right] = \log \frac{2}{(2\pi\nu)^{1/2}} - \frac{m^2}{2\nu} + \mathcal{O}\!\left(\frac{1}{\nu}\right) + \mathcal{O}\!\left(\frac{m^3}{\nu^2}\right) \tag{17}$$

and

$$\log\left[\left\{ 1 - p_k \frac{\varepsilon^{1/2}}{(2D)^{1/2}} \right\}^r \left\{ 1 + p_k \frac{\varepsilon^{1/2}}{(2D)^{1/2}} \right\}^{\nu - r} \right]$$

$$= -\frac{mp_k}{(2D)^{1/2}}\varepsilon^{1/2} - \frac{p_k{}^2}{4D}\tau + \mathcal{O}\!\left(\nu|p_k|^3\varepsilon^{3/2}\right). \tag{18}$$

Here, the orders of magnitudes of infinitesimals are

$$\frac{1}{\nu} = \mathcal{O}(\tau^{-1}\varepsilon), \quad \frac{m^3}{\nu^2} = \mathcal{O}(\tau^{-8}\varepsilon^{1/2}) \quad \text{and} \quad \nu|p_k|^3\varepsilon^{3/2} = \mathcal{O}(\tau^{-8}\varepsilon^{1/2}).$$

Applying (17) and (18) to (16), we have

$$\mathcal{G}_\alpha(t_{k+1}, p_{k+1} \,|\, t_k, p_k) = G_\mathrm{d}(t_{k+1}, p_{k+1} \,|\, t_k, p_k) \, \exp\left[\mathcal{O}\big(\tau^{-8}\varepsilon^{1/2}\big)\right]$$

$$= G_\mathrm{d}(t_{k+1}, p_{k+1} \,|\, t_k, p_k)\left(1 + \mathcal{O}\big(\tau^{-8}\varepsilon^{1/2}\big)\right),$$

so that we have completed the proof of the lemma.

Remark: By (17) and (18), we have

$$\frac{1}{2\delta}\frac{1}{2^\nu}\binom{\nu}{r} \simeq \frac{1}{(4\pi D\beta^2\tau)^{1/2}}\exp\left[-\frac{(p_{k+1}-p_k)^2}{4D\beta^2\tau}\right]$$

and

$$\left\{1 - p_k\frac{\varepsilon^{1/2}}{(2D)^{1/2}}\right\}^r\left\{1 + p_k\frac{\varepsilon^{1/2}}{(2D)^{1/2}}\right\}^{\nu-r} \simeq \exp\left[-\frac{p_k(p_{k+1}-p_k)}{2D\beta} - \frac{p_k{}^2}{4D}\tau\right],$$

the product of which agrees with the Green function for the OU-process.

Let us connect the infinitesimal fragments $X_{k,\omega} \in \mathscr{P}_k^\alpha$ for infinitesimal intervals $[t_k, t_{k+1}]$ successively, thereby constructing a $*$-path for the interval $[0, T]$.

Definition 3: *Define a $*$-path X for the whole interval $[0, t_k]$ by*

$$X = \left(X_{0,\omega_{(0)}}, X_{1,\omega_{(1)}}, \ldots, X_{k-1,\omega_{(k-1)}}\right),$$

such that

$$X_{0,\omega_{(0)}}(0) = p_0, \quad X_{j-1,\omega_{(j-1)}}(\nu\varepsilon) = X_{j,\omega_{(j)}}(0), \quad X_{k-1,\omega_{(k-1)}}(\nu\varepsilon) = p_k$$

where $j = 1, \ldots, k-1$, and a $*$-measure μ by

$$\mu(X) = \prod_{j=0}^{k-1}\mu_j(X_{j,\omega_{(j)}}).$$

The set of all $*$-paths X is denoted by $\mathscr{P}^\alpha_{(t_k,p_k|0,p_0)}$. Then, $\mathcal{G}_\alpha(t_k, p_k \,|\, 0, p_0)$ is defined by a $*$-path sum

$$\mathcal{G}_\alpha(t_k, p_k \,|\, 0, p_0) = \frac{1}{2\delta}\sum_{X \in \mathscr{P}^\alpha_{(t_k,p_k|0,p_0)}}\mu(X) \tag{19}$$

where $k = 1, 2, \ldots, N$; $t_N = T$. $\mathcal{G}_\alpha(t_k, p_k \,|\, 0, p_0)$ depends also on ε and τ, though not explicit in our notation.

We have obtained in Lemma 1 the explicit expression (14) of the $*$-path sum $\mathcal{G}_\alpha$ for one step τ, whose l-fold compositions,

$$
\begin{aligned}
G_{\mathrm{d}}(t_l, p_l \,|\, 0, p_0) \\
&= G_{\mathrm{d}}(t_l, p_l \,|\, t_{l-1}, p_{l-1}) * G_{\mathrm{d}}(t_{l-1}, p_{l-1} \,|\, t_{l-2}, p_{l-2}) * \cdots * G_{\mathrm{d}}(t_1, p_1 \,|\, 0, p_0) \\
&= \frac{1}{(4\pi D\beta^2 \tau b_l)^{1/2}} \exp\left[-\frac{(p_l - a^l p_0)^2}{4D\beta^2 \tau b_l} \right]
\end{aligned}
\tag{20}
$$

with

$$
a = 1 - \beta\tau, \quad b_l = \sum_{j=0}^{l-1} a^{2j} \quad \text{and} \quad 0 \le l \le N
$$

should be a discrete version of the Green function for the interval $[0,\, t_l]$. Here, the composition "$*$" is defined by

$$
f(p, p') * g(p', p'') = \int_{\mathbb{R}} f(p, p')\, g(p', p'')\, dp'.
$$

In fact,

$$
G_{\mathrm{d}}(t_l, p_l \,|\, 0, p_0) \simeq \frac{1}{\{2\pi D\beta(1 - e^{-2\beta t_l})\}^{1/2}} \exp\left[-\frac{(p_l - p_0 e^{-\beta t_l})^2}{2D\beta(1 - e^{-2\beta t_l})} \right]
$$

is valid if t_l and 0 are finitely separated.

In order to prove inductively that the $*$-path sum $\mathcal{G}_\alpha(t_l, p_l \,|\, 0, p_0)$ attains this expression, we check some properties of G_{d}. We first note that

$$
\begin{aligned}
G_{\mathrm{d}}(t_{k+1}, p_{k+1} \,|\, t_k, p_k)\, G_{\mathrm{d}}(t_k, p_k \,|\, 0, p_0) \\
&= \frac{b_{k+1}^{1/2}}{(4\pi D\beta^2 \tau b_k)^{1/2}} \exp\left[-\frac{b_{k+1}}{4D\beta^2 \tau b_k} \left(p_k - r(p_{k+1}) \right)^2 \right] G_{\mathrm{d}}(t_{k+1}, p_{k+1} \,|\, 0, p_0)
\end{aligned}
\tag{21}
$$

with

$$
r(p_{k+1}) = \frac{1}{b_{k+1}} (ab_k p_{k+1} + a^k p_0)
\tag{22}
$$

would become $G_{\mathrm{d}}(t_{k+1}, p_{k+1} \,|\, 0, p_0)$ if it was integrated over the whole range of p_k. However, it is not the case because we have restricted p_k to (9) discarding the $*$-paths of large momentum increment. Instead, we wish to find an interval $\left(r(p_{k+1}) - u,\, r(p_{k+1}) + u \right)$ which is large enough to give $G_{\mathrm{d}}(t_{k+1}, p_{k+1} \,|\, 0, p_0)$ up to negligible error when (21) is integrated with respect to p_k.

Lemma 2: *Define*

$$u = \frac{1}{(\beta\tau)^{1/4}} \frac{(4D\beta^2\tau b_k)^{1/2}}{b_{k+1}^{1/2}} = \mathcal{O}(\tau^{1/4}). \tag{23}$$

Then,

$$\left| \int_{R_k(u)} G_{\mathrm{d}}(t_{k+1}, p_{k+1} \,|\, t_k, p_k)\, G_{\mathrm{d}}(t_k, p_k \,|\, 0, p_0) dp_k - G_{\mathrm{d}}(t_{k+1}, p_{k+1} \,|\, 0, p_0) \right|$$
$$\leq \kappa_2\, G_{\mathrm{d}}(t_{k+1}, p_{k+1} \,|\, 0, p_0) \tag{24}$$

for some infinitesimal constant $\kappa_2 = \mathcal{O}\big(\tau^{1/4}\, e^{-1/\sqrt{\beta\tau}}\big)$, *where*

$$R_k(u) = \big(r(p_{k+1}) - u,\; r(p_{k+1}) + u\big). \tag{25}$$

Proof: The inequality (24) is simply verified by putting $A = (\beta\tau)^{-1/4}$ into a well-known inequality

$$\left| 1 - \frac{1}{\pi^{1/2}} \int_{-A}^{A} \exp(-\eta^2) d\eta \right| < \frac{1}{\pi^{1/2} A} \exp(-A^2)$$

and then changing the variable η to p_k by

$$\eta = \frac{b_{k+1}^{1/2}}{(4D\beta^2\tau b_k)^{1/2}} \left(p_k - r(p_{k+1}) \right).$$

Remark: The value $A = (\beta\tau)^{-1/4}$ in the proof above was chosen so that (40) below should hold.

Secondly, let us replace the $*$-integral in (24) by a $*$-Riemann sum.

Lemma 3: *If an internal set E contains the set $R_k(u)$ defined by (25), then*

$$\left| \sum_{p_k \in E} G_{\mathrm{d}}(t_{k+1}, p_{k+1} \,|\, t_k, p_k)\, G_{\mathrm{d}}(t_k, p_k \,|\, 0, p_0)\, 2\delta - G_{\mathrm{d}}(t_{k+1}, p_{k+1} \,|\, 0, p_0) \right|$$
$$\leq 3\kappa_2\, G_{\mathrm{d}}(t_{k+1}, p_{k+1} \,|\, 0, p_0). \tag{26}$$

Here, the constants u and κ_2 are those introduced in the preceding lemma.

Proof: Since

$$\left| \exp\left[-\frac{b_{k+1}}{4D\beta^2\tau b_k} p^2 \right] - \exp\left[-\frac{b_{k+1}}{4D\beta^2\tau b_k} (p + 2\delta)^2 \right] \right| \leq 2\delta\, \frac{b_{k+1}^{1/2}}{(2eD\beta^2\tau b_k)^{1/2}},$$

the difference between the $*$-Riemann sum and the $*$-integral over the interval $R_k(u)$ is bounded as

$$\left| \sum_{p_k \in R_k(u)} G_\mathrm{d}(t_{k+1}, p_{k+1} \,|\, t_k, p_k) \, G_\mathrm{d}(t_k, p_k \,|\, 0, p_0) \, 2\delta \right.$$
$$\left. - \int_{R_k(u)} G_\mathrm{d}(t_{k+1}, p_{k+1} \,|\, t_k, p_k) \, G_\mathrm{d}(t_k, p_k \,|\, 0, p_0) \, dp_k \right|$$
$$\leq \frac{2\delta u b_{k+1}}{(2\pi e)^{1/2} \, D\beta^2 \tau b_k} \, G_\mathrm{d}(t_{k+1}, p_{k+1} \,|\, 0, p_0)$$
$$= o(\kappa_2) \, G_\mathrm{d}(t_{k+1}, p_{k+1} \,|\, 0, p_0). \tag{27}$$

On the other hand, the $*$-Riemann sum over the complementary set $\overline{R_k}(u)$ is bounded as

$$\sum_{p_k \in \overline{R_k}(u)} G_\mathrm{d}(t_{k+1}, p_{k+1} \,|\, t_k, p_k) \, G_\mathrm{d}(t_k, p_k \,|\, 0, p_0) 2\delta$$
$$\leq \frac{4\delta b_{k+1}{}^{1/2}}{(4\pi D\beta^2 \tau b_k)^{1/2}} \, \exp\left[-\frac{b_{k+1}}{4D\beta^2 \tau b_k} u^2 \right] G_\mathrm{d}(t_{k+1}, p_{k+1} \,|\, 0, p_0)$$
$$+ \int_{\overline{R_k}(u)} G_\mathrm{d}(t_{k+1}, p_{k+1} \,|\, t_k, p_k) \, G_\mathrm{d}(t_k, p_k \,|\, 0, p_0) \, dp_k$$
$$= \{ o(\kappa_2) + \kappa_2 \} \, G_\mathrm{d}(t_{k+1}, p_{k+1} \,|\, 0, p_0). \tag{28}$$

For the equality in the last line, we have used (24). By (24), (27) and (28), we obtain

$$\left| \sum_{p_k \in E} G_\mathrm{d}(t_{k+1}, p_{k+1} | t_k, p_k) G_\mathrm{d}(t_k, p_k | 0, p_0) 2\delta - G_\mathrm{d}(t_{k+1}, p_{k+1} | 0, p_0) \right|$$
$$\leq \left| \sum_{p_k \in R_k(u)} G_\mathrm{d}(t_{k+1}, p_{k+1} | t_k, p_k) G_\mathrm{d}(t_k, p_k | 0, p_0) 2\delta - G_\mathrm{d}(t_{k+1}, p_{k+1} | 0, p_0) \right|$$
$$+ \sum_{p_k \in \overline{R_k}(u)} G_\mathrm{d}(t_{k+1}, p_{k+1} \,|\, t_k, p_k) \, G_\mathrm{d}(t_k, p_k \,|\, 0, p_0) \, 2\delta$$
$$\leq 2\{ \kappa_2 + o(\kappa_2) \} \, G_\mathrm{d}(t_{k+1}, p_{k+1} \,|\, 0, p_0)$$
$$\leq 3\kappa_2 \, G_\mathrm{d}(t_{k+1}, p_{k+1} \,|\, 0, p_0).$$

In the third line, we extended the range of summation from $\overline{R_k}(u) \cap E$ to $\overline{R_k}(u)$ using the fact that G_d is always nonnegative. Hereafter, we do not repeat this remark. We have thus completed the proof of the lemma.

Now, we have the following theorem.

Theorem 1: (i) *The $*$-path sum (19) satisfies that*

$$\mathcal{G}_\alpha(T,p\,|\,0,p_0) = \frac{1+o(1)}{\left\{2\pi D\beta(1-e^{-2\beta T})\right\}^{1/2}} \, \exp\left[-\frac{(p-p_0e^{-\beta T})^2}{2D\beta(1-e^{-2\beta T})}\right] \quad (29)$$

if p_0 is finite and $|p-p_0| = o(\tau^{-3})$, and

$$\mathcal{G}_\alpha(T,p\,|\,0,p_0) \le \frac{1+o(1)}{\left\{2\pi D\beta(1-e^{-2\beta T})\right\}^{1/2}} \, \exp\left[-\frac{(p-p_0e^{-\beta T})^2}{2D\beta(1-e^{-2\beta T})}\right] \quad (30)$$

for other (p_0,p).

(ii) *If the initial data $U(0,p) = f(p)$ is an integrable continuous function, then $U(t,p)$ defined by*

$$U(t,p) = \mathrm{st}\left(\sum_{p_0\in M} \sum_{X\in\mathscr{P}^\alpha_{(t,p|0,p_0)}} {}^*\!f(p_0)\,\mu(X) \right) \quad (31)$$

is the solution to the Fokker–Planck equation (2) with initial data $U(0,p) = f(p)$.

Remark: If we do not assume the continuity of the initial data $f(p)$, we have to replace ${}^*\!f$ in (31) by an S-integrable lifting of f to keep the conclusion (ii) (see Refs.[2,3,4] for the definition of S-integrable lifting).

Proof of Theorem 1: If (i) is proved, then (ii) is clear because the right-hand side of (29), apart from $o(1)$, is the Green function for the initial value problem of (2).

Let us prove (i). We calculate the $*$-sum

$$\mathcal{G}_\alpha(t_{k+1},p_{k+1}|0,p_0) = \sum_{p_k\in J_k} \mathcal{G}_\alpha(t_{k+1},p_{k+1}|t_k,p_k)\mathcal{G}_\alpha(t_k,p_k|0,p_0)2\delta \quad (32)$$

by replacing the right-hand side approximately by

$$\sum_{p_k\in J_k} G_{\mathrm{d}}(t_{k+1},p_{k+1}\,|\,t_k,p_k)\,G_{\mathrm{d}}(t_k,p_k\,|\,0,p_0)\,2\delta \quad (33)$$

where J_k was defined in (15), showing inductively with respect to $k = 1, 2, \ldots$ that

$$\mathcal{G}_\alpha(t_{k+1},p_{k+1}\,|\,0,p_0) = G_{\mathrm{d}}(t_{k+1},p_{k+1}\,|\,0,p_0) + \Delta^{\mathcal{G}G}_{k+1} \quad (34)$$

within a controllable error $\Delta^{\mathcal{G}G}_{k+1}$ which we shall find later.

Now, we know from Lemma 3 that (33) gives

$$\sum_{p_k \in J_k} G_{\mathrm{d}}(t_{k+1}, p_{k+1} \,|\, t_k, p_k)\, G_{\mathrm{d}}(t_k, p_k \,|\, 0, p_0)\, 2\delta$$

$$= G_{\mathrm{d}}(t_{k+1}, p_{k+1} \,|\, 0, p_0) + \Delta_{k+1}^{GG} \tag{35}$$

with

$$\left| \Delta_{k+1}^{GG} \right| \leq 3\kappa_2\, G_{\mathrm{d}}(t_{k+1}, p_{k+1} \,|\, 0, p_0) \tag{36}$$

if the range of summation J_k is large enough:

$$R_k(u) \subset J_k, \tag{37}$$

where $R_k(u)$ was defined in (25) and J_k in (15). Solving (37), or

$$\big(r(p_{k+1}) - u,\ r(p_{k+1}) + u\big) \subset \big(p_{k+1} - \alpha,\ p_{k+1} + \alpha\big), \tag{38}$$

for p_{k+1}, we find a set I_{k+1} of p_{k+1} for which (38) is valid:

$$I_{k+1} = \big\{ p_{k+1} \,|\, m_{k+1} < p_{k+1} < M_{k+1} \big\}, \quad k = 1, 2, \ldots, N-1, \tag{39}$$

where

$$m_{k+1} = \frac{a^k p_0 - b_{k+1}\alpha'}{b_{k+1} - ab_k}, \quad M_{k+1} = \frac{a^k p_0 + b_{k+1}\alpha'}{b_{k+1} - ab_k} \quad (\alpha' = \alpha - u).$$

We can now check that

$$R_k(u) \subset I_k, \tag{40}$$

which will assure that $\mathcal{G}_\alpha(t_k, p_k \,|\, 0, p_0)$ from the preceding step of the induction (32) is well approximated by $G_{\mathrm{d}}(t_k, p_k \,|\, 0, p_0)$ in (33) with an error $\Delta_k^{\mathcal{G}G}$ as shown below. To check (40), we have only to see by using (22) that

$$M_k - \big(r(p_{k+1}) + u\big) > M_k - \frac{1}{b_{k+1}}(ab_k M_{k+1} + a^k p_0) - u$$

$$= \frac{(1 - a^2)(1 - a^{2k})}{(1 + a^{2k-1})(1 + a^{2k+1})}\,(b_k \alpha' + a^{k-1} p_0) - u$$

$$> \frac{(\beta\tau)^2}{4}\,(\alpha' - |p_0|) - u > 0,$$

where we have used $1 - a^l \geq 1 - a = \beta\tau$ for the first inequality in the last line, and (C1) and (23) for the second, and similarly that the corresponding inequality for m_k holds. For $k = 1$, we define

$$I_1 = \big\{ p_1 \,|\, p_0 - \alpha < p_1 < p_0 + \alpha \big\}.$$

Let us now proceed to estimate the errors $\Delta_{k+1}^{\mathcal{G}G}$ accumulated in the process of the induction (32) from $k = 1$ through $N - 1$. For this purpose, we introduce a $*$-number A_l such that

$$\left| \Delta_l^{\mathcal{G}G} \right| \leq A_l \, G_{\mathrm{d}}(t_l, p_l \,|\, 0, p_0), \tag{41}$$

for all $p_l \in I_l$. For $p_l \notin I_l$, we anticipate that

$$0 \leq \mathcal{G}_\alpha(t_l, p_l \,|\, 0, p_0) \leq (1 + A_l) \, G_{\mathrm{d}}(t_l, p_l \,|\, 0, p_0). \tag{42}$$

For $k = 1$ to start with, we can take $A_1 = \kappa_1$ by Lemma 1 in order for (41) and (42) to hold.

Assume (41) and (42) for $l = k$. If $p_{k+1} \in I_{k+1}$, then

$$\begin{aligned}
\mathcal{G}_\alpha(t_{k+1}, &p_{k+1} \,|\, 0, p_0) \\
&= \sum_{p_k \in I_k \cap J_k} \mathcal{G}_\alpha(t_{k+1}, p_{k+1} \,|\, t_k, p_k) \, \mathcal{G}_\alpha(t_k, p_k \,|\, 0, p_0) \, 2\delta \\
&\quad + \sum_{p_k \in \overline{I_k} \cap J_k} \mathcal{G}_\alpha(t_{k+1}, p_{k+1} \,|\, t_k, p_k) \, \mathcal{G}_\alpha(t_k, p_k \,|\, 0, p_0) \, 2\delta.
\end{aligned} \tag{43}$$

By (13) and (41), the first sum S_1 in (43) can be written as

$$S_1 = \sum_{p_k \in I_k \cap J_k} (1 + f) \, G_{\mathrm{d}}(t_{k+1}, p_{k+1} \,|\, t_k, p_k) \, (1 + g) \, G_{\mathrm{d}}(t_k, p_k \,|\, 0, p_0) \, 2\delta$$

with $f = f(p_k, p_{k+1})$ and $g = g(p_k)$ satisfying

$$|f(p_k, p_{k+1})| \leq \kappa_1, \quad |g(p_k)| \leq A_k.$$

Since $R_k(u) \subset I_k \cap J_k$ by (37) and (40),

$$\left| \sum_{p_k \in I_k \cap J_k} G_{\mathrm{d}}(t_{k+1}, p_{k+1} | t_k, p_k) G_{\mathrm{d}}(t_k, p_k | 0, p_0) \, 2\delta - G_{\mathrm{d}}(t_{k+1}, p_{k+1} | 0, p_0) \right|$$

$$\leq 3\kappa_2 \, G_{\mathrm{d}}(t_{k+1}, p_{k+1} \,|\, 0, p_0)$$

holds by (26). Hence,

$$\left| S_1 - G_{\mathrm{d}}(t_{k+1}, p_{k+1} \,|\, 0, p_0) \right|$$

$$= \left| \left\{ \sum_{p_k \in I_k \cap J_k} G_{\mathrm{d}}(t_{k+1}, p_{k+1} | t_k, p_k) G_{\mathrm{d}}(t_k, p_k | 0, p_0) 2\delta - G_{\mathrm{d}}(t_{k+1}, p_{k+1} | 0, p_0) \right\} \right.$$

$$\left. + \sum_{p_k \in I_k \cap J_k} \left\{ f + (1+f)g \right\} G_{\mathrm{d}}(t_{k+1}, p_{k+1} \,|\, t_k, p_k) \, G_{\mathrm{d}}(t_k, p_k \,|\, 0, p_0) \, 2\delta \right|$$

$$\leq 3\kappa_2 \, G_{\mathrm{d}}(t_{k+1}, p_{k+1} \,|\, 0, p_0)$$

$$+ \left\{ \kappa_1 + (1 + \kappa_1) A_k \right\} (1 + 3\kappa_2) \, G_{\mathrm{d}}(t_{k+1}, p_{k+1} \,|\, 0, p_0)$$

$$\leq \left\{ 4\kappa_2 + (1 + 4\kappa_2) A_k \right\} G_{\mathrm{d}}(t_{k+1}, p_{k+1} \,|\, 0, p_0). \tag{44}$$

For the last inequality, we have used $\kappa_1 = o(\kappa_2)$. The second sum S_2 in (43) is estimated by (13), (28) and (42) as

$$S_2 \leq (1 + \kappa_1)(1 + A_k) \sum_{p_k \in \overline{I_k} \cap J_k} G_{\mathrm{d}}(t_{k+1}, p_{k+1} \,|\, t_k, p_k) \, G_{\mathrm{d}}(t_k, p_k \,|\, 0, p_0) \, 2\delta$$

$$\leq (1 + \kappa_1)(1 + A_k) \sum_{p_k \in R_k(u)} G_{\mathrm{d}}(t_{k+1}, p_{k+1} \,|\, t_k, p_k) \, G_{\mathrm{d}}(t_k, p_k \,|\, 0, p_0) \, 2\delta$$

$$\leq (1 + \kappa_1)(1 + A_k) \left\{ o(\kappa_2) + \kappa_2 \right\} G_{\mathrm{d}}(t_{k+1}, p_{k+1} \,|\, 0, p_0)$$

$$\leq 2\kappa_2 (1 + A_k) \, G_{\mathrm{d}}(t_{k+1}, p_{k+1} \,|\, 0, p_0). \tag{45}$$

By (44) and (45),

$$\left| \Delta_{k+1}^{\mathcal{G}G} \right| \leq \left\{ 6\kappa_2 + (1 + 6\kappa_2) A_k \right\} G_{\mathrm{d}}(t_{k+1}, p_{k+1} \,|\, 0, p_0). \tag{46}$$

For $p_{k+1} \notin I_{k+1}$, we have

$$\mathcal{G}_\alpha(t_{k+1}, p_{k+1} \,|\, 0, p_0)$$

$$\leq \sum_{p_k \in J_k} (1 + \kappa_1) \, G_{\mathrm{d}}(t_{k+1}, p_{k+1} \,|\, t_k, p_k) \, (1 + A_k) \, G_{\mathrm{d}}(t_k, p_k \,|\, 0, p_0) \, 2\delta$$

$$\leq (1 + \kappa_1)(1 + A_k)(1 + 3\kappa_2) \, G_{\mathrm{d}}(t_{k+1}, p_{k+1} \,|\, 0, p_0)$$

$$\leq \left\{ (1 + 4\kappa_2) + (1 + 4\kappa_2) A_k \right\} G_{\mathrm{d}}(t_{k+1}, p_{k+1} \,|\, 0, p_0). \tag{47}$$

By (46) and (47), we see that (41) and (42) also hold for $l = k + 1$ if we take

$$A_{k+1} = 6\kappa_2 + (1 + 6\kappa_2) A_k. \tag{48}$$

With initial condition $A_1 = \kappa_1$, therefore

$$A_l = (1 + 6\kappa_2)^{l-1}(\kappa_1 + 1) - 1. \tag{49}$$

We have thus obtained the results (41) and (42) with (49) for all l. Especially for $l = N$ where $t_N = T$, we have

$$A_N = (1 + 6\kappa_2)^{N-1}(\kappa_1 + 1) - 1 = o(1)$$

because $\kappa_2 = o(1/N)$ and $\kappa_1 = o(1)$. Thus,

$$\mathcal{G}_\alpha(T, p \,|\, 0, p_0) = G_\mathrm{d}(T, p \,|\, 0, p_0) \left\{ 1 + o(1) \right\}$$

for all $p \in I_N$, and

$$\mathcal{G}_\alpha(T, p \,|\, 0, p_0) \le G_\mathrm{d}(T, p \,|\, 0, p_0) \left\{ 1 + o(1) \right\}$$

for other p. Putting

$$a^N = e^{-\beta T} + o(1) \quad \text{and} \quad 2\beta\tau b_N = 1 - e^{-2\beta T} + o(1)$$

into $G_\mathrm{d}(T, p \,|\, 0, p_0)$, we obtain (29) and (30). Since the orders of the magnitude of M_N and m_N are

$$\frac{a^{N-1}p_0 \pm b_N\alpha'}{b_N - ab_{N-1}} = \frac{1+a}{1+a^{2N-1}}\left(a^{N-1}p_0 \pm b_N\alpha' \right) = \mathcal{O}(\tau^{-3}),$$

the condition $|p - p_0| = o(\tau^{-3})$ of the theorem implies $p \in I_N$, thereby completing the proof of the theorem.

We have constructed a $*$-path-space and a $*$-measure μ so far. We can extract a standard measure from μ by the Loeb measure theory as follows. Note that the theory is applicable to any bounded $*$-measure, which is the case for our probability $*$-measure μ. Let $\underline{p}$ denote $n\delta \in \mathsf{M}$ such that $n\delta \le p < (n+1)\delta$, and $\mathcal{A}$ the set of internal subsets of $\mathscr{P}^\alpha_{(T,\underline{p}|0,\underline{p_0})}$, and define μ' by

$$\mu'(E) = \frac{1}{2\delta} \sum_{X \in E} \mu(X) \quad (E \in \mathcal{A}).$$

Then, $\left\langle \mathscr{P}^\alpha_{(T,\underline{p}|0,\underline{p_0})}, \mathcal{A}, \mu' \right\rangle$ is an internal $*$-finitely additive $*$-measure space, so that we can apply the Loeb measure theory to get the following theorem.

Theorem 2: (i) *A standard completely additive measure space*

$$\left\langle \mathsf{P}_{(T,p|0,p_0)}, \mathsf{B}, m_L \right\rangle$$

is extracted from the nonstandard $\left\langle \mathscr{P}^\alpha_{(T,\underline{p}|0,\underline{p_0})}, \mathcal{A}, \mu' \right\rangle$ *such that*

$$\int_{\mathsf{P}_{(T,p|0,p_0)}} 1 \, dm_L = \mathrm{st}\left(\frac{1}{2\delta} \sum_{X \in \mathscr{P}^\alpha_{(T,\underline{p}|0,\underline{p_0})}} \mu(X) \right) \tag{50}$$

for any standard numbers p_0 and p.

(ii) *If the initial data $U(0,p) = f(p)$ is an integrable function, then $U(t,p)$ defined by*

$$U(t,p) = \int_{-\infty}^{\infty} dp_0 \int_{P_{(T,p|0,p_0)}} f(X(0)) dm_L(X)$$

is the solution to the Fokker–Planck equation (2) with initial data $U(0,p) = f(p)$.

Proof: (i) is a direct result from the Loeb measure theory. (ii) is also clear because the right-hand side of (50) is the Green function of (2) by Theorem 1 (i).

3. Path-space measure for Fokker–Planck equation with unbounded coefficient

Let us apply the technique of double strata of time with α-cutoff to a wider class of forward Fokker–Planck equation,

$$\frac{\partial}{\partial t} U(t,p) = -\frac{\partial}{\partial p} \left\{ a(p) U(t,p) \right\} + D\beta^2 \frac{\partial^2}{\partial p^2} U(t,p). \tag{51}$$

Here, $U(t,p)$ is the probability density of the momentum p of a particle whose motion $p = X(t,\omega)$ in one-dimensional momentum space satisfies the stochastic differential equation

$$dX(t,\omega) = a\big(X(t,\omega)\big) dt + dW(t,\omega). \tag{52}$$

$W(t,\omega)$ is the Brownian motion with diffusion constant $D\beta^2$.

We shall present only the overviews of the proofs, since the details have not been completed yet. The coefficient function $a(p)$ is usually assumed to have linear growth at most,

$$\big|a(p)\big| \leq M\big(1 + |p|\big),$$

in order for no path to explode to infinity in finite time. In this section, however, $a(p)$ may grow faster,

$$\big|a(p)\big| \leq M\big(1 + |p|\big)^{1+c} \quad (c > 0). \tag{53}$$

We also assume that

$$\left| \frac{d\, a(p)}{dp} \right| \leq M\big(1 + |p|\big)^c. \tag{54}$$

The upper bound of the magnitude of c shall be determined later. This order of growth causes some paths to explode in finite time (see H. P. Mckean[13]), so that the solution $U(t,p)$ to (51) may be obtained as the sum of contributions from those paths which have not exploded by time t.

3.1. *-Measure and *-path sum

By the replacements (3), (4) and (5), (51) becomes

$$U(t+\Delta t,p) = \frac{1}{2}\left\{1 - a(p+\Delta p)\frac{(\Delta t)^{1/2}}{\beta(2D)^{1/2}}\right\}U(t,p+\Delta p)$$

$$+\frac{1}{2}\left\{1 + a(p-\Delta p)\frac{(\Delta t)^{1/2}}{\beta(2D)^{1/2}}\right\}U(t,p-\Delta p). \qquad (55)$$

Here, we have chosen

$$\Delta p = (2D\beta^2\Delta t)^{1/2}.$$

The difference equation (55) shows that we have only to alter the definitions of the *-measure in the previous section by replacing (10) by

$$\mu_k(X_{k,\omega}) = \prod_{j=0}^{\nu-1}\frac{1}{2}\left\{1 + \omega(j)a(p_k)\frac{\varepsilon^{1/2}}{\beta(2D)^{1/2}}\right\}. \qquad (56)$$

With this alternation, we define a *-solution $\mathcal{U}(t,p)$ by

$$\mathcal{U}(t,p) = \sum_{p_0\in\mathsf{M}}\ \sum_{X\in\mathscr{P}^{\alpha}_{(t,p|0,p_0)}}{}^{*}\!f(p_0)\,\mu(X), \qquad (57)$$

where $f(p) = U(0,p)$ is the initial data for the differential equation.

3.2. *Outline of the proofs*

In this subsection, we use the symbol "$\simeq$" in a slightly different meaning from the definition in Sec. 1, that is, $A \simeq B$ means A is equal to B up to negligible difference. First, we prove that $\mathcal{U}(t,p)$ is near-standard so that we can define the standard function $U(t,p)$ as its standard part;

$$U(t,p) = \mathrm{st}\Big(\mathcal{U}(t,p)\Big).$$

Since the variance of the Brownian motion for time interval t is proportional to t, the cutoff-parameter α introduced in (9) should be $\tau^{1/2}$ times some infinite number. In order to control the growth of the unbounded function

$a(p)$, let us choose the infinite number smaller than any negative power of τ;

$$\alpha = \mathcal{O}\big(\tau^{1/2} \, |\log \tau|\big). \tag{58}$$

Then,

$$|p| = \mathcal{O}\big(\tau^{-1/2} |\log \tau|\big) \tag{59}$$

by (9), and hence

$$a(p) = \mathcal{O}\big(\tau^{-(1+c)/2} |\log \tau|^{1+c}\big) \tag{60}$$

and

$$\frac{d\,a(p)}{dp} = \mathcal{O}\big(\tau^{-c/2} |\log \tau|^{c}\big). \tag{61}$$

Making use of the advantage of the double strata of time that the spacing ε is much smaller than the other spacing $\tau = \nu\varepsilon$, as we have experienced in the previous section, we can use the Stirling formula as

$$\log\left[\frac{1}{2^\nu}\binom{\nu}{r}\right] \simeq \log\frac{2}{(2\pi\nu)^{1/2}} - \frac{(p_{k+1} - p_k)^2}{2\nu\delta^2},$$

and

$$\log\left[\left\{1 + a(p_k)\frac{\varepsilon^{1/2}}{\beta(2D)^{1/2}}\right\}^r \left\{1 - a(p_k)\frac{\varepsilon^{1/2}}{\beta(2D)^{1/2}}\right\}^{\nu-r}\right]$$
$$\simeq -\frac{(p_{k+1} - p_k)a(p_k)}{\beta(2D)^{1/2}\delta}\varepsilon^{1/2} - \frac{a(p_k)^2}{4D\beta^2}\tau.$$

Here, r is the number of j's such that $\omega(j) = 1$ which we have used in (16). Then, the $*$-Green function defined by the $*$-path sum is

$$\mathcal{G}_\alpha(t_{k+1}, p_{k+1} \,|\, t_k, p_k)$$
$$= \frac{1}{2\delta}\sum_{X_{k,\omega}} \mu_k(X_{k,\omega})$$
$$= \frac{1}{2\delta}\frac{1}{2^\nu}\binom{\nu}{r}\left\{1 + a(p_k)\frac{\varepsilon^{1/2}}{\beta(2D)^{1/2}}\right\}^r \left\{1 - a(p_k)\frac{\varepsilon^{1/2}}{\beta(2D)^{1/2}}\right\}^{\nu-r}$$
$$\simeq \frac{1}{2\beta(\pi D\tau)^{1/2}}\exp\left[-\frac{\big(p_{k+1} - p_k - \tau a(p_k)\big)^2}{4D\beta^2\tau}\right], \tag{62}$$

where the sum is taken over the set $\mathscr{P}^{\alpha}_{(t_{k+1}, p_{k+1} | t_k, p_k)}$.

By (62),

$$\mathcal{U}(t_{k+1}, p_{k+1})$$
$$\simeq \sum_{q \in \Lambda} \frac{1}{2\beta(\pi D\tau)^{1/2}} \exp\left[-\frac{\left(p_{k+1} - q - \tau a(q)\right)^2}{4D\beta^2\tau}\right] \mathcal{U}(t_i, q)\, 2\delta, \qquad (63)$$

where $\Lambda \subset \mathsf{M}$ is the set of q such that $\mathscr{P}^{\alpha}_{(t_{k+1}, p_{k+1} \mid t_k, q)}$ is not empty. Let $\widetilde{\Lambda}$ be the $*$-interval $\left[\min \Lambda,\, \max \Lambda\right]$, and define

$$A = \int_{\widetilde{\Lambda}} \frac{1}{2\beta(\pi D\tau)^{1/2}} \exp\left[-\frac{(p_{k+1} - q)^2}{4D\beta^2\tau}\right]\left|\frac{d\,a(q)}{dq}\right| dq,$$

which is a finite number. Assume that

$$\left|\mathcal{U}(t_k, p)\right| \le (1 + \tau A)^k. \qquad (64)$$

Then by (63),

$$\left|\mathcal{U}(t_{k+1}, p_{k+1})\right|$$

$$\le (1 + \tau A)^k \sum_{q \in \Lambda} \frac{1}{2\beta(\pi D\tau)^{1/2}} \exp\left[-\frac{\left(p_{k+1} - q - \tau a(q)\right)^2}{4D\beta^2\tau}\right] 2\delta$$

$$\simeq (1 + \tau A)^k \int_{\widetilde{\Lambda}} \frac{1}{2\beta(\pi D\tau)^{1/2}} \exp\left[-\frac{\left(p_{k+1} - q - \tau a(q)\right)^2}{4D\beta^2\tau}\right] dq$$

$$= (1 + \tau A)^k \int_{\widetilde{\Lambda}} \frac{1}{2\beta(\pi D\tau)^{1/2}} \exp\left[-\frac{\left(p_{k+1} - q'\right)^2}{4D\beta^2\tau}\right] \frac{1}{1 + \tau \frac{da(q)}{dq}}\, dq'$$

$$\simeq (1 + \tau A)^k \int_{\widetilde{\Lambda}} \frac{1}{2\beta(\pi D\tau)^{1/2}} \exp\left[-\frac{\left(p_{k+1} - q'\right)^2}{4D\beta^2\tau}\right]\left(1 - \tau \frac{da(q)}{dq}\right) dq'$$

$$\le (1 + \tau A)^k \int_{\widetilde{\Lambda}} \frac{1}{2\beta(\pi D\tau)^{1/2}} \exp\left[-\frac{\left(p_{k+1} - q'\right)^2}{4D\beta^2\tau}\right]\left(1 + \tau \left|\frac{da(q)}{dq}\right|\right) dq'$$

$$= (1 + \tau A)^{k+1},$$

and hence (64) holds for all k. Thus, we have

$$\left|\mathcal{U}(t, p)\right| \le e^{tA}, \qquad (65)$$

ascertaining $\mathcal{U}(t, p)$ to be near-standard. We note that the calculation above is valid as long as

$$\tau \left|\frac{d\,a(q)}{dq}\right| \ll 1, \qquad (66)$$

which we shall discuss later.

Let p be finite. While we have not completed the details yet, we hopefully prove that

$$U(t,p) = \mathrm{st}\Big(\mathcal{U}(t,p)\Big)$$

is differentiable and

$$\frac{\partial U(t,p)}{\partial t} \simeq \frac{1}{\tau}\Big\{\mathcal{U}(t+\tau,p) - \mathcal{U}(t,p)\Big\},$$

$$\frac{\partial U(t,p)}{\partial p} \simeq \frac{1}{2\delta}\Big\{\mathcal{U}(t,p+\delta) - \mathcal{U}(t,p-\delta)\Big\},$$

$$\frac{\partial^2 U(t,p)}{\partial p^2} \simeq \frac{1}{\delta^2}\Big\{\mathcal{U}(t,p+\delta) + \mathcal{U}(t,p-\delta) - 2\mathcal{U}(t,p)\Big\}$$

in a way similar to the one in the author's preceding paper.[10] Denote the right-hand sides by $D_t\mathcal{U}(t,p)$, $D_p\mathcal{U}(t,p)$ and $D_p^2\mathcal{U}(t,p)$, respectively. Since $\mathcal{U}(t,p)$ is defined as a $*$-path sum, its increment for interval τ is calculated as

$$\mathcal{U}(t+\tau,p) - \mathcal{U}(t,p)$$

$$\simeq \sum_{q\in\Lambda} \mathcal{U}(t,p+q)\, \frac{1}{2\beta(\pi D\tau)^{1/2}}\, \exp\left[-\frac{\big(-q-\tau a(p+q)\big)^2}{4D\beta^2\tau}\right] 2\delta$$

$$-\sum_{q\in\Lambda} \mathcal{U}(t,p)\, \frac{1}{2\beta(\pi D\tau)^{1/2}}\, \exp\left[-\frac{q^2}{4D\beta^2\tau}\right] 2\delta$$

$$\simeq I_1 + I_2,$$

where

$$I_1 = \frac{1}{2\beta(\pi D\tau)^{1/2}} \sum_{q\in\Lambda}\Big\{\mathcal{U}(t,p+q) - \mathcal{U}(t,p)\Big\}\, \exp\left[-\frac{q^2}{4D\beta^2\tau}\right] 2\delta$$

and

$$I_2 = \frac{1}{2\beta(\pi D\tau)^{1/2}} \sum_{q}\left\{\exp\left[-\frac{\big(-q-\tau a(p+q)\big)^2}{4D\beta^2\tau}\right] - \exp\left[-\frac{q^2}{4D\beta^2\tau}\right]\right\}$$

$$\times\, \mathcal{U}(t,p+q)\, 2\delta.$$

Because I_1 has $\exp\big[-q^2/(4D\beta^2\tau)\big]$, we have only to consider q such that $q = \mathcal{O}(\tau^{1/2})$, and hence we can expand $\mathcal{U}(t,p+q)$ with respect to q as

$$\mathcal{U}(t,p+q) - \mathcal{U}(t,p) \simeq q\big(D_p\mathcal{U}(t,p)\big) + \frac{q^2}{2!}\big(D_p^2\mathcal{U}(t,p)\big). \qquad (67)$$

Then,

$$I_1 \simeq \frac{1}{4\beta(\pi D\tau)^{1/2}} D_p^2 \mathcal{U}(t,p) \sum_q q^2 \exp\left[-\frac{q^2}{4D\beta^2\tau}\right] 2\delta$$

$$\simeq \tau D\beta^2 \, D_p^2 \mathcal{U}(t,p). \tag{68}$$

For I_2, let us assume that

$$\tau a(p+q) \ll q. \tag{69}$$

Then, the first exponent in I_2 can be approximated as

$$-\frac{1}{4D\beta^2\tau}\left(-q-\tau a(p+q)\right)^2 \simeq -\frac{1}{4D\beta^2\tau}\left[q^2 + 2\tau q\{a(p) + q D_p a(p)\}\right],$$

so that

$$\exp\left[-\frac{\{-q-\tau a(p+q)\}^2}{4D\beta^2\tau}\right] - \exp\left[-\frac{q^2}{4D\beta^2\tau}\right]$$

$$\simeq \exp\left[-\frac{q^2}{4D\beta^2\tau}\right]\left\{\exp\left[-\frac{q}{2D\beta^2}\{a(p) + q D_p a(p)\}\right] - 1\right\}$$

$$\simeq \exp\left[-\frac{q^2}{4D\beta^2\tau}\right]\left(-\frac{1}{2D\beta^2}\right)\left(qa(p) + q^2 D_p a(p)\right). \tag{70}$$

We expand $\mathcal{U}(t,p+q)$ as

$$\mathcal{U}(t,p+q) \simeq \mathcal{U}(t,p) + q D_p \mathcal{U}(t,p). \tag{71}$$

Putting (70) and (71) into I_2, we obtain

$$I_2 \simeq \frac{1}{2\beta(\pi D\tau)^{1/2}}\left\{a(p) D_p \mathcal{U}(t,p) + \mathcal{U}(t,p) D_p a(p)\right\}$$

$$\times \sum_q q^2 \exp\left[-\frac{q^2}{4D\beta^2\tau}\right] 2\delta$$

$$\simeq -\tau D_p\left(a(p)\mathcal{U}(t,p)\right). \tag{72}$$

By (68) and (72),

$$D_t \mathcal{U}(t,p) \simeq D\beta^2 D_p^2 \mathcal{U}(t,p) - D_p\left(a(p)\mathcal{U}(t,p)\right).$$

Taking the standard part, we obtain

$$\frac{\partial U(t,p)}{\partial t} = D\beta^2 \frac{\partial^2}{\partial p^2} U(t,p) - \frac{\partial}{\partial p}\left(a(p)U(t,p)\right),$$

showing that

$$U(t,p) = \mathrm{st}\left(\mathcal{U}(t,p)\right)$$

is the solution to the Fokker–Planck equation (51).

Recall that we have assumed

$$\tau \frac{d\,a(q)}{dq} \ll 1, \quad \text{and} \quad \tau a(p+q) \ll q \tag{73}$$

in (66) and in (69). Let us determine the magnitude of c in (53) and (54) in order for the assumption (73) to hold. It is clear by (61) that the first inequality is satisfied if

$$c < 2.$$

Let $q = \mathcal{O}(\lambda)$ with infinite λ. Then, the second inequality says that

$$\tau \lambda^{1+c} \ll \lambda \quad i.e. \quad \lambda^c \ll \tau^{-1}. \tag{74}$$

But we know that λ is atmost $\tau^{-1/2}|\log \tau|$ by (59). Hence, (74) means

$$\tau^{-c/2}|\log \tau|^c \ll \tau^{-1}$$

which is also satisfied if $c < 2$.

Thus, we have determined the conditions on $a(p)$ as

(i) $\left|a(p)\right| \le M\left(1 + |p|\right)^{1+c}$

(ii) $\left|\dfrac{d\,a(p)}{dp}\right| \le M\left(1 + |p|\right)^{c}$

for some c such that $0 < c < 2$. Let us remark that the condition $c < 2$ is imposed for the sake of the proof given above to hold, and it may be enlarged if some impovements are made in the construction of the path-space or the measure over it.

Dedication

It is my great pleasure to express on the occasion of his seventieth birthday my deep gratitude to Professor Hiroshi Ezawa, who has kindly led me to understand many things in both mathematics and physics for a greatly many years. I wish him many more years of good health and success in physics, mathematics and many others.

References

1. I.V. Girsanov, "On transforming a certain class of stochastic processes by absolutely continuous substitution of measures," Theor. Prob. Appl. **5**, 285–301 (1960).

2. A.E. Hurd and P. A. Loeb, *An Introduction to Nonstandard Real Analysis*, Pure and Appl. Math. **118**, Academic Press (1985).

3. K.D. Stroyan and J. M. Bayod, *"Foundations of infinitesimal stochastic analysis,"* Studies in Logic and the Foundations of Mathematics, North-Holland, Amsterdam (1986).

4. T. Nakamura, *"Nonstandard analysis and physics,"* in Japanese, Nihon Hyoronsha (1998).

5. H.J. Keisler, "An infinitesimal approach to stochastic analysis," Mem. Amer. Math. Soc. **48** No.297, 1–184 (1984).

6. S.A. Albeverio, J. E. Fenstad, R. J. Høegh-Krohn and T. Lindstrøm, *Nonstandard methods in stochastic analysis and mathematical physics*, Academic Press (1986).

7. S. Chandrasekhar, "Stochastic Problems in Physics and Astronomy," Rev. Mod. Phys. **15**, 1–89 (1943). pp.20–21.

8. The author recently knew that the idea of "double strata of time" had already been introduced by N. J. Cutland in 1982, *Z. Wahrscheinlichkeitstheorie verw. Gebiete, 60* 335–357, which is different from ours in the order of infinitesimals because his method used for the construction of the solution was different from ours.

9. P.A. Loeb, "Conversion from nonstandard to standard measure spaces and applications in probability theory," Trans. Amer. Math. Soc. **211**, 113–122 (1975).

10. T. Nakamura, "A nonstandard representation of Feynman's path integrals," J. Math. Phys. **32**, 457–463 (1991).

11. T. Nakamura, "Path space measures for Dirac and Schrödinger equations: Nonstandard analytical approach," J. Math. Phys. **38**(8), 4052–4072 (1997).

12. T. Nakamura, "Path space measure for the 3+1-dimensional Dirac equation in momentum space," J. Math. Phys. **41**(8), 5209–5222 (2000).

13. H.P. Mckean, *"Stochastic Integrals,"* Academic Press (1969).

Olbers' Paradox, Wireless Telephones, and Poisson Random Sets
Is the Universe Finite?

Susan Heath and Larry Shepp

Statistics Department, Rutgers U, New Brunswick

Olbers' paradox is that if the universe is either infinite in age or extent, then the night sky should be bright. A related problem exists in terms of gravity, that an infinite universe full of stars should collapse on itself. Olbers' paradox has been used to support the "big bang" hypothesis. We find there is a simpler resolution of Olbers' paradox, that perhaps ought to be considered. We show that a standard theorem on convergence of infinite series of zero-mean independent random variables, due to Kolmogorov, can be viewed as saying that the sum of all the above contributions, *allowing destructive interference*, produces a finite sum even if the universe is infinite in age and extent. Thus our explanation of Olbers' paradox, is that if one thinks in terms of waves rather than of particles, the paradox disappears.

A new model for interference noise in wireless telephony has been given in the thesis of one of us[7] which is closely related to Chandrasekhar's model for stellar gravity. If we assume that the telephones which are sending signals to the base station at the origin are at the points of a planar Poisson random set, and that the nth most distant telephone from the origin is sending a signal X_n, then the total signal at any time received at the base station is

$$T = \sum_{n=1}^{\infty} \frac{X_n}{R_n^{\gamma}}$$

where R_n is the distance, $0 < R_1 < R_2 < \dots$. The exponent, γ, is taken as 3.9 or 4 in engineering practice.

In cosmology the situation is similar but the Poisson points are in three dimensions rather than two and represent the stars or the galaxies, and T now represents either the total gravitational force on the earth or the total radiation at a frequency. In cosmology, γ is usually taken as 2 corresponding to an inverse square loss.

We will show first that under very general conditions the series for T is perfectly well convergent mathematically in both cases even if the sum extends over infinitely many summands. This seems to say that Olbers' paradox has an alternative explanation, namely that even if there are infinitely many sources of gravity and radiation the total contribution on earth has a specific finite value due to cancellation of the terms (interference). The sum of the squares of the terms above also converges although the series does not necessarily converge *absolutely* unless γ is large enough (one would need $\gamma > 3$) in 3 dimensions.

Remarkably, (a special case is due to Chandrasekhar) the distribution of T is always stable and there is only a two-parameter family of stable distributions, $S(\alpha, \beta)$, that T can possibly have. Not all stable distributions appear since $0 < \alpha < 2$ so, for example, the normal distribution is not among them.

1. Introduction

Stable distributions arise in several areas of physics and astronomy. In 1919, Holtsmark, when studying the spectral lines of stars, derived the probability distribution of the electrostatic force exerted on an atom by surrounding atoms.[8] The Holtsmark distribution is a symmetric stable law with index of stability, $\alpha = 1/2$. The Holtsmark distribution was used by Chandrasekhar to study the gravitational force acting on a star from the neighboring stars.[2] In his model, Chandrasekhar assumes the stars are distributed as a three dimensional Poisson point set, which is the same assumption we will use when considering the gravity or radiation of the stars.

Zolotarev[14] gave a more general model for these types of problems. He considered a collection of particles distributed as a Poisson set dispersed in some region, U, of n-dimensional Euclidean space. Each particle creates a "field of action"; the strength of this "field" at any point in space is given by an influence function. The examples considered in this paper fit his general model. In the case of cellular telephones the field is the radio waves transmitted by the phones, and the influence function is the power of the radio signal or the radio waves. When considering the stars, the field is the radiation emitted by the stars or the gravitational force of the stars. We will show that in both examples the sum converges to a stable distribution.

A stable distribution is characterized by four parameters. They are the index of stability, $0 < \alpha \leq 2$, the skewness parameter, $-1 \leq \beta \leq 1$, the scale parameter, σ, and the shift parameter, μ. For the examples considered in this paper $\mu = 0$. Any stable random variable with $0 < \alpha < 2$ can be represented as a convergent sum of random variables of the form $\frac{X}{\Gamma^{1/\alpha}}$ where $\{\Gamma_n\}$ are the points of a Poisson set. The series representation of an infinitely divisible random variable without a Gaussian component was first developed by Ferguson and Klass.[5] The stable distributions are a particular example of this.

2. Cellular telephones

As radio waves travel from the transmitting phone (mobile station) to the receiving antenna (base station) they scatter off buildings, the ground, *etc.* The signal received at the base station is the sum of several scattered waves, and since they have traveled different paths they arrive with varying phases. These waves interfere and this is called multi-path fading. The signal arriving at the base station consists of an in phase component, T_1, and a quadrature component, T_2. The envelope of the signal, X, is defined as $X = (T_1^2 + T_2^2)^{1/2}$. The power of the received signal is proportional to the square of the envelope of the signal. At a given distance the distribution of the envelope is most often modeled as Rayleigh or Rician (after Stephen O. Rice.) Usually, Rayleigh is used when all received waves have been reflected and Rician when there is a direct wave plus reflected waves.

The power arriving at a base station from an individual mobile station at a distance of R is $\frac{cX^2}{R^\gamma}$. Where c is a constant which depends on transmitted power, antenna height and gain, and γ depends on the terrain and is usually taken to be 3.9 or 4. Neglecting the curvature of the earth the total power received at the base station is then

$$T = \sum_{i=1}^{\infty} \frac{cX_i^2}{R_i^\gamma}.$$

Assume that the cellular telephones are located as the points of a two dimensional Poisson point set with rate λ. Then $R_1^2, R_2^2 - R_1^2, R_3^2 - R_2^2, \ldots$ are iid[a] exponential random variables with mean $\frac{1}{\pi\lambda}$, and the sequence $R_1^2, R_2^2, \ldots$ is a Poisson point set with rate $\pi\lambda$.

It follows that one way to construct a two dimensional Poisson set is to use a one dimensional Poisson set $\Gamma_1, \Gamma_2, \ldots$ with rate $\pi\lambda$ and a sequence, $\{\Theta_k\}$, of iid uniform $[0, 2\pi)$ random variables. The sequence of points whose polar coordinates are $(\Gamma_n^{1/2}, \Theta_n)$ is a two dimensional Poisson set with rate λ.[3]

The total power received at the base station is

$$T = (\pi\lambda)^{1/\alpha} \sum_{i=1}^{\infty} \frac{cW_i}{\Gamma_i^{1/\alpha}},$$

---

[a]independent and identically distributed

where $\Gamma_1, \Gamma_2, \ldots$ is a one dimensional Poisson point set with unit rate, and $W_i = X_i^2$. If $EW_i^\alpha < \infty$ then T is a stable random variable with parameters $\alpha = 2/\gamma$, $\beta = 1$ and

$$\sigma = c\left[\frac{\pi\lambda\Gamma(2-\alpha)\cos(\frac{\pi\alpha}{2})EW_1^\alpha}{1-\alpha}\right]^{1/\alpha}.$$

The following proofs are given in Ref.[13] and are included here for completeness. Since γ is usually taken to be 3.9 or 4, we will only consider the case when $0 < \alpha < 1$.

First we will show that the series $\sum_{i=1}^\infty W_i\Gamma_i^{-1/\alpha}$ converges using Kolmogorov's three series theorem and then we will show T is stable. To show convergence of the series it is enough to show that $\sum_{i=1}^\infty W_i i^{-1/\alpha} < \infty$ a.s.

(i) For $\lambda > 0$

$$\sum_{i=1}^\infty P(W_i i^{-1/\alpha} > \lambda) = \sum_{i=1}^\infty P(W_i^\alpha > i\lambda^\alpha) < \infty$$

since $EW_1^\alpha < \infty$.

(ii) For $i \geq 1$

$$E\left(W_i i^{-1/\alpha} 1(W_i i^{-1/\alpha} \leq 1)\right) \leq i^{-1/\alpha} \int_0^{i^{1/\alpha}} P(W_i > x)dx$$

and

$$\sum_{i=3}^\infty i^{-1/\alpha} \int_{3^{1/\alpha}}^{i^{1/\alpha}} P(W_1 > x)dx$$

$$= \frac{1}{\alpha}\sum_{i=3}^\infty i^{-1/\alpha} \int_3^i y^{(1-\alpha)/\alpha} P(W_1^\alpha > y)dy$$

$$\leq C\int_3^\infty P(W_1^\alpha > y)dy < \infty$$

because $EW_1^\alpha < \infty$, so

$$\sum_{i=1}^\infty E\left(W_i i^{-1/\alpha} 1(W_i i^{-1/\alpha} \leq 1)\right) < \infty.$$

(iii)

$$\sum_{i=1}^{\infty} E\left(\frac{W_i}{i^{1/\alpha}}\left(1\frac{W_i}{i^{1/\alpha}} \leq \lambda\right)\right)^2$$

$$= \sum_{i=1}^{\infty} i^{-2/\alpha} \int_0^{\infty} w^2 1\{w \leq \lambda i^{1/\alpha}\} f(w) dw$$

$$\leq C \int_0^{\infty} x^{-2/\alpha} dx \int_0^{\lambda x^{1/\alpha}} w^2 f(w) dw$$

$$= C \int_0^{\infty} w^2 f(w) dw \int_{\lambda^{-\alpha} w^\alpha}^{\infty} x^{-2/\alpha} dx$$

$$= C' \int_0^{\infty} w^\alpha f(w) dw < \infty,$$

where C and C' are positive constants.

Therefore

$$\sum_{i=1}^{\infty} \frac{W_i}{\Gamma_i^{1/\alpha}} < \infty.$$

We will now show that T is stable. Let $U_1, U_2, \ldots$ be a sequence of iid uniform (0,1) random variables which are independent of $W_1, W_2, \ldots$ and let $Y_i = \frac{W_i}{U_i^{1/\alpha}}$.

$$\lim_{\lambda \to \infty} \lambda^\alpha P(Y_i > \lambda) = EW_i^\alpha + \lim_{\lambda \to \infty} \lambda^\alpha P(W_i > \lambda) = EW_i^\alpha,$$

so then $\frac{1}{n^{1/\alpha}} \sum_{i=1}^n Y_i \to X$ where X is a stable random variable with index of stability α, $\sigma = \left(\frac{EW_1^\alpha \Gamma(2-\alpha) \cos(\pi\alpha/2)}{1-\alpha}\right)^{1/\alpha}$ and $\beta = 1$. Let $\Gamma_1, \Gamma_2 \ldots$ be a Poisson point set, the distribution of $\frac{\Gamma_1}{\Gamma_{n+1}}, \frac{\Gamma_2}{\Gamma_{n+1}}, \ldots$ given Γ_{n+1} is the distribution of n order statistics from U(0,1). The conditional distribution of $\frac{\Gamma_1}{\Gamma_{n+1}}, \frac{\Gamma_2}{\Gamma_{n+1}}, \ldots$ given Γ_{n+1} is equal to the unconditional distribution. So,

$$\frac{1}{n^{1/\alpha}} \sum_{i=1}^n Y_i = \frac{1}{n^{1/\alpha}} \sum_{i=1}^n \frac{W_i}{U_i^{1/\alpha}}$$

$$\stackrel{d}{=} \frac{1}{n^{1/\alpha}} \sum_{i=1}^n \left(\frac{\Gamma_{n+1}}{\Gamma_i}\right)^{1/\alpha} W_i$$

$$= \left(\frac{\Gamma_{n+1}}{n}\right)^{1/\alpha} \sum_{i=1}^n \frac{W_i}{\Gamma_i^{1/\alpha}} \stackrel{d}{\Rightarrow} X.$$

Γ_{n+1} is the sum of n+1 iid exponential random variables with mean 1 so then by the strong law of large numbers $P(\lim_{n\to\infty} \frac{\Gamma_{n+1}}{n} = 1) = 1$.

Consequently,

$$\left(\frac{\Gamma_{n+1}}{n}\right)^{1/\alpha} \sum_{i=1}^{n} \frac{W_i}{\Gamma_i^{1/\alpha}} \to \sum_{i=1}^{\infty} \frac{W_i}{\Gamma_i^{1/\alpha}}.$$

And

$$T = (\pi\lambda)^{1/\alpha} \sum_{i=1}^{\infty} \frac{cW_i}{\Gamma_i^{1/\alpha}} \overset{d}{=} (\pi\lambda)^{1/\alpha} cX,$$

so T has a stable distribution with

$$\sigma = c \left(\frac{EW_1^{\alpha}\pi\lambda\Gamma(2-\alpha)\cos(\pi\alpha/2)}{1-\alpha}\right)^{1/\alpha}.$$

3. Olbers' paradox and stellar gravity

Edmund Hally first raised the question as to why the sky is dark at night in 1720. It was later discussed by H.W.M. Olbers in 1823 and has become known as Olbers' paradox. Starting with the assumptions that the universe is infinitely old and infinite in extent and that the stars are uniformly distributed leads to the conclusion that the night sky should be uniformly bright. No matter which direction we look there should always be a star in our line of sight, and so the entire sky should have the same brightness as the surface of a star. Also, the energy carried by this radiation would cause the earth to have the same temperature as the surface of a star.

There have been several solutions proposed to resolve Olbers' paradox. In Olbers' time it was believed that dust and other matter absorbed the radiation en route to the earth. Generally, this is no longer accepted as a solution since the absorbing material would heat up and then radiate energy at the same rate it absorbed energy.

The expansion of the universe has also been used as a possible resolution of the paradox. The redshift, which is caused by the Doppler effect, has been observed in radiation from all galaxies. The energy from a quantum of light is given by $E = \frac{hc}{\lambda}$. The wavelength, λ, is increased by the Doppler effect, and therefore the energy of the quantum of light is decreased. Although, it is generally accepted that the expansion of the universe does decrease the energy of the light, some have argued that the decrease in energy of the radiation would not be great enough to resolve the paradox.[6]

The darkness of the night sky can be explained if the assumption that the universe is infinitely old is dropped. In the bright sky model most of the radiation would have to come from very distant sources. Due to the finite speed of light, the radiation would have to be emitted long ago, on the order of 10^{24} years.[11] However, if the universe were not that old, light from these distant galaxies would not have reached the earth yet.

Still another possible explanation for the paradox is that the lifetime of the stars is too short to produce a bright sky. Harrison[6] calculates that the time required to attain a bright sky is approximately 10^{24} years. The luminous lifetime of the stars is approximated to be only 10^8 years, which would support this resolution. Hence, the lifetime of the stars is far too short to saturate the night sky with radiation.

It would seem that another possible explanation for the paradox should be considered. If the radiation from all of the stars is summed allowing for destructive interference then the sum converges. This would allow for a universe which is infinite in age and in extent with a dark night sky. This would imply that "Rather than proving the finiteness of the universe, the Olbers' paradox proves quantum mechanics.[12]"

In the case of stellar gravity the situation is similar to Olbers' paradox, in an infinite universe with an infinite number of stars the total force exerted on the earth should be infinite. However, the gravitational force exerted from the stars in one region should be balanced by the gravitational force of stars in other regions. So, it would seem the earth should experience infinite forces in all directions, and the universe should be in a state of extreme "tension." The proposed resolutions to this problem are similar to those for Olbers' paradox.

When considering radiation, let R_i and W_i represent the distance and the magnitude of radiation from the ith most distant star. To allow for interference assign the radiation a value of $+1$ or -1 with equal probability. Assuming an inverse square loss the total radiation from all of the stars is

$$T = \sum_{i=1}^{\infty} \frac{\epsilon_i W_i}{R_i^2},$$

where $\epsilon_1, \epsilon_2, \ldots$ is an iid sequence of Rademacher variables.

The total gravitational force at a point per unit mass is

$$F = \sum_{i=1}^{\infty} \frac{GM_i R_i}{|R_i|^3},$$

where R_i is the position of the ith most distant star. The vector $\frac{R_i}{|R_i|}$ is uniformly distributed on the unit sphere and independent of $|R_i|$. We will consider the projection of the gravitational force onto a given vector v. Let

$W_i = GM_i \left| P_v \left(\frac{\boldsymbol{R}_i}{|\boldsymbol{R}_i|} \right) \right|$ and let $\epsilon_i = \cos\theta$ where θ is the angle between $\boldsymbol{v}$ and $P_v \left(\frac{\boldsymbol{R}_i}{|\boldsymbol{R}_i|} \right)$. The gravitational force in the direction of $\boldsymbol{v}$ is then

$$T = \sum_{i=1}^{\infty} \frac{\epsilon_i W_i}{R_i^2},$$

where $R_i = |\boldsymbol{R}_i|$.

We will consider the stars to be located at the points of a three dimensional Poisson set with rate λ. The sequence $R_1^3, R_2^3 - R_1^3, R_3^3 - R_2^3, \ldots$ are iid exponential random variables with mean $(\frac{4}{3}\pi\lambda)^{-1}$, and $R_1^3, R_2^3, \ldots$ is a one dimensional Poisson set with rate $\frac{4}{3}\pi\lambda$.

So,

$$T = \left(\frac{4}{3}\pi\lambda \right)^{2/3} \sum_{i=1}^{\infty} \frac{\epsilon_i W_i}{\Gamma_i^{2/3}},$$

where $\Gamma_1, \Gamma_2, \ldots$ is a Poisson set with unit rate. If $E|W_1|^{3/2} < \infty$, then T is a symmetric stable random variable with $\alpha = 3/2$ and $\sigma = 2\pi \left(\frac{2\lambda}{3} EW_1^{3/2} \right)^{2/3}$.

As before, the following proofs are given in Ref.[13] and are included for completeness. First we will show convergence of the series using Kolmogorov's three series theorem. It is enough to show that $\sum \frac{\epsilon_i W_i}{\Gamma_i^{2/3}}$ converges for each fixed sequence $\{\Gamma_i\}$. For a fixed sequence the summands $\epsilon_i \Gamma_i^{-2/3} W_i$ are independent and $C_1 i \leq \Gamma_i \leq C_2 i$ where C_1 and C_2 are positive constants.

(i) For $\lambda > 0$

$$\sum_{i=1}^{\infty} P \left(\frac{W_i}{\Gamma_i^{2/3}} > \lambda \right) = \sum_{i=1}^{\infty} P(W_i^{3/2} > \lambda^{3/2}\Gamma_i)$$

$$\leq \sum_{i=1}^{\infty} P(W_i^{3/2} > \lambda^{3/2}C_1 i) < \infty$$

since $E|W_i|^{3/2} < \infty$.

(ii)

$$\sum_{i=1}^{\infty} E \left(\frac{\epsilon_i W_i}{\Gamma_i^{2/3}} 1\left(\frac{W_i}{\Gamma_i^{2/3}} \leq \lambda \right) \right) = 0$$

since each summand equals 0.

(iii)

$$\sum_{i=1}^{\infty} E\left(\frac{\epsilon_i W_i}{\Gamma_i^{2/3}} 1\left(\frac{W_i}{\Gamma_i^{2/3}} \le \lambda\right)\right)^2$$

$$\le C_1^{-4/3} \sum_{i=1}^{\infty} i^{-4/3} \int_0^{\infty} w^2 1(w \le \lambda C_2^{2/3} i^{2/3}) f(w) dw$$

$$\le C \int_0^{\infty} x^{-4/3} dx \int_0^{\lambda C_2^{2/3} x^{2/3}} w^2 f(w) dw$$

$$= C \int_0^{\infty} w^2 f(w) dw \int_{\lambda^{-3/2} C_2^{-1} w^{3/2}}^{\infty} x^{-4/3} dx$$

$$= C' \int_0^{\infty} w^{3/2} f(w) dw < \infty,$$

where C and C' are positive constants.

Next we will show T is a stable random variable. Let $U_1, U_2, \ldots$ be a sequence of iid uniform $(0,1)$ random variables, and let $Y_i = \frac{\epsilon_i W_i}{U_i^{2/3}}$. As before, $lim_{\lambda \to \infty} \lambda^{\alpha} P(|Y_i| > \lambda) = E W_i^{\alpha}$, and so $\frac{1}{n^{2/3}} \sum_{i=1}^n Y_i \to X$, where X is stable $\alpha = 3/2$, $\sigma = (2\pi)^{1/3} (E W_1^{3/2})^{2/3}$ and $\beta = 0$.

So,

$$\frac{1}{n^{2/3}} \sum_{i=1}^n Y_i = \frac{1}{n^{2/3}} \sum_{i=1}^n \frac{\epsilon_i W_i}{U_i^{2/3}} \stackrel{d}{=} \left(\frac{\Gamma_{n+1}}{n}\right)^{2/3} \sum_{i=1}^n \frac{\epsilon_i W_i}{\Gamma_i^{2/3}} \to X.$$

And by the strong law of large numbers $P(\lim_{n \to \infty} \frac{\Gamma_{n+1}}{n} = 1)$ so then

$$\sum_{i=1}^{\infty} \frac{\epsilon_i X_i}{\Gamma^{2/3}} \stackrel{d}{=} X.$$

And $T = (\frac{4}{3}\pi\lambda)^{3/2} X$, so T has a stable distribution.

For stellar gravity, $\boldsymbol{F} = (F_x \hat{i}, F_y \hat{j}, F_z \hat{k})$, we have shown that each of $F_x, F_y,$ and F_z has a stable distribution with $\alpha = 3/2$. Since the projection of $\boldsymbol{F}$ in any direction is stable and $\alpha \ge 1$, then $\boldsymbol{F}$ is a stable vector, this follows from theorem 2.1.5(c) in Ref.[13] We will now obtain the characteristic function for $\boldsymbol{F}$,

$$E e^{i\boldsymbol{F}\cdot\boldsymbol{z}} = E e^{i|\boldsymbol{z}|\left(\boldsymbol{F}\cdot\frac{\boldsymbol{z}}{|\boldsymbol{z}|}\right)} = E e^{i|\boldsymbol{z}|T}.$$

Where T is the projection of $\boldsymbol{F}$ onto $\boldsymbol{z}$, T was shown to be a symmetric stable random variable with $\alpha = 3/2$, so

$$E e^{i|\boldsymbol{z}|T} = e^{-\sigma^{3/2}|\boldsymbol{z}|^{3/2}}$$

with

$$\sigma^{3/2} = (2\pi)^{3/2} \left(\frac{2\lambda}{3} EW_1^{3/2} \right)$$

$$= \frac{2\lambda}{3} (2\pi G)^{3/2} EM_1^{3/2} E\left(\left| P_z \frac{R_1}{|R_1|} \right|^{3/2} \right)$$

$$= \frac{4}{15} \lambda (2\pi G)^{3/2} EM_1^{3/2}.$$

The characteristic function for F is

$$Ee^{iz \cdot F} = e^{\frac{4}{15}\lambda(2\pi G)^{3/2} EM_1^{3/2}|z|^{3/2}}.$$

Perhaps experiments could be devised to test whether or not small fluctuations in the earth's gravitational pull are consistent with the stable law hypothesis. It might be of some interest to calculate the probability that the net gravitational field at a given point is *away from* the direction of the nearest Poisson star. For example, the probability that the gravitational force on earth is away from the sun is

$$P\left(\frac{GM_s}{r^2} \leq T \right) = P\left(\frac{GM_s}{\sigma r^2} \leq X \right) = P\left(\frac{M_s}{2\pi r^2 (\frac{4}{15}\lambda EM_1^{3/2})^{2/3}} \leq X \right),$$

where $T = \sigma X$, X is a stable random variable with $\alpha = 3/2$ and $\sigma = 1$, and r is the distance from the earth to the sun.

As given in Zolotarev,[14] (or Samorodnitsky and Taqqu,[13] p. 39)

$$P\left(\frac{M_s}{2\pi r^2 (\frac{4}{15}\lambda EM_1^{3/2})^{2/3}} \leq X \right)$$

$$= \frac{1}{\pi} \int_0^{\pi/2} \exp\{ -\left(\frac{M_s}{2\pi r^2 (\frac{4}{15}\lambda EM_1^{3/2})^{2/3}} \right)^3 U(\gamma) \} d\gamma,$$

where

$$U(\gamma) = \left(\frac{\sin \frac{3}{2}\gamma}{\cos \gamma} \right)^{-3} \left(\frac{\cos \frac{1}{2}\gamma}{\cos \gamma} \right).$$

It should be noted that an objection to this argument was raised in the case of radiation.[9] A light-wave is comprised of an electric field, E, and a magnetic field, B, and the intensity is proportional to the cross product of E and B. Since intensity drops off with $\gamma = 2$ then the electric field (or magnetic field) perhaps should drop off with $\gamma = 1$ not $\gamma = 2$.

4. Poisson model

For cellular telephones, as the users move form place to place, Doob[4] gives conditions under which the Poisson distribution will be preserved. The distribution of mobile phones will remain Poisson if the following conditions are satisfied. For t>0, the classes of random variables

$$\{x_i(t) - x_i(0), -\infty < i < \infty\}, \qquad \{x_i(0), -\infty < i < \infty\}$$

are mutually independent, and the random variables in the first class are mutually independent, where $x_i(t)$ represents the position of the ith mobile phone at time t.

In the case of the stars, assume that the initial position and momentum of the stars are distributed as a Poisson point set in 6 dimensions. To ensure that there is not an infinite number of stars in a finite region condition on a given energy density. For any finite region of space, the initial position and momentum of the N stars in that region can be described by a point in 6N-dimensional phase space, and the motion of the stars is described by a trajectory through phase space. It would seem that since Lebesgue measure is preserved in phase space due to Liouville's theorem, the stars should continue to have a Poisson distribution as they move. And taking the limit as the size of the region goes to infinity would seem to suggest that the stars in an infinite region should remain Poisson.

The above argument implies that an infinite homogeneous gravitating system can be in equilibrium, which according to Binney and Tremaine[1] cannot be the case. Binney and Tremaine state that the way to construct an infinite homogeneous equilibrium is to perpetrate the "Jeans' swindle," known as such because, until recently, it was believed that there is no formal justification for Jeans' relation. Kiessling[10] gives a derivation of Jeans' relation, showing that it is not a "swindle." Both Kiessling's derivation and the above argument show that it is possible to have an infinite homogeneous gravitating system in equilibrium.

5. Conclusion

We have modeled both the locations of cellular telephones and the locations of the stars as Poisson point sets and found the distribution of the total power of the radio signal and the total stellar radiation and gravity. In the case of cellular telephones we have shown that the total power received at the base station has a skewed α stable distribution where typical values for α are about .5 and $\beta = 1$. The total radiation or total gravitational force on earth has a symmetric stable distribution with $\alpha = 3/2$.

Using this Poisson model for stars and allowing for interference leads to an alternative explanation of Olbers' paradox. A universe which is infinite in age and extent can have a dark night sky since the sum of the radiation will converge.

Acknowledgments

We are grateful to Moe Win for bringing our attention to wireless which led to the Poisson model and for many useful discussions of issues on wireless. We are grateful to Tom Cover for informing us of the Olbers' paradox. We are grateful to Bill Joyce and Yuval Peres and to several physicists, who did not want to be explicitly thanked, for discussions. We are grateful to Michael Kiessling for his critical comments.

References

1. J. Binney and S. Tremaine. *Galactic Dynamics*. Princeton University Press, 1987.
2. S. Chandrasekhar. Stochastic Problems in Physics and Astronomy. *Reviews of Modern Physics*, 15, 1943.
3. D.R. Cox and V. Isham. *Point Processes*. Chapman and Hall, 1980.
4. J.L. Doob. *Stochastic Processes*. John Wiley and Sons, 1953.
5. T. Ferguson and M. Klass. A representation theorem of independent increment processes without Gaussian component. *The Annals of Mathematical Statistics*, 43, 1972.
6. E.R. Harrison. Why the sky is dark at night. *Physics Today*, 1974.
7. S. Heath. PhD thesis, Rutgers University, To be published in 2003.
8. J. Holtsmark. Uber die verbreiterung von spektrallinien. *Annalen Der Physik*, 58, 1919.
9. W. Joyce. Personal Correspondence, 2002.
10. Michael K.-H. Kiessling. The "Jeans' Swindle" A true story-mathematically speaking. *Adv. Appl. Math*, to be published in 2003.
11. J.M. Pasachoff. *Contemporary Astronomy*. W.B. Saunders Company, 1977.
12. Y. Peres. Personal Correspondence, 2002.
13. G Samorodnitsky and M.S. Taqqu. *Stable Non-Gaussian Random Processes*. Chapman and Hall, 1994.
14. V.M. Zolotarev. *One-Dimensional Stable Distributions*. American Mathematical Society, 1986.

Nonrelativistic QED at Large Momentum of Photons

Fumio Hiroshima

Department of Mathematics and Physics, Setsunan University, Osaka 572-8508, Japan
E-mail: hiroshima@mpg.setsunan.ac.jp

Dedicated to Professor Hiroshi Ezawa on the occasion of his seventieth birthday

A system of one spinless electron minimally coupled to a quantized radiation field is considered. On the Hamiltonian, H_Λ, of this system, an ultraviolet cutoff $\Lambda > 0$ and an infrared cutoff $\lambda > 0$ are imposed. We do *not* assume the existence of a ground state of the decoupled Hamiltonian. It is proven, however, that there exists Λ_* such that for all $\Lambda > \Lambda_*$, H_Λ has a unique ground state. Furthermore we investigate the asymptotic behavior of $g(\Lambda) = \inf \sigma(H_\Lambda)$ without external potentials as $\Lambda \to \infty$.

1. Introduction

It is a pleasure to dedicate this article to Professor Hiroshi Ezawa on the occasion of his seventieth birthday. He was a source of encouragement in mathematical physics. Especially he gave physical inspirations to myself in seminars in the Gakushuin University, in ICMP '97 held in Australia, and in Les Houches '98 "Ecole de Physique des Houches" in France. He also invited me as one of lecturers in the summer school of mathematical physics 2000 organized by H. Araki and himself. I wish to express my deep thanks to him for stimulating my works and for innumerable, unbounded and helpful conversations.

The aim of this paper is to consider a system of one spinless electron minimally coupled to a quantized radiation field. It is the so-called Pauli–Fierz model[24] in nonrelativistic quantum electrodynamics. The Pauli–Fierz Hamiltonian is defined as a self-adjoint operator acting in the tensor product of $L^2(\mathbb{R}^3)$ and a boson Fock space $\mathcal{F}$, $L^2(\mathbb{R}^3) \otimes \mathcal{F}$, which is given by

$$\left(\frac{1}{2m}p^2 + V\right) \otimes 1 + 1 \otimes H_{\mathrm{f}}$$

with $p \otimes 1 = -i\nabla_x \otimes 1$ replaced by $p \otimes 1 + A_\Lambda(x)$, *i.e.*,

$$\frac{1}{2m}(p \otimes 1 + A_\Lambda(x))^2 + V \otimes 1 + 1 \otimes H_{\mathrm{f}},$$

where V denotes an external potential, H_{f} the free Hamiltonian in a Fock space and $A_\Lambda(x)$ a quantized radiation field with a ultraviolet cutoff $\Lambda > 0$. Details will be seen later. Under some suitable conditions, it is established that a ground state of the Pauli–Fierz Hamiltonian exists if $\frac{1}{2m}p^2 + V$ has a ground state. See Refs.[3,8,9,10,13,18]

In this paper we take the dipole approximation, *i.e.*, we replace as

$$A_\Lambda(x) \longrightarrow 1 \otimes A_\Lambda(0),$$

and assume that on the quantized radiation field an infrared cutoff is imposed. Then the Hamiltonian under consideration is

$$H_\Lambda = \frac{1}{2m}(p \otimes 1 + 1 \otimes A_\Lambda(0))^2 + V \otimes 1 + 1 \otimes H_{\mathrm{f}}.$$

It is proven in Ref.[12] that H_Λ also has a ground state even when no infrared cutoff is imposed, if $\frac{1}{2m}p^2 + V$ has a ground state.

A more difficult, but physically reasonable, question is about enhanced binding. One may expect that a coupling with a quantized radiation field enhances binding. In this paper, we suppose that the electron is governed by a sufficiently shallow nonpositive external potential V, and then $\frac{1}{2m}p^2 + V$ has *no ground states*. We prove, however, that H_Λ has a ground state φ_{g} for sufficiently large Λ. The physical reasoning behind such a result is simple. On a heuristic level one argues that a coupling with a quantized radiation field amounts to renormalizing bare mass m to effective mass:

$$m \to m_{\mathrm{eff}} = m + \delta m(\Lambda)$$

Since effective mass $m(\Lambda)$ is increasing in Λ, a bound state could be produced for a sufficiently large Λ. Roughly speaking, H_Λ may be replaced by

$$H_\Lambda \sim H_{\mathrm{eff}} = \left(\frac{p^2}{2m_{\mathrm{eff}}} + V\right) \otimes 1 + 1 \otimes H_{\mathrm{f}}, \tag{1}$$

which binds for sufficiently large Λ. Needless to say (1) has no mathematical meaning; however we show the associated phenomena in this paper.

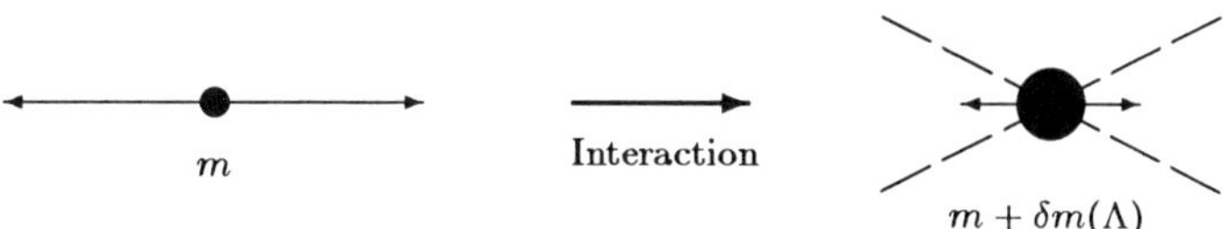

Fig. 1.   Effective mass

This kind of investigation has been done in Hiroshima and Spohn,[20] and developed by *e.g.*, Refs.[2,4,5,11] Arai and Kawano[2] proved the similar result as ours in a general framework. Catto and Hainzl,[4] Chen, Vougalter and Vugalter,[5] Hainzl, Vougalter and Vugalter[11] study the case of no dipole approximation under some conditions.

This paper is organized as follows. In Section 2 we define the Pauli–Ficrz Hamiltonian and its self-adjointness is stated. In Section 3 we diagonalize the Pauli–Fierz Hamiltonian. In Section 4 we prove the main theorem. Section 5 is devoted to showing the multiplicity and localization of a ground state. In Section 6 we investigate the asymptotic behavior of the ground state energy of H_Λ with no external potential.

2. Nonrelativistic QED

2.1. *The Pauli–Fierz Hamiltonian*

The boson Fock space $\mathcal{F}$ over $L^2(\mathbb{R}^3 \times \{1,2\})$ is defined by

$$\mathcal{F} = \mathcal{F}(L^2(\mathbb{R}^3 \times \{1,2\})) = \bigoplus_{n=0}^{\infty} \mathcal{F}^{(n)}(L^2(\mathbb{R}^3 \times \{1,2\})),$$

where

$$\mathcal{F}^{(n)}(L^2(\mathbb{R}^3 \times \{1,2\})) = \otimes_s^n L^2(\mathbb{R}^3 \times \{1,2\})$$

denotes the n-fold symmetric tensor product of $L^2(\mathbb{R}^3 \times \{1,2\})$ with

$$\otimes_s^0 L^2(\mathbb{R}^3 \times \{1,2\}) = \mathbb{C}.$$

The Fock vacuum Ω is defined by

$$\Omega = \{1, 0, 0, ...\} \in \mathcal{F}.$$

$\{a^*(k,j), a(k,j)\}_{j=1,2}$ are bose fields with commutation relations

$$[a(k,j), a^*(k',j')] = \delta_{jj'}\delta(k - k'),$$

$$[a^*(k,j), a^*(k',j')] = 0, \quad [a(k,j), a(k',j')] = 0.$$

The quantized radiation field, $A_\Lambda(x), x \in \mathbb{R}^3$, is defined by

$$A_\Lambda(x) = \sum_{j=1,2} \int \frac{\chi_\Lambda(k)}{\sqrt{2\omega(k)}} e_j(k) \left\{ a^*(k,j)e^{-ik\cdot x} + a(k,j)e^{ik\cdot x} \right\} d^3k,$$

where χ_Λ denotes the momentum cutoff given by

$$\chi_\Lambda(k) = \begin{cases} 0, & |k| < \lambda, \\ 1, & \lambda \le |k| \le \Lambda, \\ 0, & |k| > \Lambda. \end{cases}$$

We fix an infrared cutoff, $\lambda > 0$, and assume that

$$\lambda \le \Lambda.$$

$e_j(k) = (e_j^1(k), e_j^2(k), e_j^3(k)), j = 1, 2$, are polarization vectors, *i.e.*,

$$\{e_1(k), e_2(k), k/|k|\}$$

forms a right-handed system, *i.e.*,

$$e_j \cdot k = 0, \quad e_i \cdot e_j = \delta_{ij}, \quad e_1 \times e_2 = k/|k|.$$

The dispersion relation is given by

$$\omega(k) = |k|$$

and therefore the free Hamiltonian of $\mathcal{F}$ is defined by

$$H_{\mathrm{f}} = \sum_{j=1,2} \int \omega(k)a^*(k,j)a(k,j)d^3k.$$

Let us denote the spectrum (resp. discrete spectrum, essential spectrum) of a self-adjoint operator T by $\sigma(T)$ (resp. $\sigma_{\mathrm{disc}}(T)$, $\sigma_{\mathrm{ess}}(T)$). It is well known that

$$\sigma(H_{\mathrm{f}}) = [0, \infty), \quad \sigma_{\mathrm{disc}}(H_{\mathrm{f}}) = \{0\},$$

and

$$H_{\mathrm{f}}\Omega = 0.$$

The Hilbert space of our system is

$$\mathcal{H} = L^2(\mathbb{R}^3) \otimes \mathcal{F}.$$

The decoupled Hamiltonian is given by

$$H_0 = \left(\frac{1}{2m}p^2 + V\right) \otimes 1 + 1 \otimes H_{\mathrm{f}}.$$

Here m denotes the bare mass of an electron, $p = -i\nabla_x$ the momentum operator canonically conjugate to the position operator x in $L^2(\mathbb{R}^3)$ and V an external potential. Thus

$$\sigma(H_0) = \{\lambda_1 + \lambda_2 | \lambda_1 \in \sigma(\frac{1}{2m}p^2 + V),\ \lambda_2 \in \sigma(H_{\mathrm{f}})\}.$$

We take the dipole approximation, *i.e.*, we replace as

$$A_\Lambda(x) \longrightarrow 1 \otimes A_\Lambda(0).$$

Set

$$A_\Lambda = A_\Lambda(0).$$

The total Hamiltonian, H_Λ, is defined by H_0 with the replacement

$$p \otimes 1 \longrightarrow p \otimes 1 + 1 \otimes A_\Lambda,$$

i.e.,

$$H_\Lambda = \frac{1}{2m}\left(p \otimes 1 + 1 \otimes A_\Lambda\right)^2 + V \otimes 1 + 1 \otimes H_{\mathrm{f}}.$$

For notational convenience we mostly omit the tensor notation $\otimes$ unless confusions may arise, *e.g.*, we simply write as p, V, A_Λ, H_{f}, instead of $p \otimes 1$, $V \otimes 1$, $1 \otimes A_\Lambda$ and $1 \otimes H_{\mathrm{f}}$, respectively. Then H_Λ is simply written as

$$H_\Lambda = \frac{1}{2m}(p + A_\Lambda)^2 + V + H_{\mathrm{f}}.$$

Proposition 1: *Suppose that $D(p^2) \subset D(V)$ and*

$$\|Vf\| \le a\|\frac{1}{2m}p^2 f\| + b\|f\|$$

with some $0 \le a < 1$ and $b \ge 0$. Then H_Λ is self-adjoint on $D(p^2) \cap D(H_{\mathrm{f}})$ and bounded from below. Moreover it is essentially self-adjoint on any core of $\frac{1}{2m}p^2 + H_{\mathrm{f}}$.

Proof: See Refs.[15,16] $\qquad\qquad\qquad\qquad\qquad\qquad\qquad\qquad\qquad\qquad\square$

Remark 2: The Hamiltonian without the dipole approximation, *i.e.*,

$$\frac{1}{2m}(p + A_\Lambda(x))^2 + V + H_{\mathrm{f}},$$

is also self-adjoint on $D(p^2) \cap D(H_{\mathrm{f}})$. See Refs.[15,16]

2.2. *Enhanced binding and the main theorem*

In this paper we consider the case where the external potential V is sufficiently shallow. By the Lieb–Thirring inequality,[23]

$$\#\{\text{bound states of } \frac{1}{2m}p^2 + V\} \leq L_{0,3} \int |mV_-(x)|^{3/2}d^3x$$

$$(V_- : \text{ the negative part of } V)$$

with some constant $L_{0,3}$ independent of V, we see that for a sufficiently shallow nonpositive potential V, $\frac{1}{2m}p^2+V$ has no bound states. In particular it has no ground state. Thus the decoupled Hamiltonian $\frac{1}{2m}p^2 + V + H_{\mathrm{f}}$ also has no ground state. Our contribution is to show, however, that $H_\Lambda = \frac{1}{2m}(p + A_\Lambda)^2 + V + H_{\mathrm{f}}$ has a ground state for sufficiently large Λ.

Throughout this paper we assume Hypothesis (V).

Hypothesis (V)

(a) $V \in C_0^\infty(\mathbb{R}^3)$.
(b) $V \leq 0$.
(c) *There exists $\mu_0 > 1$ and $r > 0$ such that for $\mu > \mu_0$,*

$$\inf \sigma(\frac{1}{2m}p^2 + \mu V) < -r.$$

Since V is relatively compact with respect to $\frac{1}{2m}p^2$, it holds that

$$\sigma_{\mathrm{ess}}(\frac{1}{2m}p^2 + \mu V) = [0, \infty).$$

Hence $\frac{1}{2m}p^2 + \mu V$, $\mu > \mu_0$, has a ground state.

Remark 3: We do not assume the existence of ground states of $\frac{1}{2m}p^2 + V$.

The main result of this paper is as follows.

Theorem 1: *There exists Λ_* such that for $\Lambda > \Lambda_*$, H_Λ has a ground state.*

3. Canonical transformations

The following proposition is a key in this paper.

Proposition 2: *There exists a unitary operator U such that U maps $D(p^2) \cap D(H_{\mathrm{f}})$ onto itself and*

$$U^{-1}H_\Lambda U = \frac{1}{2m_{\mathrm{eff}}}p^2 + V(\cdot -K/m_{\mathrm{eff}}) + H_{\mathrm{f}} + \frac{1}{\pi}\int_{-\infty}^{\infty}\frac{t^2\|\chi_\Lambda/(t^2+\omega^2)\|^2}{m + \frac{2}{3}\|\chi_\Lambda/\sqrt{t^2+\omega^2}\|^2}dt,$$

where

$$m_{\text{eff}} = m + \frac{8\pi}{3}(\Lambda - \lambda),$$

$$K = \sum_{j=1,2} \int \frac{\chi_\Lambda(k)e_j(k)}{\sqrt{2\omega(k)}} \left\{ \frac{a^*(k,j)}{\omega(k)D_-(\omega(k)^2)} + \frac{a(k,j)}{\omega(k)D_+(\omega(k)^2)} \right\} d^3k, \quad (2)$$

and

$$D_\pm(s) = m - \frac{4\pi}{3}\left(\lim_{\epsilon \downarrow 0} \int_{|s-x|<\epsilon} \frac{\sqrt{x}\,\chi_\Lambda(\sqrt{x})}{s - x}\, dx \mp 2\pi i \chi_\Lambda(\sqrt{s})\sqrt{s} \right) \neq 0.$$

Proof: This is proven by a slight modification of Ref.[20] □

Remark 4: In Ref.[20] it is obtained that

$$m_{\text{eff}} = m + \frac{2}{3}\|\chi_\Lambda/\omega\|^2.$$

Remark 5: We see that $D_\pm(s) = m$ for $0 \leq s < \lambda^2$ or $\Lambda^2 < s$, and $\Im D_\pm(s) = 2\pi\sqrt{s}$ for $\lambda^2 \leq s \leq \Lambda^2$. In particular $D_\pm(s) \neq 0$ for $s \geq 0$.

We set

$$\varrho(\cdot,j) = (\varrho_1(\cdot,j), \varrho_2(\cdot,j), \varrho_3(\cdot,j)),$$

where

$$\varrho_\mu(k,j) = \frac{e_j^\mu(k)\chi_\Lambda(k)}{\omega^{3/2}(k)D_-(\omega(k)^2)}, \quad \mu = 1, 2, 3.$$

Then we rewrite (2), $K = (K_1, K_2, K_3)$, as

$$K_\mu = \sum_{j=1,2} \frac{1}{\sqrt{2}} \int \{\varrho_\mu(k,j)a^*(k,j) + \varrho_\mu(k,j)^*(k)a(k,j)\}\, d^3k$$

and $\varrho_\mu(\cdot,j)$ satisfies that

$$\|\omega^{n/2}\varrho_\mu(\cdot,j)\| \leq C_1\|\omega^{(n-3)/2}\chi_\Lambda\| \tag{3}$$

with some constant C_1. Let

$$\phi(f) = \sum_{j=1,2} \int \{f(k)a^*(k,j) + f^*(k)a(k,j)\}\, d^3k.$$

Then the following inequality is well known for $\Psi \in D(H_f^{1/2})$,

$$\|\phi(f)\Psi\| \leq (\|f/\omega^{1/2}\| + \|f\|)(2\|H_f^{1/2}\Psi\| + \|\Psi\|).$$

174   *F. Hiroshima*

From (3) it follows that

$$\|K_\mu \Psi\| \le C_2(\|\chi_\Lambda/\omega^2\| + \|\chi_\Lambda/\omega^{3/2}\|)(\|H_{\mathrm{f}}^{1/2}\Psi\| + \|\Psi\|) \tag{4}$$

with some constant C_2. We set

$$\delta V = V(\cdot - K/m_{\mathrm{eff}}) - V,$$

$$H_{\mathrm{eff}} = \frac{1}{2m_{\mathrm{eff}}}p^2 + V,$$

$$g(\Lambda) = \frac{1}{\pi}\int_{-\infty}^{\infty} \frac{t^2\|\chi_\Lambda/(t^2+\omega^2)\|^2}{m + \frac{2}{3}\|\chi_\Lambda/\sqrt{t^2+\omega^2}\|^2}dt,$$

and

$$\widehat{H}_\Lambda = U^{-1}H_\Lambda U = H_{\mathrm{eff}} + \delta V + H_{\mathrm{f}} + g(\Lambda). \tag{5}$$

We investigate a spectral property of $\widehat{H}_\Lambda$ instead of H_Λ in what follows.

Lemma 1: *Let Λ be such that*

$$\Lambda > \frac{3}{8\pi}(\mu_0 - 1)m.$$

Then H_{eff} has a ground state, and moreover

$$\#\{\text{bound states of } H_{\mathrm{eff}}\} \le L_{0,3}\left(m + \frac{8}{3}\pi(\Lambda - \lambda)\right)^{3/2}\int |V(x)|^{3/2}d^3x. \tag{6}$$

In particular H_{eff} has a finite number of bound states.

Proof: By Hypothesis (V),

$$H_{\mathrm{eff}} = \frac{1}{2m_{\mathrm{eff}}}p^2 + V = \frac{m}{m_{\mathrm{eff}}}\left(\frac{1}{2m}p^2 + \frac{m_{\mathrm{eff}}}{m}V\right)$$

implies that if

$$\frac{m_{\mathrm{eff}}}{m} > \mu_0, \tag{7}$$

then H_{eff} has a ground state. (7) is identical with $\Lambda > \frac{3}{8\pi}(\mu_0 - 1)m$. (6) follows from the Lieb–Thirring inequality. Then the lemma follows. $\qquad\square$

4. Existence of ground states

4.1. *Lattice approximation*

We introduce an artificial mass of photon, $\nu > 0$, and define

$$\widehat{H}_\Lambda^\nu = H_{\text{eff}} + \delta V^\nu + H_{\text{f}}^\nu + g(\Lambda),$$

where δV^ν and H_{f}^ν are defined by δV and H_{f} with ω replaced by $\omega + \nu$, respectively. Using a momentum lattice approximation we will prove that $\widehat{H}_\Lambda^\nu$ has a ground state. Let $\Gamma(l, a)$, $l = (l_1, l_2, l_3) \in \mathbb{Z}^3$, $a > 0$, be the momentum lattice with spacing $1/a$, i.e.,

$$\Gamma(l, a) = [\frac{l_1}{a}, \frac{(l_1 + 1)}{a}) \times [\frac{l_2}{a}, \frac{(l_2 + 1)}{a}) \times [\frac{l_3}{a}, \frac{(l_3 + 1)}{a}),$$

and

$$\chi_{\Gamma(l,a)}(k) = \begin{cases} 0, & k \notin \Gamma(l, a), \\ a^{3/2}, & k \in \Gamma(l, a). \end{cases}$$

For $L > 0$ we define the momentum-lattice-approximated Hamiltonian by

$$\widehat{H}_\Lambda^\nu(a, L) = H_{\text{eff}} + \delta V^\nu(a, L) + H_{\text{f}}^\nu(a, L) + g(\Lambda),$$

where

$$H_{\text{f}}^\nu(a, L) = \sum_{j=1,2} \int \sum_{|l| \le L} \chi_{\Gamma(l,a)}(k)(\omega(l) + \nu)a^*(k, j)a(k, j)d^3k,$$

$$\delta V^\nu(a, L) = V(\cdot - K^\nu(a, L)/m_{\text{eff}})$$

and $K^\nu(a, L) = (K_1^\nu(a, L), K_2^\nu(a, L), K_3^\nu(a, L))$ with

$$K_\mu^\nu(a, L) = \sum_{j=1,2} \frac{1}{\sqrt{2}} \int \sum_{|l| \le L} \chi_{\Gamma(l,a)}(k)[\varrho_\mu^\nu(l, j)a^*(k, j) + \varrho_\mu^\nu(l, j)^* a(k, j)]d^3k.$$

In a way similar to (4) we can show that

$$\|K_\mu^\nu(a, L)\Psi\|$$
$$\le C_2' \left(\left\| \frac{\sum_{|l| \le L} \chi_{\Gamma(l,a)}\chi_\Lambda(l)}{\sqrt{\omega}(\sum_{|l| \le L} \chi_{\Gamma(l,a)}\omega(l))^{3/2}} \right\| + \left\| \frac{\sum_{|l| \le L} \chi_{\Gamma(l,a)}\chi_\Lambda(l)}{(\sum_{|l| \le L} \chi_{\Gamma(l,a)}\omega(l))^{3/2}} \right\| \right)$$
$$\times(\|H_{\text{f}}^\nu(a, L)^{1/2}\Psi\| + \|\Psi\|) \tag{8}$$

with some constant C_2'. Here we used

$$c_1\|H_{\text{f}}^\nu(a, L)^{1/2}\Psi\| \le \|H_{\text{f}}^{\nu\,1/2}\Psi\| \le c_2\|H_{\text{f}}^\nu(a, L)^{1/2}\Psi\|$$

with some constants c_1 and c_2.

Lemma 2: *We have*

$$\lim_{L\to\infty}\lim_{a\to\infty}\widehat{H}^{\nu}_{\Lambda}(a,L)=\widehat{H}^{\nu}_{\Lambda}$$

in the operator norm in the sense of resolvent.

Proof: It is well known that there exists $\epsilon(a,L)$ such that

$$\|(H^{\nu}_{\mathrm{f}}(a,L)-H^{\nu}_{\mathrm{f}})\Psi\|\le \epsilon(a,L)\|H^{\nu}_{\mathrm{f}}\Psi\|$$

and

$$\lim_{L\to\infty}\lim_{a\to\infty}\epsilon(a,L)=0.$$

Moreover

$$\|(\delta V^{\nu}(a,L)-\delta V^{\nu})\Psi\|=\|(V(\cdot-K^{\nu}(a,L)/m_{\mathrm{eff}})-V(\cdot-K^{\nu}/m_{\mathrm{eff}}))\Psi\|$$

$$\le\sum_{\mu=1}^{3}\frac{\|\nabla_{\mu}V\|_{\infty}}{m_{\mathrm{eff}}}\|(K^{\nu}_{\mu}-K^{\nu}_{\mu}(a,L))\Psi\|.$$

Since

$$\|(K^{\nu}_{\mu}-K^{\nu}_{\mu}(a,L))\Psi\|$$

$$\le\sum_{j=1,2}C\left(\left\|\frac{1}{\sqrt{\omega}}\left\{\frac{\varrho^{\nu}_{\mu}(\cdot,j)}{\omega^{3/2}}-\frac{\sum_{|l|\le L}\chi_{\Gamma(l,a)}\varrho^{\nu}_{\mu}(l,j)}{(\sum_{|l|\le L}\chi_{\Gamma(l,a)}\omega(l))^{3/2}}\right\}\right\|\right.$$

$$\left.+\left\|\frac{\varrho^{\nu}_{\mu}(\cdot,j)}{\omega^{3/2}}-\frac{\sum_{|l|\le L}\chi_{\Gamma(l,a)}\varrho^{\nu}_{\mu}(l,j)}{(\sum_{|l|\le L}\chi_{\Gamma(l,a)}\omega(l))^{3/2}}\right\|\right)(\|H^{\nu\,1/2}_{\mathrm{f}}\Psi\|+\|\Psi\|)$$

$$=\epsilon(a,L)'(\|H^{\nu\,1/2}_{\mathrm{f}}\Psi\|+\|\Psi\|)$$

with some constant C, and $\epsilon(a,L)'$ satisfies that

$$\lim_{L\to\infty}\lim_{a\to\infty}\epsilon(a,L)'=0,$$

we have

$$\|(\widehat{H}^{\nu}_{\Lambda}-z)^{-1}\Psi-(\widehat{H}^{\nu}_{\Lambda}(a,L)-z)^{-1}\Psi\|$$

$$\le\|(\widehat{H}^{\nu}_{\Lambda}(a,L)-z)^{-1}\|\|(\widehat{H}^{\nu}_{\Lambda}(a,L)-\widehat{H}^{\nu}_{\Lambda})(\widehat{H}^{\nu}_{\Lambda}-z)^{-1}\Psi\|$$

$$\le\frac{(\max_{\mu}\|\nabla_{\mu}V\|_{\infty})\epsilon(a,L)'}{|\Im z|m_{\mathrm{eff}}}\left(\|H^{\nu\,1/2}_{\mathrm{f}}(\widehat{H}^{\nu}_{\Lambda}-z)^{-1}\Psi\|+\|(\widehat{H}^{\nu}_{\Lambda}-z)^{-1}\Psi\|\right)$$

$$+\frac{\epsilon(a,L)}{|\Im z|}\|H^{\nu\,1/2}_{\mathrm{f}}(\widehat{H}^{\nu}_{\Lambda}-z)^{-1}\Psi\|.$$

Since

$$\|H_{\mathbf{f}}^{\nu}(\widehat{H}_{\Lambda}^{\nu} - z)^{-1}\Psi\| \leq C'\|\Psi\|$$

with some constant C', we have

$$\|(\widehat{H}_{\Lambda}^{\nu} - z)^{-1}\Psi - (\widehat{H}_{\Lambda}^{\nu}(a, L) - z)^{-1}\Psi\| \leq \epsilon''(a, L)\|\Psi\|$$

with $\epsilon''(a, L)$ such that

$$\lim_{L\to\infty} \lim_{a\to\infty} \epsilon''(a, L) = 0.$$

Hence the lemma follows. $\qquad\square$

4.2. Spectrum of $\hat{H}_{\lambda}^{\nu}$

Let $f \in L^2(\mathbb{R}^3)$. We identify

$$\{f(l)\}_{l\in\mathbb{Z}^3} \in \ell^2(\mathbb{Z}^3)$$

with

$$\sum_{l\in\mathbb{Z}^3} f(l)\chi_{\Gamma(l,a)}(\cdot) \in L^2(\mathbb{R}^3).$$

By this identification we regard $\ell^2(\mathbb{Z}^3)$ as

$$\ell^2 = \ell^2(\mathbb{Z}^3) \subset L^2(\mathbb{R}^3).$$

Let

$$\mathcal{H}_a = L^2(\mathbb{R}^3) \otimes \mathcal{F}(\ell^2 \otimes \mathbb{C}^2).$$

By the following fundamental identification

$$\mathcal{F}(L^2(\mathbb{R}^3 \times \{1,2\})) \cong \mathcal{F}(L^2(\mathbb{R}^3) \otimes \mathbb{C}^2) = \mathcal{F}([\ell^2 \oplus \ell^{2\perp}] \otimes \mathbb{C}^2)$$

$$\cong \mathcal{F}(\ell^2 \otimes \mathbb{C}^2) \otimes \mathcal{F}(\ell^{2\perp} \otimes \mathbb{C}^2) \cong \mathcal{F}(\ell^2 \otimes \mathbb{C}^2) \otimes \left[\bigoplus_{n=0}^{\infty} \mathcal{F}^{(n)}(\ell^{2\perp} \otimes \mathbb{C}^2)\right]$$

$$= \mathcal{F}(\ell^2 \otimes \mathbb{C}^2) \otimes \left[\bigoplus_{n=1}^{\infty} \mathcal{F}^{(n)}(\ell^{2\perp} \otimes \mathbb{C}^2) \oplus \mathbb{C}\right]$$

$$\cong \left(\mathcal{F}(\ell^2 \otimes \mathbb{C}^2) \otimes \left[\bigoplus_{n=1}^{\infty} \mathcal{F}^{(n)}(\ell^{2\perp} \otimes \mathbb{C}^2)\right]\right) \oplus \mathcal{F}(\ell^2 \otimes \mathbb{C}^2),$$

we have

$$\mathcal{H} \cong \left(\mathcal{H}_a \otimes \left[\bigoplus_{n=1}^{\infty} \mathcal{F}^{(n)}(\ell^{2\perp} \otimes \mathbb{C}^2)\right]\right) \oplus \mathcal{H}_a.$$

In particular we see that

$$\mathcal{H}_a^{\perp} \cong \mathcal{H}_a \otimes \left[\bigoplus_{n=1}^{\infty} \mathcal{F}^{(n)}(\ell^{2\perp} \otimes \mathbb{C}^2) \right]. \tag{9}$$

The following lemma is easily proven.

Lemma 3: $\widehat{H}_\Lambda^\nu(a, L)$ *is reduced by* $\mathcal{H}_a$ *and, under identification (9), we have*

$$\widehat{H}_\Lambda^\nu(a, L)\Big\lceil_{\mathcal{H}_a^{\perp}} \cong \left(\widehat{H}_\Lambda^\nu(a, L)\Big\lceil_{\mathcal{H}_a} \right) \otimes 1 + 1 \otimes H_{\mathrm{f}}^\nu(a, L)\Big\lceil_{\left[\oplus_{n=1}^{\infty} \mathcal{F}^{(n)}(\ell^{2\perp} \otimes \mathbb{C}^2) \right]}.$$

In particular

$$\inf \sigma(\widehat{H}_\Lambda^\nu(a, L)\Big\lceil_{\mathcal{H}_a^{\perp}}) \geq \inf \sigma(H_{\mathrm{f}}^\nu(a, L)) + \nu.$$

Note that $H_{\mathrm{f}}^\nu(a, L)\lceil_{\mathcal{H}_a}$ has a purely discrete spectrum, *i.e.*,

$$\sigma(H_{\mathrm{f}}^\nu(a, L)\lceil_{\mathcal{H}_a}) = \sigma_{\mathrm{disc}}(H_{\mathrm{f}}^\nu(a, L)\lceil_{\mathcal{H}_a}).$$

Lemma 4: *Let* $\Psi \in D(p^2) \cap D(H_{\mathrm{f}}^{1/2})$. *Then*

(1) $\Psi \in D(\delta V^\nu)$ *and*

$$\|\delta V^\nu \Psi\| \leq \theta(\Lambda)(\|H_{\mathrm{f}}^{\nu 1/2}\Psi\| + \|\Psi\|), \tag{10}$$

where

$$\theta(\Lambda) = \frac{C\|\nabla V\|_\infty}{m_{\mathrm{eff}}}(\|\chi_\Lambda/\omega^2\| + \|\chi_\Lambda/\omega^{3/2}\|)$$

with some constant C.

(2) $\Psi \in D(\delta V^\nu(a, L))$ *and*

$$\|\delta V^\nu(a, L)\Psi\| \leq \theta(\Lambda, a, L)(\|H_{\mathrm{f}}^\nu(a, L)^{1/2}\Psi\| + \|\Psi\|), \tag{11}$$

where

$$\theta(\Lambda, a, L) = \frac{C'\|\nabla V\|_\infty}{m_{\mathrm{eff}}}$$

$$\times \left(\left\| \frac{1}{\sqrt{\omega}} \left\{ \frac{\sum_{|l| \leq L} \chi_{\Gamma(l,a)} \chi_\Lambda(l)}{(\sum_{|l| \leq L} \chi_{\Gamma(l,a)} \omega(l))^{3/2}} \right\} \right\| + \left\| \frac{\sum_{|l| \leq L} \chi_{\Gamma(l,a)} \chi_\Lambda(l)}{(\sum_{|l| \leq L} \chi_{\Gamma(l,a)} \omega(l))^{3/2}} \right\| \right)$$

with some constant C'.

Proof: We have

$$\|\delta V^\nu \Psi\| \le \sum_{\mu=1}^{3} \frac{\|\nabla_\mu V\|_\infty}{m_{\text{eff}}} \|K_\mu \Psi\|.$$

(10) follows from (4). (11) is similarly proven. $\qquad\square$

Remark 6: $\theta(\Lambda)$ and $\theta(\Lambda, a, L)$ are independent of ν.

We write $A \le B$, if $D(B) \subset D(A)$ and $(\psi, A\psi) \le (\psi, B\psi)$ for $\psi \in D(B)$.
  We have

$$|(\Psi, \delta V^\nu \Psi)| \le \theta(\Lambda) \left\{ \|\Psi\|(\|H_{\text{f}}^{\nu\,1/2}\Psi\| + \|\Psi\|) \right\}$$

$$\le \theta(\Lambda) \left(\|\Psi\|^2 + \frac{1}{2}\|\Psi\|^2 + \frac{1}{2}\|H_{\text{f}}^{\nu\,1/2}\Psi\|^2 \right) = (\Psi, \theta(\Lambda)(\frac{3}{2} + \frac{1}{2}H_{\text{f}}^\nu)\Psi).$$

Thus we conclude

$$-\theta(\Lambda) \left(\frac{1}{2}H_{\text{f}}^\nu + \frac{3}{2} \right) \le \delta V^\nu \le \theta(\Lambda) \left(\frac{1}{2}H_{\text{f}}^\nu + \frac{3}{2} \right). \tag{12}$$

Hence for $f \in C_0^\infty(\mathbb{R}^3)$,

$$\inf \sigma(\widehat{H}_\Lambda^\nu) \le (f \otimes \Omega, \widehat{H}_\Lambda^\nu f \otimes \Omega) \le (f, [H_{\text{eff}} + g(\Lambda) + \frac{3}{2}\theta(\Lambda)]f).$$

In particular, since $C_0^\infty(\mathbb{R}^3)$ is a core of H_{eff}, we have

$$\inf \sigma(\widehat{H}_\Lambda^\nu) \le \inf \sigma(H_{\text{eff}}) + \frac{3\theta(\Lambda)}{2} + g(\Lambda).$$

Similarly we have

$$-\theta(\Lambda, a, L) \left(\frac{1}{2}H_{\text{f}}^\nu(a, L) + \frac{3}{2} \right) \le \delta V^\nu(a, L) \le \theta(\Lambda, a, L) \left(\frac{1}{2}H_{\text{f}}^\nu(a, L) + \frac{3}{2} \right) \tag{13}$$

and hence

$$\inf \sigma(\widehat{H}_\Lambda^\nu(a, L)) \le \inf \sigma(H_{\text{eff}}) + \frac{3\,\theta(\Lambda, a, L)}{2} + g(\Lambda).$$

We summarize above statements in the following lemma.

Lemma 5: *We have*

$$\inf \sigma(\widehat{H}_\Lambda^\nu) \le \inf \sigma(H_{\text{eff}}) + \frac{3\,\theta(\Lambda)}{2} + g(\Lambda)$$

and

$$\inf \sigma(\widehat{H}_\Lambda^\nu(a, L)) \le \inf \sigma(H_{\text{eff}}) + \frac{3\,\theta(\Lambda, a, L)}{2} + g(\Lambda).$$

We see that as $\Lambda \to \infty$

$$\|\chi_\Lambda/\omega^l\| \sim \Lambda^{-l+\frac{3}{2}}, \quad l \neq \frac{3}{2},$$

$$\|\chi_\Lambda/\omega^l\| \sim \sqrt{\log \Lambda}, \quad l = \frac{3}{2}.$$

From this it follows that as $\Lambda \to \infty$

$$m_{\text{eff}} = m + \frac{8\pi}{3}(\Lambda - \lambda) \sim \Lambda, \tag{14}$$

$$\|\chi_\Lambda/\omega^2\| \sim \Lambda^{-1/2}, \tag{15}$$

$$\|\chi_\Lambda/\omega^{3/2}\| \sim \sqrt{\log \Lambda}. \tag{16}$$

Lemma 6: *It follows that*

$$\lim_{\Lambda \to \infty} \theta(\Lambda) = 0$$

and

$$\lim_{\Lambda \to \infty} \theta(\Lambda, a, L) = 0.$$

Proof: It follows from (14), (15), (16) and the definitions of $\theta(\Lambda)$ and $\theta(\Lambda, a, L)$. $\qquad\square$

We set

$$\Sigma = \Sigma(\Lambda) = \inf \sigma(H_{\text{eff}}).$$

It is easily seen that

$$\lim_{\Lambda \to \infty} \Sigma = \inf_x V(x) < 0. \tag{17}$$

For a self-adjoint operator M, the spectral projection of M on a Borel set $\mathcal{B} \subset \mathbb{R}$ is denoted by

$$E_{\mathcal{B}}^M.$$

Lemma 7: *Let a, L and Λ be sufficiently large such that*

$$\min\{|\Sigma|/3, 2\} > \theta(\Lambda, a, L). \tag{18}$$

Then for ν such that $|\Sigma| > 3\,\theta(\Lambda, a, L) + \nu$,

$$\widehat{H}_\Lambda^\nu(a, L)\lceil_{\mathcal{H}_a} - \inf \sigma(\widehat{H}_\Lambda^\nu(a, L)\lceil_{\mathcal{H}_a}) - \nu$$

$$\geq E_{[0,|\Sigma|)}^{H_{\text{eff}}-\Sigma} \otimes \left((1 - \frac{\theta(\Lambda, a, L)}{2})H_{\text{f}}^\nu(a, L) - 3\,\theta(\Lambda, a, L) - \nu \right).$$

Proof: In this proof we set

$$H = \widehat{H}_\Lambda^\nu(a, L)\!\restriction_{\mathcal{H}_a}$$

as an operator in $\mathcal{H}_a$ and

$$\theta = \theta(\Lambda, a, L).$$

We directly see that

$$
\begin{aligned}
H &- \inf \sigma(H) - \nu \\
&= H_{\mathrm{eff}} + \delta V^\nu(a, L) + H_{\mathrm{f}}^\nu(a, L) + g(\Lambda) - \inf \sigma(H) - \nu \\
&\geq H_{\mathrm{eff}} + \delta V^\nu(a, L) + H_{\mathrm{f}}^\nu(a, L) + g(\Lambda) - g(\Lambda) - \frac{3}{2}\theta - \Sigma - \nu \\
&\geq H_{\mathrm{eff}} + (1 - \frac{\theta}{2})H_{\mathrm{f}}^\nu(a, L) - \frac{3}{2}\theta - \frac{3}{2}\theta - \Sigma - \nu \\
&= (H_{\mathrm{eff}} - \Sigma) + (1 - \frac{\theta}{2})H_{\mathrm{f}}^\nu(a, L) - 3\theta - \nu \\
&\geq |\Sigma| E_{[|\Sigma|,\infty)}^{H_{\mathrm{eff}} - \Sigma} \otimes 1 - \theta'(E_{[0,|\Sigma|)}^{H_{\mathrm{eff}} - \Sigma} + E_{[|\Sigma|,\infty)}^{H_{\mathrm{eff}} - \Sigma}) \otimes 1 \\
&\quad + (1 - \frac{\theta}{2})(E_{[0,|\Sigma|)}^{H_{\mathrm{eff}} - \Sigma} + E_{[|\Sigma|,\infty)}^{H_{\mathrm{eff}} - \Sigma}) \otimes H_{\mathrm{f}}^\nu(a, L),
\end{aligned}
$$

where $\theta' = 3\theta + \nu$. Then

$$
H - \inf \sigma(H) - \nu \geq (|\Sigma| - \theta') E_{[|\Sigma|,\infty)}^{H_{\mathrm{eff}} - \Sigma} \otimes 1 + (1 - \frac{\theta}{2}) E_{[|\Sigma|,\infty)}^{H_{\mathrm{eff}} - \Sigma} \otimes H_{\mathrm{f}}^\nu(a, L)
$$

$$
+ E_{[0,|\Sigma|)}^{H_{\mathrm{eff}} - \Sigma} \otimes \left((1 - \frac{\theta}{2}) H_{\mathrm{f}}^\nu(a, L) - \theta' \right).
$$

Since by (18),

$$|\Sigma| - \theta' = |\Sigma| - 3\theta - \nu > 0$$

and

$$1 - \frac{\theta}{2} > 0,$$

we have

$$
H - \inf \sigma(H) - \nu \geq E_{[0,|\Sigma|)}^{H_{\mathrm{eff}} - \Sigma} \otimes \left((1 - \frac{\theta}{2}) H_{\mathrm{f}}^\nu(a, L) - \theta' \right).
$$

Thus the lemma follows. $\qquad\square$

Set

$$T = \widehat{H}_\Lambda^\nu(a, L) - \inf \sigma(\widehat{H}_\Lambda^\nu(a, L)) - \nu.$$

Define

$$\mathcal{H}_a^{(+)} = E_{[0,\infty)}^T \mathcal{H}_a,$$

$$\mathcal{H}_a^{(-)} = E_{[-\nu,0)}^T \mathcal{H}_a.$$

Lemma 8: *Suppose that* $\min\{|\Sigma|/3, 2\} > \theta(\Lambda, a, L)$. *Then for* ν *such that* $|\Sigma| > 3\theta(\Lambda, a, L) + \nu$, $T\lceil_{\mathcal{H}_a^{(-)}}$ *has a purely discrete spectrum, i.e.,*

$$\sigma(\widehat{H}_\Lambda^\nu(a, L)\Big\lceil_{\mathcal{H}_a}) \cap [\inf \sigma(\widehat{H}_\Lambda^\nu(a, L)\Big\lceil_{\mathcal{H}_a}), \ \inf \sigma(\widehat{H}_\Lambda^\nu(a, L)\Big\lceil_{\mathcal{H}_a}) + \nu)$$

$$\subset \sigma_{\mathrm{disc}}(\widehat{H}_\Lambda^\nu(a, L)\Big\lceil_{\mathcal{H}_a}).$$

Proof: Let $\{\phi_n\}_n$ be a complete orthonormal system of $\mathcal{H}_a^{(-)}$ and $\{\psi_m\}_m$ a complete orthonormal system of $\mathcal{H}_a^{(+)}$. We see that by Lemma 7,

$$0 \geq \mathrm{Tr}\, T\lceil_{\mathcal{H}_a^{(-)}} = \sum_n (\phi_n, T\phi_n) \geq \sum_n (\phi_n, T'\phi_n),$$

where

$$T' = E_{[0,|\Sigma|)}^{H_{\mathrm{eff}} - \Sigma} \otimes \left((1 - \frac{\theta(\Lambda, a, L)}{2}) H_{\mathrm{f}}^\nu(a, L) - 3\theta(\Lambda, a, L) - \nu \right).$$

Set

$$T'_- = T' E_{(-\infty,0)}^{T'}.$$

Then

$$0 \geq \mathrm{Tr}\, T\lceil_{\mathcal{H}_a^{(-)}} \geq \sum_n (\phi_n, T'_- \phi_n)$$

$$\geq \sum_n (\phi_n, T'_- \phi_n) + \sum_m (\psi_m, T'_- \psi_m) = \mathrm{Tr} T'_-.$$

Hence we obtain that

$$\left| \mathrm{Tr}\, T\lceil_{\mathcal{H}_a^{(-)}} \right| \leq \left| \mathrm{Tr}\, T'_- \right|$$

$$= \mathrm{Tr} E_{[0,|\Sigma|)}^{H_{\mathrm{eff}} - \Sigma} \times \left| \mathrm{Tr} \left((1 - \frac{\theta(\Lambda, a, L)}{2}) H_{\mathrm{f}}^\nu(a, L) - 3\theta(\Lambda, a, L) - \nu \right)_- \Big\lceil_{\mathcal{H}_a} \right|,$$

where $(\cdots)_-$ denotes the negative part of the spectrum of $(\cdots)$. Since

$$\sigma(H_{\mathrm{f}}^\nu(a, L)\lceil_{\mathcal{H}_a}) = \sigma_{\mathrm{disc}}(H_{\mathrm{f}}^\nu(a, L)\lceil_{\mathcal{H}_a})$$

and

$$\left| \mathrm{Tr}\, E^{H_{\mathrm{eff}}-\Sigma}_{[0,|\Sigma|)} \right| < \infty,$$

it follows that

$$\left| \mathrm{Tr}\, T \!\restriction_{\mathcal{H}_a^{(-)}} \right| < \infty$$

Thus the lemma follows. $\qquad\square$

Lemma 9: *Suppose that* $\min\{|\Sigma|/3, 2\} > \theta(\Lambda, a, L)$. *Then for* ν *such that* $|\Sigma| > 3\theta(\Lambda, a, L) + \nu$,

$$\sigma(\widehat{H}^\nu_\Lambda(a, L)) \cap [\inf \sigma(\widehat{H}^\nu_\Lambda(a, L)), \inf \sigma(\widehat{H}^\nu_\Lambda(a, L)) + \nu) \subset \sigma_{\mathrm{disc}}(\widehat{H}^\nu_\Lambda(a, L)).$$

Proof: We have by Lemmas 3 and 8,

$$\sigma(\widehat{H}^\nu_\Lambda(a, L)) = \sigma(\widehat{H}^\nu_\Lambda(a, L)\!\restriction_{\mathcal{H}_a^\perp}) \cup \sigma(\widehat{H}^\nu_\Lambda(a, L)\!\restriction_{\mathcal{H}_a}),$$

$$\sigma(\widehat{H}^\nu_\Lambda(a, L)\!\restriction_{\mathcal{H}_a^\perp}) \subset [\inf \sigma(\widehat{H}_\Lambda(a, L)) + \nu, \infty),$$

and

$$\sigma(\widehat{H}^\nu_\Lambda(a, L)\!\restriction_{\mathcal{H}_a}) \cap [\inf \sigma(\widehat{H}^\nu_\Lambda(a, L)\!\restriction_{\mathcal{H}_a}), \inf \sigma(\widehat{H}^\nu_\Lambda(a, L)\!\restriction_{\mathcal{H}_a}) + \nu)$$

$$\subset \sigma_{\mathrm{disc}}(\widehat{H}^\nu_\Lambda(a, L)\!\restriction_{\mathcal{H}_a}).$$

Hence the lemma follows. $\qquad\square$

Lemma 10: *Suppose that* $\min\{|\Sigma|/3, 2\} > \theta(\Lambda)$. *Then*

$$\sigma(\widehat{H}^\nu_\Lambda) \cap [\inf \sigma(\widehat{H}^\nu_\Lambda), \inf \sigma(\widehat{H}^\nu_\Lambda) + \nu) \subset \sigma_{\mathrm{disc}}(\widehat{H}^\nu_\Lambda).$$

In particular $\widehat{H}^\nu_\Lambda$ *has a ground state.*

Proof: Note that

$$\lim_{L\to\infty} \lim_{a\to\infty} \theta(\Lambda, a, L) = \theta(\Lambda)$$

Then by Lemmas 2 and 9, the lemma follows from Ref.[21] $\qquad\square$

4.3. *Binding*

The number operator N in $\mathcal{F}$ is defined by

$$N = \sum_{j=1,2} \int a^*(k,j)a(k,j)d^3k.$$

A ground state of $\widehat{H}_\Lambda^\nu$ is denoted by $\varphi_{\mathrm{g}}(\nu)$.

Lemma 11: *Suppose that* $\min\{|\Sigma|/3, 2\} > \theta(\Lambda)$. *Then for ν such that* $|\Sigma| > 3\theta(\Lambda) + \nu$,

$$\frac{\|N_{\mathrm{f}}^{1/2}\varphi_{\mathrm{g}}(\nu)\|}{\|\varphi_{\mathrm{g}}(\nu)\|} \le \frac{C\|\chi_\Lambda/\omega^{5/2}\|(\max_\mu \|\nabla_\mu V\|_\infty)}{m_{\mathrm{eff}}}$$

with some constant C.

Proof: We set

$$E = \inf \sigma(\widehat{H}_\Lambda^\nu).$$

Since

$$[\widehat{H}_\Lambda^\nu, a(k,j)] = -(\omega(k) + \nu)a(k,j) + [\delta V^\nu, a(k,j)],$$

we have

$$\widehat{H}_\Lambda^\nu a(k,j)\varphi_{\mathrm{g}}(\nu) = -(\omega(k) + \nu)a(k,j)\varphi_{\mathrm{g}}(\nu)$$

$$+ Ea(k,j)\varphi_{\mathrm{g}}(\nu) + [\delta V^\nu, a(k,j)]\varphi_{\mathrm{g}}(\nu).$$

Hence

$$(\widehat{H}_\Lambda^\nu - E + \omega(k) + \nu)a(k,j)\varphi_{\mathrm{g}}(\nu) = [\delta V^\nu, a(k,j)]\varphi_{\mathrm{g}}(\nu).$$

Here we see that

$$[\delta V^\nu, a(k,j)] = [V(\cdot - K^\nu/m_{\mathrm{eff}}), a(k,j)] = [e^{-i\frac{p\cdot K^\nu}{m_{\mathrm{eff}}}}Ve^{i\frac{p\cdot K^\nu}{m_{\mathrm{eff}}}}, a(k,j)]$$

$$= e^{-i\frac{p\cdot K^\nu}{m_{\mathrm{eff}}}}[V, e^{i\frac{p\cdot K^\nu}{m_{\mathrm{eff}}}}a(k,j)e^{-i\frac{p\cdot K^\nu}{m_{\mathrm{eff}}}}]e^{i\frac{p\cdot K^\nu}{m_{\mathrm{eff}}}}.$$

Since

$$e^{i\frac{p\cdot K^\nu}{m_{\mathrm{eff}}}}a(k,j)e^{-i\frac{p\cdot K^\nu}{m_{\mathrm{eff}}}} = a(k,j) - \frac{i}{\sqrt{2}m_{\mathrm{eff}}}p\cdot\rho^\nu(k,j),$$

it follows that

$$[\delta V^\nu, a(k,j)] = e^{-i\frac{p\cdot K^\nu}{m_{\mathrm{eff}}}}[V, -\frac{i}{\sqrt{2}m_{\mathrm{eff}}}p\cdot\rho^\nu(k,j)]e^{i\frac{p\cdot K^\nu}{m_{\mathrm{eff}}}}$$

$$= e^{-i\frac{p \cdot K^\nu}{m_{\mathrm{eff}}}} \left(\frac{1}{\sqrt{2m_{\mathrm{eff}}}} (\nabla V) \cdot \rho^\nu(k,j) \right) e^{i\frac{p \cdot K^\nu}{m_{\mathrm{eff}}}}.$$

Thus we obtain a pull-through formula:

$$a(k,j)\varphi_{\mathrm{g}}(\nu) = (\widehat{H}_\Lambda^\nu - E + \omega(k) + \nu)^{-1} \times$$

$$\times e^{-i\frac{p \cdot K^\nu}{m_{\mathrm{eff}}}} \left(\frac{1}{\sqrt{2m_{\mathrm{eff}}}} (\nabla V) \cdot \rho^\nu(k,j) \right) e^{i\frac{p \cdot K^\nu}{m_{\mathrm{eff}}}} \varphi_{\mathrm{g}}(\nu). \tag{19}$$

Using identity (19) we see that

$$(N^{1/2}\varphi_{\mathrm{g}}(\nu), N^{1/2}\varphi_{\mathrm{g}}(\nu)) = \sum_{j=1,2} \int \|a(k,j)\varphi_{\mathrm{g}}(\nu)\|^2 d^3k$$

$$= \sum_{j=1,2} \int \left\| (\widehat{H}_\Lambda^\nu - E + \omega(k) + \nu)^{-1} \times \right.$$

$$\left. \times e^{-i\frac{p \cdot K^\nu}{m_{\mathrm{eff}}}} \left(\frac{1}{\sqrt{2m_{\mathrm{eff}}}} (\nabla V) \cdot \rho^\nu(k,j) \right) e^{i\frac{p \cdot K^\nu}{m_{\mathrm{eff}}}} \varphi_{\mathrm{g}}(\nu) \right\|^2 d^3k$$

$$\leq 3 \sum_{\mu=1}^{3} \sum_{j=1,2} \int \left(\frac{1}{\omega(k)} \|\nabla_\mu V\|_\infty \right)^2 \left| \frac{1}{\sqrt{2m_{\mathrm{eff}}}} \varrho_\mu^\nu(k,j) \right|^2 d^3k \|\varphi_{\mathrm{g}}(\nu)\|^2$$

$$\leq \frac{3}{2} \sum_{\mu=1}^{3} \left(\frac{\|\nabla_\mu V\|_\infty}{m_{\mathrm{eff}}} \right)^2 \sum_{j=1,2} \|\varrho^\nu(\cdot,j)/\omega\|^2 \|\varphi_{\mathrm{g}}(\nu)\|^2$$

$$\leq C \left(\frac{(\max_\mu \|\nabla_\mu V\|_\infty)}{m_{\mathrm{eff}}} \right)^2 \|\chi_\Lambda/\omega^{5/2}\|^2 \|\varphi_{\mathrm{g}}(\nu)\|^2.$$

Hence the lemma follows. $\qquad\square$

Lemma 12: *Let P_Ω be the projection onto $\{\alpha\Omega \mid \alpha \in \mathbb{C}\}$ and*

$$Q = E_{[\Sigma+\delta,\infty)}^{H_{\mathrm{eff}}} \otimes P_\Omega$$

with some $\delta > 0$ such that

$$\delta > \frac{3}{2}\theta(\Lambda).$$

Suppose that $\min\{|\Sigma|/3, 2\} > \theta(\Lambda)$. Then for ν such that $|\Sigma| > 3\theta(\Lambda) + \nu$,

$$\frac{\|Q\varphi_{\mathrm{g}}(\nu)\|}{\|\varphi_{\mathrm{g}}(\nu)\|} \leq \sqrt{\frac{\theta(\Lambda)}{\delta - \frac{3}{2}\theta(\Lambda)}}.$$

Proof: Since $(\varphi_g(\nu), Q[\widehat{H}_\Lambda^\nu - \inf \sigma(\widehat{H}_\Lambda^\nu)]\varphi_g(\nu)) = 0$, we have

$$(\varphi_g(\nu), Q[H_{\text{eff}} - \inf \sigma(\widehat{H}_\Lambda^\nu) + g(\Lambda)]\varphi_g(\nu)) = -(\varphi_g(\nu), Q\delta V^\nu \varphi_g(\nu)).$$

The left-hand side above is estimated as

$$(\varphi_g(\nu), Q[H_{\text{eff}} - \inf \sigma(\widehat{H}_\Lambda^\nu) + g(\Lambda)]\varphi_g(\nu))$$

$$\geq [\Sigma + \delta - \inf \sigma(\widehat{H}_\Lambda^\nu) + g(\Lambda)](\varphi_g(\nu), Q\varphi_g(\nu)).$$

Note that

$$\Sigma + \delta + g(\Lambda) - \inf \sigma(\widehat{H}_\Lambda^\nu) \geq \Sigma + \delta + g(\Lambda) - \Sigma - \frac{3}{2}\theta(\Lambda) - g(\Lambda)$$

$$= \delta - \frac{3}{2}\theta(\Lambda) > 0.$$

Then

$$(\varphi_g(\nu), Q[H_{\text{eff}} - \inf \sigma(\widehat{H}_\Lambda^\nu) + g(\Lambda)]\varphi_g(\nu)) \geq (\delta - \frac{3}{2}\theta(\Lambda))\|Q\varphi_g(\nu)\|^2 > 0.$$

Moreover

$$|(\varphi_g(\nu), Q\delta V^\nu \varphi_g(\nu))| = |(\delta V^\nu Q\varphi_g(\nu), \varphi_g(\nu))| \leq \|\delta V^\nu Q\varphi_g(\nu)\|\|\varphi_g(\nu)\|$$

$$\leq \theta(\Lambda) \left(\|H_f^{1/2}Q\varphi_g(\nu)\| + \|Q\varphi_g(\nu)\| \right) \|\varphi_g(\nu)\|$$

$$= \theta(\Lambda)\|Q\varphi_g(\nu)\|\|\varphi_g(\nu)\| \leq \theta(\Lambda)\|\varphi_g(\nu)\|^2.$$

Hence we have

$$0 < (\delta - \frac{3}{2}\theta(\Lambda))\|Q\varphi_g(\nu)\|^2 \leq \theta(\Lambda)\|\varphi_g(\nu)\|^2.$$

The lemma follows.   □

4.4. *Proof of the main theorem*

We normalize $\varphi_g(\nu)$, *i.e.*, $\|\varphi_g(\nu)\| = 1$. Take a subsequence ν' such that $\varphi_g(\nu')$ weakly converges to a vector φ_g as $\nu' \to \infty$. It is established in Ref.[1] that if $\varphi_g \neq 0$, then φ_g is a ground state of H_Λ.

Proof of Theorem 1

It is enough to prove $\varphi_g \neq 0$. Note that on $\mathcal{F}$

$$N + P_\Omega \geq 1.$$

Then on $\mathcal{H}$

$$1 \otimes N + 1 \otimes P_\Omega \geq 1.$$

Hence

$$1 \otimes N + (E_{[\Sigma,\Sigma+\delta)}^{H_{\text{eff}}} + E_{[\Sigma+\delta,\infty)}^{H_{\text{eff}}}) \otimes P_\Omega$$

$$= 1 \otimes N + E_{[\Sigma,\Sigma+\delta)}^{H_{\text{eff}}} \otimes P_\Omega + Q \geq 1,$$

and

$$E_{[\Sigma,\Sigma+\delta)}^{H_{\text{eff}}} \otimes P_\Omega \geq 1 - 1 \otimes N - Q. \tag{20}$$

Suppose that $\min\{|\Sigma|/3, 2\} > \theta(\Lambda)$ and $\delta > \frac{3}{2}\theta(\Lambda)$. Then for ν' such that $|\Sigma| > 3\theta(\Lambda) + \nu'$, we have by (20),

$$(\varphi_{\text{g}}(\nu'), \left[E_{[\Sigma,\Sigma+\delta)}^{H_{\text{eff}}} \otimes P_\Omega\right]\varphi_{\text{g}}(\nu'))$$

$$\geq \|\varphi_{\text{g}}(\nu')\|^2 - (\varphi_{\text{g}}(\nu'), N\varphi_{\text{g}}(\nu')) - (\varphi_{\text{g}}(\nu'), Q\varphi_{\text{g}}(\nu'))$$

$$\geq \|\varphi_{\text{g}}(\nu')\|^2 - \frac{C\|\chi_\Lambda/\omega^{5/2}\|}{m_{\text{eff}}}(\max_\mu \|\nabla_\mu V\|_\infty)\|\varphi_{\text{g}}(\nu')\|^2$$

$$- \frac{\theta(\Lambda)}{\delta - \frac{3}{2}\theta(\Lambda)}\|\varphi_{\text{g}}(\nu')\|^2$$

$$= 1 - \frac{C\|\chi_\Lambda/\omega^{5/2}\|}{m_{\text{eff}}}(\max_\mu \|\nabla_\mu V\|_\infty) - \frac{\theta(\Lambda)}{\delta - \frac{3}{2}\theta(\Lambda)}.$$

Note that

$$\lim_{\Lambda \to \infty} \frac{\|\chi_\Lambda/\omega^{5/2}\|}{m_{\text{eff}}} = 0,$$

$$\lim_{\Lambda \to \infty} \frac{\theta(\Lambda)}{\delta - \frac{3}{2}\theta(\Lambda)} = 0.$$

Hence for sufficiently large Λ,

$$(\varphi_{\text{g}}(\nu'), \left[E_{[\Sigma,\Sigma+\delta)}^{H_{\text{eff}}} \otimes P_\Omega\right]\varphi_{\text{g}}(\nu')) > \epsilon$$

uniformly in ν' with some $\epsilon > 0$. Take $\nu' \to \infty$ on the both sides above. Since $E_{[\Sigma,\Sigma+\delta)}^{H_{\text{eff}}} \otimes P_\Omega$ is a finite rank operator, we have

$$(\varphi_{\text{g}}, \left[E_{[\Sigma,\Sigma+\delta)}^{H_{\text{eff}}} \otimes P_\Omega\right]\varphi_{\text{g}}) > \epsilon.$$

In particular $\varphi_{\text{g}} \neq 0$. Then φ_{g} is a ground state of $\widehat{H}_\Lambda$. $\qquad\square$

Remark 7: φ_{g} satisfies that $\varphi_{\mathrm{g}} \in D(N^{1/2})$. Actually in a manner similar to (19) we have

$$a(k,j)\varphi_{\mathrm{g}} = \left(H_\Lambda - \inf \sigma(H_\Lambda) + \omega(k)\right)^{-1}$$

$$\times e^{-i\frac{p\cdot K}{m_{\mathrm{eff}}}} \left(\frac{1}{\sqrt{2}m_{\mathrm{eff}}}(\nabla V) \cdot \varrho(k,j)\right) e^{i\frac{p\cdot K}{m_{\mathrm{eff}}}} \varphi_{\mathrm{g}}.$$

Using this we can obtain that

$$(N^{1/2}\varphi_{\mathrm{g}}, N^{1/2}\varphi_{\mathrm{g}}) = \sum_{j=1,2} \int \|a(k,j)\varphi_{\mathrm{g}}\|^2 d^3k < \infty.$$

5. Remarks

5.1. *Multiplicity and localization*

In Section 3 we established that H_Λ has a ground state for $\Lambda > \Lambda_*$ with some Λ_* even in the case where the decoupled Hamiltonian $\frac{1}{2m}p^2 + V + H_{\mathrm{f}}$ has no ground state. The uniqueness of ground state is established by means of a functional integral representation of e^{-tH_Λ}. There exists a unitary operator such that

$$\mathcal{U} : \mathcal{H} = L^2(\mathbb{R}^3) \otimes \mathcal{F} \longrightarrow L^2(\mathbb{R}^3 \times \mathbf{S}, dx \otimes d\mu),$$

where $d\mu$ is a Gaussian probability measure on the following direct sum of the real-valued Schwartz distribution,

$$\mathbf{S} = \mathcal{S}'_{\mathrm{real}}(\mathbb{R}^3) \times \mathcal{S}'_{\mathrm{real}}(\mathbb{R}^3) \times \mathcal{S}'_{\mathrm{real}}(\mathbb{R}^3).$$

Using the identification implemented by $\mathcal{U}$,

$$\mathcal{H} \cong L^2(\mathbb{R}^3 \times \mathbf{S}, dx \otimes d\mu),$$

we established in Ref.[14] that

$$\Theta(t) = e^{i\frac{\pi}{2}N} e^{-tH_\Lambda} e^{-i\frac{\pi}{2}N}$$

is positivity improving, *i.e.*,

$$(\Psi, \mathcal{U}\Theta(t)\mathcal{U}^{-1}\Phi) > 0$$

for $\Psi, \Phi \in L^2(\mathbb{R}^3 \times \mathbf{S}, dx \otimes d\mu)$ with $\Psi, \Phi \geq 0$ but $\Psi \not\equiv 0, \Phi \not\equiv 0$. This implies that a ground state (if it exists) of H_Λ is unique up to a phase factor.

In Section 3 we also see that $\varphi_{\mathrm{g}} \in D(N^{1/2})$. Modifying a method in Ref.[19] slightly, we can also establish that

$$\varphi_{\mathrm{g}} \in \bigcap_{k=1}^{\infty} D(N^{k/2}). \tag{21}$$

Proposition 3: *Let $\Lambda > \Lambda_*$. Then the ground state, φ_{g}, of H_Λ is unique and (21) is satisfied.*

5.2. A_Λ^2 *term*

A_Λ^2 term in H_Λ is a subjective part for large Λ. It is established in Ref.[17] that the Hamiltonian without A_Λ^2,

$$H_\Lambda' = \frac{1}{2m}p^2 + V + \frac{1}{m}p \cdot A_\Lambda + H_{\mathrm{f}},$$

is unbounded from below for Λ such that

$$\|\chi_\Lambda/\omega\|^2 > m.$$

In particular H_Λ' has no ground state for such Λ.

Proposition 4: H_Λ' *has no ground state for* $\Lambda > m/(4\pi) + \lambda$.

6. Stability of matter

Stability of matter investigated in this section are pointed out in *e.g.*, Lieb and Loss[22] and references therein. See also Fefferman, Fröhlich and Graf.[7]

6.1. *Ultraviolet cutoffs*

In this subsection we estimate the asymptotics of $g(\Lambda)$ in (5) as $\Lambda \to \infty$.

Remark 8: In the case of $V = 0$, from Proposition 5 it follows that

$$g(\Lambda) = \inf \sigma(H_\Lambda).$$

It is seen that $g(\Lambda)$ is monotonously increasing in Λ.

Since

$$\|\chi_\Lambda/\sqrt{t^2+\omega^2}\|^2 = 4\pi\left\{(\Lambda - \lambda) + t\left(\arctan\frac{\lambda}{t} - \arctan\frac{\Lambda}{t}\right)\right\},$$

and

$$\|\chi_\Lambda/(t^2+\omega^2)\|^2 = 4\pi\left\{\frac{1}{2t}\left(\arctan\frac{\Lambda}{t} - \arctan\frac{\lambda}{t}\right) + \frac{1}{2}\left(\frac{\lambda}{t^2+\lambda^2} - \frac{\Lambda}{t^2+\Lambda^2}\right)\right\},$$

we have

$$g(\Lambda) = \frac{1}{\pi}\int_{-\infty}^{\infty} \frac{2\pi t\left\{\left(\arctan\frac{\Lambda}{t} - \arctan\frac{\lambda}{t}\right) + \left(\frac{\lambda/t}{1+(\lambda/t)^2} - \frac{\Lambda/t}{1+(\Lambda/t)^2}\right)\right\}}{m + \frac{8\pi t}{3}\left\{\left(\frac{\Lambda}{t} - \frac{\lambda}{t}\right) + \left(\arctan\frac{\lambda}{t} - \arctan\frac{\Lambda}{t}\right)\right\}} dt.$$

Change the variable t to $r = \Lambda/t$. We have

$$g(\Lambda) = 4\Lambda^2 \int_0^\infty \frac{\left(\arctan r - \frac{r}{1+r^2}\right) - \left(\arctan \frac{r\lambda}{\Lambda} - \frac{\frac{r\lambda}{\Lambda}}{1+r^2\left(\frac{\lambda}{\Lambda}\right)^2}\right)}{mr + \frac{8\pi}{3}\Lambda\left\{(r - \arctan r) - (\frac{r\lambda}{\Lambda} - \arctan \frac{r\lambda}{\Lambda})\right\}} \frac{dr}{r^2}.$$

Theorem 2: *Assume that*

$$m > 8\pi\lambda/3. \tag{22}$$

Then

$$\frac{8}{3}\left(\frac{3}{8\pi}\frac{1}{m}\right)^{1/2}\frac{\pi}{2} \le \lim_{\Lambda\to\infty} \frac{g(\Lambda)}{\Lambda^{3/2}} \le \frac{8}{3}\left(\frac{9}{8\pi}\frac{1}{m}\right)^{1/2}\frac{\pi}{2}.$$

Proof: This proof is due to Ezawa.[6] We decompose

$$\frac{g(\Lambda)}{4\Lambda} = \int_0^\infty \frac{\left(\arctan r - \frac{r}{1+r^2}\right) - \left(\arctan \frac{r\lambda}{\Lambda} - \frac{\frac{r\lambda}{\Lambda}}{1+r^2\left(\frac{\lambda}{\Lambda}\right)^2}\right)}{\frac{m}{\Lambda}r + \frac{8\pi}{3}\left\{(r - \arctan r) - (\frac{r\lambda}{\Lambda} - \arctan \frac{r\lambda}{\Lambda})\right\}} \frac{dr}{r^2}$$

such as

$$\frac{g(\Lambda)}{4\Lambda} = \int_0^{1/\Lambda^{1/4}} + \int_{1/\Lambda^{1/4}}^\infty = I_1(\Lambda) + I_2(\Lambda).$$

It is enough to show that

$$\frac{2}{3}\left(\frac{3}{8\pi}\frac{1}{m}\right)^{1/2}\frac{\pi}{2} \le \lim_{\Lambda\to\infty} \frac{I_1(\Lambda)}{\sqrt{\Lambda}} \le \frac{2}{3}\left(\frac{9}{8\pi}\frac{1}{m}\right)^{1/2}\frac{\pi}{2}, \tag{23}$$

and that

$$\lim_{\Lambda\to\infty} \frac{I_2(\Lambda)}{\sqrt{\Lambda}} = 0. \tag{24}$$

Note that

$$\arctan x = \frac{x}{1+x^2} + \frac{2}{3}\frac{x^3}{(1+x^2)^2} + \frac{5}{4}\frac{3}{2}\frac{x^5}{(1+x^2)^3} + \cdots.$$

We define f and h by

$$\arctan r - \frac{r}{1+r^2} = \frac{2}{3}\frac{r^3}{(1+r^2)^2} + f(r)$$

and

$$r - \arctan r = \frac{r^3}{1+r^2} - h(r).$$

It is satisfied that $f(r) \geq 0$, $h(r) \geq 0$,

$$\lim_{r \to 0} \frac{f(r)}{r^3} = 0, \quad \lim_{r \to 0} \frac{h(r)}{r^3} = \frac{2}{3}$$

and

$$h(r) = \frac{2}{3} \frac{r^3}{(1+r^2)^2} + f(r) = \arctan r - \frac{r}{1+r^2}.$$

Let us set

$$f_\Lambda(r) = \arctan \frac{r\lambda}{\Lambda} - \frac{r\left(\frac{\lambda}{\Lambda}\right)}{1 + r^2 \left(\frac{\lambda}{\Lambda}\right)^2} > 0$$

and

$$h_\Lambda(r) = \frac{r\lambda}{\Lambda} - \arctan \frac{r\lambda}{\Lambda} > 0.$$

Then $I_1(\Lambda)$ is written as

$$I_1(\Lambda) = \int_0^{1/\Lambda^{1/4}} \frac{\frac{2}{3} + \frac{(1+r^2)^2}{r^3}[f(r) - f_\Lambda(r)]}{\frac{m}{\Lambda}(1+r^2) + \frac{8\pi}{3}r^2 - \frac{8\pi}{3}\frac{1+r^2}{r}[h(r) + h_\Lambda(r)]} \frac{dr}{1+r^2}.$$

It follows that for $0 \leq r \leq 1/\Lambda^{1/4}$,

$$\frac{8\pi}{3} \frac{1+r^2}{r}(h(r) + h_\Lambda(r)) = \frac{8\pi}{3} r^2 (1+r^2) \frac{h(r) + h_\Lambda(r)}{r^3}$$

$$\leq \frac{8\pi}{3}\left(1 + \frac{1}{\sqrt{\Lambda}}\right) r^2 \theta(\Lambda) := r^2 \delta(\Lambda),$$

where

$$\theta(\Lambda) := \sup_{0 \leq r \leq 1/\Lambda^{1/4}} \frac{h(r) + h_\Lambda(r)}{r^3}.$$

Note that

$$\lim_{r \downarrow 0} \frac{h(r) + h_\Lambda(r)}{r^3} = \frac{2}{3} + \frac{1}{3}\left(\frac{\lambda}{\Lambda}\right)^3$$

and

$$\lim_{\Lambda \to \infty} \delta(\Lambda) = \frac{8\pi}{3} \frac{2}{3}. \tag{25}$$

Moreover we have

$$\frac{(1+r^2)^2}{r^3}[f(r) - f_\Lambda(r)] \leq \sup_{0 \leq r \leq (1/\Lambda^{1/4})} \frac{(1+r^2)^2}{r^3} f(r)$$

and

$$\frac{(1+r^2)^2}{r^3}[f(r) - f_\Lambda(r)] \geq - \sup_{0 \leq r \leq (1/\Lambda^{1/4})} \frac{(1+r^2)^2}{r^3} f_\Lambda(r).$$

Set

$$\epsilon(\Lambda) := \max\left\{ \sup_{0 \leq r \leq (1/\Lambda^{1/4})} \frac{(1+r^2)^2}{r^3} f(r), \sup_{0 \leq r \leq (1/\Lambda^{1/4})} \frac{(1+r^2)^2}{r^3} f_\Lambda(r) \right\}.$$

It is trivial that

$$\lim_{\Lambda \to \infty} \epsilon(\Lambda) = 0. \tag{26}$$

Hence we have

$$\frac{(2/3) - \epsilon(\Lambda)}{1 + 1/\sqrt{\Lambda}} \int_0^{1/\Lambda^{1/4}} \frac{1}{\frac{m}{\Lambda} + \left(\frac{m}{\Lambda} + \frac{8\pi}{3}\right) r^2} dr \leq I_1(\Lambda)$$

$$\leq \frac{(2/3) + \epsilon(\Lambda)}{1 - 1/\sqrt{\Lambda}} \int_0^{1/\Lambda^{1/4}} \frac{1}{\frac{m}{\Lambda} + \left(\frac{m}{\Lambda} + \frac{8\pi}{3} - \delta(\Lambda)\right) r^2} dr.$$

Together with (25) a direct calculation yields that

$$\lim_{\Lambda \to \infty} \frac{1}{\sqrt{\Lambda}} \int_0^{1/\Lambda^{1/4}} \frac{1}{\frac{m}{\Lambda} + \left(\frac{m}{\Lambda} + \frac{8\pi}{3} - \delta(\Lambda)\right) r^2} dr$$

$$= \lim_{\Lambda \to \infty} \frac{1}{\sqrt{m(\frac{m}{\Lambda} + \frac{8\pi}{3} - \delta(\Lambda))}} \arctan \sqrt{\frac{\frac{m}{\Lambda} + \frac{8\pi}{3} - \delta(\Lambda)}{m/\sqrt{\Lambda}}}$$

$$= \left(\frac{9}{8\pi} \frac{1}{m}\right)^{1/2} \frac{\pi}{2}.$$

Similarly we have

$$\lim_{\Lambda \to \infty} \frac{1}{\sqrt{\Lambda}} \int_0^{1/\Lambda^{1/4}} \frac{1}{\frac{m}{\Lambda} + \left(\frac{m}{\Lambda} + \frac{8\pi}{3}\right) r^2} dr = \left(\frac{3}{8\pi} \frac{1}{m}\right)^{1/2} \frac{\pi}{2}.$$

Thus by (26) we have

$$\frac{2}{3} \left(\frac{3}{8\pi} \frac{1}{m}\right)^{1/2} \frac{\pi}{2} \leq \lim_{\Lambda \to \infty} \frac{1}{\sqrt{\Lambda}} I_1(\Lambda) \leq \frac{2}{3} \left(\frac{9}{8\pi} \frac{1}{m}\right)^{1/2} \frac{\pi}{2}.$$

Hence (23) follows. Next we show (24). Since

$$\left(\arctan r - \frac{r}{1+r^2}\right) - \left(\arctan \frac{r\lambda}{\Lambda} - \frac{\frac{r\lambda}{\Lambda}}{1 + (\frac{r\lambda}{\Lambda})^2}\right)$$

$$\leq \frac{2}{3}\frac{r^3}{(1+r^2)^2} + \frac{5}{4}\frac{3}{2}\frac{r^5}{(1+r^2)^2}$$

and by the assumption $m > 8\pi\lambda/3$,

$$\frac{m}{\Lambda}r + \frac{8\pi}{3}\left\{(r - \arctan r) - \left(\frac{r\lambda}{\Lambda} - \arctan\frac{r\lambda}{\Lambda}\right)\right\}$$

$$> \frac{m}{\Lambda}r - \frac{\lambda}{\Lambda}\frac{8\pi}{3}r + \frac{8\pi}{3}\left\{\frac{r^3}{1+r^2} - \frac{2}{3}\frac{r^3}{1+r^2} + \arctan\frac{r\lambda}{\Lambda}\right\}$$

$$> \frac{8\pi}{9}\frac{r^3}{1+r^2}.$$

Then

$$\lim_{\Lambda\to\infty}\frac{1}{\sqrt{\Lambda}}I_2(\Lambda) \leq \lim_{\Lambda\to\infty}\frac{1}{\sqrt{\Lambda}}\int_{1/\Lambda^{1/4}}^{\infty}\frac{\frac{2}{3}\frac{r^3}{(1+r^2)^2} + \frac{5}{4}\frac{3}{2}\frac{r^5}{(1+r^2)^2}}{\frac{8\pi}{9}\frac{r^3}{1+r^2}}\frac{dr}{r^2}$$

$$= \lim_{\Lambda\to\infty}\frac{9}{8\pi}\frac{1}{\sqrt{\Lambda}}\int_{1/\Lambda^{1/4}}^{\infty}\frac{1}{1+r^2}\left(\frac{2}{3} + \frac{15}{8}r^2\right)\frac{dr}{r^2} = 0.$$

Thus (24) follows, and then the theorem follows. $\qquad\square$

6.2. *N particle system*

We consider an N particle system. We assume simply that each particle has mass m and there is no external potential. The Hamiltonian, H_Λ^N, is defined as a self-adjoint operator acting on

$$L^2(\mathbb{R}^{3N}) \otimes \mathcal{F},$$

and is given by

$$H_\Lambda^N = \sum_{j=1}^{N}\frac{1}{2m}(p_j + A_{j\Lambda}(0))^2 + H_{\mathrm{f}},$$

where

$$A_{j\Lambda}(0) = \sum_{j'=1,2}\int\frac{\chi_{j\Lambda}(k)}{\sqrt{2\omega(k)}}e_{j'}(k)\left\{a^*(k,j') + a(k,j')\right\}d^3k.$$

Let

$$\inf\sigma(H_\Lambda^N) = g(\Lambda, N).$$

We consider the two cases such as

(1) $\chi_{j\Lambda}(k) = \chi_\Lambda(k), \quad j = 1,...,N,$

(2) $\chi_{j\Lambda}, j = 1,...,N$, are characteristic functions on closed sets in $\mathbb{R}^3$ such as

$$\mathrm{supp}\chi_{j\Lambda} \cap \mathrm{supp}\chi_{i\Lambda} \cap \{0\} = \emptyset, \quad (i \neq j).$$

We will see below that the asymptotic behavior of $g(\Lambda, N)$ as $N \to \infty$ depends on ultraviolet cutoffs. In the case of (2) we intuitively expect that

$$g(\Lambda, N) \approx N,$$

since N particles may have no interaction through quantized radiation fields. In Ref.[17] we established the following proposition.

Proposition 5: *In the case of (1),*

$$g(\Lambda, N) = \frac{N}{\pi} \int_{-\infty}^{\infty} \frac{t^2 \|\chi_\Lambda/(t^2 + \omega^2)\|^2}{m + \frac{2}{3}N\|\chi_\Lambda/\sqrt{t^2 + \omega^2}\|^2} dt,$$

in the case of (2),

$$g(\Lambda, N) = \sum_{j=1}^{N} \frac{1}{\pi} \int_{-\infty}^{\infty} \frac{t^2 \|\chi_{j\Lambda}/(t^2 + \omega^2)\|^2}{m + \frac{2}{3}\|\chi_{j\Lambda}/\sqrt{t^2 + \omega^2}\|^2} dt.$$

In the case of (1) the following theorem holds.

Theorem 3: *We assume case (1) and*

$$m > 8\pi\lambda/3. \tag{27}$$

Then

$$\frac{8}{3}\left(\frac{3}{8\pi}\frac{1}{m}\right)^{1/2}\frac{\pi}{2} \leq \lim_{\Lambda, N \to \infty} \frac{g(\Lambda, N)}{\sqrt{N}\Lambda^{3/2}} \leq \frac{8}{3}\left(\frac{9}{8\pi}\frac{1}{m}\right)^{1/2}\frac{\pi}{2}.$$

Proof: By Proposition 5 we have

$$\frac{g(\Lambda, N)}{4\Lambda} = \int_0^\infty \frac{(\arctan r - \frac{r}{1+r^2}) - \left(\arctan\frac{r\lambda}{\Lambda} - \frac{\frac{r\lambda}{\Lambda}}{1+(\frac{r\lambda}{\Lambda})^2}\right)}{\frac{m}{N\Lambda}r + \frac{8\pi}{3}\{(r - \arctan r) - (\frac{r\lambda}{\Lambda} - \arctan\frac{r\lambda}{\Lambda})\}} \frac{dr}{r^2}.$$

Then in a way similar to the proof of Theorem 3 we decompose it such as

$$\frac{g(\Lambda, N)}{4\Lambda} = \int_0^{1/(\Lambda N)^{1/4}} + \int_{1/(\Lambda N)^{1/4}}^{\infty} = I_1(\Lambda, N) + I_2(\Lambda, N),$$

and it can be seen that

$$\frac{2}{3}\left(\frac{3}{8\pi}\frac{1}{m}\right)^{1/2}\frac{\pi}{2} \leq \lim_{\Lambda,N\to\infty} \frac{I_1(\Lambda,N)}{\sqrt{N\Lambda}} \leq \frac{2}{3}\left(\frac{9}{8\pi}\frac{1}{m}\right)^{1/2}\frac{\pi}{2},$$

and that

$$\lim_{\Lambda,N\to\infty} \frac{I_2(\Lambda,N)}{\sqrt{N\Lambda}} = 0.$$

Then the theorem follows. $\qquad\square$

Remark 9: In the case of (2), if we adjust $\chi_{j\Lambda}$ such as

$$\frac{1}{\pi}\int_{-\infty}^{\infty} \frac{t^2\|\chi_{j\Lambda}/(t^2+\omega^2)\|^2}{m+\frac{2}{3}\|\chi_{j\Lambda}/\sqrt{t^2+\omega^2}\|^2}\,dt = g$$

with some constant g independent of j. Then

$$g(\Lambda,N) = Ng.$$

Acknowledgment

I thank H. Ezawa for sending his proof of Theorems 2 and 3 to myself. I thank Grant-in-Aid 13740106 for Encouragement of Young Scientists from the Ministry of Education, Science, Sports and Culture for financial support.

References

1. A. Arai and M. Hirokawa, On the existence and uniqueness of ground states of a generalized spin-boson model, *J. Funct. Anal.* **151** (1997), 455–503.
2. A. Arai and H. Kawano, Enhanced binding in a general class of quantum field models, mp-arc 02-508, preprint, 2002.
3. V. Bach, J. Fröhlich, I. M. Sigal, Spectral analysis for systems of atoms and molecules coupled to the quantized radiation field, *Commun. Math. Phys.* **207** (1999), 249–290.
4. I. Catto and C. Hainzl, Self-energy of one electron in non-relativistic QED, math-ph/0207036, preprint, 2002
5. T. Chen, V. Vougalter and S. Vugalter, The increase of binding energy and enhanced binding in non-relativistic QED, math-ph/0209062, preprint, 2002.
6. H. Ezawa, private communications.
7. C. Fefferman, J. Fröhlich, G. M. Graf, Stability of ultraviolet-cutoff quantum electrodynamics with non-relativistic matter, *Commun. Math. Phys.* **190** (1997), 309–330.
8. J. Fröhlich, On the infrared problem in a model of scalar electrons and massless, scalar bosons, *Ann. Inst. Henri Poincaré* **19** (1973), 1–103.

9. J. Fröhlich, Existence of dressed one electron states in a class of persistent models, *Fortschritte der Physik* **22** (1974), 159–198.

10. M. Griesemer, E. Lieb and M. Loss, Ground states in non-relativistic quantum electrodynamics, *Invent. Math.* **145** (2001), 557–595.

11. C. Hainzl, V. Vougalter and S. A. Vugalter, Enhanced binding in non-relativistic QED, mp-arc 01-455, preprint, 2001.

12. M. Hirokawa, Divergence of soft photon number and absence of ground state for Nelson's model, math-ph/0211051, preprint, 2002.

13. F. Hiroshima, Ground states of a model in nonrelativistic quantum electrodynamics I, *J. Math. Phys.* **40** (1999), 6209–6222.

14. F. Hiroshima, Ground states of a model in nonrelativistic quantum electrodynamics II, *J. Funct. Anal.* **41** (2000), 661-674.

15. F. Hiroshima, Essential self-adjointness of translation invariant quantum filed models for arbitrary coupling constants, *Commun. Math. Phys.* **211** (2000), 585-613.

16. F. Hiroshima, Self-adjointness of the Pauli–Fierz Hamiltonian in quantum electrodynamics for arbitrary values of coupling constants, *Ann. H. Poincaré* **3** (2002), 171–201.

17. F. Hiroshima, Observable effect and parameterized scaling limits of a model in nonrelativistic electrodynamics, *J. Funct. Anal.* **43** (2002), 1775-1795.

18. F. Hiroshima, Analysis of ground states of atoms interacting with a quantized radiation field, preprint, 2003.

19. F. Hiroshima, Localization of the number of photons of nonrelativistic QED, math-ph/0211009, to be published in *Rev. Math. Phys.*

20. F. Hiroshima and H. Spohn, Enhanced binding through coupling to a quantum field, *Ann. H. Poincaré* **2** (2001), 1159–1187.

21. R. Høegh-Krohn and B. Simon, Hypercontractive semigroup and two dimensional self-coupled Bose fields, *J. Funct. Anal.* **9** (1972), 121–180.

22. E. Lieb and M. Loss, Bound on binding energies and mass renormalization in models of quantum electrodynamics, preprint, 2001.

23. E. Lieb and W. Thirring, Inequalities for the moments of the eigenvalues of the Schrödinger Hamiltonian and their relation to Sobolev inequalities, *Studies in Mathematical Physics*, Princeton Univ. Press, 269-303, 1976.

24. W. Pauli and M. Fierz, Zur Theorie der Emission langwelliger Lichtquanten, *Nuovo Cimento* **15** (1938), 167–188, Theory of the emission of long-wave light quanta, *Early Quantum Electrodynamics*, ed. by I. Miller, Cambridge Univ. Press, 1995, 227–243.

Enhanced Binding in Models of Nonrelativistic Quantum Field Theory

Asao Arai

Department of Mathematics, Hokkaido University
Sapporo 060-0810, Japan
E-mail: arai@math.sci.hokudai.ac.jp

Dedicated to Professor Hiroshi Ezawa on the occasion of his seventieth birthday

We consider a class of models in nonrelativistic quantum field theory. This class includes the Pauli–Fierz model with the dipole approximation in nonrelativistic quantum electrodynamics. Each model in the class describes an interaction of a quantum system —typically a system of nonrelativistic quantum particles— coupled to a multi-component Bose field. We prove that, under suitable hypotheses, each model has a ground state even if the unperturbed (zero-coupling) system has no ground state.

1. Introduction

In mathematical quantum field theory, it is a fundamentally important problem to prove the existence of ground states in various models. Usually this is done under the assumption that the unperturbed system of the model under consideration has a ground state. But there may be models of quantum particles interacting with a quantum field in which the following picture is possible: the binding of quantum particles can be enhanced through coupling to the quantum field so that the coupled system has a ground state even if the uncoupled system has no ground state. In that case we say that the enhanced binding occurs in the quantum system under consideration. This picture is interesting and even attractive. From this point of view, it is important to clarify whether or not the enhanced binding *indeed* occurs in quantum field models. This problem, called the enhanced binding problem, was first discussed by Hiroshima and Spohn[1] for the Pauli–Fierz (PF) model in nonrelativistic quantum electrodynamics (QED) with the dipole approximation and then by Hainzl, Vougalter and Vugalter[2] for the PF model without the dipole approximation. In each model, the existence of the enhanced binding was established for a range of the coupling constant.

In a previous paper[3] the enhanced binding problem was considered for a general class of quantum field models, called the generalized spin-boson (GSB) model,[4,5,6] and the existence of the enhanced binding was proved under suitable hypotheses. In the GSB model, however, the interaction part of the Hamiltonian is linear in the quantum field and it contains no self-interaction terms in the quantum field. In this paper we extend the GSB model to include quadratic self-interaction terms of the quantum field and discuss the enhanced binding problem for the extended GSB model. A typical example of the extended GSB model is the PF model with the dipole approximation. We show that, under suitable hypotheses, the enhanced binding occurs in the extended GSB model.

The outline of the present paper is as follows. In Section 2 we describe the extended GSB model. The main theorems are stated in Section 3. In Section 4, we present a sketch of proof of each theorem.

2. Definition of the model

2.1. *Some notations*

In general we denote the inner product and the norm of a Hilbert space $\mathcal{X}$ by $\langle \cdot, \cdot \rangle_{\mathcal{X}}$ and $\| \cdot \|_{\mathcal{X}}$ respectively, where we use the convention that the inner product is antilinear (resp. linear) in the first (resp. second) variable. If there is no danger of confusion, then we omit the subscript $\mathcal{X}$ in $\langle \cdot, \cdot \rangle_{\mathcal{X}}$ and $\| \cdot \|_{\mathcal{X}}$. For a linear operator T on a Hilbert space, we denote its domain by $D(T)$. For a subspace $D \subset D(T)$, $T|D$ denotes the restriction of T to D. If T is densely defined, then the adjoint of T is denoted T^*. For linear operators S and T on a Hilbert space, $D(S + T) := D(S) \cap D(T)$ unless otherwise stated.

For a self-adjoint operator S on a Hilbert space, we denote its spectrum (resp. essential spectrum) by $\sigma(S)$ (resp. $\sigma_{\mathrm{ess}}(S)$) and its spectral measure by $E_S(\cdot)$. If S is bounded from below, then we set

$$E_0(S) := \inf \sigma(S) \tag{1}$$

the *ground state energy* of S. We say that S has a ground state if $E_0(S)$ is an eigenvalue of S; in this case, each non-zero vector in $\ker(S - E_0(S))$ is called a *ground state* of S.

2.2. *Bose fields*

To describe a Bose field, one uses the Boson Fock space over a separable complex Hilbert space $\mathcal{X}$:

$$\mathcal{F}_b(\mathcal{X}) := \oplus_{n=0}^{\infty} \otimes_s^n \mathcal{X}$$
$$= \left\{ \psi = \{\psi^{(n)}\}_{n=0}^{\infty} \,\Big|\, n \geq 0, \psi^{(n)} \in \otimes_s^n \mathcal{X}, \sum_{n=0}^{\infty} \|\psi^{(n)}\|^2 < \infty \right\}, \quad (2)$$

where $\otimes_s^n \mathcal{X}$ denotes the n-fold symmetric tensor product of $\mathcal{X}$ with $\otimes_s^0 \mathcal{X} := \mathbb{C}$ (the set of complex numbers).

One of basic objects on $\mathcal{F}_b(\mathcal{X})$ is the annihilation operator $a(f)$ $(f \in \mathcal{X})$ which is a densely defined closed linear operator on $\mathcal{F}_b(\mathcal{X})$ such that, for all $\psi = \{\psi^{(n)}\}_{n=0}^{\infty} \in D(a(f)^*), (a(f)^*\psi)^{(0)} = 0$ and

$$(a(f)^*\psi)^{(n)} = \sqrt{n} S_n \left(f \otimes \psi^{(n-1)} \right), \quad n \geq 1,$$

where S_n is the symmetrization operator on $\otimes^n \mathcal{X}$. The adjoint $a(f)^*$, called the creation operator, and the annihilation operator $a(g)$ $(g \in \mathcal{X})$ obey the canonical commutation relations

$$[a(f), a(g)^*] = \langle f, g \rangle_{\mathcal{X}}, \quad [a(f), a(g)] = 0, \quad [a(f)^*, a(g)^*] = 0 \quad (3)$$

for all $f, g \in \mathcal{X}$ on the dense subspace

$$\mathcal{F}_0(\mathcal{X}) := \{\psi \in \mathcal{F}_b(\mathcal{X}) \mid \text{there exists a number } n_0 \text{ such that}$$
$$\psi^{(n)} = 0 \text{ for all } n \geq n_0\}, \quad (4)$$

where $[X, Y] := XY - YX$.

We denote by $\Omega := \{1, 0, 0, \cdots\}$ the Fock vacuum in $\mathcal{F}_b(\mathcal{X})$ and introduce a subspace

$$\mathcal{F}_{\text{fin}}(\mathcal{X}) := \mathcal{L}\left(\{\Omega, a(f_1)^* \cdots a(f_n)^* \Omega | n \in \mathbb{N}, f_j \in \mathcal{X}, j = 1, \cdots, n\}\right), \quad (5)$$

where $\mathcal{L}(\{\cdot\})$ means the subspace algebraically spanned by the set $\{\cdot\}$. The subspace $\mathcal{F}_{\text{fin}}(\mathcal{X})$ is dense in $\mathcal{F}_b(\mathcal{X})$.

Let

$$\phi(f) := \frac{a(f) + a(f)^*}{\sqrt{2}}, \quad f \in \mathcal{X}, \quad (6)$$

which is called the Segal field operator. It is shown that $\phi(f)$ is essentially self-adjoint on $\mathcal{F}_0(\mathcal{X})$ (§X.7 in Ref.[7]). *We denote its closure by the same symbol $\phi(f)$.*

For every symmetric operator S on $\mathcal{X}$, one can define a symmetric operator $d\Gamma(S)$, called the second quantization of S, by

$$d\Gamma(S) := \oplus_{n=0}^{\infty} S^{(n)}, \tag{7}$$

with $S^{(0)} = 0$ and $S^{(n)}$ is the closure of

$$\left(\sum_{j=1}^{n} I \otimes \cdots \otimes \overset{j\text{th}}{\breve{S}} \otimes \cdots \otimes I \right) \Big| \otimes_{\text{alg}}^{n} D(S),$$

where I denotes identity and $\otimes_{\text{alg}}^{n}$ algebraic tensor product. If S is self-adjoint, then so is $d\Gamma(S)$.

2.3. *Quadratic operators*

In the Boson Fock space $\mathcal{F}_{\text{b}}(\mathcal{X})$, one can define, in a general manner, linear operators quadratic in the annihilation and the creation operators. Let C be a conjugation on $\mathcal{X}$, *i.e.*, C is an anti-linear, isometric mapping on $\mathcal{X}$ such that $C^2 = I$. Let K be a Hilbert–Schmidt operator on $\mathcal{X}$. Then there exist orthonormal sets $\{f_n\}_{n=1}^{M}$, $\{g_n\}_{n=1}^{M}$ in $\mathcal{X}$ and a sequence $\{\lambda_n\}_{n=1}^{M}$ of positive numbers ($M < \infty$ or M is countably infinite) such that $\sum_{n=1}^{M} \lambda_n^2 < \infty$ and

$$K = \sum_{n=1}^{M} \lambda_n \langle f_n, \cdot \rangle_{\mathcal{X}} g_n,$$

where, in the case $M = \infty$, the sum on the right hand side converges in operator norm (Theorems VI.17 and VI.22 in Ref.[8]). Using this representation of K, we define for $N \in \mathbb{N}$ linear operators $(a^*|K|a^*)_N$ and $(a|K|a)_N$ acting in $\mathcal{F}_{\text{b}}(\mathcal{X})$, with domain $\mathcal{F}_{\text{fin}}(\mathcal{X})$, by

$$(a^*|K|a^*)_N := \sum_{n=1}^{N \wedge M} \lambda_n a(Cf_n)^* a(g_n)^*, \tag{8}$$

$$(a|K|a)_N := \sum_{n=1}^{N \wedge M} \lambda_n a(f_n) a(Cg_n), \tag{9}$$

where $N \wedge M := \min\{N, M\}$ if $M < \infty$ and $N \wedge M := N$ if $M = \infty$. It is easy to see that, for all $\Psi \in \mathcal{F}_{\text{fin}}(\mathcal{X})$, the strong limit

$$(a^{\#}|K|a^{\#})\Psi := \text{s-}\lim_{N \to \infty} (a^{\#}|K|a^{\#})_N \Psi \tag{10}$$

exists, where $a^{\#}$ denotes either a^* or a. Moreover the operator $(a^{\#}|K|a^{\#})$ with domain $\mathcal{F}_{\text{fin}}(\mathcal{X})$ is closable and

$$(a^*|K|a^*)^* \supset (a|K^*|a), \qquad (a|K|a)^* \supset (a^*|K^*|a^*).$$

2.4. *An extended GSB model*

We consider, in an abstract form, a model of a quantum system S coupled to a multi-component Bose field. We denote the Hilbert space of the system S by $\mathcal{H}$, which is taken to be an arbitrary separable complex Hilbert space. In concrete realizations, S may be a system of nonrelativistic quantum particles or a quantum field system.

We assume that the Bose field is an N-component quantum field over $\mathbb{R}^d$ $(d, N \in \mathbb{N})$. Hence the one-boson Hilbert space is taken to be

$$\mathcal{K} := \oplus^N L^2(\mathbb{R}^d), \tag{11}$$

the N direct sum of $L^2(\mathbb{R}^d)$, and the Hilbert space of the Bose field is taken to be $\mathcal{F}_{\text{b}}(\mathcal{K})$. Let ω be a Borel measurable function on $\mathbb{R}^d$ such that $0 < \omega(k) < \infty$ for almost everywhere (a.e.) $k \in \mathbb{R}^d$ with respect to (w.r.t.) the Lebesgue measure on $\mathbb{R}^d$. Then ω defines a multiplication operator on $L^2(\mathbb{R}^d)$, which is nonnegative, injective and self-adjoint. We denote it by the same symbol. The function ω represents a dispersion relation of one free boson associated with the Bose field under consideration.

We define an operator

$$\widehat{\omega} := \oplus^N \omega \tag{12}$$

acting in $\mathcal{K}$.

The free Hamiltonian of the Bose filed is defined by $d\Gamma(\widehat{\omega})$ acting on $\mathcal{F}_{\text{b}}(\mathcal{K})$.

The Hilbert space of the coupled system of S and the Bose field is given by the tensor product

$$\mathcal{F} := \mathcal{H} \otimes \mathcal{F}_{\text{b}}(\mathcal{K}). \tag{13}$$

Let A be a self-adjoint operator on $\mathcal{H}$, which denotes physically the Hamiltonian of the quantum system S, and B_j $(j = 1, \cdots, J, J \in \mathbb{N})$ be a symmetric operator on $\mathcal{H}$ such that $\cap_{j=1}^{J} D(B_j)$ is dense in $\mathcal{H}$. Let $g_j \in \mathcal{K}$, $j = 1, \cdots, J$. Then the Hamiltonian of the GSB model is defined by

$$H_{\text{GSB}}(\omega, g) := A \otimes I + I \otimes d\Gamma(\widehat{\omega}) + \sum_{j=1}^{J} B_j \otimes \phi(g_j), \tag{14}$$

where $g := (g_1, \cdots, g_J)$. In the previous paper,[3] the existence of the enhanced binding of $H_{\mathrm{GSB}}(\omega, g)$ was shown under suitable hypotheses.

An extended GSB model is defined by perturbing $H_{\mathrm{GSB}}(\omega, g)$. Here we consider only a class of perturbations formed out of quadratic operators.

We say that a densely defined linear operator on a Hilbert space is Hilbert–Schmidt if it is bounded and its closure is Hilbert–Schmidt. We denote the closure by the same symbol.

In what follows, we denote by C the complex conjugation on $\mathcal{K}$. For a bounded linear operator T on $\mathcal{K}$, we define T_C by

$$T_C := CTC. \tag{15}$$

Let S, T be bounded linear operators on $\mathcal{K}$ and μ be a nonnegative Borel measurable function on $\mathbb{R}^d$ such that $0 < \mu(k) < \infty$ a.e. $k \in \mathbb{R}^d$ w.r.t. the Lebesgue measure on $\mathbb{R}^d$. We assume the following.

Hypothesis (I)

(I.1) The operators $T\widehat{\mu}S^*$ and $T\widehat{\mu}^{1/2}$ are densely defined and Hilbert–Schmidt.

(I.2) The operator $S\widehat{\mu}S^* + T_C\widehat{\mu}T_C^*$ is densely defined.

Under Hypothesis (I), we can define a symemtric operator

$$H := A \otimes I + I \otimes \bar{L} + \sum_{j=1}^{J} B_j \otimes \phi(g_j), \tag{16}$$

where $\bar{L}$ is the closure of the operator

$$L := d\Gamma(S\widehat{\mu}S^* + T_C\widehat{\mu}T_C^*) + (a|T\widehat{\mu}S^*|a) + (a|T\widehat{\mu}S^*|a)^*. \tag{17}$$

The operator H is the Hamiltonian of the extended GSB model we consider in the present paper.

Remark In applications, μ, S, T and g_j may depend on ω. If $T = 0$ and $S\widehat{\mu}S^* = \widehat{\omega}$, then $H = H_{\mathrm{GSB}}(\omega, g)$.

3. The main results

To state the main results of this paper, we formulate additional hypotheses. For this purpose, we first recall an important notion on commutativity of self-adjoint operators:

Definition 1: *We say that two self-adjoint operators S_1 and S_2 on a Hilbert space strongly commute (or S_1 strongly commutes with S_2) if* *their spectral measures commute, i.e., for all Borel sets* $I_1, I_2 \subset \mathbb{R}$, $E_{S_1}(I_1)E_{S_2}(I_2) = E_{S_2}(I_2)E_{S_1}(I_1)$.

A family of self-adjoint operators $\{S_j\}_{j=1}^n$ *on a Hilbert space is said to be strongly commuting if* S_j *strongly commutes with* S_l *for all* $j, l = 1, \cdots, n$ *with* $j \neq l$.

In what follows, we assume that A is of the form

$$A = A_0 + A_1 \tag{18}$$

with A_0 a nonnegative self-adjoint operator and A_1 a symmetric operator on $\mathcal{H}$.

We introduce $G_j \in \mathcal{K}$ $(j = 1, \cdots, J)$ by

$$G_j := S^* g_j - T^* C g_j, \tag{19}$$

where S, T and C are operators introduced in Subsection 2.4.

Hypothesis (II) $G_j, G_j/\mu^{3/2} \in \mathcal{K}$ $(j = 1, \cdots, J)$ and

$$\langle G_j(k), G_l(k)\rangle_{\mathbb{C}^N} \in \mathbb{R}, \text{ a.c. } k \in \mathbb{R}^d \quad (j, l = 1, \cdots, J).$$

Hypothesis (III) The operator A_1 is A_0-bounded, i.e., $D(A_0) \subset D(A_1)$ and there exist constants $a, b \geq 0$ such that, for all $u \in D(A_0)$,

$$\|A_1 u\| \leq a\|A_0 u\| + b\|u\|. \tag{20}$$

Hypothesis (IV) The operator A_0 strongly commutes with each B_j $(j = 1, \cdots, J)$ and

$$D(A_0) \subset \cap_{j,l=1}^J D(B_j B_l). \tag{21}$$

Moreover, there exist constants $c_j, d_j \geq 0$ such that, for all $u \in D(A_0^{1/2})$,

$$\|B_j u\| \leq c_j \|A_0^{1/2} u\| + d_j \|u\| \quad (j = 1, \cdots, J). \tag{22}$$

Hypothesis (V) The set $\{B_j\}_{j=1}^J$ is a family of strongly commuting self-adjoint operators.

Hypothesis (VI) $D(A_0) \subset \cap_{j=1}^J D(B_j A_1) \cap D(A_1 B_j)$ and $[B_j, A_1]|D(A_0)$ is bounded $(j = 1, \cdots, J)$. We denote the operator norm of $[B_j, A_1]$ by $\|[B_j, A_1]\|$.

We introduce an operator

$$R_B := \frac{1}{2} \sum_{j,l=1}^J \left\langle \frac{G_j}{\sqrt{\mu}}, \frac{G_l}{\sqrt{\mu}} \right\rangle_{\mathcal{K}} B_j B_l. \tag{23}$$

and define

$$A_{\mathrm{ren}} := A - R_B. \tag{24}$$

Under Hypotheses (II)–(IV), we have $D(A_{\mathrm{ren}}) = D(A_0)$.

Hypothesis (VII) The operator A_{ren} is self-adjoint and bounded from below.

Hypothesis (VIII)
(VIII.1) The operator T is Hilbert–Schmidt.
(VIII.2) The operators S and T satisfy the following relations:

$$S^*S - T^*T = I, \quad SS^* - T_C T_C^* = I, \tag{25}$$

$$ST^* = T_C S_C^*, \quad S^*T_C = T^* S_C. \tag{26}$$

A result on the self-adjointness of H is given by the following theorem.

Theorem 1: *Assume Hypotheses (I)–(VIII). Then H is self-adjoint and bounded from below.*

To establish an existence theorem of a ground state of H *without* the assumption that A has a ground state, we need additional conditions.

Hypothesis (IX) The function μ is continuous on $\mathbb{R}^d$ with

$$\lim_{|k|\to\infty} \mu(k) = \infty$$

and there exist constants $\gamma > 0$ and $c_0 > 0$ such that

$$|\mu(k) - \mu(k')| \le c_0 |k - k'|^\gamma (1 + \mu(k) + \mu(k')), \quad k, k' \in \mathbb{R}^d.$$

Generally speaking, the existence of a ground state of H may depend on whether

$$\mu_0 := \mathrm{ess.} \inf_{k \in \mathbb{R}^d} \mu(k) \tag{27}$$

is positive or zero,[6] where ess. inf means essential infimum. We first establish a theorem on the existence of a ground state of H in the case $\mu_0 > 0$.

For $s \ge 0$, we introduce a constant $C_s(G)$ $(G := (G_1, \cdots, G_d))$ by

$$C_s(G) := \sqrt{2} \sum_{j=1}^{J} \|[B_j, A_1]\| \left\| \frac{G_j}{\mu^s} \right\| \tag{28}$$

provided that $G_j/\mu^s \in \mathcal{K}$. We set

$$\Sigma(A_{\mathrm{ren}}) := \inf \sigma_{\mathrm{ess}}(A_{\mathrm{ren}}). \tag{29}$$

Remark If $\mu_0 > 0$, then the condition $G_j \in \mathcal{K}$ implies that $G_j/\mu^s \in \mathcal{K}$ for all $s > 0$. Hence, in this case, Hypothesis (II) is replaced by the condition that $G_j \in \mathcal{K}$ $(j = 1, \cdots, J)$ and $\langle G_j(k), G_l(k) \rangle_{\mathbb{C}^N} \in \mathbb{R}$, a.e.$k$, $j, l = 1, \cdots, J$.

Theorem 2: (Enhanced binding in the case $\mu_0 > 0$). *Consider the case* $\mu_0 > 0$. *Assume Hypotheses (I)–(IX) and that*

$$\Sigma(A_{\mathrm{ren}}) - E_0(A_{\mathrm{ren}}) > \mu_0 + \frac{1}{2}C_{3/2}(G)^2 + C_1(G). \tag{30}$$

Then H has purely discrete spectrum in the interval $[E_0(H), E_0(H) + \mu_0)$. In particular, H has a ground state.

Remark Condition (30) implies that $E_0(A_{\mathrm{ren}})$ is a discrete eigenvalue of A_{ren} and hence A_{ren} has a finite number of ground states.

Theorem 3: (Enhanced binding in the case $\mu_0 = 0$). *Consider the case* $\mu_0 = 0$. *Assume Hypotheses (I)–(IX) with $G_j/\mu^2 \in \mathcal{K}$ $(j = 1, \cdots, J)$ in addition. Suppose that*

$$\Sigma(A_{\mathrm{ren}}) - E_0(A_{\mathrm{ren}}) > \frac{1}{2}C_{3/2}(G)^2 + C_1(G). \tag{31}$$

and

$$\frac{C_1(G)^2}{[\Sigma(A_{\mathrm{ren}}) - E_0(H)]^2} + \left\{ \frac{2C_1(G)^2}{[\Sigma(A_{\mathrm{ren}}) - E_0(H)]^2} + 1 \right\} \frac{1}{2}C_2(G)^2 < 1. \tag{32}$$

Then H has a ground state.

Note that, in Theorems 2 and 3, *the existence of a ground state of A is not assumed.*

4. Sketches of proofs of the main theorems

4.1. *Proof of Theorem 1*

For each $f \in \mathcal{K}$, we define a linear operator $b(f)$ on $\mathcal{F}_{\mathrm{b}}(\mathcal{K})$ by

$$b(f) := a(Sf) + a(CTf)^*. \tag{33}$$

The following fact is well known (Theorem XI.108 in Ref.[9]).

Lemma 10: *Assume Hypothesis (VIII). Then there exists a unitary operator U on $\mathcal{F}_{\mathrm{b}}(\mathcal{K})$ such that $b(f) = U^{-1}a(f)U$, $f \in \mathcal{K}$.*

Lemma 11: *Assume Hypotheses (I) and (VIII). Then L is essentially self-adjoint and the operator equality*

$$U\overline{L}U^{-1} = d\Gamma(\widehat{\mu}) - \|T\widehat{\mu}^{1/2}\|_{\mathrm{HS}}^2 \tag{34}$$

holds, where $\|\cdot\|_{\mathrm{HS}}$ denotes Hilbert–Schmidt norm.

Proof. Similar to the proof of Theorem 4.2 in Ref.[10] ∎

Lemma 12: *Assume Hypotheses (I)–(VIII). Then*

$$(I \otimes U)H(I \otimes U)^{-1} = H_{\mathrm{GSB}}(\mu, G) - \|T\widehat{\mu}^{1/2}\|_{\mathrm{HS}}^2, \tag{35}$$

where $H_{\mathrm{GSB}}(\mu, G)$ is the operator $H_{\mathrm{GSB}}(\omega, g)$ given by (14) with ω and g replaced by μ and G respectively.

Proof. By Eq.(16) and Lemma 11, we have

$$(I \otimes U)H(I \otimes U)^{-1} = A \otimes I + I \otimes d\Gamma(\widehat{\mu}) + \sum_{j=1}^{J} B_j \otimes U\phi(g_j)U^{-1}$$
$$- \|T\widehat{\mu}^{1/2}\|_{\mathrm{HS}}^2.$$

On the other hand, we have for all $f \in \mathcal{K}$

$$a(f) = b(S^*f) - b(T^*Cf)^*, \quad a(f)^* = b(S^*f)^* - b(T^*Cf)$$

on $\mathcal{F}_0(\mathcal{K})$. This implies that

$$U\phi(g_j)U^{-1} = \phi(G_j). \tag{36}$$

Hence Eq.(35) follows. ∎

We are now ready to prove Theorem 1. We assume Hypotheses (I)–(VIII). By Eq.(35), we need only to show that $H_{\mathrm{GSB}}(\mu, G)$ is self-adjoint. But this follows from a simple application of Theorem 2.2 in Ref.[3]

4.2. *Proofs of Theorems 2 and 3*

By (35), we need only to prove the existence of a ground state of $H_{\mathrm{GSB}}(\mu, G)$. But this follows from Theorems 2.3 and 2.4 in Ref.[3]

Acknowledgments

I have been very much indebted to Professor Hiroshi Ezawa for my researches on mathematical physics since I was a graduate student. I would like to take this opportunity to express my sincere gratitude and appreciation to him for his supports, suggestions, encouragements and all that given to me in the past years. I wish that Professor Ezawa will remain healthy for a long time to come and continue to play an active role in the coming years.

This work was supported by the Grant-In-Aid 13440039 for scientific research from the Japan Society for the Promotion of Science.

References

1. F. Hiroshima and H. Spohn, Enhanced binding through coupling to a quantum field, *Ann. Henri Poincaré* **2**, 1159–1187(2001).
2. C. Hainzl, V. Vougalter and S. A. Vugalter, Enhanced binding in nonrelativistic QED, mp_arc 01-455, 2001.
3. A. Arai and H. Kawano, Enhanced binding in a general class of quantum field models, mp_02-508, 2002 or ESI preprint series 1251, 2002, to be published in *Rev. Math. Phys.*
4. A. Arai and M. Hirokawa, On the existence and uniqueness of ground states of a generalized spin-boson model, *J. Funct. Anal.* **151**, 455–503(1997).
5. A. Arai and M. Hirokawa, Ground states of a general class of quantum field Hamiltonians, *Rev. Math. Phys.* **8**, 1085–1135(2000).
6. A. Arai, M. Hirokawa and F. Hiroshima, On the absence of eigenvectors of Hamiltonians in a class of massless quantum field models without infrared cutoff, *J. Funct. Anal.* **168**, 470–497(1999).
7. M. Reed and B. Simon, *Methods of Modern Mathematical Physics Vol. II*, Academic Press, New York, 1975.
8. M. Reed and B. Simon, *Methods of Modern Mathematical Physics Vol. I*, Academic Press, New York, 1972.
9. M. Reed and B. Simon, *Methods of Modern Mathematical Physics Vol. III*, Academic Press, New York, 1979.
10. A. Arai, Noninvertible Bogoliubov transformation and instability of embedded eigenvalues, *J. Math. Phys.* **32** (1991), 1838–1846.

Recent Developments in Mathematical Methods for Models in Non-Relativistic Quantum Electrodynamics

Masao Hirokawa

Department of Mathematics, Okayama University
Okayama 700-8530, Japan
E-mail: hirokawamath.okayama-u.ac.jp

Dedicated to Professor Hiroshi Ezawa on the occasion of his 70th birthday

The mathematical concern with analysis of models in non-relativistic electrodynamics has been growing again for the last several years. So, in this paper, we introduce some recent mathematical results for models in non-relativistic electrodynamics. We demonstrate one of them to analyze the average of the total number of soft photons for hydrogen-like atoms.

1. Introduction

Professor Hiroshi Ezawa has been organizing lots of seminars for mathematical physics at Gakushuin University for a long time. The seminar is held on every Saturday, and many mathematicians and physicists attended the seminars. After the invitation letter from him, I could attend many seminars for about seven years until I left Tokyo. So, he and I became acquainted through these seminars. He used to advise me that I should become a mathematician who not only do mathematics but also do physics. And moreover, he would require that I should understand experimental physics besides. All of the experiences there (and through voyages to Paris, Brisbane, and Les Houches) affords me spiritual nourishment whenever I study mathematical physics. Thus, I dedicate this paper to Professor Hiroshi Ezawa with my pleasure, and it makes me proud.

In this paper, I would like to introduce brief outlines of our recent developments[11,12] of mathematical methods to analyze a model of non-relativistic QED in §2 and §3. Moreover, I apply the method to a non-relativistic Hamiltonian describing hydrogen-like atoms in §4.

1.1. *Hydrogen-like atoms in non-relativistic QED*

We consider hydrogen atom ($Z = 1$) or hydrogen-like ion ($Z > 1$) in non-relativistic quantum electrodynamics now (see, for example, §3 of Nishijima's textbook.[18]) We use the units, $\hbar = c = 1$. The electron of the hydrogen atom is described by the Hamiltonian

$$H_{\text{el}} := \frac{1}{2m}p^2 + V_{\text{c}}(x), \tag{1}$$

where $V_{\text{c}}(x) = -\alpha Z/|x|$ is the Coulomb potential and α, Z are positive constants. We may suppose α and Z to be the fine structure constant and atomic number, respectively.

As the field reaction, we take the quantized vector potential with the Coulomb gauge. So, we can take polarization vectors, $e_r(k) = (e_{r1}(k), e_{r2}(k), e_{r3}(k)) \in \mathbb{R}^3$, $r = 1, 2$, satisfying $e_r(k) \cdot e_s(k) = \delta_{rs}$ and $e_r(k) \cdot k = 0$. Then, radiation field (quantized vector field) $A(x) = (A_1(x), A_2(x), A_3(x))$ is given by

$$A_j(x) := \sum_{r=1,2} \int_{\mathbb{R}^3} d^3k \, \frac{\chi_\kappa(k)}{\sqrt{2|k|}} \left(e_{rj}(k)e^{-i\theta kx}a_r^\dagger(k) + e_{rj}(k)e^{i\theta kx}a_r(k) \right),$$

$$x \in \mathbb{R}^3, \; \kappa \geq 0, \; j = 1, 2, 3,$$

where $a_r(k)$ and $a_r^\dagger(k)$ are annihilation and creation operators of photons, $\chi_\kappa(k)$ a cutoff function given by

$$\chi_\kappa(k) = \begin{cases} 0, & |k| < \kappa, \\ (2\pi)^{-3/2}, & \kappa \leq |k| \leq \Lambda, \\ 0, & |k| > \Lambda \end{cases}$$

with an arbitrary infrared cutoff κ with $0 \leq \kappa \leq 1$ and an arbitrarily fixed ultraviolet cutoff Λ with $1 < \Lambda$, and θ is a positive constant. The reason for the introduction of θ is explained below. We have now $\text{div}\,A(x) = 0$ because of the Coulomb gauge. The free Hamiltonian of photons is defined by

$$H_{\text{f}} := \sum_{r=1,2} \int_{\mathbb{R}^3} d^3k \, \omega(k)a_r^\dagger(k)a_r(k),$$

where $\omega(k) = |k|$.

In non-relativistic quantum electrodynamics, the total Hamiltonian of the hydrogen atom is given by

$$H_\kappa^{\text{PF}} := \frac{1}{2m}\left(p - qA(x)\right)^2 + V_{\text{c}}(x) + H_{\text{f}} - \frac{q}{2m}\sigma B(x), \tag{2}$$

where $B(x) = \mathrm{rot}\,A(x)$ is the magnetic field, q a constant playing a role of a charge, and $\sigma = (\sigma_1, \sigma_2, \sigma_3)$ consists of the standard Pauli matrices σ_j, $j = 1, 2, 3$. H_κ^{PF} is called the *Pauli-Fierz Hamiltonian* among mathematicians.

H_κ^{PF} has the *infrared singularity* at $k = 0$ in the following sense,

$$\lim_{|k| \to 0} \frac{\lambda_{rj}(k)}{\omega(k)} = \infty \quad \text{and} \quad \frac{\lambda_{rj}}{\omega} \notin L^2(\mathbb{R}^3),$$

where

$$\lambda_{rj}(k) = \frac{\chi_0(k)}{\sqrt{2|k|}} e_{rj}(k) e^{-ikx}.$$

Considering the case $q = e$ and $\alpha = e^2/4\pi$ and following Bohr's theory of atomic structure, we can neglect $-e\sigma B(x)/2m$ when we can regard $e^2 Z$ as being very small because

$$\frac{\left\langle \frac{e}{2m}\sigma B(x) \right\rangle_{\mathrm{B}}}{\left\langle \frac{e}{m}pA(x) \right\rangle_{\mathrm{B}}} \sim e^2 Z$$

(see Nishijima's textbook[18]), where $\langle T \rangle_{\mathrm{B}}$ is the ground state expectation of an operator T. Through this omission of $-e\sigma B(x)/2m$, we have

$$H_\kappa^{\mathrm{PF}} \equiv H_\kappa^{\mathrm{PF}}(\alpha, q, \theta; \Lambda) = \frac{1}{2m}\left(p - qA(x)\right)^2 + V_{\mathrm{C}}(x) + H_{\mathrm{f}}.$$

Cosider the unitary operator U satisfying

$$U^* x U = \left(e^2/4\pi\right)^{-1} x, \quad U^* p U = \left(e^2/4\pi\right) p,$$
$$U^* a_r(k) U = \left(e^2/4\pi\right)^{-3} a_r\left(\left(e^2/4\pi\right)^{-2} k\right).$$

Then, we have

$$H_\kappa^{\mathrm{PF}}\left(1, e^3/4\pi, e^2/4\pi; (4\pi)^2 \Lambda_{\mathrm{rel}}/e^4\right)$$
$$= \left(e^2/4\pi\right)^{-2} U^* H_\kappa^{\mathrm{PF}}\left(e^2/4\pi, e, 1; \Lambda_{\mathrm{rel}}\right) U. \tag{3}$$

Actually, we introduced the parameter θ above in order to obtain Hamiltonians,

$$H_\kappa^{\mathrm{PF}}\left(1, e^3/4\pi, e^2/4\pi; (4\pi)^2 \Lambda_{\mathrm{rel}}/e^4\right) \quad \text{and} \quad H_\kappa^{\mathrm{PF}}\left(e^2/4\pi, e, 1; \Lambda_{\mathrm{rel}}\right).$$

Here we note that the two Hamiltonians are not unitarily equivalent. In the works,[5,10] the units satisfying $\alpha = 1$, $q = e^3/4\pi$, $\theta = e^2/4\pi$ were basically used (see the Hamiltonian defined in (1.5) of the paper[5]).

Moreover, the (electric) dipole approximation can be done when we can regard αZ as being so small that

$$e^{-ikx} \approx 1$$

because $\langle kx \rangle_{\mathrm{B}} \sim \alpha Z/2$.

After doing dipole approximation and neglecting $- q\sigma B(x)/2m$, we obtain the approximated Hamiltonian H_κ^{da} from H_κ^{PF} as

$$H_\kappa^{\mathrm{da}} := \frac{1}{2m}\left(p - qA(0)\right)^2 + V_{\mathrm{c}}(x) + H_{\mathrm{f}}. \tag{4}$$

It is well known that there exits a normalized ground state $\psi_\kappa^{\mathrm{da}}$ of H_κ^{da} for $0 < \kappa < \Lambda$.

We denote the number operator of photons by

$$N_{\mathrm{f}} = \sum_{r=1,2} \int_{\mathbb{R}^3} d^3k\, a_r^\dagger(k) a_r(k).$$

Then, the expectation $\langle N_{\mathrm{f}} \rangle$ of the total number of soft photons is given by

$$\langle N_{\mathrm{f}} \rangle := \lim_{\kappa \to 0} \langle \psi_\kappa^{\mathrm{da}}, N_{\mathrm{f}} \psi_\kappa^{\mathrm{da}} \rangle_{\mathcal{H}}.$$

We need frequently to remind ourselves of the following identity:

$$\langle \psi_\kappa^{\mathrm{da}}, N_{\mathrm{f}} \psi_\kappa^{\mathrm{da}} \rangle_{\mathcal{H}} = \sum_{r=1,2} \int_{\mathbb{R}^3} d^3k\, \|a_r(k)\psi_\kappa^{\mathrm{da}}\|_{\mathcal{H}}^2. \tag{5}$$

By using the pull-through formula, we get

$$a_r(k)\psi_\kappa^{\mathrm{da}} = q\,\frac{\chi_\kappa(k)}{\sqrt{2|k|}} e_r(k)\left(H_\kappa^{\mathrm{da}} - E_\kappa^{\mathrm{da}} + |k|\right)^{-1} v\psi_\kappa^{\mathrm{da}}, \tag{6}$$

where $v = (p - qA(0))/m$, and E_κ^{da} denotes the ground state energy of H_κ^{da}.

Since, generally, velocity can be given as the differential of the Heisenberg picture of position of the electron, we get

$$i\left[H_\kappa^{\mathrm{da}}, x\right] = \frac{p - qA(0)}{m} = v. \tag{7}$$

Namely, v appearing in (6) is the velocity of the electron in the system with its total energy H_κ^{da}.

Through (6) and (7) we have

$$a_r(k)\psi_\kappa^{\mathrm{da}} = iq\,\frac{\chi_\kappa(k)}{\sqrt{2|k|}} e_r(k)\left(H_\kappa^{\mathrm{da}} - E_\kappa^{\mathrm{da}} + |k|\right)^{-1}\left(H_\kappa^{\mathrm{da}} - E_\kappa^{\mathrm{da}}\right) x\psi_\kappa^{\mathrm{da}}.$$

Thus, since $\| \left(H_\kappa^{\mathrm{da}} - E_\kappa^{\mathrm{da}} + |k| \right)^{-1} \left(H_\kappa^{\mathrm{da}} - E_\kappa^{\mathrm{da}} \right) \| \leq 1$, we have

$$\sum_{r=1,2} |k|^2 \| a_r(k) \psi_\kappa^{\mathrm{da}} \|_{\mathcal{H}}^2 \leq |k|^2 \frac{q^2}{8\pi^3 |k|} \| x \psi_\kappa^{\mathrm{da}} \|_{\mathcal{H}}^2 = \frac{q^2}{8\pi^3} |k| \langle x^2 \rangle_\kappa, \qquad (8)$$

where $\langle x^2 \rangle_\kappa := \langle \psi_\kappa^{\mathrm{da}} , x^2 \psi_\kappa^{\mathrm{da}} \rangle_{\mathcal{H}}$. When we follow quantum mechanics, $\langle x^2 \rangle_\kappa$ should be almost $(m_{\mathrm{re}} \alpha Z)^{-2}$, where m_{re} is the renormalized mass of electron. M. Griesemer, E. H. Lieb, and M. Loss[10] actually proved that there exists a positive constant C independent of κ, Λ and α such that

$$\langle x^2 \rangle_\kappa = \| |x| \psi_\kappa^{\mathrm{da}} \|^2 \leq \frac{2C}{m(\alpha Z)^2}. \qquad (9)$$

It follows from this estimate that we can obtain mathematically

$$\langle N_{\mathrm{f}} \rangle \equiv \lim_{\kappa \to 0} \langle \psi_\kappa^{\mathrm{da}} , N_{\mathrm{f}} \psi_\kappa^{\mathrm{da}} \rangle_{\mathcal{H}} \leq \frac{q^2}{2\pi^2} \Lambda^2 \langle x^2 \rangle_\kappa \leq C \frac{q^2}{\pi^2 m (\alpha Z)^2} \Lambda^2. \qquad (10)$$

Since we consider only single electron now, the proof of (10) is completed by applying the first part of their proof. The analogy of the proof of (10) works for more general Hamiltonian $H(\kappa)$ describing a particle coupled to Bose field. Namely, once an operator v appears between $(H(\kappa) - E(\kappa) + |k|)^{-1}$ and a ground state in the identity (6), where $E(\kappa)$ is the ground state energy of $H(\kappa)$, the analogy works as long as v is the velocity of the particle, $v = i[H(\kappa), x]$, and x satisfies (9). To the author's best knowledge, V. Bach, J. Fröhlich, and I.M. Sigal[5] were the first to make good use of the representation (6) in order to treat the infrared problem in mathematics, though they did not write the word of velocity clearly.

In the case of atomic units ($\alpha = 1$ and $q = e^3/4\pi$ for the electric charge e), (10) has no trouble. We should, however, notice that (10) becomes a rough estimate in the case of the relativistic units ($\alpha = e^2/4\pi$ and $q = e$). Because q^2/α^2 becomes $16\pi^2 e^{-2}$ in this case. We cannot avoid this appearance of α^2 in the denominator because of the physical reason written above as far as we estimate $\langle x^2 \rangle_\kappa$. So, the point to observe is how we can estimate $\langle N_{\mathrm{f}} \rangle_\kappa$ avoiding that of $\langle x^2 \rangle_\kappa$. In our recent work, we make a device to avoid such an appearance of α^2 in the denominator of the upper bound in (6) (see Corollary 7.4 in our work.[11]) One of the main purpose of this paper is to explain the mathematical method to achieve it (see §4 for H_κ^{PF} and also §3 for Nelson's model). Incidentally, Griesemer, Lieb, and Loss use the atomic units ($\alpha = 1$ and $q = e^3/4\pi$). In this units, $\Lambda = (4\pi)^2 \Lambda_{\mathrm{rel}}/e^4$ for the ultraviolet cutoff $\Lambda = \Lambda_{\mathrm{rel}}$ in the relativistic units ($\alpha = e^2/4\pi$ and $q = e$). So, it is very important that they can prove the existence of a ground state for arbitrary ultraviolet cutoff as well as arbitrary coupling constant.[10]

By (10), if we follow the way shown in papers,[2,5] we at least obtain that *there exists a ground state ψ_0 of H_0^{da} for sufficiently small $|q|$*, and $\psi_0 \in D(N_{\mathrm{f}}^{1/2})$. Lemma 4.9 in the work[2] shows a general method to prove existence of a ground state. By applying the Lemma 4.9 to a general Hamiltonian H describing an electron interacting with a quantized field, we can prove the existence of the ground state for sufficiently small coupling constant $|q|$ as long as we can show that there exists positive constnats κ_i, c_i, and γ_i, $i = 1,2$, such that $\langle N_{\mathrm{f}} \rangle_\kappa \leq c_1 |q|^{\gamma_1}$ for all κ with $0 < \kappa < \kappa_1$ and $\langle Q \rangle_\kappa \leq c_2 |q|^{\gamma_2}$ for all κ with $0 < \kappa < \kappa_2$, where Q is an operator defined such that (34) below holds. So, it is important to investigate the q-dependence of $\langle N_{\mathrm{f}} \rangle \equiv \lim_{\kappa \to 0} \langle N_{\mathrm{f}} \rangle_\kappa$. On the other hand, we expect that there is no ground state when $\langle N_{\mathrm{f}} \rangle = \infty$, which is still one of the open problems in mathematics except some models (see (vH1), (vH2) below, and §2).

As a brief remark, for H_κ^{PF} without $-q\sigma B(x)/2m$, the plane wave e^{-ikx} appears between $(H_\kappa^{\mathrm{PF}} - E_\kappa^{\mathrm{PF}} + |k|)^{-1}$ and v in the right hand side of (6). Then, we have to face with the non-commutativity between e^{-ikx} and v, which is treated by the identity,

$$e^{-ikx} \left[H_\kappa^{\mathrm{PF}}, x \right] = \left[H_\kappa^{\mathrm{PF}}, e^{-ikx}x \right] - \left[H_\kappa^{\mathrm{PF}}, e^{-ikx} \right] x.$$

The identity is useful because we can calculate $\left[H_\kappa^{\mathrm{PF}}, e^{-ikx} \right]$ concretely.

Thus, although the Pauli-Fierz Hamiltonian has the infrared singularity at $k = 0$, the average value of the total number of soft photons converges, and we can realize that the existence of a ground state of the Pauli-Fierz Hamiltonian does not depend on whether the dipole approximation is done or not. It is important that (i) $A(x)^2$ (resp. $A(0)^2$) in H_κ^{PF} (resp. H_κ^{da}) causes the appearance of the velocity of the electron in the representation (6) using the pull-through formula and that (ii) the term $pA(x)$ (resp. $pA(0)$) in H_κ^{PF} (resp. H_κ^{da}) determines the form of the velocity of the electron by following its definition. For H_κ^{PF} (resp. H_κ^{da}) the velocity is given by mv-momentum.[8]

Although we have to fix a gauge when we do the second quantization, we meet physical difference if we employ the approximation which breaks the gauge invariance of the energy in quantum mechanics. We linearize H_κ^{PF} and get a Hamiltonian

$$H_\kappa^{\mathrm{GSB}} := H_{\mathrm{el}} - \frac{q}{m}pA(x) + H_{\mathrm{f}}, \tag{11}$$

which is an example of the generalized spin-boson model[2] without dipole approximation. For H_κ^{GSB}, the velocity of the electron is still given by mv-

momentum as $i\,[H_\kappa^{\mathrm{GSB}}\,,\,x] = (p - qA(x))/m = v$. However, by the pull-through formula, we have

$$a_r(k)\psi_\kappa^{\mathrm{GSB}} = q\,\frac{\chi_\kappa(k)}{\sqrt{2|k|}}\,e_r(k)\cdot(H_\kappa^{\mathrm{GSB}} - E_\kappa^{\mathrm{GSB}} + |k|)^{-1}\,\frac{p}{m}\,\psi_\kappa^{\mathrm{GSB}},$$

where $\psi_\kappa^{\mathrm{GSB}}$ is a normalized ground state of H_κ^{GSB}. We note that p-momentum[8] divided by the mass, p/m, appearing in the right hand side of the above expression, is the velocity of electron which does not feel the quantized vector field $A(x)$, i.e., $i\,[H_{\mathrm{el}} + H_{\mathrm{f}}\,,\,x] = p/m$. In this case, the existence or absence might be influenced by whether the dipole approximation is done or not, which is still controversial.

The Hamiltonian H_κ^{N} of the *Nelson's model* [17] is given by replacing the quantized vector field $A(x)$ by the scalar Bose field $\phi_\kappa(x)$ in H_κ^{GSB}, and moreover, by omitting p from the term $pA(x)$. So, the Hamiltonian for Nelson's model with $m = 1$ reads

$$H_\kappa^{\mathrm{N}} = \frac{1}{2}p^2 + V_{\mathrm{c}}(x) + H_{\mathrm{f}} + q\phi_\kappa(x), \tag{12}$$

where

$$H_{\mathrm{f}} = \int_{\mathbb{R}^3} d^3k\,\omega(k)a^\dagger(k)a(k), \tag{13}$$

$$\phi_\kappa(x) = \int_{\mathbb{R}^3} d^3k\,\frac{\chi_\kappa(k)}{\sqrt{2\omega(k)}}\left(e^{ikx}a(k) + e^{-ikx}a^\dagger(k)\right). \tag{14}$$

Then, the velocity is given by $v = p/m$, however the operator sitting on the place of v in (6) is the plane wave. More precisely, we get

$$a(k)\psi_\kappa = -q\frac{\chi_\kappa(k)}{\sqrt{2|k|}}\left(H_\kappa^{\mathrm{N}} - E_\kappa^{\mathrm{N}} + |k|\right)^{-1} e^{-ikx}\psi_\kappa. \tag{15}$$

Although mathematicians call H_κ^{N} Nelson's Hamiltonian now, this model was already studied as a model for polaron before they called it so. By works,[6,12] the average of the total number of soft photons diverges logarithmically:

$$\langle N_{\mathrm{f}}\rangle \equiv \lim_{\kappa\to 0}\langle\psi_\kappa,\,N_{\mathrm{f}}\psi_\kappa\rangle_{\mathcal{H}} \sim \mathrm{const.}\,\lim_{\kappa\to 0}\log\frac{\Lambda}{\kappa}, \tag{16}$$

where the existence of ψ_κ for $\kappa > 0$ is well-known. And moreover, as we expect it, *for every q with $q \neq 0$, $H_N = H_0^N$ has no ground state in the standard state space $\mathcal{H}$.* For confining external potential such as quantum harmonic oscillator, the same statement is proved by Lörinczi *et al.*[16] Recently, C. Gérard and J. Dereziński also succeeded in proving the absence of ground state for Nelson's Hamiltonian with $\kappa = 0$ and a certain external potential (see their work[7] announced in Gérard's paper.[9])

In order to see the divergence (16), the following simplification helps us. We apply the dipole approximation to H_κ^{N} now. Then, we obtain the Hamiltonian

$$H_\kappa^{\mathrm{N,da}} = \frac{1}{2}p^2 + V_{\mathrm{c}}(x) + H_{\mathrm{f}} + q\phi_\kappa(0) = H_{\mathrm{el}} + H_\kappa^{\mathrm{vH}},$$

where $H_\kappa^{\mathrm{vH}} = H_{\mathrm{f}} + q\phi_\kappa(0)$ is the Hamiltonian of the van Hove model (or fixed source model). For H_κ^{vH} it is well known that

(vH1) H_0^{vH} has no ground state in $\mathcal{F}$.

(vH2) $\langle N_{\mathrm{f}} \rangle \equiv \lim_{\kappa \to 0} \langle \psi_\kappa^{\mathrm{vH}}, N_{\mathrm{f}} \psi_\kappa^{\mathrm{vH}} \rangle = \infty$, where $\psi_\kappa^{\mathrm{vH}}$ is a normalized ground state of H_κ^{vH} for $0 < \kappa$.

(vH3) $\lim_{\kappa \to 0} \sigma(H_\kappa^{\mathrm{vH}}) = \sigma(H_0^{\mathrm{vH}})$, where $\sigma(T)$ denotes the spectrum for an operator T.

As a matter of fact, we get

$$a(k)\psi_\kappa^{\mathrm{N,da}} = -q\frac{\chi_\kappa(k)}{\sqrt{2|k|}} \left(H_\kappa^{\mathrm{N,da}} - E_\kappa^{\mathrm{N,da}} + |k| \right)^{-1} \psi_\kappa^{\mathrm{N,da}} = -q\frac{\chi_\kappa(k)}{\sqrt{2|k|}\,|k|} \psi_\kappa^{\mathrm{N,da}},$$

where $\psi_\kappa^{\mathrm{N,da}} = \psi_{\mathrm{el}} \otimes \psi_\kappa^{\mathrm{vH}}$ is a unique ground state of $H_\kappa^{\mathrm{N,da}}$ and ψ_{el} the unique ground state of H_{el}. Then, for $H_\kappa^{\mathrm{N,da}}$ we get

$$\langle N_{\mathrm{f}} \rangle_\kappa \equiv \langle \psi_\kappa^{\mathrm{N,da}}, N_{\mathrm{f}} \psi_\kappa^{\mathrm{N,da}} \rangle = \frac{q^2}{4\pi^2} \log \frac{\Lambda}{\kappa}.$$

So, the problem for H_κ^{N} is to derive the cause of the infrared catastrophe, explained above, from e^{-ikx}. Then, the rest purpose of this paper is to explain a mathematical proof for (16) and the absence of a ground state (see §2).

2. Infrared catastrophe of Nelson's model

For the sake of simplicity, we set $m = 1$ throughout. As explained in §1, we consider a particle coupled with a massless scalar Bose field. In momentum representation the creation and annihilation operators of the field satisfy the standard CCR,

$$[a(k), a^\dagger(k')] = \delta(k - k'), \quad [a(k), a(k')] = 0, \quad [a^\dagger(k), a^\dagger(k')] = 0, \quad k, k' \in \mathbb{R}^3.$$

The field and its energy are given by (14) and (13), respectively, and the number operators is defined by

$$N_{\mathrm{f}} = \int_{\mathbb{R}^3} d^3k\, a^\dagger(k)a(k).$$

They act on the symmetric Fock space, $\mathcal{F}$, over $L^2 = L^2(\mathbb{R}^3)$. For zero mass bosons the dispersion relation is given by $\omega(k) = |k|$.

We set

$$\lambda_{\kappa,x}(k) = \frac{\chi_\kappa(k)}{\sqrt{2\omega(k)}}\, e^{-ikx} \quad \text{for } \forall k, x \in \mathbb{R}^3.$$

Then,

$$\phi_\kappa(x) = a^\dagger(\lambda_{\kappa,x}) + a(\lambda_{\kappa,x}),$$

where

$$a^\dagger(f) = \int_{\mathbb{R}^3} d^3k\, a^\dagger(k) f(k), \qquad a(f) = \int_{\mathbb{R}^3} d^3k\, a(k) f(k)^*$$

for every $f \in L^2(\mathbb{R}^3)$. Here, as for the definition of $a(f)$, we follow the mathematical way. We note that for $\kappa = 0$ the *infrared singularity* at $k = 0$:

$$\lim_{|k| \to 0} \frac{\lambda_{0,x}(k)}{\omega(k)} = \infty \quad \text{and} \quad \frac{\lambda_{0,x}}{\omega} \notin L^2(\mathbb{R}^3).$$

The cutoff Nelson Hamiltonian H_κ^{N} defined in (12) reads

$$H_\kappa^{\mathrm{N}} = H_{\mathrm{at}} + H_{\mathrm{f}} + q\phi_\kappa(x),$$

acting on $\mathcal{H} = L^2(\mathbb{R}^3) \otimes \mathcal{F}$, where $H_{\mathrm{at}} = p^2/2 + V_{\mathrm{c}}(x)$. We define a non-negative free Hamiltonian by $H_0 = H_{\mathrm{at}} - E_1 + H_{\mathrm{f}}$. We set $H_{\mathrm{N}} := H_0^{\mathrm{N}}$.

Applying Proposition 2.1 in Arai's paper[1] to H_κ^{N}, H_κ^{N} *is self-adjoint with* $D(H_\kappa^{\mathrm{N}}) = D(H_0) \equiv D(p^2) \cap D(H_{\mathrm{f}})$ *for all* q, $0 < \kappa < \Lambda$. Moreover, H_κ^{N} *is bounded from below for arbitrary values of* q, *and* H_κ^{N} *is essentially self-adjoint on each core for* H_0.

We define a positive constant q_Λ by

$$-E_1 = \frac{q_\Lambda^2}{2} \int_{\mathbb{R}^3} d^3k\, |\lambda_{0,0}(k)|^2 \frac{k^2}{\omega(k) + k^2/2}.$$

We note that q_Λ is independent of κ. Fix $\Lambda > 0$. Then, Theorem 1 in Spohn's paper[21] says that H_κ^{N} *has a unique ground state* ψ_κ *for every* κ, q *with* $0 < \kappa < \Lambda$ *and* $|q| < q_\Lambda$.

In our situation, we have for $\kappa > 0$ $\lambda_{\kappa,x}/\omega \in L^2(\mathbb{R}^3)$, which is called *infrared regularity* condition, and $\lambda_{0,x}/\omega \notin L^2(\mathbb{R}^3)$ for $\kappa = 0$, which is called *infrared singularity* condition (see our papers[3,4]).

We derive the following identity from the pull-through formula as done in Lemma 4.1 of the paper.[12] *If* H_κ^{N}, $0 \le \kappa < \Lambda$, *has a ground state* ψ_κ, *then*

$$a(k)\psi_\kappa = \sum_{j=1}^{5} I_j(k)\psi_\kappa, \qquad 0 \le \kappa < \Lambda, \tag{17}$$

as an identity in $L^2(\mathbb{R}^3, d^3k; \mathcal{H})$ with

$$I_1(k) = -q\Xi_1(k)e^{-ikx},$$
$$\Xi_1(k) = \frac{\chi_\kappa(k)}{\sqrt{2\omega(k)}\,\omega(k)},$$
$$I_2(k) = -iq\Xi_2(k)\left(H_\kappa^N - E_\kappa^N + \omega(k)\right)^{-1}\left(H_\kappa^N - E_\kappa^N\right)e^{-ikx}x,$$
$$\Xi_2(k) = \frac{\chi_\kappa(k)k}{\sqrt{2\omega(k)}\,\omega(k)},$$
$$I_3(k) = -iq\Xi_3(k)\left(H_\kappa^N - E_\kappa^N + \omega(k)\right)^{-1}pe^{-ikx}x,$$
$$\Xi_3(k) = \frac{\chi_\kappa(k)k^2}{\sqrt{2\omega(k)}\,\omega(k)},$$
$$I_4(k) = -iq\Xi_4(k)\left(H_\kappa^N - E_\kappa^N + \omega(k)\right)^{-1}e^{-ikx}x,$$
$$\Xi_4(k) = \frac{\chi_\kappa(k)k^2 k}{2\sqrt{2\omega(k)}\,\omega(k)},$$
$$I_5(k) = q\Xi_5(k)\left(H_\kappa^N - E_\kappa^N + \omega(k)\right)^{-1}e^{-ikx},$$
$$\Xi_5(k) = \frac{\chi_\kappa(k)k^2}{2\sqrt{2\omega(k)}\,\omega(k)}.$$

It is easy to check that $I_j\psi_0$, $j = 2, 3, 4, 5$, are $L^2(\mathbb{R}^3)$-functions of k, but it is clear that $I_1\psi_0$ is IR-catastrophic. Moreover, (17) implies that *for every q with $0 < |q| < q_\Lambda$, there exist a positive constant K_0 and $M_{q,\alpha,\Lambda}$ such that*

$$\frac{q^2}{8\pi^2}\left(\log\frac{\Lambda}{\kappa}\right) - M_{q,\alpha,\Lambda} \le \langle N_f\rangle_\kappa \le \frac{q^2}{2\pi^2}\left(\log\frac{\Lambda}{\kappa}\right) + 2M_{q,\alpha,\Lambda} \tag{18}$$

for $\kappa < K_0$ (see Theorem 2.4 in my paper,[12]) which is an implementation of (16). The inequality (18) is proved, by combining the identity (17) and mathematical techniques developed in the recent works.[5,10,11] They are so-called *strict positivity of binding energy* and *spatial localization* with respect to ground state(s). The importance of strict positivity of binding energy was showed for H_κ^{PF} by Griesemer, Lieb, and Loss[10] first. Let $E_\kappa^{N\,V=0} = \inf\sigma\left(H_\kappa^{N\,V=0}\right)$, where the superscript means that in H_κ^N we replace the

Coulomb potential V_{C} by just 0. The (positive) binding energy is defined by

$$E_{\kappa}^{\mathrm{bin}} = E_{\kappa}^{N\,V=0} - E_{\kappa}^{N}.$$

Following the way to prove Theorem 3.1 in their work,[10] we can show that $E_{\kappa}^{\mathrm{bin}} > 0$ with a slight revision like Proposition 4.4 in our work.[11] The statement is the following. *Fix κ with $0 \le \kappa < \Lambda$ and suppose that H_{κ}^{N} has a ground state ψ_{κ}. Then,*

$$E_{\kappa}^{\mathrm{bin}} \ge -E_1 > 0.$$

This is the strict positivity of binding energy.

On the other hand, a useful proof of the spatial localization was also introduced by Griesemer, Lieb, and Loss.[10] We can revise it so that it can meet our case. Let us fix κ with $0 \le \kappa < \Lambda$ and suppose that H_{κ}^{N} has a ground state ψ_{κ} again. Here we take a differentiable and non-negative function g and positive number R satisfying $\sup_{R/2 < |x|} |\nabla g(x)|^2 < \infty$, and moreover, we set $G_R(x) = \chi_R(|x|)g(x)$, where $\chi_R(r) = 0$ for $r < R/2$ and $\chi_R(r) = 1$ for $r > R$ with linear interpolation. Then, Lemma 3.9 in our work[12] says that $\psi_{\kappa} \in D(g)$ *and*

$$\|g\psi_{\kappa}\|_{\mathcal{H}}^2$$

$$\le \left\{ \sup_{|x| \le R} |g(x)|^2 + \left(-E_1 - \sup_{R < |x|} |V_{\mathrm{C}}(x)| \right)^{-1} \sup_x |\nabla G_R(x)|^2 \right\} \|\psi_{\kappa}\|_{\mathcal{H}}^2$$

holds for sufficiently large R so that $-E_1 > \sup_{R < |x|} |V_{\mathrm{C}}(x)|$. This inequality is the spatial localization with respect to ground state(s). As its corollary, we have

$$\langle x \rangle_{\kappa} \le \left\{ R^2 + 9 \left(-E_1 - \sup_{R < |x|} |V_{\mathrm{C}}(x)| \right)^{-1} \right\}.$$

Also, by adding the argument in our work[4] to the above results, we can show *the absence of a ground state for $H_N \equiv H_0^N$* (see Theorem 2.5 in my work.[12]) The argument uses the reduction to absurdity. More precisely, suppose now that H_N has a ground state ψ_0, which is an assumption of the reduction to absurdity. Then we can show that $\langle \phi, a(k)\psi_0 \rangle_{\mathcal{H}}$ is a $L^2(\mathbb{R}^3)$-function of $k \in \mathbb{R}^3$ for $\phi \in D(N_{\mathrm{f}})$, by using the Riesz' lemma and some other techniques in functional analysis. On the other hand, we know that $\langle \phi, I_j(k)\psi_0 \rangle_{\mathcal{H}}$, $j = 2, 3, 4, 5$, are also $L^2(\mathbb{R}^3)$-functions of $k \in \mathbb{R}^3$ because of the Schwarz inequality and the fact that $I_j(k)\psi_0$, $j = 2, 3, 4, 5$, are $L^2(\mathbb{R}^3)$-functions of k. Thus, $\langle \phi, I_1(k)\psi_0 \rangle_{\mathcal{H}}$ has to be also a $L^2(\mathbb{R}^3)$-function of $k \in \mathbb{R}^3$ by the identity (17), which implies the contradiction that ψ_0 has to be 0 because $\langle \phi, I_1(k)\psi_0 \rangle_{\mathcal{H}}$ is actually IR-catastrophic.

3. Nelson's model without IR & UV cutoffs

3.1. *Change of representation*

In our work[11] we consider two particles. One of them is infinitely heavy and has charge qZ, and the other one has charge q.

We start with mathematical formulation to explain our results. The infinitely heavy nucleus is nailed down at the origin. The other particle has bare mass $m = 1$. Its position is denoted by x, momentum by $p = -i\nabla_x$. Remember we set $\hbar = 1, c = 1$ throughout. The two particles are coupled through a massless scalar Bose field.

The Nelson Hamiltonian for this situation reads

$$H_{\mathrm{N},\kappa} = \frac{1}{2}p^2 + H_{\mathrm{f}} + qZ\phi_\kappa(0) + q\phi_\kappa(x), \tag{19}$$

acting in the state space

$$\mathcal{H} := L^2(\mathbb{R}^3) \otimes \mathcal{F}.$$

It is easy to show that $H_{\mathrm{N},\kappa}$ *is self-adjoint on the domain* $D(p^2) \cap D(H_{\mathrm{f}})$ *and bounded from below for arbitrary values of* qZ by applying the Kato-Rellich theorem.

Unfortunately $H_{\mathrm{N},\kappa}$ is both infrared and ultraviolet divergent: As $\check{\chi}_\kappa(x) \to \delta(x)$, the anticipated ground state does not lie in $\mathcal{H}$ and the ground state energy tends to $-\infty$. These divergences are of a mild nature however and can be unraveled through the unitary Gross transformation e^{-T} which is generated by

$$T = -q \int_{\mathbb{R}^3} d^3k \, \frac{\chi_\kappa(k)}{\sqrt{2\omega(k)}}$$
$$\times \left\{ \left(\beta(k)e^{ikx} + \frac{Z}{\omega(k)} \right) a(k) - \left(\beta(k)e^{-ikx} + \frac{Z}{\omega(k)} \right) a^\dagger(k) \right\}$$

with $\beta(k) = (\omega(k) + |k|^2/2)^{-1}$.

We obtain

$$\left\{ e^T H_{\mathrm{N},\kappa} e^{-T} + q^2 \int_{\mathbb{R}^3} d^3k \frac{\chi_\kappa(k)^2}{2\omega(k)} \left(\beta(k) + \frac{Z^2}{\omega(k)} \right) \right\} = H_{\kappa\Lambda} + V_{\kappa\Lambda},$$

where

$$H_{\kappa\Lambda} = H_{\mathrm{at}} + H_{\mathrm{f}} + q(pA_{\kappa\Lambda} + A_{\kappa\Lambda}{}^{*}p) + \frac{q^2}{2}\left(A_{\kappa\Lambda}{}^{2} + 2A_{\kappa\Lambda}{}^{*}A_{\kappa\Lambda} + A_{\kappa\Lambda}{}^{*2}\right),$$

$$(20)$$

and

$$H_{\mathrm{at}} = \frac{1}{2}p^2 - \frac{q^2 Z}{4\pi|x|}.$$

After the transformation, the Coulomb potential is derived as an effective potential. We have to note the appearance of q^2 in the numerator of the effective potential comes from the Gross transformation, so in this case we have to treat both, q^2 in the numerator of the Coulomb potential and q in front of the interaction of the field, at the same time.

The Gross transformation generates an extra potential

$$V_{\kappa\Lambda} = q^2 Z \int_{\mathbb{R}^3} d^3k \left((2\pi)^{-3} - \chi_{\kappa}(k)^2\right) \omega(k)^{-2} e^{ikx}.$$

Since $\lim_{\kappa\to 0,\Lambda\to\infty} V_{\kappa\Lambda} = 0$ in the $L^2(\mathbb{R}^3)$-norm, we consider only $H_{\kappa\Lambda}$ for the purpose that we remove both cutoffs, infrared and ultraviolet. Then, following the way Nelson did,[17] we can show that *there exist a strictly positive constant* q_{UV} *and self-adjoint operators* $H_{\kappa\infty}$ *and* H *such that for every* q *with* $|q| < q_{UV}$

(N1) $H_{\kappa\Lambda}$ *converges to* $H_{\kappa\infty}$ *as* $\Lambda \to \infty$ *in the norm resolvent sense,*
(N2) $H_{\kappa\infty}$ *converges to* H *as* $\kappa \to 0$ *in the norm resolvent sense.*

(See Theorem 1.2 in our work.[11]) We have to remember that $H_{\mathrm{N}} \equiv H_0^{\mathrm{N}} \equiv \lim_{\kappa\to 0} H_{\kappa}^{\mathrm{N}}$ (removed the infrared cutoff) does not have any ground state. Applying the same treatment to $H_{\mathrm{N},\kappa}$, we know that $H_{\mathrm{N},0} \equiv \lim_{\kappa\to 0} H_{\mathrm{N},\kappa}$ does not have any ground state either. In such a case, following the Bloch-Nordsieck theorem in physics, we have to change the representation of the model. Actually, e^{-T} is not unitary as $\kappa = 0$, which is proved in Arai's paper.[1] So, "$H_{N,0}$" *and* "$H_{0\Lambda} + (extra\ potential) + (self\text{-}energy)$" are not *unitarily equivalent. Thus, we can say that* H *is Nelson's model with a changed representation.*

In this section, we explain how we can show broadly that *there exist positive constants* $D_{\mathrm{N},n}$, $n = 0,1,2,3$, *such that*

$$\langle N_{\mathrm{f}} \rangle \leq q^2 D_{\mathrm{N},0} \left(D_{\mathrm{N},1} + D_{\mathrm{N},2} \log q + D_{\mathrm{N},3}(\log q)^2\right) \qquad (21)$$

for sufficiently small $|q|$, of which precise proof is in the paper.[11] We show the algorithm to solve the problem in §4 for H_{κ}^{PF}.

3.2. *Strategy for estimate of photon number*

Our $H_{\kappa\Lambda}$ has a similar form to H_κ^{PF}. We can say that $H_{\kappa\Lambda}$ has the normal ordered interaction of H_κ^{PF}. As noted in §3.1, q^2 appears in the numerator of the Coulomb potential in $H_{\kappa\Lambda}$. Thus, the same trouble occurs as explained at (10) in §1. So, we work out our strategy to treat this problem.

We denote x, p, and v with their components by $x = (x_1, x_2, x_3)$, $p = (p_1, p_2, p_3)$, and $v = (v_1, v_2, v_3)$, respectively.

We have for $0 < \kappa$

$$[H_{\kappa\Lambda}, a(k)]\psi_{\kappa\Lambda} = (H_{\kappa\Lambda} - E_{\kappa\Lambda})a(k)\psi_{\kappa\Lambda}, \tag{22}$$

where $\psi_{\kappa\Lambda}$ is a normalized ground state of $H_{\kappa\Lambda}$ with its energy $E_{\kappa\Lambda}$. The existence of $\psi_{\kappa\Lambda}$ is well-known for $0 < \kappa < \Lambda$. On the other hand, we can calculate the commutator directly and get

$$[H_{\kappa\Lambda}, a(k)] = -\omega(k)a(k) - qh_x(k)v\psi_{\kappa\Lambda}, \tag{23}$$

where v is the velocity of the particle, *i.e.*,

$$v = p + qA_{\kappa\Lambda} + qA_{\kappa\Lambda}{}^* = i[H_{\kappa\Lambda}, x], \tag{24}$$

and

$$h_x(k) = 2^{-1/2}k\beta(k)e^{-ikx}\omega(k)^{-1/2}\chi_\kappa(k).$$

Remember that $m = 1$ now. Combining (22) and (23), we have the so-called *pull-through formula*

$$(H_{\kappa\Lambda} - E_{\kappa\Lambda})a(k)\psi_{\kappa\Lambda} = -\omega(k)a(k)\psi_{\kappa\Lambda} - qh_x(k)v\psi_{\kappa\Lambda},$$

which implies

$$a(k)\psi_{\kappa\Lambda} = -q(H_{\kappa\Lambda} - E_{\kappa\Lambda} + \omega(k))^{-1}h_x(k)v\psi_{\kappa\Lambda}. \tag{25}$$

Remember

$$\langle N_{\mathrm{f}}\rangle_\kappa = \int_{\mathbb{R}^3} d^3k\, \|a(k)\psi_{\kappa\Lambda}\|_{\mathcal{H}}^3.$$

Then, we divide the momentum space into the low energy infrared (IR) region, $|k| < 1$, and the high energy ultraviolet (UV) region, $|k| \geq 1$. Because different kinds of inequality are required of us according to the individual regions. Then, $\langle N_{\mathrm{f}}\rangle_\kappa$ is divided as

$$\langle N_{\mathrm{f}}\rangle_\kappa = \int_{|k|<1} d^3k\, \|a(k)\psi_{\kappa\Lambda}\|_{\mathcal{H}}^2 + \int_{|k|\geq 1} d^3k\, \|a(k)\psi_{\kappa\Lambda}\|_{\mathcal{H}}^2.$$

When we consider the UV region ($|k| \geq 1$), we do not need to worry about infrared singularities. Also, when we consider the IR region ($|k| < 1$), we have only to pay attention to the infrared singularity.

3.3. *UV region*

As for $H_{\kappa\Lambda}$, it is not so difficult to treat UV region actually. This is one of the big differences between the Nelson model and Pauli-Fierz model. However, non-commutativity between the particle momentum p and the plane wave, *i.e.*, $[p\,,\,e^{ikx}] = ke^{ikx}$, introduces k, which is not good for the treatment of ultraviolet problem.

Then, we obtain that *for* $q < q_{UV}$ *there exists a positive constant C independent of* κ, Λ *such that*

$$\| (H_{\kappa\Lambda} - E_{\kappa\Lambda} + |k|)^{-1}\, h_x(k)v\psi_{\kappa\Lambda}\| \leq C\frac{h_0(k)}{|k|^{1/2}}.$$

As its corollary, we have *for* $q < q_{UV}$

$$\int_{|k|\geq 1} d^3k\, \|a(k)\psi_{\kappa\Lambda}\|_{\mathcal{H}}^2 \leq \text{const. } q^2.$$

3.4. *Energy scale renormalization*

In this subsection, we show briefly a decomposition of $\|a(k)\psi_{\kappa\Lambda}\|_{\mathcal{H}}$, $|k| < 1$, following a repeatable algorithm to avoid the appearance of q^2 in the denominator in the upper bound as explained at (10). The precise contents is introduced for H_{κ}^{PF} in §4.

From (25) we have

$$a(k)\psi_{\kappa\Lambda} = - q\gamma(k)\left(H_{\kappa\Lambda} - E_{\kappa\Lambda} + \omega(k)\right)^{-1} I_0\psi_{\kappa\Lambda},$$

where

$$\gamma(k) = \frac{\beta(k)\chi_\kappa(k)}{\sqrt{2|k|}},$$

$$I_0 = e^{-ikx}v. \tag{26}$$

For f_j dependent only on $k \in \mathbb{R}^3$, we have

$$\left[H_{\kappa\Lambda}\,,\,e^{if_jx_j}\right] = f_je^{if_jx_j}v + \frac{f_j^2}{2}e^{if_jx_j}, \tag{27}$$

where we used $[p\,,\,e^{if_jx_j}] = f_je^{if_jx_j}$. Thus, we can represent the velocity by

$$v = \frac{1}{f_j}e^{-if_jx}\left[H_{\kappa\Lambda}\,,\,e^{if_jx_j}\right] - \frac{f_j}{2}, \qquad j = 1, 2, 3. \tag{28}$$

Similarly, for $g = (g_1, g_2, g_3)$ with components g_j dependent only on $k \in \mathbb{R}^3$, we get

$$\left[H_{\kappa\Lambda}, e^{igx}\right] = \frac{1}{2} g^2 e^{igx} + e^{igx} gv. \tag{29}$$

Let $\{f^{[n]}\}_{n=1}^\infty$ be a given scale of $\mathbb{R}^3$-valued functions on $\mathbb{R}^3$. We define

$$g^{[j,n_\ell]} = k_\ell + \delta_{j\ell} \sum_{m=1}^{n_l} f_j^{[m]}, \quad j = 1,2,3, \quad \ell = 1,2,3,$$

and set $g^{[j,n]} = (g^{[j,n_1]}, g^{[j,n_2]}, g^{[j,n_3]})$ and $g^{[j,0]} = (k_1, k_2, k_3)$. Note that

$$e^{-ig^{[j,n]}\cdot x} e^{-if_j^{[n+1]} x_j} = e^{-ig^{[j,n+1]}\cdot x}. \tag{30}$$

Since

$$\begin{aligned}
\left[H_{\kappa\Lambda}, e^{-ig^{[j,n-1]}\cdot x}\right] &= \left[H_{\kappa\Lambda}, e^{-ig^{[j,n]}\cdot x} e^{if_j^{[n]} x_j}\right] \\
&= \left[H_{\kappa\Lambda}, e^{-ig^{[j,n]}\cdot x}\right] e^{if_j^{[n]} x_j} + e^{-ig^{[j,n]}\cdot x}\left[H_{\kappa\Lambda}, e^{if_j^{[n]} x_j}\right],
\end{aligned}$$

we obtain that

$$\begin{aligned}
\left[H_{\kappa\Lambda}, e^{if_j^{[n]} x_j}\right] &\\
= e^{ig^{[j,n]}\cdot x}\left[H_{\kappa\Lambda}, e^{-ig^{[j,n-1]}\cdot x}\right] &- e^{ig^{[j,n]}\cdot x}\left[H_{\kappa\Lambda}, e^{-ig^{[j,n]}\cdot x}\right] e^{if_j^{[n]} x_j}. \tag{31}
\end{aligned}$$

Using these identities, we obtain the following decomposition:

$$I_0 \psi_{\kappa\Lambda} = \sum_{\nu=0}^3 \Psi_{1\nu} + I_1 \psi_{\kappa\Lambda} + \Psi_1^{er},$$

where

$$\Psi_{11} = \sum_{j=1}^3 \frac{k_j}{f_j^{[1]}} \left(H_{\kappa\Lambda} - E_{\kappa\Lambda}\right) e^{-ig^{[j,0]}\cdot x} \psi_{\kappa\Lambda},$$

$$\Psi_{12} = -\frac{1}{2} \sum_{j=1}^3 \frac{k_j}{f_j^{[1]}} |g^{[j,1]}|^2 e^{-ig^{[j,0]}\cdot x} \psi_{\kappa\Lambda},$$

$$\Psi_{13} = \sum_{\substack{j=1 \\ }}^3 \sum_{\substack{\ell=1 \\ \ell \neq j}}^3 \frac{k_j k_\ell}{f_j^{[1]}} e^{-ig^{[j,1]}\cdot x} v_\ell e^{if_j^{[1]} x_j} \psi_{\kappa\Lambda},$$

and

$$I_1 = \sum_{j=1}^{3} \frac{k_j \left(k_j + f_j^{[1]} \right)}{f_j^{[1]}} e^{-ig^{[j,1]} \cdot x} v_j,$$

$$\Psi_1^{er} = \sum_{j=1}^{3} \frac{k_j \left(k_j + f_j^{[1]} \right)}{f_j^{[1]}} e^{-ig^{[j,1]} \cdot x} v_j e^{if_j^{[1]} x_j} \left(1 - e^{-if_j^{[1]} x_j} \right) \psi_\kappa^{\mathrm{PF}}.$$

Therefore we obtain that

$$I_0 \psi_k = \sum_{\nu=0}^{3} \Psi_{1\nu} + I_1 \psi_k + \Psi_1^{er}. \tag{32}$$

The scale parameter $f^{[1]}$ is chosen so that $\gamma(k) \| (H_{\kappa\Lambda} - E_{\kappa\Lambda} + |k|)^{-1} \Psi_{1\nu} \|_{\mathcal{H}}$, $\nu = 0, 1, 2, 3$, can become $L^2(\mathbb{R}^3)$-functions of k for $f^{[1]}$. We call the map from $I_0 \psi_\kappa$ to $I_1 \psi_\kappa$ the first iteration $\mathcal{R}_1$. We repeat the same procedure as applied to $I_1 \psi_{\kappa\Lambda}$. Then, we obtain

$$I_1 \psi_{\kappa\Lambda} = \sum_{\nu=0}^{3} \Psi_{2\nu} + I_2 \psi_{\kappa\Lambda},$$

where

$$\Psi_{21} = \sum_{j=1}^{3} \frac{k_j \left(k_j + f_j^{[1]} \right)}{f_j^{[1]} f_j^{[2]}} \left(H_{\kappa\Lambda} - E_{\kappa\Lambda} \right) e^{-ig^{[j,1]} \cdot x} \psi_{\kappa\Lambda},$$

$$\Psi_{22} = -\frac{1}{2} \sum_{j=1}^{3} \frac{k_j \left(k_j + f_j^{[1]} \right)}{f_j^{[1]} f_j^{[2]}} |g^{[j,2]}|^2 e^{-ig^{[j,1]} \cdot x} \psi_{\kappa\Lambda},$$

$$\Psi_{23} = \sum_{j=1}^{3} \sum_{\substack{\ell=1 \\ \ell \neq j}}^{3} \frac{k_j \left(k_j + f_j^{[1]} \right) k_\ell}{f_j^{[1]} f_j^{[2]}} e^{-ig^{[j,2]} \cdot x} v_\ell e^{if_j^{[2]} x_j} \psi_{\kappa\Lambda},$$

and

$$I_2 = \sum_{j=1}^{3} \frac{k_j \left(k_j + f_j^{[1]} \right) \left(k_j + f_j^{[1]} + f_j^{[2]} \right)}{f_j^{[1]} f_j^{[2]}} e^{-ig^{[j,2]} \cdot x} v_j e^{if_j^{[2]} x_j}.$$

Again $f^{[2]}$ is chosen so that $\gamma(k) \| (H_{\kappa\Lambda} - E_{\kappa\Lambda} + |k|)^{-1} \Psi_{2\nu} \|_{\mathcal{H}}$, $\nu = 0, 1, 2, 3$, can become $L^2(\mathbb{R}^3)$-functions of k for $f^{[2]}$. We call the map from $I_1 \psi_{\kappa\Lambda}$ to $I_2 \psi_{\kappa\Lambda}$ the second iteration $\mathcal{R}_2$. In particular the iterations may be repeated ad infinitum. However, we stop our iteration here.

We choose our energy scale parameters as follows: for fixed ε with $1/2 < \varepsilon < 1$, we set

$$f_j^{[1]}(k) = |k|^\varepsilon, \quad f_j^{[2]}(k) = -f_j^{[1]}(k), \quad j = 1, 2, 3. \tag{33}$$

Then, it follows from direct estimate that $\gamma(k)\|(H_{\kappa\Lambda} - E_{\kappa\Lambda} + |k|)^{-1}\Psi_{n\nu}\|_{\mathcal{H}}$ are $L^2(\mathbb{R}^3)$-functions of k for $f^{[n]}$, $n = 1, 2$, and $\nu = 0, 1, 2, 3$. Moreover, fortunately $\gamma(k)\|(H_{\kappa\Lambda} - E_{\kappa\Lambda} + |k|)^{-1}I_2\psi_{\kappa\Lambda}\|_{\mathcal{H}}$ is also a $L^2(\mathbb{R}^3)$-function for $f^{[2]}$. As we expected that the velocity makes a buffer from the infrared singularity, we obtain that *there exists δ with $0 < \delta < 1/2$ and a positive constant C_{IR} independent of κ, Λ such that for every q with $|q| < q_{UV}$ and arbitrary κ, Λ with $0 < \kappa < \Lambda < \infty$*

$$
\begin{aligned}
I_{\kappa\Lambda}(k) \\
:= \left\|(H_{\kappa\Lambda} - E_{\kappa\Lambda} + |k|)^{-1}h_x(k)v\psi_{\kappa\Lambda} - \gamma(k)(H_{\kappa\Lambda} - E_{\kappa\Lambda} + |k|)^{-1}\Psi_1^{er}\right\|_{\mathcal{H}} \\
\leq C_{IR}|\gamma(k)|\,|k|^\delta
\end{aligned}
$$

holds in the IR region. Here, actually, we have only to set $\delta := 1 - \varepsilon$, and we note that $I_{\kappa\Lambda} \in L^2(\mathbb{R}^3)$ with

$$
\begin{aligned}
\int_{|k|<1} d^3k \, |I_{\kappa\Lambda}(k)|^2 &\leq \frac{C_{IR}^2}{4\pi^2} \int_0^1 |k|^2 d|k|\,|\gamma(|k|)|^2|k|^{2\delta} \\
&\leq \frac{C_{IR}^2}{4\pi^2} \int_0^1 d|k|\,|k|^{2\delta-1} < \infty.
\end{aligned}
$$

On the other hand, we can estimate the error term Ψ_1^{er} as follows: *there exist constants c_0, c_1, c_2, c_3 independent of κ, Λ such that for all q with $|q| < q_{UV}$ and $qZ < 4\pi$, it holds*

$$
\begin{aligned}
q^2 \int_{|k|<1} d^3k\,|\gamma(k)|^2\|(H_{\kappa\Lambda} - E_{\kappa\Lambda} + |k|)^{-1}\Psi_1^{er}\|_{\mathcal{H}}^2 \\
\leq c_0 q^2\left\{c_1 + c_2 \log q^2 Z + c_3\left(\log q^2 Z\right)^2\right\}.
\end{aligned}
$$

Thus, $\gamma(k)\|(H_{\kappa\Lambda} - E_{\kappa\Lambda} + |k|)^{-1}\Psi_1^{er}\|_{\mathcal{H}}$ is also a $L^2(\mathbb{R}^3)$-function of k for $f^{[1]}$.

3.5. *IR region*

The two estimates obtained in the previous subsection, we obtain that *there are positive constants C_0, C_1, C_2, C_3 independent of κ, Λ such that for every*

q *with* $|q| < q_{UV}$ *and arbitrary* κ, Λ *with* $0 < \kappa < \Lambda < \infty$,

$$\int_{|k|<1} d^3k \, \|a(k)\psi_{\kappa\Lambda}\|_{\mathcal{H}}^2 \leq C_0 q^2 \left[C_1 + C_2 \log q + C_3 (\log q)^2 \right].$$

The reason why our estimate can avoid the appearance of q^2 in the denominator of the upper bound such as (10) is our method $\mathcal{R}_n$ can reduce the estimate of $\langle x^2 \rangle_\kappa$ appearing in (10) to that of $\langle \log(3 + |x|) \rangle_\kappa$. And, $\langle \log(3 + |x|) \rangle_\kappa$ comes from the estimate of the error term Ψ_1^{er}. We should also mention that in Bach, Fröhlich, and Sigal's work[5] there is a hint of how to introduce $\left\{ f^{[n]} \right\}_n$. It was very small at a glance, but a very big clue for us.

3.6. *Existence of a ground state of H*

We can prove the existence of a ground state of H at least for sufficiently small $|q|$, by using Lemma 4.9 in our paper.[2]

It follows from (N1) and (N2) that

(N1′) $\lim\limits_{\Lambda \to \infty} \inf \sigma(H_{\kappa\Lambda}) = \inf \sigma(H_{\kappa\infty})$.

(N2′) $\lim\limits_{\kappa \to 0} \inf \sigma(H_{\kappa\infty}) = \inf \sigma(H) =: E$.

On the other hand, since we take normalized ground state ψ_κ for every κ, Λ, we can take the weak limit

$$\psi := \text{w-}\lim_{\kappa \to 0} \left(\text{w-}\lim_{\Lambda \to \infty} \psi_{\kappa\Lambda} \right).$$

Then, the Lemma 4.9 in the paper says that if $\psi \neq 0$, then ψ is a ground state of H with the ground state energy E, and moreover, $\psi \in D(N_{\mathrm{f}}^{1/2})$. So, we have only to show $\psi \neq 0$, which is not so easy in fact.

Let P_{at} and P_Ω be orthogonal projections onto the space spanned by ψ_{at} and Ω, respectively, where ψ_{at} is the unique ground state of H_{at} and Ω the Fock vacuum. We set $P := P_{\mathrm{at}} \otimes P_\Omega$ and $Q := (1 - P_{\mathrm{at}}) \otimes P_\Omega$. It is easy to show that

$$\langle \psi_{\kappa\Lambda} \, , \, P\psi_{\kappa\Lambda} \rangle_{\mathcal{H}} \geq 1 - \langle \psi_{\kappa\Lambda} \, , \, N_{\mathrm{f}}\psi_{\kappa\Lambda} \rangle_{\mathcal{H}} - \langle \psi_{\kappa\Lambda} \, , \, Q\psi_{\kappa\Lambda} \rangle_{\mathcal{H}}. \tag{34}$$

The second term in the right hand side of (34) is treated by (21). That is, we get

$$\langle \psi_{\kappa\Lambda} \, , \, P\psi_{\kappa\Lambda} \rangle_{\mathcal{H}} \geq 1 - q^2 D_{\mathrm{N},0} \left(D_{\mathrm{N},1} + D_{\mathrm{N},2} \log q + D_{\mathrm{N},3}(\log q)^2 \right)$$
$$- \langle \psi_{\kappa\Lambda} \, , \, Q\psi_{\kappa\Lambda} \rangle_{\mathcal{H}}. \tag{35}$$

So, we need to estimate $\langle \psi_{\kappa\Lambda} , Q\psi_{\kappa\Lambda}\rangle_{\mathcal{H}}$ in the above. If we had invoked the standard way to estimate it such as in the paper,[2] we would have

$$|\langle \psi_{\kappa\Lambda} , Q\psi_{\kappa\Lambda}\rangle_{\mathcal{H}}| \le C \left(\frac{4\pi}{q^2 Z}\right)^2 \times q^2$$

for a positive constant C. So, we have to be faced with the same trouble as in (10). Then, we make a device again to avoid such a trouble.

In our work,[11] we consider changes of units. We set $\varrho := q^2 Z\lambda_1/4\pi$ with a positive parameter λ_1. Then, for $\tau \in \mathbb{R}$ we define a change of variables as follows:

For $\Psi = \left\{ \Psi^{(0)}, \Psi^{(1)}, \Psi^{(2)}, \cdots , \Psi^{(n)}, \cdots , \right\} \in \mathcal{H}$, we set

$$(U_\tau\Psi)^{(n)}(x, k_1, \cdots , k_n) = \varrho^{3\tau/2}\varrho^{-3n\tau}\Psi^{(n)}\left(\varrho^\tau x, \varrho^{-2\tau}k_1, \cdots , \varrho^{-2\tau}k_n\right). \tag{36}$$

Then, $U_{\tau_1}U_{\tau_2} = U_{\tau_1+\tau_2}$, $U_0 = I$, and $x, p, a(k)$ transform as

$$U_\tau^* x U_\tau = \varrho^{-\tau} x, \;\; U_\tau^* p U_\tau = \varrho^\tau p, \, U_\tau^* a(k)U_\tau = \varrho^{-3\tau} \, a(\varrho^{-2\tau}k). \tag{37}$$

Namely, we make a unitary group to connect continuously several units.

We define Hamiltonians $H_{\kappa\Lambda}^{(\tau)}$ by

$$H_{\kappa\Lambda}^{(\tau)} := \varrho^{-2\tau}U_\tau^* H_{\kappa\Lambda}U_\tau.$$

Then, we have

$$H_{\kappa\Lambda}^{(\tau)} = H_{\mathrm{at}}^{(\tau)} + H_{\mathrm{f}} + H_{\mathrm{I}}^{(\tau)},$$

where

$$H_{\mathrm{at}}^{(\tau)} := \frac{1}{2}p^2 - \frac{\varrho^{1-\tau}}{\lambda_1|x|}$$

with the ground state $\psi_{\mathrm{at}}^{(\tau)}(x) = e^{-\varrho^{1-\tau}|x|}$ and the ground state energy $E_{\mathrm{at}}^{(\tau)} = \inf \sigma \left(H_{\mathrm{at}}^{(\tau)}\right) = -\varrho^{2(1-\tau)}/2\lambda_1^2$, and moreover,

$$H_{\mathrm{I}}^{(\tau)} := q\varrho^\tau(pA_{\kappa\Lambda}^{(\tau)} + A_{\kappa\Lambda}^{(\tau)\,*}p) + \frac{q^2\varrho^{2\tau}}{2} \left(A_{\kappa\Lambda}^{(\tau)\,2} + 2A_{\kappa\Lambda}^{(\tau)\,*} A_{\kappa\Lambda}^{(\tau)} + A_{\kappa\Lambda}^{(\tau)\,*2}\right),$$

$$A_{\kappa\Lambda}^{(\tau)} := \int_{\mathbb{R}^3} d^3k \, \frac{\chi_\kappa(\varrho^{2\tau}k)}{\sqrt{2|k|}}k\beta_\tau(k)a(k)e^{i\varrho^\tau kx},$$

$$\beta_\tau(k) := \left(|k| + \varrho^{2\tau}|k|^2/2\right)^{-1}.$$

We set $Q^{(\tau)} := (1 - P_{\mathrm{at}}^{(\tau)}) \otimes P_\Omega$. Then, we obtain

$$\langle \psi_{\kappa\Lambda} , Q\psi_{\kappa\Lambda}\rangle_{\mathcal{H}} \le \frac{8\lambda_1^{2\tau}}{3} \left(\frac{4\pi}{q^2 Z}\right)^{2(1-\tau)} \langle Q^{(\tau)}\psi_{\kappa\Lambda}^{(\tau)} , H_{\mathrm{I}}^{(\tau)}\psi_{\kappa\Lambda}^{(\tau)}\rangle_{\mathcal{H}}. \tag{38}$$

(see (8.11) in the paper.[11]) Here, we note that $\langle Q^{(\tau)}\psi_{\kappa\Lambda}^{(\tau)}, H_{\mathrm{I}}^{(\tau)}\psi_{\kappa\Lambda}^{(\tau)}\rangle_{\mathcal{H}}$ includes the information about dependence of q of the cloud which the electron dresses.

By applying some techniques of old constructive field theory, we can find the information when $\lambda_1 = 4\pi/Z$ and $3/4 < \tau \le 1$, and we obtain that there exist a positive constants C_i^{Q}, $i = 1, 2$, and positive function Θ of q such that

$$\Theta(q) \sim C_1^{\mathrm{Q}}|q| \quad \text{as } |q| \to 0$$

and

$$|\langle Q^{(\tau)}\psi_{\kappa\Lambda}^{(\tau)}, H_{\mathrm{I}}^{(\tau)}\psi_{\kappa\Lambda}^{(\tau)}\rangle_{\mathcal{H}}| \le C_2^{\mathrm{Q}}\Theta(q) \tag{39}$$

(see (8.16), (8.17), and Lemma 8.2 in the work.[11]) Hence it follows from (38) and (39) that there exists a positive constant C_{Q} such that

$$\langle \psi_{\kappa\Lambda}, Q\psi_{\kappa\Lambda}\rangle_{\mathcal{H}} \le C_{\mathrm{Q}}|q|^{4\tau-3} \tag{40}$$

for sufficiently small $|q|$. Combining (35) and (40), we obtain

$$\langle \psi_{\kappa\Lambda}, P\psi_{\kappa\Lambda}\rangle_{\mathcal{H}} \ge 1 - q^2 D_{\mathrm{N},0}\left(D_{\mathrm{N},1} + D_{\mathrm{N},2}\log q + D_{\mathrm{N},3}(\log q)^2\right) - C_{\mathrm{Q}}|q|^{4\tau-3},$$

which implies that there exists a positive constant $\widetilde{C}_{\mathrm{Q}}$ independent of κ, Λ such that

$$|\langle \psi_{\kappa\Lambda}, \psi_{\mathrm{at}} \otimes \Omega\rangle_{\mathcal{H}}|_{\mathcal{H}}^2 = \langle \psi_{\kappa\Lambda}, P\psi_{\kappa\Lambda}\rangle_{\mathcal{H}} \ge \widetilde{C}_{\mathrm{Q}} > 0$$

for sufficiently small $|q|$. Taking the weak limit $\Lambda \to \infty$ and $\kappa \to 0$, we have $\langle \psi, \psi_{\mathrm{at}} \otimes \Omega\rangle_{\mathcal{H}} \ne 0$. Therefore, we obtain $\psi \ne 0$ and that ψ *is a ground state of H with the ground state energy E, and moreover, $\psi \in D(N_{\mathrm{f}}^{1/2})$.*

4. Total number of soft photons for H_κ^{PF}

For H_κ^{PF}, we set $q = e$ and $\alpha = e^2/4\pi$ for electric charge $e > 0$. Moreover, we set $m = 1$ for the sake of simplicity. So, θ in $A(x)$ is just 1 now, *i.e.*, $\theta = 1$. We consider H_κ^{PF} without dipole approximation and $-e\sigma B(x)/m$.

Let $a^\dagger(f)$ and $a(f)$, $f \in W := L^2(\mathbb{R}^3) \oplus L^2(\mathbb{R}^3)$, be the smeared creation and annihilation operators acting in $\mathcal{F}$, the symmetric Fock space over W, respectively. We set $a_1^\sharp(f) := a^\sharp(f \oplus 0)$ and $a_2^\sharp(f) := a^\sharp(0 \oplus f)$, where $a^\sharp$ stands for a or $a^\dagger$. Then, $a_r(f)$ and $a_r^\dagger(f)$, $r = 1, 2$, satisfy the canonical commutation relation

$$\left[a_r(f), a_s^\dagger(g)\right] = \delta_{rs} \int_{\mathbb{R}^3} d^3k\, f(k)^* g(k),$$

$$\left[a_r^\dagger(f),\, a_s^\dagger(g)\right] = 0 = \left[a_r(f),\, a_s(g)\right]$$

on the finite particle subspace $\mathcal{F}_{\text{fin}}$ of $\mathcal{F}$ for $f, g \in W$. In this paper, we used the kernel representation already, $a^\natural(k)$, for $a^\natural(f)$, *i.e.*,

$$a(f) = \int_{\mathbb{R}^3} d^3k\, f(k)^* a(k), \quad a^\dagger(f) = \int_{\mathbb{R}^3} d^3k\, f(k) a^\dagger(k)$$

according to the custom in mathematics.

Then, we can realize H_κ^{PF} as

$$H_\kappa^{\text{PF}} := \frac{1}{2}\left(p - qA(x)\right)^2 + V_{\text{c}}(x) + H_{\text{f}}$$

as an operator acting in the state space

$$\mathcal{H} := W \otimes \mathcal{F} = L^2(\mathbb{R}^3; \mathbb{C}^2) \otimes \mathcal{F}.$$

Self-adjointness of the above H_κ^{PF}, $0 \leq \kappa < \Lambda$, is studied in Hiroshima's work,[15] and existence of ground state is investigated by Bach, Fröhlich, and Sigal[5] and Griesemer, Lieb, and Loss.[10] We denote by ψ_κ^{PF} the normalized ground state of H_κ^{PF}.

The purpose in this section is to show that *there exist positive constants* $D_{\text{PF},n}$, $n = 0, 1, 2, 3$, *such that*

$$\langle N_{\text{f}} \rangle \leq e^2 D_{\text{PF},0}\left(D_{\text{PF},1} + D_{\text{PF},2}\log e + D_{\text{PF},3}(\log e)^2\right)$$

for sufficiently small e.

4.1. *Strategy for estimate of photon number*

We employ the same way as we do for the Nelson Hamiltonian $H_{\kappa\Lambda}$. So, we start with

$$\left[H_\kappa^{\text{PF}},\, a_r(k)\right]\psi_\kappa^{\text{PF}} = \left(H_\kappa^{\text{PF}} - E_\kappa^{\text{PF}}\right)a_r(k)\psi_\kappa^{\text{PF}}. \tag{41}$$

On the other hand, using $\operatorname{div} A(x) = 0$, we can calculate the commutator directly and get

$$\left[H_\kappa^{\text{PF}},\, a_r(k)\right] = -\omega(k)a_r(k) + e\frac{\chi_\kappa(k)}{\sqrt{2|k|}}e_r(k)e^{-i\gamma kx}v, \tag{42}$$

where v is the velocity of the electron, *i.e.*,

$$v = \frac{p - eA(x)}{m} = i\left[H_\kappa^{\text{PF}},\, x\right]. \tag{43}$$

Combining (41) and (42), we have the *pull-through formula*

$$\left(H_\kappa^{\text{PF}} - E_\kappa^{\text{PF}}\right)a_r(k)\psi_\kappa^{\text{PF}} = -\omega(k)a_r(k)\psi_\kappa^{\text{PF}} + e\frac{\chi_\kappa(k)}{\sqrt{2|k|}}e_r(k)e^{-i\gamma kx}v,$$

which implies

$$a_r(k)\psi_\kappa^{\mathrm{PF}} = e\,\frac{\chi_\kappa(k)}{\sqrt{2|k|}}\,e_r(k)\,(H_\kappa^{\mathrm{PF}} - E_\kappa^{\mathrm{PF}} + \omega(k))^{-1}\,e^{-i\gamma kx}\,v\psi_\kappa^{\mathrm{PF}}. \tag{44}$$

In the same way as in the case of $H_{\kappa\Lambda}$, we divide the momentum space, $k \in \mathbb{R}^3$, into IR and UV regions. As for H_κ^{PF}, however, there is no problem about UV region because we fix the ultraviolet cutoff Λ. In fact, the removal of Λ for H_κ^{PF} has not been showed in mathematics so far. It remains to be proved.

Thus, we use the identity

$$\langle N_{\mathrm{f}}\rangle_\kappa = \sum_{r=1,2}\int_{|k|<1} d^3k\,\|a_r(k)\psi_\kappa^{\mathrm{PF}}\|_{\mathcal{H}}^2 + \sum_{r=1,2}\int_{1\leq|k|<\Lambda} d^3k\,\|a_r(k)\psi_\kappa^{\mathrm{PF}}\|_{\mathcal{H}}^2.$$

Set

$$H_0 := H_{\mathrm{el}} + H_{\mathrm{f}}, \qquad H_I := H_\kappa^{\mathrm{PF}} - H_0.$$

By, for instance, (2.9) and (2.10) in Hiroshima's work,[14] there exists a positive constant C_1 such that

$$\|H_I\varphi\|_{\mathcal{H}} \leq eC_1\|(H_0 + 1)\varphi\|_{\mathcal{H}} \tag{45}$$

for $\varphi \in C_0^\infty(\mathbb{R}^3)\widehat{\otimes}\mathcal{F}_{\mathrm{fin}}$, where $\widehat{\otimes}$ means the algebraic tensor product and $\mathcal{F}_{\mathrm{fin}}$ is a subspace of $\mathcal{F}$ consisting of finite number of particles. Although we can show self-adjointness of H_κ^{PF} for restricted e by applying the Kato-Rellich theorem to (45), Hiroshima proved self-adjointness of H_κ^{PF} for arbitrary e.[15]

We denote by $H_\kappa^{V=0}$ the Hamiltonian given by replacing $V_{\mathrm{c}}(x)$ by just 0. As defined in the case of H_κ^{N}, we also define the (positive) binding energy E_κ^{bin} by $E_\kappa^{\mathrm{bin}} := E_\kappa^{V=0} - E_\kappa^{\mathrm{PF}}$. Then, Griesemer, Lieb, and Loss[10] proved that $E_\kappa^{\mathrm{bin}} \geq -E_1$. So, we get $E_\kappa^{\mathrm{PF}} \leq E_1 + E_\kappa^{V=0}$, which implies that

$$E_\kappa^{\mathrm{PF}} \leq \frac{e^2}{8m\pi^2}\left(\Lambda^2 - \frac{me^2Z^2}{4}\right).$$

On the other hand, by the diamagnetic inequality (*e.g.* see Theorem 5.1 in Ref.[13]), we get $E_1 \leq E_\kappa^{\mathrm{PF}}$, which implies

$$-\frac{me^4Z^2}{32\pi^2} \leq E_\kappa^{\mathrm{PF}}.$$

Thus, set

$$E(e) := \max\left\{\frac{me^4Z^2}{32\pi^2},\,\frac{e^2}{8m\pi^2}\left(\Lambda^2 - \frac{me^2Z^2}{4}\right)\right\},$$

then we obtain

$$|E_\kappa^{\mathrm{PF}}| \le E(e). \tag{46}$$

By (45) and (46), we have the estimate

$$\|(H_\kappa^{\mathrm{PF}} - E_\kappa^{\mathrm{PF}} + 1)\varphi\|_{\mathcal{H}} \ge \|(H_0 + 1)\|_{\mathcal{H}} - \|H_I\varphi\|_{\mathcal{H}} - |E_\kappa^{\mathrm{PF}}|\,\|\varphi\|_{\mathcal{H}}$$
$$\ge (1 - C_*(e))\,\|(H_{0+1})\varphi\|_{\mathcal{H}}$$

for $\varphi \in C_0^\infty(\mathbb{R}^3)\widehat{\otimes}\mathcal{F}_{\mathrm{fin}}$, where

$$C_*(e) := eC_1 + E(e).$$

From now on, we assume that we take such small charge e that

$$C_*(e) < 1 \tag{47}$$

holds.

Then, under the assumption (47), we have $\|(H_0+1)(H_\kappa^{\mathrm{PF}} - E_\kappa^{\mathrm{PF}} + 1)^{-1}\| \le C_{\mathrm{PF}}(e)$, where

$$C_{\mathrm{PF}}(e) := (1 - C_*(e))^{-1}.$$

It is easy to check $(H_\kappa^{\mathrm{PF}} - E_\kappa^{\mathrm{PF}} + 1)(H_\kappa^{\mathrm{PF}} - E_\kappa^{\mathrm{PF}} + \epsilon)^{-1} \le \epsilon^{-1}$ if $\epsilon < 1$; ≤ 1 if $\epsilon \ge 1$. So, combining these inequalities, we have

$$\|(H_0 + 1)(H_\kappa^{\mathrm{PF}} - E_\kappa^{\mathrm{PF}} + \epsilon)^{-1}\| \le \frac{C_{\mathrm{PF}}(e)}{\min\{\epsilon, 1\}}. \tag{48}$$

In the same way, using Theorem X.18(a) in Ref.[20], we have

$$\|(H_0 + 1)^{1/2}(H_\kappa^{\mathrm{PF}} - E_\kappa^{\mathrm{PF}} + \epsilon)^{-1/2}\| \le \frac{\widetilde{C}_{\mathrm{PF}}(e)}{\min\{\sqrt{\epsilon}, 1\}}. \tag{49}$$

As written above, since we fix the ultraviolet cutoff for H_κ^{PF}, it is easy to show that *there exists a positive constant C_0 independent of κ with $0 \le \kappa < \Lambda$ such that*

$$\sum_{r=1,2} \int_{1 \le |k| < \Lambda} d^3k\, \|a_r(k)\psi_\kappa^{\mathrm{PF}}\|_{\mathcal{H}}^2 \le C_0 e^2. \tag{50}$$

4.2. *Energy scale renormalization*

We also introduce energy scale parameters $\{f^{[n]}\}_n$ for H_κ^{PF} and take the procedure $\mathcal{R}_n$. Here we explain the reason why we thought of the method with energy scale parameters. When we ignore the non-commutativity between $(H_\kappa^{\mathrm{PF}} + E_\kappa^{\mathrm{PF}} + |k|)^{-1}$ and $e^{-ikx}e_r(k)v$, we have a rough estimate

$$\int_{|k|<1} d^3k\, \|a_r(k)\|_{\mathcal{H}}^2 \le \mathrm{const.} \int_\kappa^1 d|k|\, \frac{1}{|k|} \sim -\log\kappa,$$

and we meet logarithmically divergence $\lim_{\kappa \to 0} \log \kappa = \infty$ in the right hand side. So, we introduced the energy scale parameters $\{f^{[n]}\}_n$.

By (44), we have

$$a_r(k)\psi_\kappa^{\mathrm{PF}} = e\gamma(k)\left(H_\kappa^{\mathrm{PF}} - E_\kappa^{\mathrm{PF}} + \omega(k)\right)^{-1} I_{r0}\psi_\kappa^{\mathrm{PF}},$$

where

$$\gamma(k) := \frac{\chi_\kappa(k)}{\sqrt{2|k|}}, \qquad I_{r0} = e^{-ikx}e_r(k)v. \tag{51}$$

In the case of $H_{\kappa\Lambda}$, I_0 includes k, but $\gamma(k)$ includes $|k|^{-1}$ in $\beta(k)$. As for H_κ^{PF}, I_{r0} includes $e_r(k)$ instead of k, but $\gamma(k)$ does not include $|k|^{-1}$.

For f_j dependent only on $k \in \mathbb{R}^3$, we have

$$\left[H_\kappa^{\mathrm{PF}}, e^{if_jx_j}\right] = f_j e^{if_jx_j}v + \frac{f_j^2}{2}e^{if_jx_j}, \tag{52}$$

where we used $\left[p, e^{if_jx_j}\right] = f_j e^{if_jx_j}$. Thus, we can represent the velocity as

$$v = \frac{1}{f_j}e^{-if_jx}\left[H_\kappa^{\mathrm{PF}}, e^{if_jx_j}\right] - \frac{f_j}{2}, \qquad j = 1, 2, 3. \tag{53}$$

Similarly, for $g = (g_1, g_2, g_3)$ with components g_j dependent only on $k \in \mathbb{R}^3$, we get

$$\left[H_\kappa^{\mathrm{PF}}, e^{igx}\right] = \frac{1}{2}g^2 e^{igx} + e^{igx}gv. \tag{54}$$

We employ the same energy scale parameters $\{f^{[n]}\}_{n=1}^\infty$ and other notations as those for $H_{\kappa\Lambda}$.

Since

$$[H_\kappa^{\mathrm{PF}}, e^{-ig^{[j,n-1]}\cdot x}] = [H_\kappa^{\mathrm{PF}}, e^{-ig^{[j,n]}\cdot x}e^{if_j^{[n]}x_j}]$$
$$= [H_\kappa^{\mathrm{PF}}, e^{-ig^{[j,n]}\cdot x}]e^{if_j^{[n]}x_j} + e^{-ig^{[j,n]}\cdot x}[H_\kappa^{\mathrm{PF}}, e^{if_j^{[n]}x_j}],$$

we obtain

$$[H_\kappa^{\mathrm{PF}}, e^{if_j^{[n]}x_j}]$$
$$= e^{ig^{[j,n]}\cdot x}[H_\kappa^{\mathrm{PF}}, e^{-ig^{[j,n-1]}\cdot x}] - e^{ig^{[j,n]}\cdot x}[H_\kappa^{\mathrm{PF}}, e^{-ig^{[j,n]}\cdot x}]e^{if_j^{[n]}x_j}. \tag{55}$$

$\underline{\mathcal{R}_1}$: In the same way as for $H_{\kappa\Lambda}$, we infer

$$I_{r0}\psi_\kappa^{\mathrm{PF}} = \sum_{j=1}^{3} e^{-ig^{[j,0]}\cdot x}e_{rj}(k)\left(\frac{1}{f_j^{[1]}}e^{-if_j^{[1]}x_j}[H_\kappa^{\mathrm{PF}}, e^{if_j^{[1]}x_j}] - \frac{1}{2}f_j^{[1]}\right)\psi_\kappa^{\mathrm{PF}}$$
$$= I_{r11}\psi_\kappa^{\mathrm{PF}} + \Psi_{r10}, \tag{56}$$

where

$$I_{r11} = \sum_{j=1}^{3} \frac{e_{rj}(k)}{f_j^{[1]}} e^{-ig^{[j,1]}\cdot x} [H_\kappa^{\mathrm{PF}}, e^{if_j^{[1]}x_j}],$$

$$\Psi_{r10} = -\frac{1}{2} \sum_{j=1}^{3} e_{rj}(k) f_j^{[1]} e^{-ig^{[j,0]}\cdot x} \psi_\kappa^{\mathrm{PF}}.$$

Moreover, we get

$$I_{r11}\psi_\kappa^{\mathrm{PF}}$$
$$= \sum_{j=1}^{3} \frac{e_{rj}(k)}{f_j^{[1]}} [H_\kappa^{\mathrm{PF}}, e^{-ig^{[j,0]}\cdot x}]\psi_\kappa^{\mathrm{PF}} - \sum_{j=1}^{3} \frac{e_{rj}(k)}{f_j^{[1]}} [H_\kappa^{\mathrm{PF}}, e^{-ig^{[j,1]}\cdot x}]e^{if_j^{[1]}x_j}\psi_\kappa^{\mathrm{PF}}$$
$$= \Psi_{r11} + \Psi_{r12} + \widetilde{\Psi}_{r13}, \tag{57}$$

where

$$\Psi_{r11} = \sum_{j=1}^{3} \frac{e_{rj}(k)}{f_j^{[1]}} \left(H_\kappa^{\mathrm{PF}} - E_\kappa^{\mathrm{PF}}\right) e^{-ig^{[j,0]}\cdot x}\psi_\kappa^{\mathrm{PF}},$$

$$\Psi_{r12} = -\frac{1}{2} \sum_{j=1}^{3} \frac{e_{rj}(k)}{f_j^{[1]}} |g^{[j,1]}|^2 e^{-ig^{[j,0]}\cdot x}\psi_\kappa^{\mathrm{PF}},$$

$$\widetilde{\Psi}_{r13} = \sum_{j=1}^{3} \frac{e_{rj}(k)}{f_j^{[1]}} e^{-ig^{[j,1]}\cdot x} \left(g^{[j,1]}v\right) e^{if_j^{[1]}x_j}\psi_\kappa^{\mathrm{PF}}.$$

Moreover we decompose $\widetilde{\Psi}_{r13}$ as

$$\widetilde{\Psi}_{r13} = \sum_{j=1}^{3} \frac{e_{rj}(k)}{f_j^{[1]}} e^{-ig^{[j,1]}\cdot x} g_j^{[j,1]} v_j e^{if_j^{[1]}x_j}\psi_\kappa^{\mathrm{PF}}$$
$$+ \sum_{j=1}^{3}\sum_{\substack{\ell=1 \\ \ell\neq j}}^{3} \frac{e_{rj}(k)}{f_j^{[1]}} e^{-ig^{[j,1]}\cdot x} g_\ell^{[j,1]} v_\ell e^{if_j^{[1]}x_j}\psi_\kappa^{\mathrm{PF}}$$
$$= \widetilde{I}_{r1}\psi_\kappa + \Psi_{r13},$$

where

$$\widetilde{I}_{r1} = \sum_{j=1}^{3} \widetilde{I}_{r1}^{j}, \quad \widetilde{I}_{r1}^{j} := \frac{e_{rj}(k)\left(k_j + f_j^{[1]}\right)}{f_j^{[1]}} e^{-ig^{[j,1]}\cdot x} v_j e^{if_j^{[1]}x_j},$$

$$\Psi_{r13} = \sum_{j=1}^{3}\sum_{\substack{\ell=1 \\ \ell\neq j}}^{3} \frac{e_{rj}(k)k_\ell}{f_j^{[1]}} e^{-ig^{[j,1]}\cdot x} v_\ell e^{if_j^{[1]}x_j}\psi_\kappa^{\mathrm{PF}},$$

and we used the definition of $g^{[j,1]}$.

Set

$$I_{r1} = \sum_{j=1}^{3} \widetilde{I}_{r1}^{j} e^{-if_j^{[1]} x_j} = \sum_{j=1}^{3} \frac{e_{rj}(k)\left(k_j + f_j^{[1]}\right)}{f_j^{[1]}} e^{-ig^{[j,1]}\cdot x} v_j.$$

Then the error term Ψ_{r1}^{er} is defined as

$$\Psi_{r1}^{\mathrm{er}} := \left(\widetilde{I}_{r1} - I_{r1}\right) \psi_\kappa^{\mathrm{PF}} = \sum_{j=1}^{3} \widetilde{I}_{r1}^{j} \left(1 - e^{-if_j^{[1]} x_j}\right) \psi_\kappa^{\mathrm{PF}}.$$

Therefore we obtain

$$I_{r0}\psi_\kappa^{\mathrm{PF}} = \sum_{\nu=0}^{3} \Psi_{r1\nu} + I_{r1}\psi_\kappa^{\mathrm{PF}} + \Psi_{r1}^{\mathrm{er}}. \tag{58}$$

$\underline{\mathcal{R}_2}$: We get

$$I_{r1}\psi_\kappa^{\mathrm{PF}} = \sum_{j=1}^{3} \frac{e_{rj}(k)\left(k_j + f_j^{[1]}\right)}{f_j^{[1]}} e^{-ig^{[j,1]}\cdot x}$$

$$\times \left\{ \frac{1}{f_j^{[2]}} e^{-if_j^{[2]} x_j} [H_\kappa^{\mathrm{PF}}, e^{if_j^{[2]} x_j}] - \frac{1}{2} f_j^{[2]} \right\} \psi_\kappa^{\mathrm{PF}} \tag{59}$$

$$= I_{r21}\psi_\kappa^{\mathrm{PF}} + \Psi_{r20}, \tag{60}$$

where

$$I_{r21} = \sum_{j=1}^{3} \frac{e_{rj}(k)\left(k_j + f_j^{[1]}\right)}{f_j^{[1]} f_j^{[2]}} e^{-ig^{[j,2]}\cdot x} [H_\kappa^{\mathrm{PF}}, e^{if_j^{[2]} x_j}],$$

$$\Psi_{r20} = -\frac{1}{2} \sum_{j=1}^{3} \frac{e_{rj}(k)\left(k_j + f_j^{[1]}\right) f_j^{[2]}}{f_j^{[1]}} e^{-ig^{[j,1]}\cdot x} \psi_\kappa^{\mathrm{PF}}.$$

By (30) and (55) we obtain

$$I_{r21}\psi_\kappa^{\mathrm{PF}} = \sum_{j=1}^{3} \frac{e_{rj}(k)\left(k_j + f_j^{[1]}\right)}{f_j^{[1]} f_j^{[2]}} (H_\kappa^{\mathrm{PF}} - E_\kappa^{\mathrm{PF}}) e^{-ig^{[j,1]}\cdot x} \psi_\kappa^{\mathrm{PF}}$$

$$- \sum_{j=1}^{3} \frac{e_{rj}(k)\left(k_j + f_j^{[1]}\right)}{f_j^{[1]} f_j^{[2]}} [H_\kappa^{\mathrm{PF}}, e^{-ig^{[j,2]}\cdot x}] e^{if_j^{[2]} x_j} \psi_\kappa^{\mathrm{PF}}$$

$$= \Psi_{r21} + \Psi_{r22} + \widetilde{\Psi}_{r23},$$

where

$$\Psi_{r21} = \sum_{j=1}^{3} \frac{e_{rj}(k)\left(k_j + f_j^{[1]}\right)}{f_j^{[1]} f_j^{[2]}} \left(H_\kappa^{\mathrm{PF}} - E_\kappa^{\mathrm{PF}}\right) e^{-ig^{[j,1]}\cdot x} \psi_\kappa^{\mathrm{PF}},$$

$$\Psi_{r22} = -\frac{1}{2} \sum_{j=1}^{3} \frac{e_{rj}(k)\left(k_j + f_j^{[1]}\right)}{f_j^{[1]} f_j^{[2]}} |g^{[j,2]}|^2 e^{-ig^{[j,1]}\cdot x} \psi_\kappa^{\mathrm{PF}},$$

$$\widetilde{\Psi}_{r23} = \sum_{j=1}^{3} \frac{e_{rj}(k)\left(k_j + f_j^{[1]}\right)}{f_j^{[1]} f_j^{[2]}} e^{-ig^{[j,2]}\cdot x} \left(g^{[j,2]}v\right) e^{if_j^{[2]}x_j} \psi_\kappa^{\mathrm{PF}}.$$

Moreover $\widetilde{\Psi}_{r23}$ is decomposed as $\widetilde{\Psi}_{r23} = \widetilde{I}_{r2}\psi_\kappa^{\mathrm{PF}} + \Psi_{r23}$, where

$$\widetilde{I}_{r2} = \sum_{j=1}^{3} \frac{e_{rj}(k)\left(k_j + f_j^{[1]}\right)\left(k_j + f_j^{[1]} + f_j^{[2]}\right)}{f_j^{[1]} f_j^{[2]}} e^{-ig^{[j,2]}\cdot x} v_j e^{if_j^{[2]}x_j},$$

$$\Psi_{r23} = \sum_{j=1}^{3} \sum_{\substack{\ell=1 \\ \ell \neq j}}^{3} \frac{e_{rj}(k)\left(k_j + f_j^{[1]}\right) k_\ell}{f_j^{[1]} f_j^{[2]}} e^{-ig^{[j,2]}\cdot x} v_\ell e^{if_j^{[2]}x_j} \psi_\kappa^{\mathrm{PF}}.$$

Here we used the definition of $g_\ell^{[j,2]}$. Therefore, we obtain

$$I_{r1}\psi_\kappa^{\mathrm{PF}} = \sum_{\nu=0}^{3} \Psi_{r2\nu} + \widetilde{I}_{r2}\psi_\kappa^{\mathrm{PF}}. \tag{61}$$

Although the iterations may be repeated ad infinitum, we stop our iteration here and set

$$I_{r2} = \widetilde{I}_{r2}.$$

So, there is no error term from the second step.

4.3. *IR-safety*

We have only to treat IR region, $|k| < 1$, because there is no ultraviolet problem as written above. We have to show that $\gamma(k)(H_\kappa^{\mathrm{PF}} - E_\kappa^{\mathrm{PF}} + |k|)^{-1}\Psi_{rn\nu}$, $r = 1,2; n = 1,2; \nu = 0,1,2,3$, and $\gamma(k)(H_\kappa^{\mathrm{PF}} - E_\kappa^{\mathrm{PF}} + |k|)^{-1}I_{r2}\psi_\kappa^{\mathrm{PF}}$ are $L^2(\mathbb{R}^3)$-functions of k .

    Set

$$B(k) := \left(H_\kappa^{\mathrm{PF}} - E_\kappa^{\mathrm{PF}} + |k|\right)^{-1}.$$

We choose the same energy scale parameters as those for $H_{\kappa\Lambda}$: for fixed ε with $1/2 < \varepsilon < 1$, we set

$$f_j^{[1]}(k) = |k|^\varepsilon, \quad f_j^{[2]}(k) = -f_j^{[1]}(k), \quad j = 1, 2, 3. \tag{62}$$

Then, the following inequality

$$|k_j + f_j^{[1]}| \le |k_j| + |f_j^{[1]}| \le (|k| + |k|^\varepsilon) \le 2|k|^\varepsilon \tag{63}$$

holds, since $|k| < 1$.

Since we have

$$B(k)\left(I_{r0}\psi_\kappa^{\mathrm{PF}} - \Psi_1^{er}\right) = \sum_{n=1}^{2}\sum_{\nu=0}^{3} B(k)\Psi_{rn\nu} + B(k)I_{r2}\psi_\kappa^{\mathrm{PF}}, \tag{64}$$

we estimate each term in (64). Using (62), $B(k)\Psi_{rn\nu}$ is estimated from above. Let us first consider $B(k)\Psi_{r1\nu}$. We have

$$\|B(k)\Psi_{r10}\|_{\mathcal{H}} \le |k|^{\varepsilon-1} \le |k|^{-\varepsilon} \tag{65}$$

since $|k| < 1$ and $-\varepsilon < -1/2 < \varepsilon - 1$. And, we get

$$\|B(k)\Psi_{r11}\|_{\mathcal{H}} \le 3|k|^{-\varepsilon} \tag{66}$$

and

$$\|B(k)\Psi_{r12}\|_{\mathcal{H}} \le \frac{1}{2|k|}\sum_{j=1}^{3}\frac{1}{|f_j^{[1]}|}|g^{[j,1]}|^2.$$

Note that

$$\sum_{j=1}^{3}\frac{1}{|f_j^{[1]}|}|g^{[j,1]}|^2 \le |k|^{-\varepsilon}\sum_{j=1}^{3}|g^{[j,1]}|^2.$$

Let $\mathcal{P}_3^+$ be the set of even permutations of $(1, 2, 3)$. Then it is easy to show that

$$\sum_{j=1}^{3}|g^{[j,1]}|^2 \le 18|k|^{2\varepsilon}. \tag{67}$$

by (63). Hence it follows from $|k| < 1$ that

$$\|B(k)\Psi_{r12}\|_{\mathcal{H}} \le 9|k|^\varepsilon \le 9|k|^{-\varepsilon}. \tag{68}$$

Moreover, since we can show the same lemma as Lemma 7.1 in our work,[11] we know that there is a positive constant C_2 independent of κ such that

$$\|B(k)\Psi_{r13}\|_{\mathcal{H}} \le C_2|k|^{-\varepsilon}. \tag{69}$$

Next we estimate $B(k)\Psi_{r2\nu}$. By (63) we get

$$\|B(k)\Psi_{r20}\|_{\mathcal{H}} \leq 3|k|^{\varepsilon-1} \leq 3|k|^{-\varepsilon}, \tag{70}$$

$$\|B(k)\Psi_{r21}\|_{\mathcal{H}} \leq 6|k|^{-\varepsilon} \tag{71}$$

and

$$\|B(k)\Psi_{r22}\|_{\mathcal{H}} \leq \frac{1}{2|k|} \sum_{j=1}^{3} \frac{|k_j + f_j^{[1]}|}{|f_j^{[1]}|\,|f_j^{[2]}|} |g^{[j,2]}|^2.$$

Note that $f_j^{[1]} = -f_j^{[2]}$. Therefore by (63)

$$\sum_{j=1}^{3} \frac{|k_j + f_j^{[1]}|}{|f_j^{[1]}|\,|f_j^{[2]}|} |g^{[j,2]}|^2 \leq 2|k|^{-\varepsilon} \sum_{j=1}^{3} |g^{[j,2]}|^2$$

and in addition

$$\sum_{j=1}^{3} |g^{[j,2]}|^2 = 3|k|^2. \tag{72}$$

Since $|k| < 1$, we have

$$\|B(k)\Psi_{r22}\|_{\mathcal{H}} < 3|k|^{2-\varepsilon} < 3|k|^{-\varepsilon}. \tag{73}$$

Moreover, since we can show the same lemma as Lemma 7.1 in our work,[11] we know that there is a positive constant C_3 independent of κ such that

$$\|B(k)\Psi_{r23}\|_{\mathcal{H}} \leq C_3|k|^{-\varepsilon}. \tag{74}$$

Finally let us consider $I_{r2}\psi_\kappa^{\mathrm{PF}}$. Note that $f_j^{[1]} = -f_j^{[2]}$. By the same lemma as Lemma 7.1 in our work[11] and (63) we can show that there exists a positive constant C_4 independent of κ such that

$$\|B(k)I_{r2}\psi_\kappa^{\mathrm{PF}}\|_{\mathcal{H}} \leq C_4|k|^{-\varepsilon}. \tag{75}$$

Set $\delta = 1 - \varepsilon$. Then $0 < \delta < 1/2$. By (65)–(75) we can find a positive constant C_5 independent of κ such that

$$\|B(k)I_{r0}\psi_\kappa^{\mathrm{PF}} - B(k)\Psi_{r1}^{\mathrm{er}}\|_{\mathcal{H}} \leq C_5|k|^{\delta-1}.$$

Hence it follows that $\gamma(k)(H_\kappa^{PF} - E_\kappa^{PF} + |k|)^{-1}\Psi_{rn\nu}$, $r = 1, 2$; $n = 1, 2$; $\nu = 0, 1, 2, 3$, and $\gamma(k)(H_\kappa^{PF} - E_\kappa^{PF} + |k|)^{-1}I_{r2}\psi_\kappa^{PF}$ *are $L^2(\mathbb{R}^3)$-functions of k, and* moreover, *for every q under condition (47) and arbitrary κ with $0 \le \kappa < \Lambda$,*

$$
\begin{aligned}
I_{PF}(k) := \Bigg\| &\frac{\chi_\kappa(k)}{\sqrt{2|k|}} e_r(k)(H_\kappa^{PF} - E_\kappa^{PF} + |k|)^{-1} v \psi_\kappa^{PF} \\
&- \frac{\chi_\kappa(k)}{\sqrt{2|k|}} e_r(k)(H_\kappa^{PF} - E_\kappa^{PF} + \omega(k))^{-1} \Psi_{r1}^{er} \Bigg\|_{\mathcal{H}} \\
\le\ & C_5 \left| \frac{\chi_\kappa(k)}{\sqrt{2|k|}} \right| |k|^{\delta-1}
\end{aligned}
$$

holds. Here $\delta - 1$ appears as the power of $|k|$, but this appearance of -1 does not make a trouble, considering the Jacobian when we employ change of variables from Euclidean coordinates to the polar coordinates. Namely,

$$
\int_{|k|<1} d^3k\, |I_{\mathrm{PF}}(k)|^2 \le \frac{C_5^2}{4\pi^2} \int_0^1 |k|^2 d|k|\, \frac{|k|^{2\delta-2}}{|k|} = \frac{C_5^2}{4\pi^2} \int_0^1 d|k|\, |k|^{2\delta-1} < \infty.
$$

4.4. *Estimate of error terms*

The estimate of error terms Ψ_{r1}^{er}, $r = 1, 2$, remains.

We use the same ε as in the previous subsection 4.3, *i.e.*, $1/2 < \varepsilon < 1$. First, by (63) and using $\psi_\kappa^{\mathrm{PF}} = |k| B(k) \psi_\kappa^{\mathrm{PF}}$ and $(H_{\mathrm{f}}+1)^{-1/2}(H_{\mathrm{f}}+1)^{1/2} = I$, we have

$$
\begin{aligned}
\Big\| B(k) \widetilde{I}_{r1}^j &\left(1 - e^{-i f_j^{[1]} x_j} \right) \psi_\kappa^{\mathrm{PF}} \Big\|_{\mathcal{H}} \\
&\le 2 \| B(k)^{1/2} e^{-i g^{[j,1]} \cdot x} v_j (H_{\mathrm{f}} + 1)^{-1/2} \| \\
&\qquad \times \| \left(e^{i f_j^{[1]} x_j} - 1 \right) (H_{\mathrm{f}} + 1)^{1/2} B(k)^{1/2} \psi_\kappa^{\mathrm{PF}} \|_{\mathcal{H}}.
\end{aligned} \tag{76}
$$

Since $\left[v_j \,,\, e^{-i g^{[j,1]} \cdot x} \right] = \left[p_j \,,\, e^{-i g^{[j,1]} \cdot x} \right] = -g_j^{[j,1]} e^{-i g^{[j,1]} \cdot x}$, we have

$$
\begin{aligned}
\| B(k)^{1/2} &e^{-i g^{[j,1]} \cdot x} v_j (H_{\mathrm{f}} + 1)^{-1/2} \| \\
&\le \| B(k)^{1/2} v_j (H_{\mathrm{f}} + 1)^{-1/2} \| + |g_j^{[j,1]}| \, \| B(k)^{1/2} \| \, \| (H_{\mathrm{f}} + 1)^{-1/2} e^{-i g^{[j,1]} \cdot x} \| \\
&\le \| B(k)^{1/2} v_j (H_{\mathrm{f}} + 1)^{-1/2} \| + 2|k|^{\varepsilon-1/2}.
\end{aligned} \tag{77}
$$

On the other hand, we get

$$
\begin{aligned}
\| B(k)^{1/2} &v_j (H_{\mathrm{f}} + 1)^{-1/2} \| \\
&\le \| B(k)^{1/2} (H_0 + 1)^{1/2} \| \left(\| (H_0 + 1)^{-1/2} p_j \| + e \| (H_0 + 1)^{-1/2} A_{rj}(x) \| \right).
\end{aligned}
$$

Hence it follows from (49) that there exists a positive constant C_6 such that

$$
\| B(k)^{1/2} v_j (H_{\mathrm{f}} + 1)^{-1/2} \| \le C_6 |k|^{-1/2}. \tag{78}
$$

By (76)–(78), we obtain

$$\|B(k)\widetilde{I}_{r1}^{j}\left(1 - e^{-if_{j}^{[1]}x_{j}}\right)\psi_{\kappa}^{\mathrm{PF}}\|_{\mathcal{H}}$$

$$\leq 2C_{7}|k|^{-1/2}\|\left(e^{i|k|^{\epsilon}x_{j}} - 1\right)(H_{\mathrm{f}} + 1)^{1/2}B(k)^{1/2}\psi_{\kappa}^{\mathrm{PF}}\|_{\mathcal{H}}, \qquad (79)$$

where

$$C_{7} := C_{6} + 2.$$

Then, by (79) and using the change of variable, we have

$$\int_{|k|<1} d^{3}k\,|\gamma(k)|^{2}\|B(k)\Psi_{r1}^{\mathrm{er}}\|_{\mathcal{H}}^{2}$$

$$\leq \frac{4\pi}{2(2\pi)^{3}}\int_{0}^{1}|k|^{2}d|k|\,\frac{1}{|k|}$$

$$\times 3\sum_{j=1}^{3}\|\left(H_{\kappa}^{\mathrm{PF}} - E_{\kappa}^{\mathrm{PF}} + |k|\right)^{-1}\widetilde{I}_{r1}^{j}\left(1 - e^{-if_{j}^{[1]}x_{j}}\right)\psi_{\kappa}^{\mathrm{PF}}\|_{\mathcal{H}}^{2}$$

$$\leq \frac{3}{\pi^{2}}C_{7}^{2}\int_{0}^{1}d|k|$$

$$\times \sum_{j=1}^{3}\|\left(e^{i|k|^{\epsilon}x_{j}} - 1\right)(H_{\mathrm{f}} + 1)^{1/2}\left(H_{\kappa}^{\mathrm{PF}} - E_{\kappa}^{\mathrm{PF}} + |k|\right)^{-1/2}\psi_{\kappa}^{\mathrm{PF}}\|_{\mathcal{H}}^{2}$$

$$\leq \frac{3}{\pi^{2}}C_{7}^{2}\int_{0}^{1}d|k|\,S(|k|), \qquad (80)$$

where $S : [0,1] \to [0,\infty)$ is defined by

$$S(t) = \begin{cases} \displaystyle\sum_{j=1}^{3}\left\|t^{-1/2}\left(e^{it^{\epsilon}x_{j}} - 1\right)(H_{\mathrm{f}} + 1)^{1/2}\psi_{\kappa}^{\mathrm{PF}}\right\|_{\mathcal{H}}^{2} & \text{for } 0 < t \leq 1, \\[2em] 0 & \text{for } t = 0. \end{cases} \qquad (81)$$

We note the following points: (i) $S(t)$ can be written as an integral over x according to

$$S(t) = \sum_{j=1}^{3}\int_{\mathbb{R}^{3}} d^{3}x\left\|t^{-1/2}\left(e^{it^{\epsilon}x_{j}} - 1\right)(H_{\mathrm{f}} + 1)^{1/2}\psi_{\kappa}^{\mathrm{PF}}\right\|_{\mathcal{F}}^{2}(x).$$

(ii) By the work,[10] we know $\psi_{\kappa}^{\mathrm{PF}} \in D(x^{2}) \cap D(H_{\mathrm{f}})$. Moreover x and $(H_{\mathrm{f}} + 1)^{1/2}$ are (strongly) commutable in the sense of the definition on p.271 in the book.[19] Thereby it can be proved that $\psi_\kappa \in D\left(x\left(H_{\mathrm{f}}+1\right)^{1/2}\right)$. Thus, we have

$$S(t) = t^{2(\varepsilon-1/2)} \sum_{j=1}^{3} \left\| t^{-\varepsilon}\left(e^{it^\varepsilon x_j}-1\right)\left(H_{\mathrm{f}}+1\right)^{1/2}\psi_\kappa^{\mathrm{PF}} \right\|_{\mathcal{H}}^{2} \to 0$$

as $t \to 0$. It is easy to see that $S(t)$ is continuous on $(0,1]$. Thus, $S(t)$ is continuous on $[0,1]$, and

$$\int_0^1 dt\, S(t)$$

$$= \int_{\mathbb{R}^3} d^3x\, \|(H_{\mathrm{f}}+1)^{1/2}\psi_\kappa^{\mathrm{PF}}\|_{\mathcal{F}}^2(x) \sum_{j=1}^{3} \int_0^1 dt\, \left| \frac{e^{it^\varepsilon x_j}-1}{\sqrt{t}} \right|^2$$

$$= \frac{2}{\varepsilon} \int_{\mathbb{R}^3} d^3x\, \|(H_{\mathrm{f}}+1)^{1/2}\psi_\kappa^{\mathrm{PF}}\|_{\mathcal{F}}^2(x) \sum_{j=1}^{3} \int_0^{|x_j|} ds\, \frac{1-\cos s}{s} \qquad (82)$$

by Fubini's theorem. So, in the same way as in our work,[11] we set $R = 10^2$ and estimate $\int_0^1 dt\, S(t)$ as

$$\int_0^1 dt\, S(t) \le C_8\left(\gamma_E + \log 15 + \frac{91}{30}\right) + C_9\|(H_{\mathrm{f}}+1)^{1/2}\chi_R\sqrt{3+|x|}\psi_\kappa^{\mathrm{PF}}\|_{\mathcal{H}}^2$$

$$\le C_8\left(\gamma_E + \log 15 + \frac{91}{30}\right) + C_{10}\left(\log e\right)^2 + C_{11}\log e \qquad (83)$$

for positive constants C_8, C_9, C_{10}, C_{11} independent of κ.

Therefore, we obtain that *there exist positive constants C_{12}, C_{13}, C_{14}, C_{15} such that*

$$e^2 \int_{|k|<1} d^3k\, |\gamma(k)|^2 \|(H_\kappa^{PF} - E_\kappa^{PF} + |k|)^{-1}\Psi_{r1}^{\mathrm{er}}\|_{\mathcal{H}}^2$$

$$\le C_{12}e^2\left\{C_{13} + C_{14}\log e^2 + C_{15}\left(\log e^2\right)^2\right\}.$$

References

1. A. Arai, *Rev. Math. Phys.* **13** (2001) 1075.
2. A. Arai and M. Hirokawa, *J. Funct. Anal.* **151** (1997) 455.
3. A. Arai and M. Hirokawa, *Rev. Math. Phys.* **12** (2000) 1085.
4. A. Arai, M. Hirokawa, and F. Hiroshima, *J. Funct. Anal.* **168** (1999) 470.

5. V. Bach, J. Fröhlich, and I. M. Sigal, *Commun. Math. Phys.* **207** (1999) 249.

6. V. Betz, F. Hiroshima, J. Lőrinczi, R. A. Minlos and H. Spohn, *Rev. Math. Phys.* **14** (2002) 173.

7. J. Dereziński and C. Gérard, Scattering theory of infrared-divergent Pauli-Fierz Hamiltonian, in preparation.

8. R. P. Feynman, R. B. Leighton, and M. Sands, *The Feynman lectures on physics, Vol.III*, Addison-Wesley Publishing Company, 1965.

9. C. Gérard, *Rev. Math. Phys.* **14** (2002) 1165.

10. M. Griesemer, E. H. Lieb, and M. Loss, *Invent. Math.* **145** (2001) 557.

11. M. Hirokawa, F. Hiroshima, and H. Spohn, Ground state for point particles interacting through a massless scalar bose field, arXiv:math-ph/0211050, 2002.

12. M. Hirokawa, Divergence of soft photon number and absence of ground state for Nelson's Model, arXiv:math-ph/0211051, 2002.

13. F. Hiroshima, *Rev. Math. Phys* **9** (1997) 489.

14. F. Hiroshima, *J. Math. Phys* **40** (1999) 6209.

15. F. Hiroshima, *Ann. H. Poincaré* **3** (2002) 171.

16. J. Lőrinczi, R. A. Minlos and H. Spohn, *Ann. Henri Poincaré* **3** (2002) 269.

17. E. Nelson, *J. Math. Phys.***5** (1964) 1190.

18. K. Nishijima, *Relativistic Quantum Mechanics* (in Japanese), Baihukan, 2001.

19. M. Reed and B. Simon, *Methods of Modern Mathematical Physics I*, Academic Press, 1980.

20. M. Reed and B. Simon, *Methods of Modern Mathematical Physics II*, Academic Press, 1980.

21. H. Spohn, *Lett. Math. Phys.* **44** (1998) 9.

Localization in Quantum Field Theory

Shigeaki Nagamachi

Department of Applied Physics & Mathematics, Faculty of Engineering,
The University of Tokushima,
Tokushima 770-8506, Japan
E-mail: shigeaki@pm.tokushima-u.ac.jp

Erwin Brüning

Department of Mathematics & Applied Mathematics,
University of Durban-Westville,
Private Bag X54001, Durban 4000, South Africa
E-mail: ebruning@pixie.udw.ac.za

Dedicated to Professor Hiroshi Ezawa on the occasion
of his seventieth birthday

Hyperfunctions are considered to be the most general generalized functions which have the localization property. So, the quantum field theory formulated in terms of Fourier hyperfunctions, called hyperfunction quantum field theory (HFQFT), is the most general framework for local quantum field theories. The authors' works about HFQFT are reviewed and a new attempt is presented in which the authors formulate a relativistic quantum field theory with a fundamental length in terms of tempered ultra-hyperfunctions. Although, tempered ultra-hyperfunctions have no standard localization properties, a model for such theories can be constructed which offers many familiar features.

1. Introduction

A detailed motivation for the general theory of quantized relativistic fields can be found in several textbooks, for instance Refs.[5,47] Here we highlight some basic points.

The theory of quantized fields has its roots in the quantum theory of M.

Planck (1900).[40] For the solution of the problem of black body radiation, he introduced the universal quantum of action to the theory of electromagnetic fields. Nonrelativistic quantum mechanics is established by W. Heisenberg (1925)[20] and E. Schrödinger (1926).[42] M. Born, W. Heisenberg and P. Jordan (1926)[6] realized the quantization of electromagnetic fields, and P.A.M. Dirac (1927)[12] quantized electromagnetic fields in interaction with a material system. But P. Ehrenfest noticed soon that the theory had to lead to infinities. The existence of the positron is suggested by the theory of P.A.M. Dirac (1928).[13,14] The discovery of the positron by C.D. Andersen (1932) established the theory of quantum electrodynamics which treats the behavior of electron, positron and electromagnetic fields. But this theory still has to lead to infinities, and these difficulties are (partially) removed by S. Tomonaga (1946)[48], H.A. Bethe (1947)[2] and J. Schwinger (1948)[43] using the subtraction formalism in perturbation theory. R.P. Feynman (1948)[18] developed the method of path integral which simplifies the calculation and F.J. Dyson (1949)[15] derives Feynman's prescription from Tomonaga–Schwinger theory. Thus the prescription of the subtraction formalism in electrodynamics was completely worked out. Those theories in which the infinity of each term in the perturbation series can be subtracted consistently are called renormalizable theories. But for these renormalizable theories, there are still doubts about summability of the series in perturbation theory. In electrodynamics, the expansion parameter (the coupling constant) is small and the sum of the first few terms gives an amazing agreement with experiments, but there are no proofs about the convergence of the series. In some theory of strong interaction the parameter is greater than 1, and this subtraction formalism does not work. The main question is: What is hidden behind these formal infinite series? To this question, one answer is given by the formalism of A.S. Wightman and L. Gårding (1964),[50] which is a mathematically rigorous study of quantum fields. Since the structure of the theory is axiomatic, the theory is called axiomatic quantum field theory. The axioms formulate the basic physical postulates (*e.g.*, relativistic invariance of the state space, spectral property, unique existence of a vacuum state, Poincaré-covariance of the fields, locality or micro causality, *etc.*, and some technical postulate: temperedness of the fields, *etc.*) in a mathematical language. The axioms show the way to reconcile the principles of quantum mechanics and those of special relativity. Quantum field theory is also important as effective theory in low-energy approximation to a deeper theory like a string theory. From this perspective, quantum field theory with a fundamental length becomes interesting.

In Section 2, we present Wightman's axioms W.I – W.VII, and the axioms w.1 – w.7 for the n-point functionals (*i.e.*, n-fold vacuum expectation values of the quantum field), often called Wightman distributions. Some basic results are mentioned. Among those, the reconstruction theorem (Theorem 1) shows that these two sets of axioms are equivalent.

These axioms are consistent and seem to characterize local relativistic quantum fields well. But up to now, there are no theories which satisfy Wightman's axioms and have nontrivial scattering matrices in 4-dimensional space-time. This fact can be considered to indicate that Wightman's axioms might be too strong. There are some reasons (see Ref.[49]) and many attempts to weaken axiom W.IV of temperedness which is a technical one and is believed to exclude nonrenormalizable theories. One of the most famous attempts is Jaffe's work.[26] However, on page 425 of Ref.[4] one finds:

> In conclusion we remark that Jaffe's formulation of the problem is not the most general. One might try (in the spirit of the work of Martineau (1963), for example) to define local properties of generalized functions (in particular the notion of support of a generalized function) even when the space of test functions does not include functions of compact support. The hope is that in this general framework one might arrive at a class of generalized functions satisfying weaker conditions than (15.53–54); in particular, the generalized functions which increase at infinity no faster than any linear exponential $|f(p)| \leq C_\epsilon \exp \epsilon |p|$ for all $\epsilon > 0$ might be included.

All this is realized in hyperfunction quantum field theory (HFQFT) of S. Nagamachi and N. Mugibayashi.[34] Among the class of generalized functions, hyperfunctions are considered to be the most general ones which have the localization property, and using them, one can formulate local quantum field theory. Section 2 introduces Wightman theory. At the end of Section 2 and Section 3 a brief discussion of the Euclidean reformulation of Wightman theory is presented.

In Section 3, Fourier hyperfunctions are explained and the axioms of HFQFT are presented. The test functions of Fourier hyperfunctions are analytic functions and consequently they have no bounded supports. Nevertheless we can define the concept of support for Fourier hyperfunctions by which we can formulate the locality (local commutativity) and the spectral property in HFQFT.

The two point function $D_m^{(-)}(x - y) = \langle \phi(x)\phi(y) \rangle$ of the free field with mass m has a local singularity of the form

$$\frac{1}{[(x-y)^2]^{(s-1)/2}}$$

in $s > 1$ space dimension. Thus an infinite power series in this quantity will have an essential singularity at the origin, and therefore not be tempered. However these power series can converge in the sense of Fourier hyperfunctions. Thus (Wick ordered) entire functions of free fields find a natural formulation in HFQFT and thus provide a huge class of additional explicit models of quantum fields which are far more complex than the class of explicit models in standard QFT. Certain 'formally interacting' models can be treated rigorously in HFQFT, but not in standard QFT.[37] Using this observation we indicate in Section 4 the proof of existence of a field ρ which satisfies all the axioms of HFQFT and does not admit test-functions with bounded support.

There are two main approaches to construct a net of local algebras of observables from tempered quantum fields (see Refs.[1,22,3] for a survey of these and other attempts). One way to define local algebras $\{\mathcal{M}(O)\}$ for bounded open sets O in $\mathbb{R}^4$, is to define $\mathcal{M}(O)$ as the set of all 'bounded functions' of self-adjoint field operators $\Phi(f)$ for $f \in \mathcal{S}(\mathbb{R}^4)$ with supp $f \subset O$. Since the test functions of the Fourier hyperfunctions are analytic functions, there are no nonzero functions with compact supports. Therefore we can not define local algebras for hyperfunction quantum fields by this method. Another definition of local algebras is that $\mathcal{M}(O)$ is the set of all bounded operators which commute weakly with all field operators $\Phi(f)$ with supp $f \subset O'$ where $O' = \text{int} \{x \in \mathbb{R}^4; (x-y)^2 < 0, \forall y \in O\}$. S. Nagamachi and E. Brüning[33] show that this definition has a counter part in hyperfunction quantum field theory. In Section 5, we recall how to construct local observable algebras from HFQFT.

In the standard tempered field theory, the Borchers algebra plays an important role. Since the test-function space of tempered fields has functions with compact supports, the Borchers algebra is a local algebra. But the Borchers algebra in HFQFT is not local. Nevertheless Borchers algebra plays some interesting role in HFQFT too, since it has the property of asymptotic Abelianness (see Ref.[9]).

Although the axioms of HFQFT are considered to be the most general (weakest) among those of local quantum field theories, there are still no nontrivial models which satisfy the axioms of HFQFT. Therefore, we try to weaken the axiom of locality. In string theory, it is said that there is a length $\ell > 0$ such that one cannot distinguish events which occur in a smaller distance than ℓ (see Ref.[41]). How can we take this condition into account? In Wightman's framework, in terms of the field operators $A(x)$, locality is expressed by the condition that the commutator of the field operators vanishes for space-like separated points, *i.e.*, by $A(x)A(y) - A(y)A(x) = 0$ for all $x, y \in \mathbb{R}^4$ with $(x-y)^2 < 0$. Even if one replaces this condition by the apparently weaker condition that this commutator vanishes for all points $x, y \in \mathbb{R}^4$ which are space-like separated and are in a certain distance of at least $\ell > 0$ form each other, *i.e.*, $(x - y)^2 < -\ell^2 < 0$, one can prove that this commutator vanishes for all space-like separated points by using the other axioms of the theory (see for instance Theorem 19.3 of Ref.[4]). The situation is the same in hyperfunction quantum field theory (see Ref.[8]). The reason is that the n-point functionals in both theories are boundary values of holomorphic functions in $\mathrm{Im}\,(z_{i+1} - z_i) \in V_+$ and as such, distributions and hyperfunctions have the same type of localization property, expressed by the fact that both form a sheaf over space-time (see for instance Ref.[27]).

Thus, in order to formulate a quantum field theory with a fundamental length, a space of generalized functions different from both Schwartz distributions and Fourier hyperfunctions has to be used. By a variety of reasons, the problem to formulate relativistic quantum field theory in terms of generalized functions which are different from Schwartz distributions and/or Fourier hyperfunctions, has been addressed in the past. Some prominent articles about this and related problems are found in the Refs.[26,23,10,11,17,16,44,45,46] In these articles, roughly, it is argued that the resulting theory is nonlocalizable if it is formulated in terms of generalized functions for which the underlying test-function space does not contain test-functions of compact support. Clearly, in order to formulate a relativistic quantum field theory with a fundamental length, a different type of generalized functions has to be used.

The following simple example will provide some insight into the properties of the class of generalized functions which could be used in theories with fundamental length. Let $\{a_n\}$ be a sequence of real numbers. The support of the distributions $\sum_{n=0}^{N} a_n \delta^{(n)}(x)$ is $\{0\}$, but $\sum_{n=0}^{\infty} a_n \delta^{(n)}(x)$ does not define a distribution unless all except finitely many of the coefficients a_n vanish. If $\lim_{n \to \infty} [n! a_n]^{1/n} = 0$, then $\sum_{n=0}^{\infty} a_n \delta^{(n)}(x)$ converges to a hyperfunction. The support of this limit is the same as the supports of the approximating sums, *i.e.*, the hyperfunction has the support $\{0\}$.

Now consider the case $a_n = a^n/n!$ and take the underlying space of test-functions into account. If $f(z)$ is a holomorphic function in $|\operatorname{Im} z| < \ell$, then, for $|a| < \ell$, we have

$$\left\langle \sum_{n=0}^{\infty} \frac{a^n}{n!} \delta^{(n)}(x), f(x) \right\rangle = \sum_{n=0}^{\infty} \frac{(-a)^n}{n!} f^{(n)}(0)$$
$$= f(0-a) = \langle \delta(x+a), f(x) \rangle,$$

that is, as an equation for functionals defined on the function space $\mathcal{T}(T(-\ell,\ell))$ whose elements are holomorphic functions in $T(-\ell,\ell) = \mathbb{R} + i(-\ell,\ell) \subset \mathbb{C}$, we have

$$\sum_{n=0}^{\infty} \frac{a^n}{n!} \delta^{(n)}(x) = \delta(x+a).$$

The sequence of generalized functions $S_N = \sum_{n=0}^{N} \frac{a^n}{n!} \delta^{(n)}(x)$ with support $\{0\}$ converges (weakly, in the dual space of $\mathcal{T}(T(-\ell,\ell))$) to the generalized function $\delta(x+a)$ with support $\{-a\}$, as $N \to \infty$. However, if $|a| > \ell$, then this sequence does not converge in $\mathcal{T}(T(-\ell,\ell))'$.

This phenomenon can be understood as follows. If $|a| < \ell$, then elements in $\mathcal{T}(T(-\ell,\ell))'$ do not distinguish between the points $\{0\}$ and $\{-a\}$, but if $|a| > \ell$ then elements in $\mathcal{T}(T(-\ell,\ell))'$ can distinguish between the points $\{0\}$ and $\{-a\}$. Since $|a| < \ell$ is arbitrary, one can say that elements in $\mathcal{T}(T(-\ell,\ell))'$ do not distinguish between points which are separated by less than ℓ. Spaces of functionals with this property have been studied under the name of ultra-hyperfunctions in Refs.[19,29]

In the final section of this paper we propose to formulate relativistic quantum field theory with a fundamental length ℓ in terms of tempered ultra-hyperfunctions, *i.e.*, those ultra-hyperfunctions which admit the Fourier transform as an isomorphism of topological vector spaces. Section 6 gives the definition and the basic properties of this space and axioms of quantum field theory with a fundamental length. Obviously one would like to have examples of ultra-hyperfunction quantum fields. We show that the field $A(x) =: e^{g\phi(x)^2} :$ defined by the free neutral scalar field $\phi(x)$ satisfies all our axioms of a relativistic quantum field theory with a fundamental length.

2. Wightman's axioms

In Wightman's scheme, the concept of a relativistic quantum field $\phi^{(\kappa)}$ of type κ plays a fundamental role. Such a field, for example a scalar, tensor or spinor field, has a finite number of Lorentz components $\phi_j^{(\kappa)}$ $(j = 1, \ldots, r_\kappa)$.

The field components $\phi_j^{(\kappa)}(x)$ are operator-valued generalized functions, i.e.,

$$\phi_j^{(\kappa)}(f) = \int \phi_j^{(\kappa)}(x) f(x) d^4x$$

are densely defined linear operators in a complex Hilbert space $\mathcal{H}$. They are not assumed to be bounded.

We recall Wightman's axioms.

W.I **Relativistic invariance of the state space**: There is a complex Hilbert space $\mathcal{H}$ with positive metric in which a unitary representation $U(a, A)$ of the Poinaré spinor group $\mathcal{P}_0$ acts. $(a, A) \mapsto U(a, A)$ is weakly continuous.

W.II **Spectral property**: The spectrum of the energy-momentum operator P which generates the translations in this representation, i.e., $e^{iaP} = U(a, 1)$, is contained in the closed forward light cone
$\bar{V}_+ = \{p = (p^0, \ldots, p^3) \in \mathbb{R}^4; p^0 \geq |\boldsymbol{p}|\}$.

W.III **Existence and uniqueness of the vacuum**: In $\mathcal{H}$ there exists unit vector Φ_0 (also denoted by $|0\rangle$ and called the vacuum vector) which is unique up to a phase factor and which is invariant under all space-time translations $U(a, 1)$, $a \in \mathbb{R}^4$.

W.IV **Fields and temperedness**: The components $\phi_j^{(\kappa)}$ of the quantum field $\phi^{(\kappa)}$ are operator-valued generalized functions $\phi_j^{(\kappa)}(x)$ over the Schwartz space $\mathcal{S}(\mathbb{R}^4)$ with common dense domain $\mathcal{D}$; i.e., for all $\Psi \in \mathcal{D}$ and all $\Phi \in \mathcal{H}$,

$$\mathcal{S}(\mathbb{R}^4) \ni f \to (\Phi, \phi_j^{(\kappa)}(f)\Psi) \in \mathbb{C}$$

is a tempered distribution. It is supposed that the vacuum vector Φ_0 is contained in $\mathcal{D}$ and that $\mathcal{D}$ is taken into itself under the action of the operators $\phi_j^{(\kappa)}(f)$ and $U(a, A)$, i.e.,

$$\phi_j^{(\kappa)}(f)\mathcal{D} \subset \mathcal{D}, \; U(a, A)\mathcal{D} \subset \mathcal{D}.$$

Moreover it is supposed that there exist indices $\bar{\kappa}, \bar{j}$ such that $\phi_{\bar{j}}^{(\bar{\kappa})}(\bar{f}) \subset \phi_j^{(\kappa)}(f)^*$ where * indicates the Hilbert space adjoint of the operator in question.

W.V Poincaré-covariance of the fields: According to the type of the field, there is a finite dimensional real or complex matrix representation $V^{(\kappa)}(A)$ of $SL(2, \mathbb{C})$ such that

$$U(a, A)\phi_j^{(\kappa)}(x)U(a, A)^{-1} = \sum_\ell V_{j,\ell}^{(\kappa)}(A^{-1})\phi_\ell^{(\kappa)}(\Lambda(A)x + a),$$

i.e., for any $f \in \mathcal{S}(\mathbb{R}^4)$ and $\Psi \in \mathcal{D}$,

$$U(a, A)\phi_j^{(\kappa)}(f)U(a, A)^{-1}\Psi = \sum_\ell V_{j,\ell}^{(\kappa)}(A^{-1})\phi_\ell^{(\kappa)}(f_{(a,A)})\Psi,$$

where $f_{(a,A)}(x) = f(\Lambda(A)^{-1}(x - a))$. We have $V^{(\kappa)}(-1) = \pm 1$. If $V^{(\kappa)}(-1) = 1$, then the field is called a tensor field. If $V^{(\kappa)}(-1) = -1$, then the field is called a spinor field.

W.VI Locality, local commutativity or microcausality: Any two field components $\phi_j^{(\kappa)}(x)$ and $\phi_\ell^{(\kappa')}(y)$ either commute or anti-commute at space-like separated points x and y: If f and g have space-like separated supports then, for all $\Psi \in \mathcal{D}$, $[\phi_j^{(\kappa)}(f), \phi_\ell^{(\kappa')}(g)]_{\mp}\Psi = 0$, *i.e.*,

$$[\phi_j^{(\kappa)}(x), \phi_\ell^{(\kappa')}(y)]_{\mp} = 0 \quad \text{for } (x - y)^2 < 0.$$

W.VII Cyclicity of the vacuum: The set $\mathcal{D}_0$ of finite linear combinations of vectors of the form

$$\phi_{j_1}^{(\kappa_1)}(f_1) \cdots \phi_{j_n}^{(\kappa_n)}(f_n)\Phi_0, \quad f_j \in \mathcal{S}(\mathbb{R}^4) \ (n = 0, 1, \dots)$$

is dense in $\mathcal{H}$.

The vacuum expectation values or n-point functionals of the field operators are often called Wightman functions or Wightman distributions. They are defined as follows ($n \geq 1$, $1 \leq \ell \leq n - 1$):

$$\mathcal{S}(\mathbb{R}^4)^n \ni (f_1, \dots, f_n) \to \left(\Phi_0, \phi_{j_1}^{(\kappa_1)}(f_1) \cdots \phi_{j_n}^{(\kappa_n)}(f_n)\Phi_0\right)$$
$$= \left(\phi_{\bar{j}_\ell}^{(\bar{\kappa}_\ell)}(\bar{f}_\ell) \cdots \phi_{\bar{j}_1}^{(\bar{\kappa}_1)}(\bar{f}_1)\Phi_0, \phi_{j_{\ell+1}}^{(\kappa_{\ell+1})}(f_{\ell+1}) \cdots \phi_{j_n}^{(\kappa_n)}(f_n)\Phi_0\right) \in \mathbb{C}$$

is clearly multilinear and separately continuous by W.IV. Therefore there exists a tempered distribution $w_{j_1 \dots j_n}^{(\kappa_1 \dots \kappa_n)}(x_1, \dots, x_n)$, called the n-point *Wightman distribution*, such that

$$\left(\Phi_0, \phi_{j_1}^{(\kappa_1)}(f_1) \cdots \phi_{j_n}^{(\kappa_n)}(f_n)\Phi_0\right) = \langle w_{j_1 \dots j_n}^{(\kappa_1 \dots \kappa_n)}, f_1 \otimes \cdots \otimes f_n \rangle$$
$$= \langle w_{j_1 \dots j_n}^{(\kappa_1 \dots \kappa_n)}(x_1, \dots, x_n), f_1(x_1) \cdots f_n(x_n) \rangle,$$

where $\langle \cdot, \cdot \rangle$ denotes the duality between distributions and test functions and $\otimes$ the tensor product for test functions.

The set of axioms for the fields allow to deduce a number of interesting properties of the Wightman distributions. Among these properties are the properties [w.1] – [w.7] listed below. They are distinguished by the fact that allows to reconstruct the fields, as we will learn later.

w.1 **Admissible nature of a singularity and growth**: The n-point Wightman function

$$w_{j_1\ldots j_n}^{(\kappa_1\ldots\kappa_n)}(x_1,\ldots,x_n) = \langle 0|\phi_{j_1}^{(\kappa_1)}(x_1)\cdots\phi_{j_n}^{(\kappa_n)}(x_n)|0\rangle$$

is a tempered distribution.

w.2 **Property of adjoint**:

$$\overline{w_{j_1\ldots j_n}^{(\kappa_1\ldots\kappa_n)}(x_1,\ldots,x_n)} = w_{j_n\ldots j_1}^{(\bar{\kappa}_n\ldots\bar{\kappa}_1)}(x_n,\ldots,x_1).$$

w.3 **Positive definiteness**: For any finite system $f = \{f_{j_1\ldots j_m}^{(k_1\ldots k_m)}\}$ of functions $f_{j_1\ldots j_n}^{(k_1\ldots k_n)}(x_1,\ldots,x_n) \in \mathcal{S}(\mathbb{R}^{4n})$ we have

$$\sum_{m,n=0}^{\infty} \sum_{\substack{k_1\ldots k_m\\ j_1\ldots j_m}} \sum_{\substack{k_1'\ldots k_n'\\ j_1'\ldots j_n'}} \left\langle w_{\bar{j}_m\ldots\bar{j}_1 j_1'\ldots j_n'}^{(\bar{k}_m\ldots\bar{k}_1 k_1'\ldots k_n')}(x_m,\ldots,x_1,x_1',\ldots,x_n'), \right.$$

$$\left. \overline{f_{j_1\ldots j_m}^{(k_1\ldots k_m)}(x_1,\ldots,x_m)} f_{j_1'\ldots j_n'}^{(k_1'\ldots k_n')}(x_1',\ldots,x_n') \right\rangle \;\geq\; 0.$$

w.4 **Poincaré-covariance**:

$$\sum_{\ell_1\ldots\ell_n} \left\{ w_{\ell_1\ldots\ell_n}^{(\kappa_1\ldots\kappa_n)}(\Lambda(A)x_1 + a,\ldots,\Lambda(A)x_n + a) \right.$$

$$\left. \times V_{j_1\ell_1}^{(\kappa_1)}(A^{-1})\cdots V_{j_n\ell_n}^{(\kappa_n)}(A^{-1}) \right\} = w_{j_1\ldots j_n}^{(\kappa_1\ldots\kappa_n)}(x_1,\ldots,x_n).$$

w.5 **Locality**: Whenever $(x_\ell - x_{\ell+1})^2 < 0$, then

$$w_{j_1\ldots j_n}^{(\kappa_1\ldots\kappa_n)}(x_1,\ldots,x_n)$$

$$= \pm w_{j_1\ldots j_{\ell-1},j_{\ell+1},j_\ell\ldots j_n}^{(\kappa_1\ldots\kappa_{\ell-1},\kappa_{\ell+1},\kappa_\ell\ldots\kappa_n)}(x_1,\ldots,x_{\ell-1},x_{\ell+1},x_\ell,\ldots,x_n).$$

w.6 **Cluster property**: For any space-like vector $a \in \mathbb{R}^4$, the following relation holds (in the sense of convergence of generalized functions) for

$\lambda \to \infty$:

$$w_{j_1 \ldots j_n}^{(\kappa_1 \ldots \kappa_n)}(x_1, \ldots, x_\ell, x_{\ell+1} + \lambda a, \ldots, x_n + \lambda a)$$
$$\to w_{j_1 \ldots j_\ell}^{(\kappa_1 \ldots \kappa_\ell)}(x_1, \ldots, x_\ell) w_{j_{\ell+1} \ldots j_n}^{(\kappa_{\ell+1} \ldots \kappa_n)}(x_{\ell+1}, \ldots, x_n).$$

w.7 **Spectral property**: The support of the Fourier transform $\tilde{w}_{j_1 \ldots j_n}^{(\kappa_1 \ldots \kappa_n)}$ $(p_1, \ldots, p_n)$ of $w_{j_1 \ldots j_n}^{(\kappa_1 \ldots \kappa_n)}(x_1, \ldots, x_n)$ is contained in

$$\{(p_1, \ldots, p_n) ; \; p_k + p_{k+1} + \ldots + p_n \in \bar{V}_+, \; k = 1, 2, \ldots, n\}.$$

The announced reconstruction of the fields from the Wightman distributions reads as follows:

Theorem 1: (Wightman's reconstruction theorem) *Given a system of tempered distributions* $\{w_{j_1 \ldots j_n}^{(\kappa_1 \ldots \kappa_n)}(x_1, \ldots, x_n); \; n = 1, 2, \ldots\}$ *which satisfy the conditions w1 – w7, there exist a separable Hilbert space* $\mathcal{H}$, *a unitary representation* $U(a, A)$ *of the Poinaré spinor group* $\mathcal{P}_0$ *and quantum fields* $\phi^{(\kappa)}$ *which satisfy the Wightman axioms W.I – W.VII and whose vacuum expectation values are the given set of distributions, i.e.,*

$$w_{j_1 \ldots j_n}^{(\kappa_1 \ldots \kappa_n)}(x_1, \ldots, x_n) = \langle 0 | \phi_{j_1}^{(\kappa_1)}(x_1) \cdots \phi_{j_n}^{(\kappa_n)}(x_n) | 0 \rangle.$$

Two other fundamental results which are relevant for our later discussion are:

Theorem 2: *In Wightman's theory (in which Axioms W.I – W.VII hold) the field operators form an irreducible system, that is, any bounded operator* C *in* $\mathcal{H}$ *satisfying*

$$(\phi_j^{(\kappa)}(f)^* \Phi, C\Psi) = (\Phi, C\phi_j^{(\kappa)}(f)\Psi)$$

for all κ, j, *and* $\Phi, \Psi \in \mathcal{D}$, $f \in \mathcal{S}(\mathbb{R}^4)$, *must be a multiple of the identity operator.*

Theorem 3: (Reeh–Schlieder) *For any non-empty open set* $\mathcal{O} \subset \mathbb{R}^4$, *the set of finite linear combinations of vectors of the form* $\phi_{j_1}^{(\kappa_1)}(f_1) \cdots \phi_{j_n}^{(\kappa_n)}(f_n) \Phi_0$ $(n = 0, 1, \ldots)$, *where* $\operatorname{supp} f_j \subset \mathcal{O}$, *is dense in* $\mathcal{H}$.

Condition w.4 implies that there exist generalized functions $W_{j_1\ldots j_n}^{(\kappa_1\ldots\kappa_n)}(\xi_1,\ldots,\xi_{n-1}) \in \mathcal{S}(\mathbb{R}^{4(n-1)})$ such that

$$w_{j_1\ldots j_n}^{(\kappa_1\ldots\kappa_n)}(x_1,\ldots,x_{n-1}) = W_{j_1\ldots j_n}^{(\kappa_1\ldots\kappa_n)}(x_2 - x_1,\ldots,x_n - x_{n-1}).$$

Taking the spectral condition w.7 into account and applying the Fourier Laplace transformation for tempered distributions we get

Theorem 4: *The Wightman distribution* $W_{j_1\ldots j_n}^{(\kappa_1\ldots\kappa_n)}(\xi_1,\ldots,\xi_{n-1})$ *is the boundary value in* $\mathcal{S}(\mathbb{R}^{4(n-1)})'$ *of a function* $W_{j_1\ldots j_n}^{(\kappa_1\ldots\kappa_n)}(\zeta_1,\ldots,\zeta_{n-1})$ *which is holomorphic in the tube* $T_+^{n-1} = (\mathbb{R}^4 + iV_+)^{n-1}$. $w_{j_1\ldots j_n}^{(\kappa_1\ldots\kappa_n)}(x_1,\ldots,x_n)$ *is the boundary value of a function* $w_{j_1\ldots j_n}^{(\kappa_1\ldots\kappa_n)}(z_1,\ldots,z_n)$ *holomorphic in the tube* $\mathcal{T}_+^n$:

$$\mathcal{T}_+^n = \{(z_1,\ldots,z_n) \in \mathbb{C}^{4n}; \mathrm{Im}\,(z_{k+1} - z_k) \in V_+, k = 1,\ldots,n-1\}.$$

With each point $x = (x^1,\ldots,x^4) = (\boldsymbol{x},x^4) \in E = \mathbb{R}^4$, we associate the point $x' \in \mathbb{C}^4$ according to the rule

$$x' = (ix^4, \boldsymbol{x}).$$

Then, in the set

$$E_<^n = \{(x_1,\ldots,x_n) \in E^n; x_1^4 < x_2^4 < \ldots < x_n^4\},$$

we can define the function $s_{j_1\ldots j_n}^{(\kappa_1\ldots\kappa_n)}(x_1,\ldots,x_n)$, called the Schwinger function, by

$$s_{j_1\ldots j_n}^{(\kappa_1\ldots\kappa_n)}(x_1,\ldots,x_n) = w_{j_1\ldots j_n}^{(\kappa_1\ldots\kappa_n)}(x_1',\ldots,x_n').$$

The Schwinger functions can be extended to the set of non-exceptional Euclidean points

$$E_{\neq}^n = \{(x_1,\ldots,x_n) \in E^n; x_j \neq x_k \text{ for } j \neq k\}.$$

Remark We can derive the axioms e.1 – e.6 of Schwinger functions from w.1 – w.7 (see Ref.[38]). But they do not imply the axioms w.1 – w.7. In order to be able to reconstruct the axioms of Wightman functions from e.1 – e.6, HFQFT plays an important role (see Section 3).

3. Hyperfunction quantum field theory

3.1. *Fourier hyperfunctions*

Hyperfunctions are the elements of the cohomology group $H^n_{\mathbb{R}^n}(U, \mathcal{O})$ with coefficients in the sheaf $\mathcal{O}$ of holomorphic functions on $\mathbb{C}^n = \mathbb{R}^n + i\mathbb{R}^n$, where U is a neighborhood of $\mathbb{R}^n$ in $\mathbb{C}^n$. They are represented as finite sums of boundary values of certain analytic functions. There are no test-functions for hyperfunctions. But localization and thus supports are defined for hyperfunctions. Just as distributions they have a 'good notion' of *localization* (which agrees with the known localization properties of distribution if applied to them). However in general hyperfunctions do not admit a canonical definition of Fourier transform as an isomorphism. Some 'growth restrictions at infinity' are needed for this. The space of Fourier hyperfunctions is stable under Fourier transformation. In order to define Fourier hyperfunctions, we introduce the radial compactification D^n of $\mathbb{R}^n$. It is defined in a natural way as follows: Let S^{n-1}_∞ be the $(n-1)$-dimensional sphere at infinity, which is homeomorphic to the unit sphere $S^{n-1} = \{x \in \mathbb{R}^n; |x| = 1\}$ by the mapping $x \to x_\infty$, where the point $x_\infty \in S^{n-1}_\infty$ lies on the ray connecting the origin with the point $x \in S^{n-1}$. The set $\mathbb{R}^n \cup S^{n-1}_\infty$ equipped with its 'natural topology' (a fundamental system of neighborhoods of x_∞ is given by all open cones of arbitrary vertex generated by an arbitrary open neighborhoods of x_∞ in S^{n-1}_∞) is denoted by D^n. The space of Fourier hyperfunction then is defined by the cohomology group $H^n_{D^n}(U, \tilde{\mathcal{O}})$ with coefficients in the sheaf $\tilde{\mathcal{O}}$ of slowly increasing holomorphic functions on $Q^n = D^n + i\mathbb{R}^n$, where U is a neighborhood of D^n in Q^n. In contrast to hyperfunctions, Fourier hyperfunctions can be defined as continuous linear functionals on a certain space of test-functions.

In fact, we have the following isomorphism

$$H^n_{D^n}(U, \tilde{\mathcal{O}}) \cong \underset{\sim}{\mathcal{O}}(D^n)'.$$

Here $\underset{\sim}{\mathcal{O}}(D^n)$ is the space of rapidly decreasing holomorphic functions, *i.e.*, $\underset{\sim}{\mathcal{O}}(D^n) \ni f \Leftrightarrow \exists\, m \in \mathbb{N}$ such that $f(z)$ is analytic in $|\text{Im } z| < 1/m$ and

$$\sup_{|\text{Im } z| < 1/m} |f(z)| e^{|z|/m} < \infty.$$

For a closed subset $K \subset D^n$, the space $H^n_K(U, \tilde{\mathcal{O}})$ of Fourier hyperfunctions with supports in K is isomorphic to the dual space $\underset{\sim}{\mathcal{O}}(K)'$ of $\underset{\sim}{\mathcal{O}}(K)$, *i.e.*,

$$H^n_K(U, \tilde{\mathcal{O}}) \cong \underset{\sim}{\mathcal{O}}(K)', \text{ for any neighborhood } U \text{ of } K \text{ in } Q^n.$$

The space $\underset{\sim}{\mathcal{O}}(K)$ is defined as the inductive limit of the Banach spaces $\mathcal{O}_c^m(U_m)$:

$$\underset{\sim}{\mathcal{O}}(K) = \text{ind} \lim_{m \to \infty} \mathcal{O}_c^m(U_m),$$

where $\{U_m, m \in \mathbb{N}\}$ is a fundamental sequence of neighbourhoods of K in $\boldsymbol{Q}^n = \boldsymbol{D}^n + i\mathbb{R}^n$, and $\mathcal{O}_c^m(U_m)$ is the Banach space of functions f analytic in $U_m \cap \mathbb{C}^n$ and continuous on $\bar{U}_m \cap \mathbb{C}^n$ such that

$$\|f\|_m = \sup_{z \epsilon U_m \cap \mathbb{C}^n} |f(z)|e^{|z|/m}$$

is finite.

We can understand easily why elements $\mu \in \underset{\sim}{\mathcal{O}}(K)'$ are considered to have supports contained in K, in analogy to the case of distributions. It is well known that the space of distributions with compact support in Ω can be characterized as a dual space, *i.e.*, as the space $\mathcal{E}(\Omega)'$. We can understand this fact as follows: $T(f)$ must have a definite value for any $f \in \mathcal{E}(\Omega)$ whatever singular behavior f has near the boundary $\partial\Omega$ of Ω. For that, the functional T must avoid these 'singular points' of the test function at the boundary, *i.e.*, the support of T must be contained in some compact subset K of Ω. For $\mu \in \underset{\sim}{\mathcal{O}}(K)'$, $\mu(f)$ must have a definite value for any $f \in \underset{\sim}{\mathcal{O}}(K)$ whatever singular behaviour f has out side of K. So, the support must be contained in K.

It is easily seen that $\underset{\sim}{\mathcal{O}}(K_1) \subset \underset{\sim}{\mathcal{O}}(K_2)$ if $K_1 \supset K_2$. The following theorem is important for this characterization of the support of Fourier hyperfunctions through continuity properties.

Theorem 5: (Theorem of T. Kawai[28]) *For a compact set $K \subset \boldsymbol{D}^n$, $\mathcal{O}(\boldsymbol{D}^n)$ is dense in $\mathcal{O}(K)$.*

Here we collect several results about the relations between tempered distributions and Fourier hyperfunctions proved in Nagamachi and Brüning[32].

Proposition 1: *The embedding e of the test function space $\underset{\sim}{\mathcal{O}}(\boldsymbol{D}^n)$ for Fourier hyperfunctions into the test function space $\mathcal{S}(\mathbb{R}^n)$ for tempered distributions is continuous and has a dense range.*

Corollary 1: *There is an injection $\iota : \mathcal{S}'(\mathbb{R}^n) \to \underset{\sim}{\mathcal{O}}(\boldsymbol{D}^n)'$ of the space of tempered distributions into the space of Fourier hyperunctions on $\mathbb{R}^n$.*

This corollary shows that a tempered distribution can naturally be considered as a Fourier hyperfunction. Furthermore, if a Fourier hyperfunction T on $\mathbb{R}^n$, *i.e.*, a continuous linear functional on $\underset{\sim}{\mathcal{O}}(\boldsymbol{D}^n)$, is continuous with respect to the topology induced by $\mathcal{S}(\mathbb{R}^n)$ on $\underset{\sim}{\mathcal{O}}(\boldsymbol{D}^n)$, then this Fourier hyperfunction can be considered as a tempered distribution. Since tempered distributions can be considered as Fourier hyperfunctions, one can define the support of a tempered distribution T in two ways, firstly in the sense of tempered distributions through the use of compactly supported test functions and secondly in the sense of Fourier hyperfunctions. The following proposition establishes the relation between these two concepts of localization.

Proposition 2: *If $K \subset \mathbb{R}^n$ is the support of a tempered distribution $T \in \mathcal{S}'(\mathbb{R}^n)$ and if $L \subset \boldsymbol{D}^n$ is the support of T considered as a Fourier hyperfunction, i.e., the support of $\iota(T) \in \underset{\sim}{\mathcal{O}}(\boldsymbol{D}^n)'$, then one has*

$$K = L \cap \mathbb{R}^n.$$

3.2. *Hyperfunction quantum field theory*

In order to formulate locality of quantum fields, we need vector-valued Fourier hyperfunctions. The space $L(\underset{\sim}{\mathcal{O}}(\boldsymbol{D}^n), \mathcal{H})$ of vector-valued Fourier hyperfunctions is defined in Ref.[24] (see also Refs.[30,31]) For a vector-valued Fourier hyperfunction, the support is defined. $L(\underset{\sim}{\mathcal{O}}(K), \mathcal{H})$ is the space of Fourier hyperfunctions with support contained in K. Now we can give the axioms W.I$'$ – W.VII$'$ of HFQFT.

First we replace the test-function space $\mathcal{S}(\mathbb{R}^4)$ of tempered distributions appearing in axioms W.I – W.VII by the test-function space $\underset{\sim}{\mathcal{O}}(\boldsymbol{D}^4)$ of Fourier hyperfunctions. Then it follows from the axiom W.IV that $\phi_j^{(\kappa)}(x)\phi_\ell^{(\kappa')}(y)\Psi$ is a vector-valued Fourier hyperfunction. Only the axiom W.VI needs further modification. Let $V = \{(x,y) \in \mathbb{R}^{4\cdot2}; (x-y)^2 > 0\}$. Then the axiom W.VI of locality has the following form.

W.VI$'$ (Locality). The support of $[\phi_j^{(\kappa)}(x)\phi_\ell^{(\kappa')}(y)\Psi \mp \phi_\ell^{(\kappa')}(y)\phi_j^{(\kappa)}(x)\Psi]$ is contained in the closure $\bar{V}$ of V in $\boldsymbol{D}^{4\cdot2}$, i.e.,

$$\operatorname{supp}[\phi_j^{(\kappa)}(x)\phi_\ell^{(\kappa')}(y)\Psi \mp \phi_\ell^{(\kappa')}(y)\phi_j^{(\kappa)}(x)\Psi] \subset \bar{V}$$
$$\Leftrightarrow \phi_j^{(\kappa)}(x)\phi_\ell^{(\kappa')}(y)\Psi \mp \phi_\ell^{(\kappa')}(y)\phi_j^{(\kappa)}(x)\Psi = 0, \text{ for } (x-y)^2 < 0.$$

The axioms w.1$'$ – w.7$'$ of Wightman hyperfunction are derived from W.I$'$ – W.VII$'$ and are obtained as follows: Replace $\mathcal{S}(\mathbb{R}^{4n})$ in the axioms w.1 – w.7 by $\underset{\sim}{\mathcal{O}}(\boldsymbol{D}^{4n})$. The axiom w.1 needs further modification:

w.1′ (Admissible nature of a singularity and growth). The n-point Wightman function

$$w^{(\kappa_1\dots\kappa_n)}_{j_1\dots j_n}(x_1,\dots,x_n) = \langle 0|\phi^{(\kappa_1)}_{j_1}(x_1)\cdots\phi^{(\kappa_n)}_{j_n}(x_n)|0\rangle$$

is a Fourier hyperfunction.

The equivalence of the axioms W.I′ − W.VII′ of HQFT and the axioms w.1′ − w.7′ is proved in the Theorem 6.1 of Ref.[34], *i.e.*, the modified Theorem 1 holds where the axioms W.I − W.VII and w.1 − w.7 are replaced by W.I′ − W.VII′ and w.1′ − w.7′ respectively.

3.3. *Euclidean reformulation*

Briefly, Fourier hyperfunctions of type II are defined as follows. Let $\boldsymbol{D}^{2n} = \mathbb{C}^n \cup S^{2n-1}_\infty$ be the radial compactification of $\mathbb{C}^n$, where S^{2n-1}_∞ is the sphere at infinity. The test-function space of type II Fourier hyperfunctions is denoted by $\underset{\approx}{\mathcal{O}}(\boldsymbol{D}^n)$. Let K be a closed set in $\boldsymbol{D}^n$. Then the space $\underset{\approx}{\mathcal{O}}(K)$ is defined as the inductive limit of the Banach spaces $\mathcal{O}^m_c(U_m)$:

$$\underset{\approx}{\mathcal{O}}(K) = \text{ind}\lim_{m\to\infty}\mathcal{O}^m_c(U_m),$$

where $\{U_m; m\in\mathbb{N}\}$ is a fundamental sequence of neighborhoods of K in $\boldsymbol{D}^{2n}$, and $\mathcal{O}^m_c(U_m)$ is the Banach space of functions f analytic in $U_m\cap\mathbb{C}^n$ and continuous on $\bar{U}_m\cap\mathbb{C}^n$ such that

$$\|f\|_m = \sup_{z\in U\cap\mathbb{C}^n}|f(z)|e^{|z|/m} < \infty.$$

For $n=1$ and $K=\boldsymbol{D}$, we can choose U_m as follows:

$$U_m\cap\mathbb{C} = \{z\in\mathbb{C}; |\text{Im } z| < (1+|\text{Re } z|)/m\},$$

whereas for $\underset{\sim}{\mathcal{O}}(\boldsymbol{D})$ we can choose

$$U_m\cap\mathbb{C} = \{z\in\mathbb{C}; |\text{Im } z| < 1/m\}.$$

Since $\underset{\approx}{\mathcal{O}}(\boldsymbol{D}^n)\subset\underset{\sim}{\mathcal{O}}(\boldsymbol{D}^n)$, the space $\underset{\approx}{\mathcal{O}}(\boldsymbol{D}^n)'$ of Fourier hyperfunctions of type II is bigger than the space $\underset{\sim}{\mathcal{O}}(\boldsymbol{D}^n)'$ of Fourier hyperfunctions. Fourier hyperfunctions of type II are natural in the context of relating Wightman hyperfunctions to a given set Schwinger functions, since the Schwinger function $S^{(\kappa_1\dots\kappa_n)}_{j_1\dots j_n}(\xi_1,\dots,\xi_{n-1}) = s^{(\kappa_1\dots\kappa_n)}_{j_1\dots j_n}(x_1,\dots,x_n)$ can be analytically continued in the variables ξ^4_j to

$$V_k = \{z\in\mathbb{C}; |\text{Im } z| < k\,\text{Re } z\}$$

$k = 1, 2, \ldots$, and are bounded there. Thus the Wightman hyperfunction $W^{(\kappa_1 \ldots \kappa_n)}_{j_1 \ldots j_n}(\xi'_1, \ldots, \xi'_{n-1})$ is well defined as a boundary value of the holomorphic function

$$W^{(\kappa_1 \ldots \kappa_n)}_{j_1 \ldots j_n}(\zeta'_1, \ldots, \zeta'_{n-1}) = S^{(\kappa_1 \ldots \kappa_n)}_{j_1 \ldots j_n}(\zeta_1, \ldots, \zeta_{n-1}),$$

where $\zeta'_j = (i\zeta^4_j, \zeta_j)$ (see Refs.[35,36]).

Then one can give the axioms w.1'' – w.7'' of Wightman Fourier hyperfunctions of type II by replacing $\mathcal{O}(\boldsymbol{D}^{4n})$ with $\mathcal{O}(\boldsymbol{D}^{4n})$.

Theorem 6: *The axioms e.1, e.2 – e.6 imply a modified axioms w.1'' – w.7'' for hyperfunction quantum fields. Moreover, there are modified axioms e.1'' – e.6'' which are weaker than e.1 – e.6 and equivalent to w.1'' – w.7''.*

4. Local fields without localized test-functions

Take a neutral free scalar field $\phi(x)$ of mass $m \geq 0$ and a sequence of real numbers a_n satisfying

$$\lim_{n \to \infty} \left(\frac{|a_n|^2}{n!} \right)^{1/n} = 0. \tag{1}$$

Then it is shown in Refs.[37,33] that

$$\rho(x) = \sum_{n=0}^{\infty} a_n \frac{: \phi(x)^n :}{n!} \tag{2}$$

is a well defined hyperfunction quantum field in the Fock space of the field ϕ. Introducing the two point function of the field ϕ as

$$\big(\Phi_0, \phi(x)\phi(y)\Phi_0\big) = D^{(-)}_m(x - y)$$

the two point function of the field ρ is

$$\big(\Phi_0, \rho(x)\rho(y)\Phi_0\big) = \sum_{n=0}^{\infty} \frac{|a_n|^2}{n!} [D^{(-)}_m(x - y)]^n.$$

For $m = 0$, the Fourier transform $G_n(p)$ of $[D^{(-)}_0(x)]^n$ for $n \geq 2$ is

$$G_n(p) = (2\pi)^{1-2n} 4^{1-n}(n - 1)^{-1}[(n - 2)!]^{-2}(p^2)^{n-2}\theta(p_0)\theta(p^2)$$

(see Ref.[37]). Introduce the Fourier hyperfunction

$$G(p) = \sum_{n=2}^{\infty} \frac{|a_n|^2}{n!} G_n(p) = \sum_{n=2}^{\infty} c_{n-2}(p^2)^{n-2}\theta(p_0)\theta(p^2),$$

i.e., for $n \geq 2$, we set

$$c_{n-2} = |a_n|^2 / [n!(2\pi)^{2n-1} 4^{n-1}(n-1)\{(n-2)!\}^2]. \tag{3}$$

Let $(L_k)_{k \in \mathbb{N}}$ be an increasing sequence of positive numbers such that $L_0 = 1$ and

$$k \leq L_k, \ L_{k+1} \leq C L_k$$

for some constant C. We denote by $\mathcal{C}^L(\mathbb{R})$ the set of all complex valued functions on the real line which are infinitely often differentiable, *i.e.*, $u \in \mathcal{C}^\infty(\mathbb{R})$, such that for every compact set $K \subset \mathbb{R}$ there is a constant C_K (depending on K in general) satisfying

$$|D^q u(x)| \leq C_K (C_K L_q)^q, \qquad \forall x \in K, \quad q = 0, 1, \dots .$$

Consider the case

$$L_0 = L_1 = 1, L_2 = 2 \text{ and } L_k = k(\log k)^s, \quad (0 < s \leq 1) \text{ for } k \geq 3. \tag{4}$$

Then

$$\sum_{k=0}^{\infty} \frac{1}{L_k} = \infty.$$

According to Theorem 1.3.8 of Ref.[21], for such a sequence, the space $\mathcal{C}^L(\mathbb{R})$ is a space of quasi-analytic functions, that is, if $\phi \in \mathcal{C}^L(\mathbb{R})$ satisfies $\phi^{(j)}(x) = 0$ for every j at some $x \in \mathbb{R}$, then $\phi = 0$ in $\mathbb{R}$. There are no non-trivial functions with compact support in $\mathcal{C}^L(\mathbb{R})$.

The following proposition is crucial. It says: If all the absolute moments of a function $\phi : \mathbb{R} \to \mathbb{C}$ are finite and satisfy a certain growth restriction with respect to the order of the moments, then its Fourier transform is quasi-analytic.

Proposition 3: *Let $(L_k)_{k \in \mathbb{N}}$ be a sequence satisfying condition (4) and let ϕ be a function on $\mathbb{R}$ with the following property: There are constants A and C such that*

$$\int_{-\infty}^{\infty} |x^k \phi(x)| dx \leq C(A L_k)^k \qquad \forall k = 0, 1, 2, \dots .$$

Then the Fourier transform $\psi(\sigma)$ of $\phi(x)$ is quasi-analytic.

This proposition is the technical core of the proof of our main result in this section. It is applied as follows: For a real-valued function $f \in \mathcal{S}(\mathbb{R}^4)$ denote, for each $p \in \mathbb{R}^3$,

$$\tilde{f}_{\boldsymbol{p}}(p_0) = \tilde{f}(p_0, \boldsymbol{p}) = \tilde{f}(p) = (2\pi)^{-2} \int e^{-ip\cdot x} f(x) dx.$$

If $\int \tilde{W}(p) |\tilde{f}(p)|^2 dp$ is convergent, then $|\tilde{f}_{\boldsymbol{p}}(p_0)|^2$ satisfies the basic hypothesis of this proposition

$$\int_{-\infty}^{\infty} |p_0|^n |\tilde{f}_{\boldsymbol{p}}(p_0)|^2 dp_0 \leq M(AL_n)^n$$

for $A = \sqrt{2}$, some $p \in \mathbb{R}^3$ and $n = 0, 1, \ldots$.

Theorem 7: *Let ϕ be a free neutral field of mass $m = 0$ with vacuum state Φ_0 and $(L_n)_{n\in\mathbb{N}}$ a sequence of real numbers satisfying the condition (4). For this sequence define the hyperfunction quantum field ρ by Equation (2) where the coefficients a_n are determined by the elements L_n of this sequence according to equations $c_n = L_n^{-2n}$ and (3). Then this hyperfunction quantum field ρ does not admit real-valued test functions $f \in \mathcal{S}(\mathbb{R}^4)$ of compact support and no complex-valued test functions $h \in \mathcal{S}(\mathbb{R}^4)$ of compact support such that $\rho(h)\Phi_0$ and $\rho(h^*)\Phi_0$ are well defined vectors in the state space of ρ.*

For further details, see Ref.[32]

5. Local observables

Introduce the vector-valued Fourier hyperfunction for an Hermitian field $\Phi(x)$

$$\Phi_n(x_1, \ldots, x_n) = \Phi(x_1) \cdots \Phi(x_n)\Phi_0$$

and denote the Fourier transform of $\Phi_n(x_1, \ldots, x_n)$ by $\tilde{\Phi}_n(p_1, \ldots, p_n)$. Now, in appropriately chosen variables

$$(q_1, \ldots, q_n) = \chi^{-1}(p_1, \ldots, p_n), \quad q_k = \sum_{j=k}^{n} p_j, \quad k = 1, \ldots, n$$

the Fourier hyperfunction $\tilde{\Phi}_n$ has convenient support properties. Consider

$$\tilde{Z}_n = \tilde{\Phi}_n \circ \chi_n.$$

Then we have (see Ref.[8])

$$\operatorname{supp} \tilde{Z}_n(q_1, \ldots, q_n) \subset \bar{V}_+^n.$$

If $\operatorname{Im}\zeta_k \in V_+$, then

$$q \to \exp(i\sum_{k=1}^{n}\langle\zeta_k, q_k\rangle) = \psi_{\zeta_1,\ldots,\zeta_n}(q_1,\ldots,q_n)$$

belongs to the space $\underset{\sim}{\mathcal{O}}(\bar{V}_+^n)$ and therefore

$$Z_n(\zeta_1,\ldots,\zeta_n) = \tilde{Z}_n(\psi_{\zeta_1,\ldots,\zeta_n})$$

is holomorphic in T_+^n, where

$$T_+^n = \left\{(\zeta_1,\ldots,\zeta_n) \in \mathbb{C}^{4n} \; ; \; \operatorname{Im}\zeta_k \in V_+, \text{ for } k = 1,\ldots,n\right\}$$

and

$$\Phi_n(x_1,\ldots,x_n) = Z_n(x_1, x_2 - x_1,\ldots,x_n - x_{n-1}) = Z_n(\xi_1,\ldots,\xi_n). \quad (5)$$

$Z_n(\xi_1,\ldots,\xi_n)$ is the Fourier hyperfunction which is defined as the boundary value of $Z_n(\zeta_1,\ldots,\zeta_n)$ on T_+^n. It is also the Fourier transform of the Fourier hyperfunction $\tilde{Z}_n$ introduced above. It follows that the Fourier hyperfunction (5) is the boundary value of the analytic function

$$\Phi_n(z_1,\ldots,z_n) = Z_n(z_1, z_2 - z_1,\ldots,z_n - z_{n-1}) = Z_n(\zeta_1,\ldots,\zeta_n)$$

in $\{(z_1,\ldots,z_n) \in \mathbb{C}^{4n}; (z_1, z_2 - z_1,\ldots,z_n - z_{n-1}) \in T_+^n\}$. The condition W.V of Lorentz covariance implies for any Lorentz transformation Λ

$$U(\Lambda)\Phi_n(z_1,\ldots,z_n) = \Phi_n(\Lambda z_1,\ldots,\Lambda z_n) = Z_n(\Lambda\zeta_1,\ldots,\Lambda\zeta_n) \quad (6)$$

for all $(z_1,\ldots,z_n)$ such that $(z_1, z_2 - z_1,\ldots,z_n - z_{n-1}) \in T_+^n$. The following theorem is the counter part of the Reeh–Schlieder theorem (Theorem 3) in hyperfunction quantum field theory and plays an important role in the definition of local observable algebras.

Theorem 8: *Let* U_n $(n = 1, 2,\ldots)$ *be open non-empty sets in* $\mathbb{R}^{4n}$ *and* u *a vector of* $\mathcal{H}$. *If* $(u, Z_n(\xi_1,\ldots,\xi_n)) = 0$ *(respectively* $(Z_n(\xi_1,\ldots,\xi_n), u) = 0$*)* *in* U_n *as Fourier hyperfunctions for* $n = 0, 1, 2,\ldots$ *(we use here* $Z_0 = \Phi_0$*),* *then* $u = 0$.

The following theorem says that TCP-Theorem holds also in hyperfunction quantum field theory.

Theorem 9: *In hyperfunction quantum field theory, there is an anti-unitary operator* θ *on* $\mathcal{H}$ *with the properties*

$$\theta\Phi_0 = \Phi_0, \quad \theta\Phi(f_1)\cdots\Phi(f_n)\Phi_0 = \Phi(f_1^-)\cdots\Phi(f_n^-)\Phi_0,$$

where $f_k^-(z) = \overline{f_k(-\bar{z})}$. *This anti-unitary operator θ is called TCP-operator and satisfies*

$$\theta^2 = I, \quad \theta U(a, \Lambda)\theta = U(-a, \Lambda) \quad \text{on } \mathcal{H};$$

$$\theta \Phi(f)\theta = \Phi(f^-) \quad \text{on } \mathcal{D}.$$

Now consider the following 1-parameter subgroup of the Lorentz group

$$v_t = \begin{pmatrix} \cosh t \; \sinh t & & 0 \\ \sinh t \; \cosh t & & \\ & & 1 \; 0 \\ 0 & & 0 \; 1 \end{pmatrix} \in G.$$

Then

$$V(t) = U(v_t)$$

is a 1-parameter group of unitary operators in the Hilbert space $\mathcal{H}$ which has a self-adjoint generator L. Thus, by functional calculus, we have an analytic continuation of $V(t)$ to $V(\tau)$, $\tau \in \mathbb{C}$. Relation (6) implies

$$V(t)Z_n(\zeta_1, \dots, \zeta_n) = Z_n(\zeta_1(t), \dots, \zeta_n(t))$$

where $\zeta(t) = v_t \zeta$, for all $t \in \mathbb{R}$ and all $(\zeta_1, \dots, \zeta_n) \in T_+^n$. Recall that Z_n is holomorphic on T_+^n, and introduce the spatial wedges

$$W_\pm = \{x \in \mathbb{R}^4; \pm x^1 > |x^0|\}.$$

Then we have the following proposition.

Proposition 4: *As identities between analytic functions in*

$$\xi = (\xi_1, \dots, \xi_n) \in W_\pm^n$$

the following holds: Define

$$J = U(0, R)\theta$$

where θ is the CPT-operator of Theorem 9 and R the Euclidean rotation by π around the x^1-axis. If $\Psi \in \operatorname{dom} V(\pm i\pi)$, then

$$(V(\pm i\pi)\Psi, Z_n(\xi_1, \dots, \xi_n)) = (\Psi, JZ_n(\xi_1, \dots, \xi_n)).$$

Denote $\mathcal{D}_\pm = \operatorname{dom} V(\pm i\pi)$ and define

$$\mathcal{A}_\pm = \{X \in \mathcal{B}(\mathcal{H}); \ X\Phi_0 \in \mathcal{D}_\pm\},$$

$$\mathcal{V}(W_\pm) = \{X \in \mathcal{A}_\pm; \ V(\pm i\pi)X\Phi_0 = JX^*\Phi_0\}.$$

Proposition 4 implies:

Proposition 5: *An operator $X \in \mathcal{A}_\pm$ belongs to the set $\mathcal{V}(W_\pm)$, i.e., it satisfies the equation*

$$V(\pm i\pi)X\Phi_0 = JX^*\Phi_0$$

if, and only if, for all $n = 1, 2, \ldots$

$$(X^*\Phi_0, Z_n(\xi_1, \ldots, \xi_n))_{|W_\mp^n} = (Z_n(\xi_1, \ldots, \xi_n), X\Phi_0)_{|W_\mp^n}.$$

For $u, v \in \mathcal{D}_0$, $n = 1, 2, \ldots$, and $X \in \mathcal{B}(\mathcal{H})$ define the Fourier hyperfunction

$$\Psi_{X;n,u,v}(x_1, \ldots, x_n) = (X^*u, \Phi(x_1)\cdots\Phi(x_n)v) - (\Phi(\bar{x}_n)\cdots\Phi(\bar{x}_1)u, Xv).$$

If this Fourier hyperfunction vanishes on an open set O^n, $O \subset \boldsymbol{D}^4$, i.e., if $\Psi_{X;n,u,v}|O^n = 0$, then this means intuitively that the operator X commutes weakly with the field operators localized on O. To any open nonempty set O in $\boldsymbol{D}^4$ we assign the following set of bounded linear operators on the state space $\mathcal{H}$,

$$\mathcal{L}'_w(O) = \{X \in \mathcal{B}(\mathcal{H}); \ \Psi_{X;n,u,v}|O^n = 0, \ \forall u, v \in \mathcal{D}_0, \ \forall n \in \mathbb{N}\}.$$

We study in some detail the sets $\mathcal{L}'_w(W_\mp)$, and relate them in particular to the sets $\mathcal{V}(W_\pm)$. The following result follows from Propositon 5.

Proposition 6: *In a hyperfunction quantum field theory the following relations hold for the spaces introduced above:*

(i) $\mathcal{L}'_w(W_\mp) \cap \mathcal{A}_\pm \subset \mathcal{V}(W_\pm)$.
(ii) $J\mathcal{L}'_w(W_\pm)J = \mathcal{L}'_w(W_\mp)$
(iii) $V(t)\mathcal{L}'_w(W_\pm)V(t)^* = \mathcal{L}'_w(W_\pm)$ *for all $t \in \mathbb{R}$.*

In general, $\mathcal{K}_\pm = \mathcal{L}'_w(O) \cap \mathcal{A}_\pm \cap \mathcal{A}_\pm^*$ is only a *-invariant subspace but not a *-algebra of bounded operators on the state space $\mathcal{H}$. For the special case $O = W_\pm$ we define the *-algebra $\mathcal{M}(W_\pm)$ as follows.

$$\mathcal{M}(W_\pm) = \mathcal{K}_{alg}(W_\pm) = \{C \in \mathcal{K}_\pm; \ C\mathcal{K}_\pm \subset \mathcal{K}_\pm, \ \mathcal{K}_\pm C \subset \mathcal{K}_\pm\}.$$

Then we have the following theorem.

Theorem 10: *If in a hyperfunction quantum field theory the space $\mathcal{M}(W_\pm)\Phi_0$ is dense in the state space $\mathcal{H}$, then the following assignment*

$$W_\pm \longrightarrow \mathcal{M}(W_\pm)$$

is local, i.e.,

$$\mathcal{M}(W_\pm) \subset \mathcal{M}(W_\mp)'.$$

Next we consider the set of all wedge-like regions

$$\mathcal{W} = \{W \subset \mathbb{R}^4; \ W = g \cdot W_+, \ g \in G\} = G \cdot W_+.$$

Assign to the wedge $W \in \mathcal{W}$ the set of operators $\mathcal{M}(W)$ defined by

$$\mathcal{M}(W) = U(g)\mathcal{M}(W_+)U(g)^{-1}, \quad W = g \cdot W_+.$$

Proposition 7: *The assignment of operator *-algebras $\mathcal{M}(W)$ to wedges $W \in \mathcal{W}$ is causal in the sense that*

$$\mathcal{M}(W) \subset \mathcal{M}(W')'$$

for all wedges $W \in \mathcal{W}$.

The following argument is the same as in standard tempered field theory. Recall that a double cone in Minkowski space is the intersection of all wedges that contain it. Accordingly we assign a *-algebra $\mathcal{M}(D)$ of bounded operators on $\mathcal{H}$ to a double cone D and a *-algebra $\mathcal{M}(D')$ to the space-like complement D' of D according to the following formulas:

$$\mathcal{M}(D) = \bigcap_{D \subset W \in \mathcal{W}} \mathcal{M}(W), \quad \mathcal{M}(D') = \{ \bigcup_{D' \supset W \in \mathcal{W}} \mathcal{M}(W)\}''.$$

Then by definition we have

$$\mathcal{M}(D_1) \subset \mathcal{M}(W) \subset \mathcal{M}(D_2')$$

for any triplet (D_1, D_2, W), with $W \in \mathcal{W}$ and where D_1, D_2 are double cones such that $D_1 \subset W \subset D_2'$. Thus, to any net of double cones we can assign a net of local observable algebras:

Theorem 11: *Let a hyperfunction quantum field theory be given. Assume that $\mathcal{M}(W_+)\Phi_0$ is dense in the state space $\mathcal{H}$, then the net $\{\mathcal{M}(D)\}$ assigned to double cones D satisfies the condition of causality (local commutativity)* *in the sense that for any pair of double cones D_1, D_2 with $D_1 \subset D_2'$ the operator algebra $\mathcal{M}(D_1)$ is contained in the commutant of the operator algebra $\mathcal{M}(D_2)$:*

$$\mathcal{M}(D_1) \subset \mathcal{M}(D_2)'.$$

6. Axioms for quantum field theories with fundamental length

6.1. *Ultra-hyperfunctions*

The discussion in the Introduction demonstrates that the framework of standard quantum fields and the framework of hyperfunction quantum fields do not allow a fundamental length in the formulation of the condition of local commutativity. Thus, for such a theory, a new framework has to be used. We argue that quantum fields in terms of ultra-hyperfunctions is a suitable framework. First we give the definition of tempered ultra-hyperfunctions. For a subset A of $\mathbb{R}^n$, we denote by $T(A) = \mathbb{R}^n + iA \subset \mathbb{C}^n$ the tubular set with base A. For a convex compact set K of $\mathbb{R}^n$, $\mathcal{T}_b(T(K))$ is, by definition, the space of all continuous functions f on $T(K)$ which are holomorphic in the interior of $T(K)$ and satisfy

$$\|f\|^{T(K),j} = \sup\{|z^p f(z)|; z \in T(K), |p| \le j\} < \infty, \quad j = 0, 1, \dots$$

where $p = (p_1, \dots, p_n)$ and $z^p = z_1^{p_1} \cdots z_n^{p_n}$. $\mathcal{T}_b(T(K))$ is a Fréchet space with the semi-norms $\|f\|^{T(K),j}$. If $K_1 \subset K_2$ are two compact convex sets, we have the canonical injections:

$$\mathcal{T}_b(T(K_2)) \to \mathcal{T}_b(T(K_1)). \tag{7}$$

Let O be a convex open set in $\mathbb{R}^n$. We define

$$\mathcal{T}(T(O)) = \lim_{\leftarrow} \mathcal{T}_b(T(K_1)),$$

where K_1 runs through the convex compact sets contained in O and the projective limit is taken following the restriction mappings (7).

Definition 1: *A **tempered ultra-hyperfunction** is by definition a continuous linear functional on $\mathcal{T}(T(\mathbb{R}^n))$.*

The gauge functional h_K of a compact convex set $K \subset \mathbb{R}^n$ is defined by

$$h_K(x) = \sup\{\langle x, \xi \rangle; \xi \in K\}.$$

For a convex compact set K of $\mathbb{R}^n$, define $H_b(\mathbb{R}^n; K)$ as the space of all C^∞ functions f on $\mathbb{R}^n$ which satisfy

$$\|f\|_{K,j} = \sup\{\exp(h_K(x))|D^p f(x)|; |p| \le j\} < \infty, \ j = 0, 1, \ldots .$$

$H_b(\mathbb{R}^n; K)$ is a Fréchet space with the semi-norms $\|f\|_{K,j}$. If $K_1 \subset K_2$ are two compact convex sets, we have the canonical injections:

$$H_b(\mathbb{R}^n; K_2) \to H_b(\mathbb{R}^n; K_1). \tag{8}$$

Furthermore, for a convex open set O in $\mathbb{R}^n$ we define

$$H(\mathbb{R}^n; O) = \lim_{\leftarrow} H_b(\mathbb{R}^n; K_1),$$

where K_1 runs through the convex compact sets contained in O and the projective limit is taken following the restriction mappings (8).

Proposition 8: *The Fourier transformation* $f \mapsto \tilde{f} \equiv \mathcal{F}f$,

$$\tilde{f}(p) = (2\pi)^{-n/2} \int_{\mathbb{R}^n} f(z) e^{i\langle p, z \rangle} \, dz$$

is a topological isomorphism between $\mathcal{T}(T(O))$ *and* $H(\mathbb{R}^n; O)$.

In order to be able to formulate the concept of extended locality, we introduce a space $\mathcal{T}(U)$ as follows.

Let U be an open set in $T(A)$ for some compact set A in $\mathbb{R}^n$. Let K_m be the closure in $\mathbb{C}^n$ of the set

$$\{z \in \mathbb{C}^n; \operatorname{dist}(z, U^c) > 1/m\}.$$

For a closed set K_m of $\mathbb{C}^n$, $\mathcal{T}_b(K_m)$ is, by definition, the space of all continuous functions f on K_m which are holomorphic in the interior of K_m and satisfy

$$\|f\|^{K_m,j} = \sup\{|z^p f(z)|; z \in K_m, |p| \le j\} < \infty, \ j = 0, 1, \ldots .$$

$\mathcal{T}_b(K_m)$ is a Fréchet space with the seminorms $\|f\|^{K_m,j}$. If $l < m$, we have the canonical injections:

$$\mathcal{T}_b(K_m) \to \mathcal{T}_b(K_l). \tag{9}$$

Finally, $\mathcal{T}(U)$ is defined as the projective limit of the spaces $\mathcal{T}_b(K_m)$ along the restriction mappings (9)

$$\mathcal{T}(U) = \lim_{\leftarrow} \mathcal{T}_b(K_m).$$

6.2. *Quantum fields with a fundamental length*

Relativistic quantum field theory with a fundamental length is formulated in terms of ultra-hyperfunctions. Here we restrict ourselves to the case of a neutral scalar field and begin in the spirit of our paper[8] by defining such a field as a relativistic quantum field over the test function space $E = \mathcal{T}(T(\mathbb{R}^4))$ for ultra-hyperfunctions. While most conditions only need minor modifications, the condition of local commutativity has to be formulated in such a way as to take the existence of a fundamental length into account.

A **relativistic quantum field theory with fundamental length** l or an **ultra-hyperfunction quantum field theory** is defined through the set of axioms U.I – U.VII. U.I, U.II and U.III have the same form as W.I, W.II and W.III in Section 2 respectively, and U.V and U.VII are obtained by replacing $\mathcal{S}(\mathbb{R}^4)$ of W.V and W.VII in Section 2 respectively with $\mathcal{T}(T(\mathbb{R}^4))$. Thus we only need to formulate conditions U.IV and U.VI.

U.IV The field A is an operator-valued generalized function $A(x)$ over the space $\mathcal{T}(T(\mathbb{R}^4))$ with common dense domain $\mathcal{D}$ for all the operators $A(f)$, *i.e.*, for all $\Psi \in \mathcal{D}$ and all $\Phi \in \mathcal{H}$

$$\mathcal{T}(T(\mathbb{R}^4)) \ni f \to (\Phi, A(f)\Psi) \in \mathbb{C}$$

is a tempered ultra-hyperfunction. It is supposed that the vacuum vector Φ_0 is contained in $\mathcal{D}$ and that $\mathcal{D}$ is taken into itself under the action of the operators $A(f)$ and $U(a, A)$, *i.e.*,

$$A(f)\mathcal{D} \subset \mathcal{D}, \quad U(a, A)\mathcal{D} \subset \mathcal{D}.$$

Moreover it is supposed that $A(\bar{f}) \subset A(f)^*$ for $\bar{f}(z) = \overline{f(\bar{z})}$.

U.VIa The functional

$$\mathcal{T}(T(\mathbb{R}^4)) \times \mathcal{T}(T(\mathbb{R}^4)) \ni f_1 \otimes f_2 \to (\Phi, A(f_1)A(f_2)\Psi)$$

can be extended continuously to $\mathcal{T}(T(L^\ell))$ in some Lorentz frame, for arbitrary elements Φ, Ψ in the domain of the field operators $A(f)$, where

$$T(L^\ell) = \{(z_1, z_2) \in \mathbb{C}^{4\cdot2}; |\text{Im } z_1 - \text{Im } z_2| < \ell\}.$$

U.VIb *Extended local commutativity:* The carrier of the functional

$$(f_1, f_2) \to (\Phi, A(f_1)A(f_2)\Psi) - (\Phi, A(f_2)A(f_1)\Psi)$$

on $\mathcal{T}(T(\mathbb{R}^4)) \times \mathcal{T}(T(\mathbb{R}^4))$ is contained in the set

$$W^\ell = \{(z_1, z_2) \in \mathbb{C}^{4\cdot2}; z_1 - z_2 \in V^\ell\},$$

where

$$V^\ell = \{z \in \mathbb{C}^4; \exists x \in V, |\mathrm{Re}\, z - x| + |\mathrm{Im}\, z| < \ell\}$$

is a complex neighborhood of light cone V, *i.e.*, this functional can be extended continuously to $\mathcal{T}(W^\ell)$.

In the light of our discussion of the elementary example given in Introduction, Axiom U.VIa expresses the fact that, in a quantum field theory with fundamental length ℓ, field operators $A(x_1)$ and $A(x_2)$ at two distinct points x_1 and x_2 can only be distinguished if the distance between the two points x_1 and x_2 is greater than ℓ.

Axiom U.VIb (extended local commutativity) is the counterpart of the locality condition in standard quantum field theory, respectively hyperfunction quantum field theory.

Clearly, the localization properties of tempered ultra-hyperfunction are very different from those of Fourier hyperfunctions and Schwartz distributions, but the spectral condition is not so much different from that of Schwartz distributions because the Fourier transformation of tempered ultra-hyperfunctions are distributions (see Proposition 8).

In analogy to standard quantum field theory we single out a set of properties of these vacuum expectation values which actually characterizes an ultra-hyperfunction quantum field theory up to isomorphisms.

u.1 **Ultra-hyperfunction property**: $\mathcal{W}_0 = 1$, and for all $n = 1, 2, \ldots$,

$$\mathcal{W}_n \in \mathcal{T}(T(\mathbb{R}^{4n}))'.$$

u.2 **Hermiticity**: $\mathcal{W}_n(f^*) = \overline{\mathcal{W}_n(f)}$, for all $f \in \mathcal{T}(T(\mathbb{R}^{4n})) \equiv E(n)$, where $f^*(z_1, \ldots, z_n) = \overline{f(\bar{z}_n, \ldots, \bar{z}_1)}$.

u.3 **Positivity**: For any finite set $f_0, f_1, \ldots, f_N$ of test functions such that $f_0 \in \mathbb{C}$, $f_n \in \mathcal{T}(T(\mathbb{R}^{4n}))$ for $1 \leq n \leq N$, one has

$$\sum_{m,n=0}^{N} \mathcal{W}_{m+n}(f_m^* \otimes f_n) \geq 0.$$

u.4 **Relativistic covariance**: $\mathcal{W}_n(f) = \mathcal{W}_n(f_{(a,\Lambda)})$ for all $(a, \Lambda) \in \mathcal{P}_0$, all $f \in \mathcal{T}(T(\mathbb{R}^{4n}))$, and all $n = 1, 2, \ldots$.

u.5 **Extended locality**: For all $n = 2, 3, \ldots$ and all $i = 1, \ldots, n-1$ denote

$$L_i^\ell = \{x = (x_1, \ldots, x_n) \in \mathbb{R}^{4n}; |x_i - x_{i+1}| < \ell\},$$

$$W_i^\ell = \{(z_1, \ldots, z_n) \in \mathbb{C}^{4n}; z_i - z_{i+1} \in V_\ell\}.$$

Then $\mathcal{W}_n \in \mathcal{T}(T(\mathbb{R}^{4n}))'$ belongs to $\mathcal{T}(T(L_i^\ell))'$ and $\mathcal{W}_n \circ c_i^n$ belongs to $\mathcal{T}(W_i^\ell)'$, where

$$(\mathcal{W}_n \circ c_i^n)(f) = \mathcal{W}_n(c_i^n(f))$$

and

$$c_i^n(f)(x_1, \dots, x_n)$$
$$= f(x_1, \dots, x_i, x_{i+1}, \dots, x_n) - f(x_1, \dots, x_{i+1}, x_i, \dots, x_n).$$

u.6 **Cluster property**: For any space-like vector $a \in \mathbb{R}^4$ and any $g_n \in E(n)$ introduce, for all $\lambda > 0$,

$$g_{n,\lambda}(x_1, \dots, x_n) = g_n(x_1 - \lambda a, \dots, x_n - \lambda a).$$

Then

$$\mathcal{W}_{m+n}(f_m \otimes g_{n,\lambda}) \to \mathcal{W}_m(f_m)\mathcal{W}_n(g_n)$$

as $\lambda \to \infty$ for every $f_m \in E(m)$ and $g_n \in E(n)$.

u.7 **Spectral property**: For the Fourier transform $\tilde{\mathcal{W}}_n \in H(\mathbb{R}^{4n}; \mathbb{R}^{4n})'$ of $\mathcal{W}_n$, there exists $\tilde{W}_{n-1} \in H(\mathbb{R}^{4(n-1)}; \mathbb{R}^{4(n-1)})'$ such that

$$\tilde{\mathcal{W}}_n \circ \chi_n(q_0, \dots, q_{n-1}) = \delta(q_0)\tilde{W}_{n-1}(q_1, \dots, q_{n-1})$$

and $\operatorname{supp} \tilde{W}_{n-1} \subset \Sigma^{n-1}$.

Again we can reconstruct the fields from the vacuum expectation values.

Theorem 12: *To a given sequence $(\mathcal{W}_n)_{n\in\mathbb{N}}$ of tempered ultra-hyperfunctions satisfying the conditions u.1 – u.7, there corresponds a neutral scalar field $A(f)$ which obeys all the axioms U.1 – U.VII and has the given tempered ultra-hyperfunctions as vacuum expectation values. The field A is unique up to isomorphisms.*

6.3. *Models of quantum fields with fundamental length*

Naturally, the important question arises whether there are models of ultra-hyperfunction quantum fields. We prove the existence of such models by verifying the axioms u.1 – u.7 and then apply the reconstruction theorem 12. Here we discuss only the basic ideas. Details are given in Ref.[7]

The model is given by a Wick power series of a free field. Let $(a_n^{(i)})_{n\in\mathbb{N}}$ be a sequence of real numbers and

$$\rho^{(i)}(x) = \sum_{n=0}^{\infty} a_n^{(i)} \frac{: \phi(x)^n :}{n!}.$$

Then we have the following theorem.

Theorem 13: (Theorem A.1 of Ref.[25]) *As a formal power series the following identity holds*

$$\left(\Phi_0, \rho^{(1)}(x_1)\cdots\rho^{(n)}(x_n)\,\Phi_0\right) = \sum_{r_{ij}=0;1\leq i<j\leq n}^{\infty} \frac{A(R)T^R}{R!},$$

where

$$r_{ij} = r_{ji}, \quad r_{ii} = 0,$$

$$A(R) = \prod_{j=1}^{n} a_{R_j}^{(j)}, \quad R_i = \sum_{j=1}^{n} r_{ij}, \quad R! = \prod_{1\leq i<j\leq n} (r_{ij})!,$$

$$T^R = \prod_{1\leq i<j\leq n} (t_{ij})^{r_{ij}}, \quad t_{ij} = \left(\Phi_0, \phi(x_i)\phi(x_j)\,\Phi_0\right) = D_m^{(-)}(x_i - x_j).$$

In particular

$$\left(\Phi_0, \rho^{(i)}(x)\rho^{(i)}(y)\Phi_0\right) = \sum_{n=0}^{\infty} \frac{a_n^{(i)2}}{n!} D_m^{(-)}(x - y)^n$$

with

$$D_m^{(-)}(x) = (2\pi)^{-3} \int_{\mathbb{R}^3} [2\omega(\boldsymbol{k})]^{-1} e^{-i\omega(\boldsymbol{k})x^0} e^{i\boldsymbol{p}\cdot\boldsymbol{x}} d\boldsymbol{k},$$

$k \cdot x = k^0 x^0 - \boldsymbol{k}\cdot\boldsymbol{x}$, *and* $\omega(\boldsymbol{k}) = \sqrt{\boldsymbol{k}^2 + m^2}$.

We apply these results to a special Wick power series. Consider the series

$$\rho(x) =: e^{g\phi(x)^2} := \sum_{n=0}^{\infty} g^n \frac{: \phi(x)^{2n} :}{n!} = \sum_{n=0}^{\infty} g^n \frac{(2n)!}{n!} \frac{: \phi(x)^{2n} :}{(2n)!},$$

then, by Theorem 13,

$$\left(\Phi_0, \rho(x)\rho(y)\Phi_0\right) = \sum_{n=0}^{\infty} \left(g^n \frac{(2n)!}{n!}\right)^2 \frac{1}{(2n)!} D_m^{(-)}(x - y)^{2n}.$$

Since

$$(1-x)^{-\alpha} = 1 + \alpha x + \frac{\alpha(\alpha+1)}{2!}x^2 + \ldots + \frac{\alpha(\alpha+1)\cdots(\alpha+n-1)}{n!}x^n + \ldots,$$

and for $\alpha = 1/2$

$$\frac{\alpha(\alpha+1)\cdots(\alpha+n-1)}{n!} = \frac{(2n)!}{4^n n!}\frac{1}{n!},$$

we have

$$\left(\Phi_0, \rho(x)\rho(y)\Phi_0\right) = [1 - 4g^2 D_m^{(-)}(x-y)^2]^{-1/2}.$$

One shows that the n-point functions also converge as tempered ultra-hyperfunctions. Furthermore the following identity holds:

$$\left(\Phi_0, \rho(x_1)\cdots\rho(x_n)\Phi_0\right) = \sqrt{\det A},$$

where A is the $n \times n$ symmetric matrix whose entries $a_{j,k}$ are given by

$$a_{j,k} = a_{k,j} = -2g D_m^{(-)}(x_j - x_k)$$

for $j < k$ and $a_{j,j} = 1$. This can be seen as follows. The equation

$$(2\pi)^{-1/2}\int e^{itp}e^{-t^2/2}dt = e^{-p^2}$$

can be considered as an equation for two power series of p:

$$(2\pi)^{-1/2}\int \sum_{n=0}^{\infty}[(itp)^n/n!]e^{-t^2/2}dt = \sum_{n=0}^{\infty}(-p^2)^n/n!,$$

and as a formal series we have

$$(2\pi)^{-1/2}\int \sum_{n=0}^{\infty}[: (ith\phi(x))^n : /n!]e^{-t^2/2}dt = \sum_{n=0}^{\infty} : (-(h\phi(x))^2)^n : /n!$$

which can be written as

$$(2\pi)^{-1/2}\int : e^{ith\phi(x)} : e^{-t^2/2}dt =: e^{-(h\phi(x))^2} : .$$

Let $\psi^{(j)}(x) =: e^{it_j h\phi(x)} :$. Then

$$\left(\Phi_0, \psi^{(1)}(x_1)\cdots\psi^{(n)}(x_n)\Phi_0\right) = \exp\left\{\sum_{1 \leq j < k \leq n} -t_j t_k h^2 D_m^{(-)}(x_j - x_k)\right\}$$

(see Corollary 3.2 of Ref.[37]). Let $h = i\sqrt{2g}$. Then $\rho(x) =: e^{-(h\phi(x))^2/2}:$ and we have

$$\left(\Phi_0, \rho(x_1)\cdots\rho(x_n)\Phi_0\right)$$

$$= (2\pi)^{-n/2}\int \exp\left\{\sum_{1\leq j<k\leq n} -t_j t_k h^2 D_m^{(-)}(x_j - x_k) - \sum_{j=1}^{n} t_j^2/2\right\}dt_1\ldots dt_n$$

$$= \sqrt{\det A}.$$

$\det A$ has the form of $1 + P(a_{j,k})$, where $P(a_{j,k})$ is a polynomial in the entries $a_{j,k}$'s.

Now we consider the case of $m = 0$ for simplicity. Then

$$D_0^{(-)}(x) = \lim_{\epsilon\to+0}(2\pi)^{-1}[(x^0 - i\epsilon)^2 - x^2]^{-1},$$

and we can find $\ell > 0$ such that if $\epsilon \geq \ell$

$$|(2\pi)^{-1}[(x^0 - i\epsilon)^2 - x^2]^{-1}| < 1/(2g).$$

From this expression, for a suitable choice of $|\text{Im } z_j|$, we have $|P(a_{j,k})| < 1$ and $\sqrt{\det A}$ defines a tempered ultra-hyperfunctions. This allows to show that the system of n point functions constructed above, satisfies the axioms u.1 – u.7 (see Ref.[7]).

References

1. H. Baumgärtel and M. Wollenberg, *Causal Nets of Operator Algebras.* Akademie Verlag, Berlin, 1992.
2. H.A. Bethe, The electromagnetic shift of energy levels. Phys. Rev. **72** 339, 1947.
3. J.J. Bisognano and E.H. Wichmann, On the duality condition for a Hermitian scalar field. J. Math. Phys. **16** 985–1007, 1975.
4. N.N. Bogolubov, A.A. Logunov and I.T. Todorov, *Introduction to Axiomatic Quantum Field Theory.* W. A. Benjamin, London–Amsterdam–Don Mills, Ontario–Sydney–Tokyo, 1975.
5. N.N. Bogolubov, A.A. Logunov, A.I. Oksak and I.T. Todorov, *General Principles of Quantum Field Theory,* volume 10 of *Mathematical Physics and Applied Mathematics.* Kluwer Academic Publishers, Dordrecht, Boston, London, 1990.
6. M. Born, W. Heisenberg and P. Jordan, Zur Quantenmechanik, II. Z. Phys. **35** 557, 1926.
7. E. Brüning and S. Nagamachi, Relativistic quantum field theory with a fundamental length. Preprint.
8. E. Brüning and S. Nagamachi, Hyperfunction quantum field theory: Basic structural results. J. Math. Phys. **30** 2340–2359, 1989.

9. E. Brüning and S. Nagamachi, Closure of field operators, asymptotic Abelianness and vacuum structure in hyperfunction quantum field theory. J. Math. Phys. **39** 14, 1998.

10. F. Constantinescu and J.G. Taylor, Equivalence between non-localizable and local fields. Commun. Math. Phys. **30** 211–227, 1973.

11. F. Constantinescu and J.G. Taylor, Causality and non-localizable fields. J. Math. Phys. **15** 824–830, 1974.

12. P.A.M. Dirac, The quantum theory of emission and absorption of radiation. Proc. Roy. Soc. **114** 243, 1927.

13. P.A.M. Dirac, The quantum theory of the electron. Proc. Roy. Soc. **117** 610, 1928.

14. P.A.M. Dirac, Quantized singularities in the electromagnetic field. Proc. Roy. Soc. **133** 60, 1931.

15. F.J. Dyson, The radiation theories of Tomonaga, Schwinger and Feynman. Phys. Rev. **75** 486, 1736, 1949.

16. V.Ya Fainberg and M.A. Soloviev, Nonlocalizability and and asymptotical commutativity. Theor. Math. Phys. **93** 1439–1449, 1992.

17. V.Ya Fainberg and M.A. Soloviev, How can local properties be described in field theories without strict locality? Ann. Phys. **113** 421–447, 1978.

18. R.P. Feynman, A relativistic cut-off for classical electrodynamics. Phys. Rev. **74** 939, 1430, 1948.

19. M. Hasumi, Note on the n-dimensional tempered ultra-distributions. Tohoku Math. J. **13** 94–104, 1961.

20. W. Heisenberg, Über quantentheoretische Umdeutung kinematischen und mechanischen Beziehungen, Z. f. Phys. **33** 879, 1925.

21. L. Hörmander *The Analysis of Linear Partial Differential Operators I, Grundlehren der mathematischen Wissenschaften* **256**. Springer-Verlag, Berlin, Heidelberg, New York, Tokyo, 1983.

22. S.S. Horuzhy, *Introduction to Algebraic Quantum Field Theory*, Kluwer Academic Publishers, Dordrecht, Boston, London, 1990.

23. M.Z. Iofa and V.Ya. Fainberg, Wightman formulation for a nonlocalizable field theory. I. Soviet Physics JETP **29**(5) 880–886, 1969.

24. Y. Ito and S. Nagamachi, On the theory of vector valued Fourier hyperfunctions. J. Math. Tokushima Univ. **9** 1–33, 1975.

25. A.M. Jaffe, Entire functions of the free field. Ann. Phys. **32** 127–156, 1965.

26. A.M. Jaffe, High-energy behavior in quantum field theory. I. Strictly localizable fields. Phys. Rev. **158**(5) 1454–1461, 1967.

27. A. Kaneko, *Introduction to Hyperfunctions, Mathematics and Its Applications* (Japanese Series). Kluwer Academic Publishers, Dordrecht, Boston, London, 1988.

28. T. Kawai, On the theory of Fourier hyperfunctions and its applications to partial differential equations with constant coefficients. J. Fac. Sci. Univ. Tokyo, Sect. I.A **17** 467–517, 1970.

29. M. Morimoto, Theory of tempered ultrahyyperfunctions I. Proc. Japan Acad. **51** 81–87, 1975.

30. S. Nagamachi, The theory of vector valued Fourier hyperfunctions of mixed

type. I. Publ. RIMS, Kyoto Univ. **17** 25–63, 1981.

31. S. Nagamachi, The theory of vector valued Fourier hyperfunctions of mixed type. II. Publ. RIMS, Kyoto Univ. **17** 65–93, 1981.

32. S. Nagamachi and E. Brüning, Hyperfunction quantum field theory: Localized fields without localized test functions. to appear in Lett. Math. Phys.

33. S. Nagamachi and E. Brüning, Hyperfunction quantum field theory: Analytic structure, modular aspects, and local observable algebras. J. Math. Phys. **42**(1) 1–31, 2001.

34. S. Nagamachi and N. Mugibayashi, Hyperfunction quantum field theory. Commun. Math. Phys. **46** 119–134, 1976.

35. S. Nagamachi and N. Mugibayashi, Hyperfunction quantum field theory II. Euclidean Greens Functions. Commun. Math. Phys. **49** 257–275, 1976.

36. S. Nagamachi and N. Mugibayashi, Quantum field theory in terms of Fourier hyperfunctions. Publ. RIMS. Kyoto Univ. Supplement **12** 309–341, 1977.

37. S. Nagamachi and N. Mugibayashi, Hyperfunctions and renormalization. J. Math. Phys. **27** 832–839, 1986.

38. K. Osterwalder and R. Schrader, Axioms for Euclidean Greens functions II. Commun. Math. Phys. **42** 281–305, 1975.

39. W. Pauli, Paul Ehrenfest, Naturwissenschaften **21** 841, 1933.

40. M. Planck, Zur Theorie des Gesetzes der Energieverteilung in Normalspektrum. Verh. Deutsch. Phys. Ges. **2** 237, 1900.

41. J. Polchinski, *String Theory I*. Cambridge University Press, Cambridge, 1998.

42. E. Schrödinger, Quantisierung als Eigenwertproblem (Dritte Mitteilung). Ann. d. Phys. **80** 437, 1926.

43. J. Schwinger, Quantum electrodynamics, I. Phys. Rev. **74** 1439, 1948.

44. M.A. Soloviev, Beyond the theory of hyperfunctions, in *Developments in Mathematics. The Moscow School*, V. Arnold, M. Monastyrsky ed.. Chapman/Hall, 1993.

45. M.A. Soloviev, Wick-ordered entire functions of indefinite metric free field. Lett. Math. Phys. **41** 265–277, 1997.

46. M.A. Soloviev, A uniqueness theorem for distributions and its applications to nonlocal quantum field theory. J. Math. Phys. **39** 2635–2642, 1998.

47. R.F. Streater and A.S. Wightman, *PCT, Spin and Statistics, and All That*. Benjamin, New York, 1964.

48. S. Tomonaga, On a relativistically invariant formulation of the quantum theory of wave fields I. Prog. Theoret. Phys. **1** 27, 1946.

49. A.S. Wightman, The choice of test functions in quantum field theory, in *Mathematical Analysis and its Applications, Advances in Mathematical Supplementary Studies* **7**(B) 769–791, L. Nachbin ed.. Academic Press, New York London Toronto Sydney San Francisco, 1981.

50. A.S. Wightman and L. Gårding, Fields as operator-valued distributions in relativistic quantum theory. Arkiv för Fysik **28** 129–184, 1964.

Remarks on the Commutator of Quark Mass Matrices

Makoto Kobayashi

Institute of Particle and Nuclear Studies, KEK
and
Department of Particle and Nuclear Study,
The Graduate University of Advanced Study,
1-1 Oho, Tsukuba, Ibaraki 305-0801, Japan
E-mail: kobayath@post.kek.jp

The commutator of u- and d-type quark mass matrices is discussed. An exact formula for the eigenvalue equation is presented in terms of the physical parameters. Expressions for the mass matrices are also derived with the representation in which the commutator is diagonal.

1. Introduction

The standard model of elementary particles has succeeded in explaining most observed phenomena. Nevertheless, it is thought that the standard model is not the ultimate understanding of the fundamental laws of the microscopic world. One of the reasons for this sentiment is the fact that the standard model contains many parameters, in particular, all of the quark masses and the flavor mixing angles in the quark current of the weak interactions. The standard model cannot predict the values of these parameters from first principles.

As is well known, these masses and mixing angles come from Yukawa couplings of Higgs boson to quarks. Since the gauge interactions of the standard model do not discriminate the generation of quarks, any linear combination of the quark fields of different generations has an equal right as a basic field to describe the Lagrangian. This situation can be expressed as an $SU(3)_L \times SU(3)_U \times SU(3)_D$ transformation of the quark fields, where L, U and D indicate the transformation of the left- handed doublet quarks, the right-handed u-type quarks and the right-handed d- type quarks, respectively. Under the transformation, the quark mass matrices change their shape, but the values of the physical masses and mixing angles do not change.

It seems natural to assume that the generations have their origin in yet unknown mechanisms belonging to a level more fundamental than that of the standard model, and such mechanisms distinguish the generations in contrast to the gauge interactions of the standard model. Then we can expect that there are preferred directions in the $SU(3)_L \times SU(3)_U \times SU(3)_D$ space which single out special representations of the quark mass matrices. It can be also expected that there will be some kind of regularities for the mass matrices in such special representations. With this expectation, various kinds of ansatz on the regularities of the quark mass matrices have been discussed. So far, however, we have not obtained much insight on the origin of the generations from this approach.

In this note, we try to understand the structure of the quark mass matrices from a somewhat different direction. We will pay a close attention to the commutator of up- and down-type quark mass matrices. The commutator carries partial information on the disparity of two matrices which causes the quark mixing. A role of the commutator is already known: Yarlskog pointed out that CP violation in the standard model is proportional to the determinant of the commutator.[1] In the next section, we will fully investigate the $SU(3)_L \times SU(3)_U \times SU(3)_D$ invariant properties of the commutator. An exact expression for the eigenvalue equation of the commutator will be given in terms the quark masses and the mixing parameters. The representation in which the commutator is diagonal will be discussed in the final section. The expressions for the mass matrices in such representations will be given.

2. Commutator of quark mass matrices

In the standard model with a single Higgs doublet, the most general Yukawa coupling of the Higgs boson is given by

$$\mathcal{L}_{\text{Yukawa}} = \sum_{i,j=1}^{3} \{ f_{ij}^{u} \bar{q}_{L}^{i} \tilde{\Phi} u_{R}^{j} + f_{ij}^{d} \bar{q}_{L}^{i} \Phi d_{R}^{j} \} + h.c. \,,$$

where i and j represent the generation of the quarks. The elements of mass matrices of u- and d-type quarks are related to the Yukawa coupling constants as follows

$$U_{ij} = -\frac{1}{\sqrt{2}} f_{ij}^u v ,$$

$$D_{ij} = -\frac{1}{\sqrt{2}} f_{ij}^d v ,$$

where v is a vacuum expectation value of the Higgs boson field.

Under the following redefinition of the quark field with respect to the generation indices

$$q_L' = V_L q_L , \qquad V_L \in \mathrm{SU}(3)_L$$

$$u_R' = V_U u_R , \qquad V_U \in \mathrm{SU}(3)_U$$

$$d_R' = V_D d_R , \qquad V_U \in \mathrm{SU}(3)_D$$

the mass matrices are transformed as

$$U \to V_L U V_U^\dagger ,$$

$$D \to V_L D V_D^\dagger ,$$

In the following, we consider the hermitian matrices defined by

$$\mathcal{U} \equiv UU^\dagger ,$$
$$\mathcal{D} \equiv DD^\dagger ,$$

which are invariant under the $\mathrm{SU}(3)_U \times \mathrm{SU}(3)_D$ transformations.

Since $\mathcal{U}$ and $\mathcal{D}$ are independent matrices in general, we cannot diagonalize them at the same time. In the $\mathcal{U}$-diagonal representation, they can be expressed as

$$\mathcal{U}_u = \begin{pmatrix} u_1 & 0 & 0 \\ 0 & u_2 & 0 \\ 0 & 0 & u_3 \end{pmatrix} = \begin{pmatrix} m_u^2 & 0 & 0 \\ 0 & m_c^2 & 0 \\ 0 & 0 & m_t^2 \end{pmatrix} ,$$

$$\mathcal{D}_u = V \begin{pmatrix} d_1 & 0 & 0 \\ 0 & d_2 & 0 \\ 0 & 0 & d_3 \end{pmatrix} V^\dagger = V \begin{pmatrix} m_d^2 & 0 & 0 \\ 0 & m_s^2 & 0 \\ 0 & 0 & m_b^2 \end{pmatrix} V^\dagger ,$$

where V is a unitary matrix, often called the CKM matrix, which describes the flavor mixing in the charged weak current interactions. The suffixes attached to $\mathcal{U}$ and $\mathcal{D}$ indicate that they are the matrices in a specific representation.

Now we consider the commutator of $\mathcal{U}$ and $\mathcal{D}$, in the generic representation:

$$i\mathcal{C} \equiv [\mathcal{U}, \mathcal{D}] ,$$

where i is attached to make $\mathcal{C}$ hermitian. Since $\mathcal{C}$ is traceless, $SU(3)_L$ invariant properties of $\mathcal{C}$ are characterized by

$$c_2 = \operatorname{tr}\mathcal{C}^2$$

and

$$c_3 = \operatorname{tr}\mathcal{C}^3 = 3\det\mathcal{C}.$$

It is well-known that $\det\mathcal{C}$ is a measure of CP violation in the flavor mixing. If and only if $\det\mathcal{C}$ is non-zero, CP symmetry is violated in the standard model, barring the strong CP violation due to the θ vacuum. In terms of c_2 and c_3, the eigenvalue equation for $\mathcal{C}$ is expressed as

$$x^3 - \frac{1}{2}c_2 x - \frac{1}{3}c_3 = 0.$$

If CP is conserved, one of the eigenvalues is zero, and the non-zero eigenvalues are given by $\pm\sqrt{c_2/2}$.

c_2 and c_3 should be expressed as functions of 10 physical parameters, namely, 6 quark masses and 4 mixing angles. In the rest of this section, we will obtain those functions explicitly. Using the $\mathcal{U}$-diagonal representation, we have

$$c_2 = -2\sum_{i,j,k,l=1}^{3} u_i u_j d_k d_l V_{ik} V_{jl} V_{il}^* V_{jk}^* + 2\sum_{i,k=1}^{3} u_i^2 d_k^2 V_{ik} V_{ik}^*$$

and

$$c_3 = -6\sum_{i,j,k,l=1}^{3} u_i^2 u_j d_k^2 d_l \operatorname{Im}V_{ik} V_{jl} V_{il}^* V_{jk}^*.$$

We note that $\operatorname{Im}V_{ik} V_{jl} V_{il}^* V_{jk}^*$ is antisymmetric with respect to (ij) and (kl).

As for the parameterization of V, we take the following exactly unitary Wolfenstein-type one[2]:

$$V = \begin{pmatrix} D_1 & \lambda & A\lambda^3(\rho - i\eta) \\ -\lambda\dfrac{D_2 + A^2\lambda^4(\rho + i\eta)}{D_1} & D_2 & A\lambda^2 \\ A\lambda^3\dfrac{D_2 - (\rho + i\eta)D_0^2}{D_1 D_3} & -A\lambda^2\dfrac{D_2 + \lambda^2(\rho + i\eta)}{D_3} & D_3 \end{pmatrix},$$

where

$$D_1 = \sqrt{1 - \lambda^2 - A^2\lambda^6(\rho^2 + \eta^2)}\,,$$

$$D_2 = \frac{-A^2\lambda^6\rho + \sqrt{D_1^2 D_3^2 - A^4\lambda^{12}\eta^2}}{1 - A^2\lambda^6(\rho^2 + \eta^2)}\,,$$

$$D_3 = \sqrt{1 - A^2\lambda^4 - A^2\lambda^6(\rho^2 + \eta^2)}\,,$$

$$D_0 = \sqrt{1 - \lambda^2 - A^2\lambda^4 - A^2\lambda^6(\rho^2 + \eta^2)}\,.$$

The usual form of the Wolfenstein parameterization is obtained by expanding D_i's with respect to λ.

In order to obtain explicit forms of c_2 and c_3 as functions of the physical parameters, it is convenient to note that they are invariant under the following replacement of u_i and d_i:

$$u_i \to u_i - u_0\,,$$
$$d_i \to d_i - d_0\,,$$

where u_0 and d_0 are arbitrary numbers. In particular, if we choose

$$u_0 = u_3\,,$$
$$d_0 = d_1\,,$$

the derivation becomes very simple in the above representation of V.

The result for c_3 is already known:

$$c_3 = -6A^2\lambda^6\eta D_2\Delta\,,$$

where

$$\Delta = (u_1 - u_2)(u_2 - u_3)(u_3 - u_1)(d_1 - d_2)(d_2 - d_3)(d_3 - d_1)\,.$$

If CP is conserved, $\eta = 0$ and therefore c_3 vanishes. The following expression for c_2 is newly obtained in the present paper:

$$\begin{aligned}
c_2 = {}& -4(u_1 - u_3)^2(d_2 - d_1)(d_3 - d_1)A^2\lambda^6(\rho^2 + \eta^2) \\
& -4(u_2 - u_3)^2(d_2 - d_1)(d_3 - d_1)D_2^2 A^2\lambda^4 \\
& -4(u_1 - u_3)(u_2 - u_3)(d_2 - d_1)^2 D_2^2\lambda^2 \\
& -4(u_1 - u_3)(u_2 - u_3)(d_3 - d_1)^2 A^4\lambda^{10}(\rho^2 + \eta^2) \\
& -8(u_1 - u_3)(u_2 - u_3)(d_2 - d_1)(d_3 - d_1)D_2 A^2\lambda^6\rho \\
& +2(u_1 - u_3)^2(d_2 - d_1)^2(\lambda^2 - \lambda^4) \\
& +2(u_1 - u_3)^2(d_3 - d_1)^2(A^2\lambda^6(\rho^2 + \eta^2) - A^4\lambda^{12}(\rho^2 + \eta^2)^2) \\
& +2(u_2 - u_3)^2(d_2 - d_1)^2(D_2^2 - D_2^4) \\
& +2(u_2 - u_3)^2(d_3 - d_1)^2(A^2\lambda^4 - A^4\lambda^8)\,.
\end{aligned}$$

So far all the expressions are exact. There is, however, a large hierarchy among the actual values of the quark masses, and also the value of λ is relatively small, $\lambda \simeq 0.22$. We find that the leading contribution to c_2 comes from the last term and we have

$$c_2 \simeq 2u_3^2 d_3^2 A^2 \lambda^4 \,,$$

as a good approximation in evaluating an actual value of c_2.

3. Diagonalization

In the $\mathcal{U}$-diagonal representation, $\mathcal{C}$ can be written as

$$\mathcal{C}_u = \begin{pmatrix} 0 & -i\gamma & i\beta^* \\ i\gamma^* & 0 & -i\alpha \\ -i\beta & i\alpha^* & 0 \end{pmatrix} \,,$$

where

$$\alpha = (u_2 - u_3)\sum_{i=1}^{3} V_{2i} d_i V_{3i}^* \,,$$
$$\beta = (u_3 - u_1)\sum_{i=1}^{3} V_{3i} d_i V_{1i}^* \,,$$
$$\gamma = (u_1 - u_2)\sum_{i=1}^{3} V_{1i} d_i V_{2i}^* \,.$$

Similarly we can consider the $\mathcal{D}$-diagonal representation, in which

$$\mathcal{U}_d = V^\dagger \begin{pmatrix} u_1 & 0 & 0 \\ 0 & u_2 & 0 \\ 0 & 0 & u_3 \end{pmatrix} V \,,$$

$$\mathcal{D}_d = \begin{pmatrix} d_1 & 0 & 0 \\ 0 & d_2 & 0 \\ 0 & 0 & d_3 \end{pmatrix}$$

and

$$\mathcal{C}_d = \begin{pmatrix} 0 & -i\tilde{\gamma} & i\tilde{\beta}^* \\ i\tilde{\gamma}^* & 0 & -i\tilde{\alpha} \\ -i\tilde{\beta} & i\tilde{\alpha}^* & 0 \end{pmatrix} \,,$$

where

$$\tilde{\alpha} = (d_3 - d_2)\sum_{i=1}^{3} V_{i3} u_i V_{i2}^* \,,$$

$$\tilde{\beta} = (d_1 - d_3)\sum_{i=1}^{3} V_{i1} u_i V_{i3}^* \,,$$

$$\tilde{\gamma} = (d_2 - d_1)\sum_{i=1}^{3} V_{i2} u_i V_{i1}^* \,.$$

We note that c_2 and c_3 can be expressed as

$$c_2 = 2(|\alpha|^2 + |\beta|^2 + |\gamma|^2) = 2(|\tilde{\alpha}|^2 + |\tilde{\beta}|^2 + |\tilde{\gamma}|^2)$$

and

$$c_3 = -6\,\mathrm{Im}(\alpha\beta\gamma) = -6\,\mathrm{Im}(\tilde{\alpha}\tilde{\beta}\tilde{\gamma}) \,.$$

These representations are treating either $\mathcal{U}$ or $\mathcal{D}$ in a special manner. However, since there are no fundamental differences between $\mathcal{U}$ and $\mathcal{D}$, it is likely that they look even in a representation which gives some insight on their origin. Thus we are interested in those representations which treat $\mathcal{U}$ and $\mathcal{D}$ on an equal footing. In this sense, it is tempting to investigate the $\mathcal{C}$-diagonal representation as an example of such representations.

In the following, we consider the $\mathcal{C}$-diagonal representation only for the CP conserving case by putting $\eta = 0$. In this case, V is real and orthogonal and all of $\alpha, \beta, \gamma, \tilde{\alpha}, \tilde{\beta}$ and $\tilde{\gamma}$ are also real. We note that there is the following relation when V is orthogonal:

$$\begin{pmatrix} \tilde{\alpha} \\ \tilde{\beta} \\ \tilde{\gamma} \end{pmatrix} = V^\dagger \begin{pmatrix} \alpha \\ \beta \\ \gamma \end{pmatrix} \,.$$

We also note that as long as we apply orthogonal transformations to the $\mathcal{U}$-diagonal representation, $\mathcal{C}$ remains as a pure real and antisymmetric matrix, so that the diagonal elements vanish. This means that the mass matrices become complex in the $\mathcal{C}$-diagonal representation, even if CP is conserved.

A unitary transformation which diagonalizes $\mathcal{U}_u$ can be obtained by solving the equations for the eigenvector of $\mathcal{U}_u$. There is, however, arbitrariness in the choice of the phase factors of the eigenvectors. If we choose the unitary matrix as

$$
\mathcal{W} = \begin{pmatrix}
\dfrac{-\beta\alpha - i\chi\gamma}{\chi\sqrt{2(\alpha^2+\gamma^2)}} & \dfrac{\alpha}{\chi} & \dfrac{-\beta\alpha + i\chi\gamma}{\chi\sqrt{2(\alpha^2+\gamma^2)}} \\[3mm]
\dfrac{\alpha^2+\gamma^2}{\chi\sqrt{2(\alpha^2+\gamma^2)}} & \dfrac{\beta}{\chi} & \dfrac{\alpha^2+\gamma^2}{\chi\sqrt{2(\alpha^2+\gamma^2)}} \\[3mm]
\dfrac{-\beta\gamma + i\chi\alpha}{\chi\sqrt{2(\alpha^2+\gamma^2)}} & \dfrac{\gamma}{\chi} & \dfrac{-\beta\gamma - i\chi\alpha}{\chi\sqrt{2(\alpha^2+\gamma^2)}}
\end{pmatrix}
$$

with

$$
\chi = \sqrt{c_2/2},
$$

then we have

$$
\mathcal{W}^\dagger \mathcal{C}_u \mathcal{W} = \begin{pmatrix}
\chi & 0 & 0 \\
0 & 0 & 0 \\
0 & 0 & -\chi
\end{pmatrix}.
$$

With this choice of the unitary transformation to the $\mathcal{C}$-diagonal representation, $\mathcal{U}_u$ is transformed as

$$
\mathcal{U}_c = \mathcal{W}^\dagger \mathcal{U}_u \mathcal{W}.
$$

The expressions for the elements of $\mathcal{U}_c$ are slightly complicated, but calculation is straightforward and we have

$$
(\mathcal{U}_c)_{11} = (\mathcal{U}_c)_{33}
$$
$$
= u_1 \frac{\alpha^2\beta^2 + \chi^2\gamma^2}{2\chi^2(\alpha^2+\gamma^2)} + u_2 \frac{(\alpha^2+\gamma^2)^2}{2\chi^2(\alpha^2+\gamma^2)} + u_3 \frac{\gamma^2\beta^2 + \chi^2\alpha^2}{2\chi^2(\alpha^2+\gamma^2)},
$$
$$
(\mathcal{U}_c)_{22} = u_1 \frac{\alpha^2}{\chi^2} + u_2 \frac{\beta^2}{\chi^2} + u_3 \frac{\gamma^2}{\chi^2},
$$
$$
(\mathcal{U}_c)_{21} = (\mathcal{U}_c)_{12}^* = (\mathcal{U}_c)_{32} = (\mathcal{U}_c)_{23}^*
$$
$$
= u_1 \frac{-\alpha^2\beta - i\chi\alpha\gamma}{\chi^2\sqrt{2(\alpha^2+\gamma^2)}} + u_2 \frac{\beta(\alpha^2+\gamma^2)}{\chi^2\sqrt{2(\alpha^2+\gamma^2)}} + u_3 \frac{-\gamma^2\beta + i\chi\alpha\gamma}{\chi^2\sqrt{2(\alpha^2+\gamma^2)}},
$$
$$
(\mathcal{U}_c)_{31} = (\mathcal{U}_c)_{13}^*
$$
$$
= u_1 \frac{(-\alpha\beta - i\chi\gamma)^2}{2\chi^2(\alpha^2+\gamma^2)} + u_2 \frac{(\alpha^2+\gamma^2)^2}{2\chi^2(\alpha^2+\gamma^2)} + u_3 \frac{(-\gamma\beta + i\chi\alpha)^2}{2\chi^2(\alpha^2+\gamma^2)}.
$$

The calculation of $\mathcal{D}_c$ is more involved. $\mathcal{D}_c$ can be written as

$$
\mathcal{D}_c = \mathcal{W}^\dagger \mathcal{D}_u \mathcal{W} = \tilde{\mathcal{W}}^\dagger \mathcal{D}_d \tilde{\mathcal{W}},
$$

where we defined $\tilde{\mathcal{W}}$ as

$$\tilde{\mathcal{W}} \equiv V^\dagger \mathcal{W}.$$

Since $\tilde{\mathcal{W}}$ diagonalizes $\mathcal{C}_d$ which is the commutator in the $\mathcal{D}$-diagonal representation, each column of $\tilde{\mathcal{W}}$ must be a eigenvector of $\mathcal{C}_d$. However the choice of the phase factors of the eigenvectors is no longer arbitrary. After some calculations, we have

$$\tilde{\mathcal{W}} = \begin{pmatrix} e^{i\theta} \dfrac{-\tilde{\beta}\tilde{\alpha} - i\chi\tilde{\gamma}}{\chi\sqrt{2(\tilde{\alpha}^2 + \tilde{\gamma}^2)}} & \dfrac{\tilde{\alpha}}{\chi} & e^{-i\theta} \dfrac{-\tilde{\beta}\tilde{\alpha} + i\chi\tilde{\gamma}}{\chi\sqrt{2(\tilde{\alpha}^2 + \tilde{\gamma}^2)}} \\[2em] e^{i\theta} \dfrac{\tilde{\alpha}^2 + \tilde{\gamma}^2}{\chi\sqrt{2(\tilde{\alpha}^2 + \tilde{\gamma}^2)}} & \dfrac{\tilde{\beta}}{\chi} & e^{-i\theta} \dfrac{\tilde{\alpha}^2 + \tilde{\gamma}^2}{\chi\sqrt{2(\tilde{\alpha}^2 + \tilde{\gamma}^2)}} \\[2em] e^{i\theta} \dfrac{-\tilde{\beta}\tilde{\gamma} + i\chi\tilde{\alpha}}{\chi\sqrt{2(\tilde{\alpha}^2 + \tilde{\gamma}^2)}} & \dfrac{\tilde{\gamma}}{\chi} & e^{-i\theta} \dfrac{-\tilde{\beta}\tilde{\gamma} - i\chi\tilde{\alpha}}{\chi\sqrt{2(\tilde{\alpha}^2 + \tilde{\gamma}^2)}} \end{pmatrix},$$

where the phase factor θ satisfies

$$\sin\theta = \frac{(\alpha V_{32} + \gamma V_{12})\chi}{\sqrt{(\alpha^2 + \gamma^2)(\tilde{\alpha}^2 + \tilde{\gamma}^2)}}.$$

Now we can calculate the elements of $\mathcal{D}_c$ with the use of $\tilde{\mathcal{W}}$:

$$(\mathcal{D}_c)_{11} = (\mathcal{D}_c)_{33}$$

$$= d_1 \frac{\tilde{\alpha}^2\tilde{\beta}^2 + \chi^2\tilde{\gamma}^2}{2\chi^2(\tilde{\alpha}^2 + \tilde{\gamma}^2)} + d_2 \frac{(\tilde{\alpha}^2 + \tilde{\gamma}^2)^2}{2\chi^2(\tilde{\alpha}^2 + \tilde{\gamma}^2)} + d_3 \frac{\tilde{\gamma}^2\tilde{\beta}^2 + \chi^2\tilde{\alpha}^2}{2\chi^2(\tilde{\alpha}^2 + \tilde{\gamma}^2)},$$

$$(\mathcal{D}_c)_{22} = d_1 \frac{\tilde{\alpha}^2}{\chi^2} + d_2 \frac{\tilde{\beta}^2}{\chi^2} + d_3 \frac{\tilde{\gamma}^2}{\chi^2},$$

$$(\mathcal{D}_c)_{21} = (\mathcal{D}_c)_{12}^* = (\mathcal{D}_c)_{32} = (\mathcal{D}_c)_{23}^*$$

$$= e^{i\theta} \left(d_1 \frac{-\tilde{\alpha}^2\tilde{\beta} - i\chi\tilde{\alpha}\tilde{\gamma}}{\chi^2\sqrt{2(\tilde{\alpha}^2 + \tilde{\gamma}^2)}} + d_2 \frac{\tilde{\beta}(\tilde{\alpha}^2 + \tilde{\gamma}^2)}{\chi^2\sqrt{2(\tilde{\alpha}^2 + \tilde{\gamma}^2)}} + d_3 \frac{-\tilde{\gamma}^2\tilde{\beta} + i\chi\tilde{\alpha}\tilde{\gamma}}{\chi^2\sqrt{2(\tilde{\alpha}^2 + \tilde{\gamma}^2)}} \right),$$

$$(\mathcal{D}_c)_{31} = (\mathcal{D}_c)_{13}^*$$

$$= e^{2i\theta} \left(d_1 \frac{(-\tilde{\alpha}\tilde{\beta} - i\chi\tilde{\gamma})^2}{2\chi^2(\tilde{\alpha}^2 + \tilde{\gamma}^2)} + d_2 \frac{(\tilde{\alpha}^2 + \tilde{\gamma}^2)^2}{2\chi^2(\tilde{\alpha}^2 + \tilde{\gamma}^2)} + d_3 \frac{(-\tilde{\gamma}\tilde{\beta} + i\chi\tilde{\alpha})^2}{2\chi^2(\tilde{\alpha}^2 + \tilde{\gamma}^2)} \right).$$

The above expressions are exact but too complicated to see the structure of the mass matrices. The following very crude approximation may be useful to have some idea about the $\mathcal{C}$-diagonal representation: If we put $u_1 = u_2 = d_1 = d_2 = 0$, which reflects more or less the physical situation, the mass matrices look like

$$\mathcal{U}_c \simeq \begin{pmatrix} \frac{u_3}{2} & 0 & -\frac{u_3}{2} \\ 0 & 0 & 0 \\ -\frac{u_3}{2} & 0 & \frac{u_3}{2} \end{pmatrix}$$

and

$$\mathcal{D}_c \simeq \begin{pmatrix} \frac{d_3}{2} & 0 & -e^{-2i\theta}\frac{d_3}{2} \\ 0 & 0 & 0 \\ -e^{2i\theta}\frac{d_3}{2} & 0 & \frac{d_3}{2} \end{pmatrix} .$$

In the present approximation, θ satisfies

$$\sin\theta \simeq A\lambda^2 \sqrt{1 + \lambda^2\rho^2} .$$

The fact that phase factors appear only in $\mathcal{D}_c$ is a consequence of the present choice of the phase convention in $\mathcal{W}$. The phase factors will change in a different choice of the phase convention, but we cannot eliminate the phases from both $\mathcal{U}_c$ and $\mathcal{D}_c$.

As we have noticed, the mass matrices are inevitably complex in the $\mathcal{C}$-diagonal representation, even if CP is conserved. It seems important, therefore, to extend this analysis to the CP violating case, because it is an interesting question how CP violation behaves in a necessarily complex representation of the mass matrices. It is also interesting to incorporate the lepton sector into this kind of analysis. It is hoped that increasing experimental information on the masses and the mixing angles of the neutrinos will provide some hints on the origin of the generations.

Acknowledgments

I would like to express my gratitude to Prof. Hiroshi Ezawa for continuous encouragement. I learned from him the pleasure of physics, in particular in my graduate school days.

References

1. C. Jarlskog, *Phys. Rev. Letters* **55**, 1039 (1985).
2. M. Kobayashi, *Progr. Theor. Phys.* **92**, 287 (1994).

Probing Extra Dimensions with Neutrino Oscillations

C.S. Lam

Department of Physics, McGill University
3600 University St., Montreal, Q.C., Canada H3A 2T8
E-mail: Lam@physics.mcgill.ca

In the braneworld scenario, gravity and neutrino oscillation can both be used to detect the presence of an extra dimension. We argue that neutrino oscillation is particularly suitable if the size of the extra dimension is small, in which case the signature for the extra dimension is the disappearance of active neutrino fluxes into the bulk, caused by the destructive interference from the Kaluza–Klein states. We discuss a class of models to illustrate this general feature.

1. Introduction

This article is dedicated to Prof. Hiroshi Ezawa on the occasion of his seventieth birthday. I met Hiroshi in the early 1960's, at the University of Maryland. Right away it was clear that I could learn much from him, both in physics and in mathematics. What I did not realize until later was his administrative talent and his superb quality in leadership, both amply demonstrated in his illustrious career. I would like to take this opportunity to wish Hiroshi a happy birthday, a happy retirement, and many many happy returns.

The possible existence of extra (spatial) dimensions beyond our three was first suggested by Theodor Kaluza in 1919, later modified by Oskar Klein in 1926. When superstring came along, consistency requires it to live in six extra dimensions. Unfortunately, there is no experimental evidence to date for the presence of these extra dimensions. That may be due to the smallness of the extra dimensions, too small even for the largest accelerators to see. The lack of Kaluza–Klein excited states up to about 1 TeV places an upper bound on the size of the extra dimensions to be about 10^{-19} m.

Inspired by the discovery of higher-dimensional D-branes in non-perturbative string theories,[1] where open strings are trapped, a *braneworld scenario* was proposed[2] in which the Standard-Model (SM) particles are confined to our three dimensional world, called a 3-brane. Only SM singlets such as gravitons and right-handed neutrinos may leave our world to roam in the extra-dimensional bulk. In that scenario, SM particles have no Kaluza–Klein (KK) excited states simply because they cannot get into the bulk, so the upper bound of 10^{-19} m placed on the size of the extra dimensions is no longer valid. In order to probe the presence and the property of the extra dimensions, we must use either gravity, or the right-handed neutrinos ν_R.

A deviation from the inverse-square law of gravity can be used to detect the presence of an extra dimension. In a $3+n$ dimensional world, the surface area of a sphere of radius r is proportional to r^{2+n}. Hence the gravitational force between two masses m_1 and m_2 is given by $G_n m_1 m_2 / r^{2+n}$, where G_n is the Newtonian gravitational constant in n extra dimensions. This is the case when r is less than the size R of the extra dimensions. Otherwise, the sphere is squashed in the extra dimensions to a size R, so the surface area of the squashed object is now proportional to $r^2 R^n$. The resulting gravitational force $G_n(\kappa/R^n)/r^2$ once again obeys the inverse-square law, where κ is a computable geometrical factor. If a deviation from the inverse-square law is detected experimentally at $r < R$, then R marks the size of the extra dimension. Present experiments found no deviation down to about 0.2 mm,[3] so R must be smaller than that. However, it could be as large as 0.1 mm. If so, and if there are at least two extra dimensions, then $G_n = G_0 R_n/\kappa$ is large enough for strong gravitational effects to be seen at TeV energies. This possibility led to a lot of excitement and many papers.

What if the size of the extra dimension is much smaller than 0.1 mm, or, there is only one extra dimension with such a large size? In that case gravity remains weak, and it is powerless to yield any information in the near future on extra dimensions.

What about the other probe, the right-handed neutrinos ν_R? They are assumed to be absent in the SM. In any case, they are SM singlets, hence *sterile*, in the sense that they experience none of the SM forces. How can they be detected even when they are present?

The answer is 'mass'. If ν_R's exist, they would probably produce neutrino masses through the Dirac-mass coupling $\overline{\nu_R}\nu_L$ with the left-handed neutrinos ν_L. A good indication of the presence of a right-handed neutrino is therefore the presence of a neutrino mass.

No mass has been detected in the tritium β-decay experiments. This places an upper bound of 2.2 eV on the mass[4] of the electron–antineutrino, $\bar{\nu}_e$. However, the pioneering neutrino experiments led by Davis and by Koshiba, the beautiful data coming from Super-Kamiokande, SNO, and KamLAND,[5] support an oscillation explanation for the missing neutrinos they detected. It demands at least two of three neutrinos to have non-zero masses. Specifically, if M_1, M_2, M_3 are the masses of the three mass eigenstates, then $\Delta M_\odot \equiv M_2^2 - M_1^2 \simeq (7.5 \times 10^{-3} \text{ eV})^2$, and $\Delta M_{atm}^2 \equiv |M_3^2 - M_2^2| \simeq (50 \times 10^{-3} \text{ eV})^2$. There are also astrophysical evidence suggesting the neutrino masses to be bounded above by 0.23 eV, if they are degenerate.[6]

Unlike the quarks, oscillation experiments discover that neutrinos mix strongly among themselves. Three rotation angles and one phase angle are needed to describe the mixing of three left-handed fermions. In the case of quarks, all three rotation angles are small. In the case of neutrinos, two of them (θ_{12} and θ_{23}) are large and one of them (θ_{13}) is small.

Now that we know the neutrinos have a mass, we shall assume the right-handed neutrinos to exist. Where do we find them? Neutrino mixing is large but quark mixing is small; neutrino masses are small but quark masses are at least a million times larger. These differences suggest that neutrinos are quite different from the quarks. Since the left-handed quarks and the left-handed neutrinos behave in much the same way under the SM, that indicates the right-handed quarks and the right-handed neutrinos are very different. Right-handed quarks are found at the SM energies, this difference may be telling us that the right-handed neutrinos should be found elsewhere.

Where? The popular scenario is to assume the right-handed neutrinos to live at very high energies. Through the seesaw mechanism,[7] this scenario explains the smallness of the neutrino mass, though extra assumptions are needed to explain the large mixing of neutrinos this way.

With the braneworld scenario, there is another possibility. Quarks are confined to the 3-brane we live in, but ν_R's are free to roam in the bulk. That distinction might give rise to the qualitative difference between quarks and neutrinos. In the rest of this article, we shall examine this possibility more closely.

Each neutrino in the bulk yields an infinite tower of KK neutrinos in four dimensions, which we shall refer to as the *bulk neutrinos*, or the bulk states. These neutrinos are non-chiral, containing both the left-handed and the right-handed components. Since they originate from the bulk, they are sterile.

To proceed further, let us keep two general questions in mind. First, can the braneworld scenario explain the difference between quarks and neutrinos in a natural way? That is important because that is the *raison d'être* for going into extra dimensions. Second, do the data on neutrino oscillations even allow the extra-dimensions to exist? The second question is relevant because solar and atmospheric neutrino data demand the mixing with sterile neutrinos to be small, but in the braneworld scenario all the bulk neutrinos are necessarily sterile.

The answer to these questions depends to some extent on the size R of the extra dimension. In most of the recent literature,[8] the size is assumed to be large. A size of $R = 0.1$ mm corresponds to a characteristic energy of 2×10^{-3} eV, putting it in the right ball park of the neutrino masses. Using perturbation theory, a *weak* coupling between the left-handed brane (the SM) neutrinos and the right-handed bulk neutrinos can be shown to produce a small neutrino mass as well as a small mixing with the sterile neutrinos. It however does not explain the large neutrino mixing in a natural way, though that can be arranged.

What if the size of the extra dimensions is much smaller than 0.1 mm? Then the weak-coupling assumption explains nothing, so neutrino oscillation is no longer a useful probe for extra dimensions. Neither is gravity in that case. We are then back to the unenviable position of having no way to tell the presence or absence of an extra dimension.

That however is based on the assumption that the brane-bulk coupling is weak. We shall argue that it is more natural to expect that coupling to be *strong*, not weak. In that case, things are quite different, and neutrino oscillation can provide useful information even when the extra dimensions are small.

Strong coupling with the bulk may induce a large mixing between the brane neutrinos, thereby explaining naturally why neutrino mixing can be large while quark mixing is small. This answers the first question posted before in the affirmative. But why then is one of the three neutrino mixing angles small, instead of being all large? We shall show that the smallness of that particular mixing angle is intimately related to the smallness of the mass-gap ratio $\Delta M_\odot^2 / \Delta M_{atm}^2$.

What about the second question? With a strong coupling surely we expect a large amount of sterile neutrinos showing up in the the solar and atmospheric data, contrary to observation. Is there any way out of this fatal problem?

There is. The details will be discussed in the rest of this article, but let us summarize here what is involved.

The trick is to introduce an extra sterile neutrino on the brane, so that most of the mixings are between the sterile brane and bulk neutrinos, and not between the active and sterile neutrinos. In the strong coupling limit, the introduction of this sterile brane neutrino is *forced on us* by the dynamics; it is not arbitrary and artificial. This is so because one of the brane neutrinos is always absorbed by the KK tower of bulk neutrinos in the strong coupling limit. In other words, if we start with f flavor neutrinos in the brane, in the strong coupling limit there are only $f-1$ mass eigenstates left on the brane.

Therefore, in order to have two separate mass gaps needed to explain the solar and atmospheric neutrino experiments, we need to have three mass eigenstates in the brane, and hence four flavor states to start out with. We know from the Z^0 width that there are only three active neutrinos, ν_e, ν_μ, and ν_τ, so the fourth one must be sterile. We shall denote it by ν_s.

With the extra dimensions small, the small mass of these neutrinos can no longer be explained in the usual way,[8] so seesaw or some other mechanism must be invoked. In what follows we shall not ask the origin of these small masses, we simply introduce four parameters m_a to describe the Majorana masses of the flavor neutrinos on the brane.

All masses from now on are understood to be measured in some unit U, so the parameters m_a and later on d_a are dimensionless.

In addition to the four brane neutrinos, a minimal model would consist of a single massless bulk neutrino in five spacetime dimensions. It decomposes into an infinite tower of four-dimensional KK neutrinos, to be called the *bulk neutrinos*, or bulk states. With U properly chosen, the spectrum of this infinite tower can be taken to be the set of all integers.

A Dirac mass coupling of the type $d\overline{\nu_R}\nu_L + h.c.$ is introduced to couple the left-handed brane neutrinos ν_L to the right-handed bulk neutrinos ν_R. Since the KK tower comes from a single neutrino in the bulk, there is only one coupling constant d_a per brane neutrino.

If the Dirac masses d_a are comparable to the Dirac mass of any of the charged fermions, and if the Majorana masses m_a are comparable to the neutrino masses, then d_a is larger than m_b by more than a million times. Defining the overall coupling strength to be $d^2 = \sum_{a=1}^4 d_a^2$, and letting $e_a = d_a/d$, it is therefore likely that Nature is operating in the strong-coupling regime, where $d \gg m_a, e_a, 1$.

In summary, there are eight real parameters in the minimal theory. Four

m_a's, and four d_a's. The flavor neutrinos are the states when $d = 0$; the mass eigenstates in the strong-coupling limit are the neutrinos when $d \to \infty$.

To get the mass eigenstates for $d \neq 0$, we have to diagonalize an infinite dimensional matrix, whose rows and columns are labeled by the four flavor neutrinos on the brane, and the infinite number of flavor neutrinos of the bulk. The neat thing is that the eigenvalues and the eigenvectors of this infinite dimensional matrix take on a very simple form in the strong coupling limit.

As mentioned before, one brane neutrino is absorbed into the KK tower of bulk neutrinos. Taking this into account, the final mass spectrum in the strong coupling limit is as follows. The whole bulk spectrum is rigidly shifted by half a unit, so that the masses are now half integers. The three mass eigenvalues M_1, M_2, M_3 on the brane are sandwiched between the four Majorana masses m_a of the flavor neutrinos. Namely, if we order the parameters according to $m_1 < m_2 < m_3 < m_4$, and $M_1 < M_2 < M_3$, then $m_1 < M_1 < m_2 < M_2 < m_3 < M_3 < m_4$. We shall refer to this inequality as the *ordering relation*. It is a crucial feature of the strong coupling model.

In the strong coupling limit, we are left with seven free parameters, four m_a's and three independent e_a's (because $\sum_{a=1}^{4} e_a^2 = 1$). It turns out that we can replace them by the four m_a's and the three M_i's, *provided* the ordering relation is maintained. Technically this is quite important because it is much simpler to deal with the latter set of parameters than the former set.

The eigenvalues, and hence the unitary overlapping matrix between the flavor and the mass eigenstates, can also be worked out.

One can then compute the probability amplitude of a flavor neutrino ν_a oscillating into a flavor neutrino ν_b, after traversing a distance L. To do so, we must decompose the flavor neutrino ν_a into a linear combination of the eigenstates, because it is these normal states that propagate with a definite frequency. After a distance L, the mass eigenstates must all be converted back into the flavor state ν_b to get the probability amplitude.

When d is large, the flavor states on the brane have only a tiny overlap with each of the bulk eigenstates. Nevertheless, since there are an infinite number of bulk eigenstates that a strongly coupled flavor brane neutrino can reach into, the total effect of the bulk is not negligible. The contribution from the infinite number of bulk states destructively interfere with one another, so completely that any active neutrino that oscillates into the bulk will not be able to come back. In other words, the bulk acts like an absorber to the active neutrinos in the brane.

Oscillations that go through the three brane eigenstates act just like ordinary oscillations without the presence of extra dimensions.

In other words, the effect of the extra dimensions is to cause part of the oscillating flux of active neutrinos to be lost in the bulk. This then is the signature of the presence of an extra dimension that we should look for.

We can now describe the physical significance of the seven parameters in the theory. M_1, M_2, M_3 are the eigenmasses of the three active brane neutrinos. Since the mass is actually $M_i U$, it is only these products that can be determined experimentally. Different values of U corresponds to different size R of the extra dimension, because U was chosen to make the flavor mass of the bulk neutrinos to be an integer. Since U alone cannot be determined from the experiment, neither can R.

This is not to say that we can use this method to probe an extra dimension no matter how small it is. The strong-coupling requirement $d \gg 1$ places a limit how small $R \sim U^{-1}$ can be, if we assume the Dirac mass $dU \sim d/R$ to be comparable to the Dirac mass of the charged fermions. The precise value of course depends on what we use for d. Let us illustrate it with two extremes. If d/R is the electron mass 0.5 MeV, then we need to have $R \gg 4 \times 10^{-13}$ m. If it is the top quark mass 175 GeV, then we can go down to an $R \gg 10^{-18}$ m.

The other four parameters, m_1, m_2, m_3, m_4 can be used to fit the three mixing angles, and the amount of absorption by the bulk. Note that there is only one free parameter to describe the potential absorption for any ν_a oscillating into any ν_b, so there are predictions that can be potentially falsified.

No absorption has been detected in the present data. Refined and precise data in the future may. It that happens, it is a good indication that an extra dimension exists.

In the minimal medel, the absence of absorption can be achieved by letting $m_4 \to \infty$. In that limit the mixing between the active neutrinos ν_e, ν_μ, ν_τ and the sterile neutrino ν_s also disappears. Then we revert to a situation indistinguishable from the case without extra dimensions. In principle, there is actually a way that might tell them apart, because the three parameters m_1, m_2, m_3 in the minimal model are constrained by the neutrino masses M_1 and M_2 through the ordering relation. As such we may not be able to use them to fit the three experimentally measured mixing angles. In reality these constraints are fulfilled in the fit, so these two cases become indistinguishable.

When M_1 approaches M_2, the ordering constraint requires $M_1 = m_2 =$

M_2. This pinching of m_2 implies $\theta_{13} = 0$. So in this model, one obtains the interesting prediction that $\Delta M_\odot^2 = 0$ implies $\theta_{13} = 0$. The smallness of θ_{13} is then related to the smallness of the mass gap ratio $\Delta M_\odot^2 / \Delta M_{atm}^2$.

This minimal model in five spacetime dimensions can be substantially generalized without changing any of the crucial features discussed above.

These descriptions and conclusions will be put into mathematical formulas in the next few sections. In Sec. 2, the mass matrix, its eigenvalue, and its eigenvectors of the minimal model are examined. They are then used to compute the oscillating amplitude in Sec. 3. Generalization beyond the minimal model will be discussed in Sec. 4.

2. The minimal model and its solution

The mathematical solution of the minimal model will be sketched here. For more details, please consult Refs.[9,10]

This model contains four flavor neutrinos in the brane, and a single massless flavor neutrino in a five dimensional spacetime. The latter decomposes into a KK tower of bulk neutrinos with integer masses, and the former are each given a Majorana mass m_a ($a = 1, 2, 3, 4$). The brane neutrinos are coupled to the bulk neutrinos by a Dirac-mass coupling, with coupling strengths d_a. We assume all masses to be expressed in some common unit U, so that the eight real parameters m_a and d_a are dimensionless. Direct coupling between the brane neutrinos is assumed to be absent, and CP violation is ignored in this simple model.

The symmetric mass matrix of this model is

$$\mathcal{M} = \begin{pmatrix} m & D \\ D^T & B \end{pmatrix}, \tag{1}$$

where $m = \mathrm{diag}(m_1, m_2, m_3, m_4)$ is the 4×4 mass matrix of the brane neutrinos, and $B = \mathrm{diag}(0, +1, -1, +2, -2, +3, -3, \cdots)$ is the infinite dimensional mass matrix of the bulk neutrinos. The coupling between the two is supplied by D, a $4 \times \infty$ matrix in which every element of the ith row is equal to d_a.

The eigenvalue equation for the mass matrix is

$$\mathcal{M} \begin{pmatrix} w \\ v \end{pmatrix} = \lambda \begin{pmatrix} w \\ v \end{pmatrix}, \tag{2}$$

where w is a 4-dimensional column vector with components w_a, and v is an ∞-dimensional column vector with components v_n. λ is the mass eigenvalue.

In component form, (2) reads

$$m_a w_a + d_a A = \lambda w_a, \tag{3}$$

$$b + (Bv)_n = \lambda v_n, \tag{4}$$

where

$$A = \sum_n v_n,$$

$$b = \sum_{a=1}^{4} d_a w_a. \tag{5}$$

We shall choose the normalization of the eigenvectors by setting $b = 1$.

The eigenvector components can be solved from (3) and (4) to be

$$v_n = \frac{1}{\lambda - n},$$

$$w_a = A \frac{d_a}{\lambda - m_a} = (Ad) \frac{e_a}{\lambda - m_a}, \tag{6}$$

where

$$d^2 = \sum_{a=1}^{4} d_a^2,$$

$$e_a \equiv d_a / d, \qquad \Rightarrow$$

$$1 = \sum_{a=1}^{4} e_a^2. \tag{7}$$

The constant A may now be computed to be

$$A = \sum_n v_n = \sum_m \frac{1}{\lambda - m} = \frac{\pi}{\tan(\pi\lambda)}. \tag{8}$$

The eigenvalue equation is obtained from (8) and (6) and the normalization condition $b = 1$ to be

$$1 = \sum_{a=1}^{4} d_a w_a = Ad^2 \sum_{a=1}^{4} \frac{e_a^2}{\lambda - m_a}, \qquad \Rightarrow$$

$$\frac{1}{\pi} \tan(\pi\lambda) = d^2 \sum_{a=1}^{4} \frac{e_a^2}{\lambda - m_a} \equiv d^2 r(\lambda). \tag{9}$$

Let us solve this equation for the flavor eigenvalue ($d = 0$), and for the mass eigenvalue in the strong coupling limit ($d \to \infty$). For $d = 0$, it follows from (9) that $\tan(\pi\lambda) = 0$, which implies $\lambda \in \mathbb{Z}$, unless $\lambda = m_a$ for some

a. These are the expected eigenvalues because they are simply the matrix elements of the diagonal matrix $\mathcal{M}$ when $d = 0$.

For $d \to \infty$, we should have $\tan(\pi\lambda) = \infty$, which implies $\lambda = \mathbb{Z} + \frac{1}{2}$, unless $r(\lambda) = 0$. The former are the bulk eigenvalues, and the solutions of the latter are the brane eigenvalues M_i ($i = 1, 2, 3$). Since $r(\lambda)$ approaches $\pm\infty$ when $\lambda \to m_a + 0^{\pm}$, there is one zero of $r(\lambda)$ between each successive pairs of m_a's. In other words, the ordering relation $m_1 < M_1 < m_2 < M_2 < m_3 < M_3 < m_4$ mentioned in the Introduction is obeyed.

If d is large but not infinite, the eigenvalues will shift somewhat, but they are still bounded between consecutive m_a's or consecutive n's.

We will now show that the three independent parameters e_a^2 may be replaced by the three independent parameters M_i^2, provided the ordering relation holds. The argument is based on the simple observation that $r(\lambda)$ is a meromorphic function of λ, with four simple poles occuring at $\lambda = m_a$, and three zeros occuring at $\lambda = M_i$. Moreover, $r(\lambda)$ approaches $1/\lambda$ when $|\lambda| \to \infty$. Hence we can write $r(\lambda) = \prod_{i=1}^{3}(\lambda - M_a)/\prod_{a=1}^{4}(\lambda - m_a)$. The residue at $\lambda = m_b$ is then

$$e_b^2 = \prod_{i=1}^{3}(m_b - M_i)/\prod_{a \neq b}(m_b - m_a). \tag{10}$$

This formula determines e_a^2 once m_a and M_i are known. To keep $e_a^2 > 0$, the ordering relation has to be obeyed.

Let U_λ be the normalized eigenvector, with components $U_{a\lambda} = w_a/N$ and $U_{n\lambda} = v_n/N$. The norm N^2 of the original eigenvector (w_a, v_n) is given by $N^2 = (Ad)^2 s + T$, where

$$s = \frac{1}{(Ad)^2} \sum_{a=1}^{4} w_a^2 = \sum_{a=1}^{4} \frac{e_a^2}{(\lambda - m_a)^2},$$

$$T = \sum_n v_n^2 = \frac{1}{(\lambda - n)^2}. \tag{11}$$

3. Oscillation amplitude

Using (6) and (11), we can calculate the transition amplitude $\mathcal{A}_{ab}$ from a brane neutrino of flavor b and energy E (measured in units of U), to a brane neutrino of flavor a after it has traversed a distance $L = 2E\tau$ (measured in units of U^{-1}). The transition amplitude is determined by the formula

$$\mathcal{A}_{ab}(\tau) = \sum_\lambda U_{a\lambda}^* U_{b\lambda} e^{-i\lambda^2 \tau} \equiv \mathcal{A}_{ab}^{S}(\tau) + \mathcal{A}_{ab}^{K}(\tau), \tag{12}$$

where $\mathcal{A}^S$ is the contribution from the brane eigenvalues $\lambda = M_1, M_2, M_3$, and $\mathcal{A}^K$ is the contribution from the bulk eigenvalues $\lambda \in \mathbb{Z} + \frac{1}{2}$.

When $d \to \infty$, the quantities $v_n, w_a/Ad, s$ and T are all of order 1, so the magnitude of w_a is determined by Ad and the magnitude of N^2 is determined by $(Ad)^2$. According to (9), $Ad = 1/(dr)$. For bulk eigenvalues, $r = O(1)$, so $Ad = O(1/d)$. This implies $N^2 \simeq T$ and $U_{a\lambda} = O(1/d)$. In that case the bulk components of an eigenvector are much larger than the brane components. For brane eigenvalues, $A = O(1)$, hence $Ad = O(d)$ and $w_a = O(d)$. In that case the brane components of an eigenvector dominate and $N^2 \simeq (Ad)^2 s$.

Let us denote the large-d value of U_{aM_i} by V_{ai}, the value of s at $\lambda = M_i$ by s_i, and $1/(M_i - m_a)$ by x_{ai}. Then

$$V_{ai} = e_a x_{ai}/\sqrt{s_i} \quad (1 \le a \le 4,\ 1 \le i \le 3), \tag{13}$$

and

$$\mathcal{A}^S_{ab}(\tau) = \sum_{i=1}^{3} V^*_{ai} V_{bi} e^{-iM_i^2 \tau}. \tag{14}$$

As it stands, V is a 4×3 matrix, but we can make it into a square 4×4 matrix by letting the last column to be $V_{af} = e_a$. The meaning of this last column will be discussed later. Note that we can write V_{a4} in the same form as the other V_{ai}, namely, $V_{a4} = e_a x_{a4}/\sqrt{s_4}$, provided we let $\lambda = \infty$.

The resulting 4×4 matrix V can be shown to be real orthogonal. It depends on the seven parameters, m_a and M_i.

We have shown in (10) how to express e_b^2 in terms of these parameters. Similarly, it can be shown that

$$s_i = -\prod_{k \ne i}^{3}(M_i - M_k)/\prod_{c=1}^{4}(M_i - m_c). \tag{15}$$

We turn to the contribution from the bulk eigenvalues. Since $U_{a\lambda}$ is unitary, it follows from (12) that $\mathcal{A}_{ab}(0) = \delta_{ab}$, hence

$$\mathcal{A}^K_{ab}(0) = \delta_{ab} - \mathcal{A}^S_{ab}(0). \tag{16}$$

Using (14) and the unitarity of the matrix V, we conclude that

$$\mathcal{A}^S_{ab}(0) = \delta_{ab} - V^*_{af} V_{bf} = \delta_{ab} - e_a e_f. \tag{17}$$

Therefore

$$\mathcal{A}^K_{ab}(0) = e_a e_b. \tag{18}$$

The contribution from the bulk eigenvalues can also be obtained directly from (12) and the paragraph following that equation to be

$$\mathcal{A}_{ab}^{K}(\tau) = \sum_{\lambda \in \mathbb{Z}+\frac{1}{2}} \frac{1}{(dr)^2 T} \frac{e_a e_b}{(\lambda - m_a)(\lambda - m_b)} e^{-i\lambda^2 \tau} \equiv e_a e_b F(\tau). \quad (19)$$

Both r and T are of order 1 as $d \to \infty$, so the contribution from each bulk eigenvalue to the sum is $O(1/d^2)$. Since there are an infinite number of bulk eigenvalues, the total contribution to the sum in (19) is not necessarily zero. In fact, we know from (18) that $F(0) = 1$ even at an infinite d.

It can be shown that $F(\tau) = g(K^2 \tau)$, where $K^2 = d^2(1 + \pi^2 d^2)$, and that

$$g(x) \equiv \frac{1}{\pi} \int_{-\infty}^{\infty} \frac{e^{-iu^2 x}}{u^2 + 1} \quad (20)$$

is zero at $x = 0$, and decrease to 0 like $(1-i)/\sqrt{2\pi x}$ for large x. This means that $\mathcal{A}_{ab}^{K}(\tau) = e_a e_b g(K^2 \tau)$ is zero whenever $\tau > 0$, in the limit $d \to \infty$. This function describes the absorption of the active neutrino flux into the bulk.

Therefore, in the strong coupling limit, we end up with

$$\mathcal{A}_{ab}(\tau) = \mathcal{A}_{ab}^{S}(\tau) = \sum_{i=1}^{3} V_{ai}^{*} V_{bi} e^{-iM_i^2 \tau}. \quad (21)$$

In the limit $m_4 \to \infty$, it follows from (10) that $e_4^2 \to 1$, and hence from (7) that $e_b \to 0$ for $b = 1, 2, 3$. Since the matrix V is real orthogonal, and $e_4 = V_{44}$, it also follows that $V_{a4} = 0$ for $a = 1, 2, 3$. As a result, the active neutrinos ν_e, ν_μ, ν_τ do not mix with the sterile neutrino ν_s, and the active neutrinos do not get absorbed by the bulk.

If furthermore $M_1 = M_2$, then the ordering relation forces $m_2 = M_1$. In that case $V_{23} = 0$. This means the mixing angle $\theta_{13} = 0$ if we identify the $a = 2$ flavor neutrino with ν_e. Hence a vanishing $\Delta M_\odot^2 / \Delta M_{atm}^2$ implies a vanishing θ_{13}.

4. Generalization of the minimal model

The final result (21) remains valid for almost all B in (1). This is reasonable because (21) does not depend on the property of the absorptive bulk, which B affects. For a detailed argument, please consult Ref.[11]

This research is supported by NSERC and FRNT.

References

1. J. Polchinski, Phys. Rev. Lett. 75 (1995) 4724.
2. N. Arkani-Hamed, S. Dimopoulos, and G. Dvali, Phys. Lett. B429 (1998) 263; Phys. Rev. D59 (1999) 086004; I. Antoniadis, N. Arkani-Hamed, S. Dimopoulos, and G. Dvali, Phys. Lett. B436 (1998) 257; L. Randall and R. Sundrum, Phys. Rev. Lett. 83(1999) 3370, 4690.
3. J.C. Long, A.B. Churnside, and J.C. Price; C.D. Hoyle et. al., Phys. Rev. Lett. 86 (2001) 1418.
4. J. Bonn *et al.*, Nucl. Phys. B (Proc. Suppl.) 91 (2001) 273.
5. http://www-sk.icrr.u-tokyo.ac.jp/doc/sk/;
   http://www.sno.phy.queensu.ca/;
   http://www.awa.tohoku.ac.jp/html/KamLAND/.
6. H.V. Peiris *et al.*, http://arxiv.org/ps/astro-ph/0302225.
7. M.Gell-Mann, P.Ramond, R.Slansky, in 'Supergravity' (North-Holland, Amsterdam,1979); T.Yanagida, in 'Proc.of the workshop on unified Theory and Baryon Number of the Universe', (KEK, Japan,1979); R.N. Mohapatra, G. Senjanovic, Phys. Rev. Lett. 44 (1980) 912.
8. N. Arkani-Hamed, S. Dimopoulos, G. Dvali, and J. March-Russell, hep-ph/9811448; K.R. Dienes, E. Dudas, and T. Gherghetta, Nucl. Phys. B557 (1999) 25; R.N. Mohapatra and A. Pérez-Lorenzana, Nucl. Phys. B576 (2000) 466; K.R. Dienes and I. Sarcevic, Phys. Lett. B500 (2001) 133; A. Lukas, P. Ramond, A. Romanino, and G.G. Ross, Phys. Lett. B495 (2000) 136; D.O. Caldwell, R.N. Mohapatra, and S. J. Yellin, Phys. Rev. Lett. 87 (2001) 041601, hep-ph/0101043, hep-ph/0102279; N. Cosme, J.-M. Frere, Y. Gouverneur, F.-S. Ling, D. Monderen, V. Van Elewyck, Phys. Rev. D63 (2001) 113018; R. Barbieri, P. Creminelli, and A. Strumia, Nucl. Phys. B585 (2000) 28; R.N. Mohapatra and A. Pérez-Lorenzana, Nucl. Phys. B593 (2001) 451.
9. C.S. Lam and J.N. Ng, Phys. Rev. D64 (2001) 113006 (http://arxiv.org/ps/hep-ph/0104129).
10. C.S. Lam, Phys. Rev. D65 (2002) 053009 (http://arxiv.org/ps/hep-ph/0110142).
11. D. Charuchittipan and C.S. Lam, to be published.

BPS Wall in $\mathcal{N} = 2$ SUSY Nonlinear Sigma Model with Eguchi–Hanson Manifold

Masato Arai

Institute of Physics, AS CR, 182 21, Praha 8, Czech Republic

Masashi Naganuma

*Department of Physics, Tokyo Institute of Technology,
Tokyo 152-8551, Japan*

Muneto Nitta

*Department of Physics, Purdue University,
West Lafayette, IN 47907-1396, USA*

Norisuke Sakai

*Department of Physics, Tokyo Institute of Technology,
Tokyo 152-8551, Japan*

BPS wall solutions are obtained for $\mathcal{N} = 2$ supersymmetric nonlinear sigma model with Eguchi–Hanson target manifold in a manifestly supersymmetric manner. The model is constructed by a massive hyper-Kähler quotient method both in the $\mathcal{N} = 1$ superfield and in the $\mathcal{N} = 2$ superfield (harmonic superspace). We describe the model in simple terms and give relations between various parameterizations which are useful to describe the model and the solution. Some more details can be found in our previous paper[1] [hep-th/0211103]. This article is dedicated to Professor Hiroshi Ezawa on the occasion of his seventieth birthday.

1. Introduction

Supersymmetry (SUSY) has been a most promising guiding principle to construct realistic unified models beyond the standard model.[2] In recent years there have been vigorous studies on models with extra dimensions,[3,4] where our world is assumed to be realized on an extended topological defects such as domain walls or various branes. Supersymmetry can be combined with this *brane-world scenario* and helps the construction of the extended topological defects.

Solitons saturating an energy bound, called the BPS bound,[5,6] have played a crucial role also in non-perturbative studies of supersymmetric (SUSY) field theories in four dimensions.[7] BPS domain walls are topological solitons of co-dimension one, which depend on one spatial coordinate and connect two SUSY vacua. Since they preserve half of the original SUSY, they are called $\frac{1}{2}$ BPS states. Such BPS domain walls were well studied in various models with global $\mathcal{N} = 1$ SUSY in four dimensions.[8] Non-BPS multi-wall solutions were also studied to understand the SUSY breaking mechanism on the brane due to the coexistence of the other brane.[9−11] More recently we have constructed an exact BPS wall solution as well as non-BPS multi-wall solutions in the supergravity theory in four dimensions.[12] The intersections or junctions of domain walls preserve $\frac{1}{4}$ of the original SUSY and have been discussed in $\mathcal{N} = 1$ models in four dimensions.[13−16]

In order to consider models with extra dimensions, we need to discuss supersymmetric theories in spacetime with dimensions higher than four. They should have at least eight supercharges. The simplest field theory with eight SUSY is based on hypermultiplets containing only scalar and spinor as physical fields. Recently we have formulated $\frac{1}{2}$ BPS domain walls in an eight SUSY model in four dimensions.[1] Moreover we have also succeeded in constructing the $\frac{1}{2}$ BPS wall consistently in five-dimensional supergravity.[17] Before discussing the SUSY five-dimensional theories, it is useful to consider models with eight SUSY in four dimensions without gravity.

The rest of our paper is organized as follows. Sec. 2 explains how to obtain nonlinear sigma models of hypermultiplets with eight SUSY. Secs. 3, 4 and 5 are devoted to $\mathcal{N} = 1$ superfield formulation of the model. In Sec. 3, we present the model using the $U(1)$ gauge field. We give the bosonic part of the action and eliminate auxiliary fields in the Wess–Zumino gauge. The hyper-Kähler quotient method in terms of the so-called moment map becomes very clear. In Sec. 4, we eliminate auxiliary superfields in the superfield level, taking a gauge compatible with SUSY rather than the Wess–Zumino gauge. This has the advantage because we obtain the Lagrangian in terms of independent superfields. In Sec. 5, we use the $O(2)$ gauge field to formulate the model instead of the $U(1)$ gauge field. Sec. 6 is devoted to a brief review of harmonic superspace formalism (HSF). In Sec. 7, we formulate the model in HSF and eliminate auxiliary fields in the Wess–Zumino gauge. In Sec. 8, the constraints are solved by independent fields.

We close our paper by Sec. 9, in which the BPS equation and the domain wall solution are given.

2. $\mathcal{N} = 2$ Model with hypermultiplets in 4-dimensions

If we take two free chiral scalar supermultiplets ϕ and χ with a complex mass term m between them, they together become a free massive hypermultiplet with $\mathcal{N} = 2$ SUSY[18]

$$
\begin{aligned}
\mathcal{L} &= [\phi^*\phi + \chi^*\chi]_{\theta^2\bar\theta^2} + (m\,[\chi\phi]_{\theta^2} + \text{c.c.}) \\
&= -\partial_\mu\phi^*\partial^\mu\phi - \partial_\mu\chi^*\partial^\mu\chi + F_\phi^*F_\phi + F_\chi^*F_\chi \\
&\quad + (m(F_\phi\chi + F_\chi\phi) + \text{c.c.}) + \text{fermionic terms} \\
&= -\partial_\mu\phi^*\partial^\mu\phi - \partial_\mu\chi^*\partial^\mu\chi - |m|^2\,(\phi^*\phi + \chi^*\chi) \\
&\quad + \text{fermionic terms} ,
\end{aligned}
\tag{1}
$$

where complex conjugate is denoted as c.c. , and the scalar components are denoted by the same letter as the superfields[b] ϕ, χ. Since four real scalar fields $\mathrm{Re}\phi$, $\mathrm{Im}\phi$, $\mathrm{Re}\chi$, and $\mathrm{Im}\chi$ are symmetric, we can form three complex fields using any one of the fields, say $\mathrm{Re}\phi$ with any one of the other three fields : $\mathrm{Re}\phi + i\mathrm{Re}\chi$, and $\mathrm{Re}\phi + i\mathrm{Im}\chi$ beside the ordinary $\mathrm{Re}\phi + i\mathrm{Im}\phi = \phi$. These three complex structures are completely symmetric and serve as a characterization of $\mathcal{N} = 2$ SUSY for hypermultiplets. It has been shown that any nonlinear sigma model consisting of hypermultiplets should have a triplet of complex structures, and the target manifold should be hyper-Kähler[20] (HK) in contrast to Kähler of the $\mathcal{N} = 1$ SUSY nonlinear sigma model.[21]

Theories with eight SUSY are so restrictive that the nontrivial interactions require the nonlinearity of kinetic term (nonlinear sigma model) if there are only hypermultiplets. In order to obtain a wall solution, we need to have a nontrivial potential. In the case of $\mathcal{N} = 2$ SUSY nonlinear sigma model containing only hypermultiplets, one can introduce a nontrivial potential which is the square of the Killing vector acting on the HK manifold multiplied by a mass parameter. Moreover the Killing vector has to be holomorphic with respect to the three complex structures (tri-holomorphic).[22] These models are called "massive HK nonlinear sigma models".

Let us now explain a mechanism to obtain a nontrivial potential as a Sherk–Schwarz reduction[23] from six or five dimensions.[24] It is usually

---

[b]We follow mostly the notation of Ref.[19], except that $\mu, \nu, \ldots$ denote space time in four dimensions.

best to start from a model in spacetime with maximal dimensions which is allowed by the postulated number of SUSY charges. In the present case of eight SUSY, we should consider hypermultiplets in six dimensions. Let us first illustrate the dimensional reduction by a free massless hypermultiplet in six dimensions, since a mass term is forbidden by SUSY. If two (one) spatial dimensions are compactified, a nonvanishing momentum in these compactified dimensions gives a complex (real) mass term resulting in Eq.(1). This mass parameter gives rise to a central term $Z = -i(\partial_5 + i\partial_6)$ in the $\mathcal{N} = 2$ SUSY algebra in four dimensions. In the case of a nonlinear sigma model with the target space metric g_{ij*} in six dimensions, the kinetic term (of bosonic part of the Lagrangian) reads

$$\mathcal{L} = -g_{ij*}\partial_M \phi^i \partial_N \phi^{j*} \eta^{MN}, \tag{2}$$

where $\eta^{MN} = \text{diag.}(-1,1,1,1,1,1)$, $M, N = 0, \cdots, 5$. Then we can twist the boundary condition for the compactified directions, say x^4 and x^5 using a Killing vector $k^i(\phi, \phi^*)$ for the isometry of the target space metric g_{ij*}

$$-i(\partial_5 + i\partial_6)\phi^i = \mu k^i(\phi, \phi^*), \quad \mu \in \mathbb{C}. \tag{3}$$

Then we obtain a nontrivial potential term $V(\phi)$ from the Lagrangian (2)

$$\mathcal{L} = -g_{ij*}\partial_\mu \phi^i \partial_\nu \phi^{j*} \eta^{\mu\nu} - V(\phi, \phi^*), \tag{4}$$

$$V(\phi, \phi^*) = |\mu|^2 g_{ij*} k^i(\phi, \phi^*) k^{j*}(\phi, \phi^*). \tag{5}$$

This is the Sherk–Schwarz dimensional reduction. Since theories with eight SUSY have three complex structures, the Killing vector k^i has to be holomorphic with respect to all three complex structures (tri-holomorphic).

Many target space metrics can be embedded in higher dimensional flat space as illustrated by a sphere embedded in three dimensions in Fig.1. The nonlinear sigma models with these target space metrics can be realized by giving a constraint on hypermultiplets with minimal kinetic terms. One of the most convenient methods to impose the constraint is to introduce a vector multiplet without a kinetic term. If we integrate the vector multiplet, it acts as a Lagrange multiplier field to produce a constraint on hypermultiplets, resulting in a curved target manifold such as the Eguchi–Hanson manifold. In this process, the mass term automatically becomes a nontrivial potential which is a square of a Killing vector corresponding to the isometry of the resulting curved target space. Since the gauge field serves to identify the gauge orbit, the introduction of the gauge field without a kinetic term gives a quotient manifold. In particular, we call the method as hyper-Kähler quotient method, when the resulting manifold is hyper-Kähler. In

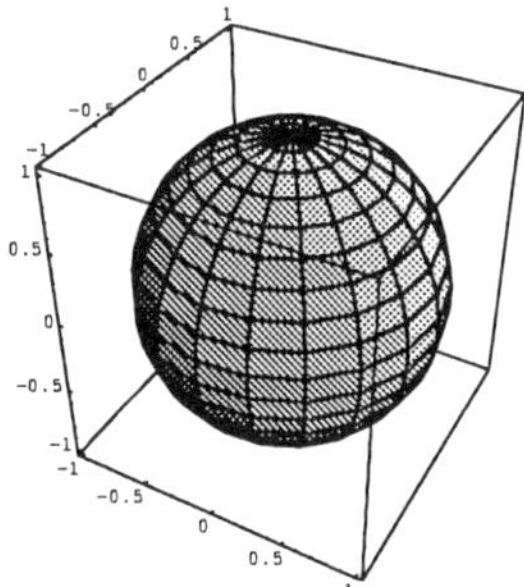

Fig. 1.   Sphere as a constrained subspace embedded in three-dimensional flat space.

our SUSY case, the curved manifold is a result of the constraint coming from integrating the auxiliary fields in the gauge supermultiplet as well as from the gauge orbit quotient. Since we also have a mass term as a central extension of the SUSY algebra, this procedure is called the *massive hyper-Kähler quotient* method. In this way, we can understand the potential term as the square of the tri-holomorphic Killing vector and the mass parameter multiplying the potential as the central extension of the $\mathcal{N} = 2$ SUSY algebra in four dimensions.

There have been a number of works to study nonlinear sigma models with eight supercharges.[20,22,24−36] The massive nonlinear sigma model with nontrivial Kähler metric as target space was studied, and BPS equations and BPS solutions such as walls and junctions were obtained.[25−31] Multi domain walls solution was also obtained and the dynamics of those walls was examined.[28,29] Single or parallel domain walls in such models preserve $\frac{1}{2}$ SUSY,[25,28] whereas their intersections preserve $\frac{1}{4}$ SUSY.[26] In most papers, nonlinear sigma models were studied in terms of component fields. However, it is often useful to maintain as much SUSY as possible.[37] Harmonic superspace formalism (HSF)[38] is most suited to maintain the SUSY maximally, but there has been relatively few attempt to formulate the BPS equations and to obtain BPS solutions in the HSF[39] until our recent work.[1]

Our nonlinear sigma model can most easily be obtained by a quotient method in terms of a $U(1)$ vector multiplet without a kinetic term. In $\mathcal{N} = 1$ formalism, the massless HK sigma model on $T^*\mathbf{C}P^n$, namely cotangent bundle over $\mathbf{C}P^n$, was obtained as the HK quotient.[34,35] The massive HK quotient was obtained in component level.[29] The massless model with the Eguchi–Hanson target manifold $(T^*\mathbf{C}P^1)^{43}$ has been constructed in $\mathcal{N} = 2$ formalism,[40] and its central extension was analysed.[42] In order to obtain also a potential term, we need to perform a quotient method for a massive hypermultiplet charged under the $U(1)$ vector multiplet.[1] When we are writing this up this work, another interesting work appeared discussing various wall and flux tube solutions in similar $\mathcal{N} = 2$ models.[48]

3. Eguchi–Hanson nonlinear sigma model in $\mathcal{N} = 1$ superfields in the $U(1)$ basis

The massive hyper-Kähler quotient method for the massive Eguchi–Hanson[43] nonlinear sigma model requires two hypermultiplets : ϕ, χ as doublets. The doublet $\phi(\chi)$ has charge $+1(-1)$ under an $\mathcal{N} = 2$ $U(1)$ vector multiplet (V, Σ) without a kinetic term which serves as a Lagrange multiplier constraining hypermultiplets to form a four-dimensional (in terms of number of real degrees of freedom) target manifold. Here V and Σ are vector and chiral superfields in $\mathcal{N} = 1$ superfield formalism, respectively. Representing two doublets ϕ, χ as column vectors, the action is given in terms of $\mathcal{N} = 1$ superfields as[1]

$$\mathcal{L} = \left[e^V \phi^\dagger \phi + e^{-V} \chi^\dagger \chi - cV \right]_{\theta^2 \bar{\theta}^2} + \left(\left[\Sigma(\chi^T \phi - b) + \frac{\mu}{2} \chi^T \sigma_3 \phi \right]_{\theta^2} + \text{c.c.} \right), \tag{6}$$

where we have absorbed a common mass of hypermultiplets into the field Σ and denote μ as a complex parameter for the mass splitting. The electric and magnetic Fayet–Iliopoulos (FI) parameters are denoted as $c \in \mathbb{R}, b \in \mathbb{C}$. We see below that these parameters become the value of the triplet of the moment map for the $U(1)$ gauge symmetry.

In the limit of $\mu = 0$, the model has a global (flavor) symmetry $U(2) = SU(2) \times U(1)_A$ defined by

$$\phi \to \phi' = g\phi, \quad \chi \to \chi' = g^*\chi, \quad \Sigma \to \Sigma' = \Sigma, \qquad g \in U(2), \tag{7}$$

and, in the case of $b = 0$, it has the additional $U(1)_D$ symmetry

$$\phi \to \phi' = e^{i\theta_D}\phi, \qquad \chi \to \chi' = e^{i\theta_D}\chi, \qquad \Sigma \to \Sigma' = e^{-2i\theta_D}\Sigma. \tag{8}$$

The $U(2)$ symmetry is consistent with the $\mathcal{N} = 2$ SUSY, but the $U(1)_D$ symmetry is only with the manifest $\mathcal{N} = 1$ SUSY. The mass splitting parameter breaks this $U(2)[\times U(1)_D]$ global symmetry (for $b = 0$) down to $U(1) \times U(1) \in U(2)$ defined by $\phi \to e^{i\theta_1 \mathbf{1} + i\theta_2 \sigma_3}\phi$, $\chi \to e^{-i\theta_1 \mathbf{1} - i\theta_2 \sigma_3}\chi$, with Σ unchanged. The $U(1)$ subgroup parametrized by θ_1 is gauged. Since this mass splitting parameter affects only the potential term without affecting the kinetic term, the curved target manifold has Killing vectors for the isometry $SU(2)[\times U(1)_D$ for $b = 0]$. The $U(1)_A$ symmetry of

$U(2) = SU(2) \times U(1)_{\mathrm{A}}$ is gauged away and is absent in the target manifold. Only the Killing vectors for the isometry $SU(2)$ are consistent with the $\mathcal{N} = 2$ SUSY and hence tri-holomorphic. The Killing vector for the $U(1)_{\mathrm{D}}$ isometry for $b = 0$ is holomorphic, but not tri-holomorphic. Since we have introduced the mass parameter through the σ_3 generator, we will eventually obtain the potential term which is a square of the tri-holomorphic Killing vector corresponding to σ_3, after eliminating the vector multiplet.

In the Wess–Zumino gauge, the bosonic action is given by

$$\mathcal{L}_{\mathrm{boson}} = \mathcal{L}_{\mathrm{kin}} + \mathcal{L}_{\mathrm{constr}} + \mathcal{L}_{\mathrm{pot}}, \tag{9}$$

$$
\begin{aligned}
\mathcal{L}_{\mathrm{kin}} &= -\left|\left(\partial_\mu + \frac{i}{2}v_\mu\right)\phi\right|^2 - \left|\left(\partial_\mu - \frac{i}{2}v_\mu\right)\chi\right|^2 \\
&= -|\partial_\mu\phi|^2 - |\partial_\mu\chi|^2 + \frac{i}{2}v^\mu\left(\phi^\dagger\overleftrightarrow{\partial_\mu}\phi - \chi^\dagger\overleftrightarrow{\partial_\mu}\chi\right) \\
&\quad - \frac{1}{4}v^\mu v_\mu\left(\phi^\dagger\phi + \chi^\dagger\chi\right),
\end{aligned} \tag{10}
$$

where $\phi^\dagger\overleftrightarrow{\partial_\mu}\phi \equiv \phi^\dagger(\partial_\mu\phi) - (\partial_\mu\phi^\dagger)\phi$ and

$$\mathcal{L}_{\mathrm{constr}} = \frac{D}{2}\left(\phi^\dagger\phi - \chi^\dagger\chi - c\right) + \left(F_\Sigma\left(\chi^T\phi - b\right) + \mathrm{c.c.}\right), \tag{11}$$

$$
\begin{aligned}
\mathcal{L}_{\mathrm{pot}} &= F_\phi^\dagger F_\phi + F_\chi^\dagger F_\chi \\
&\quad + \left(F_\chi^T\left(\Sigma\mathbf{1} + \frac{\mu}{2}\sigma_3\right)\phi + \chi^T\left(\Sigma\mathbf{1} + \frac{\mu}{2}\sigma_3\right)F_\phi + \mathrm{c.c.}\right) \\
&\equiv -V(\phi,\chi,\Sigma).
\end{aligned} \tag{12}
$$

The equation of motion for gauge field v_μ in (9) allows us to write gauge field v_μ in terms of scalar fields

$$v_\mu = \frac{i\left(\phi^\dagger\overleftrightarrow{\partial_\mu}\phi - \chi^\dagger\overleftrightarrow{\partial_\mu}\chi\right)}{\phi^\dagger\phi + \chi^\dagger\chi}. \tag{13}$$

If we eliminate the gauge field by this algebraic equation of motion, we obtain the kinetic term for hypermultiplets as

$$\mathcal{L}_{\mathrm{kin}} = -|\partial_\mu\phi|^2 - |\partial_\mu\chi|^2 + \frac{\left|i\left(\phi^\dagger\overleftrightarrow{\partial_\mu}\phi - \chi^\dagger\overleftrightarrow{\partial_\mu}\chi\right)\right|^2}{4(\phi^\dagger\phi + \chi^\dagger\chi)}. \tag{14}$$

If we integrate the Lagrange multiplier fields D and F_Σ in (11), we obtain two constraints

$$\phi^\dagger \phi - \chi^\dagger \chi = c, \qquad \chi^T \phi = b. \tag{15}$$

The Lagrangian (12) gives algebraic equations of motion for the auxiliary fields F_ϕ, F_χ

$$F_\phi^\dagger = -\chi^T \left(\Sigma \mathbf{1} + \frac{\mu}{2}\sigma_3\right), \qquad F_\chi^\dagger = -\phi^T \left(\Sigma \mathbf{1} + \frac{\mu}{2}\sigma_3\right). \tag{16}$$

After eliminating the auxiliary fields F_ϕ, F_χ by these algebraic equations of motion, we obtain the potential term

$$
\begin{aligned}
V(&\phi, \chi, \Sigma) \\
&= |F_\phi|^2 + |F_\chi|^2 \\
&= \left|\left(\Sigma \mathbf{1} + \frac{\mu}{2}\sigma_3\right)\phi\right|^2 + \left|\left(\Sigma \mathbf{1} + \frac{\mu}{2}\sigma_3\right)\chi\right|^2 \\
&= \left(\left|\frac{\mu}{2}\right|^2 + |\Sigma|^2\right)\left(|\phi|^2 + |\chi|^2\right) + \left(\frac{\mu^*}{2}\Sigma + \frac{\mu}{2}\Sigma^*\right)\left(\phi^\dagger \sigma_3 \phi + \chi^\dagger \sigma_3 \chi\right) \\
&= \left|\frac{\mu}{2}\right|^2 \left[|\phi|^2 + |\chi|^2 - \frac{\left(\phi^\dagger \sigma_3 \phi + \chi^\dagger \sigma_3 \chi\right)^2}{|\phi|^2 + |\chi|^2}\right].
\end{aligned}
\tag{17}
$$

In the last line, we have eliminated the scalar field Σ in the $\mathcal{N} = 2$ vector multiplet (V, Σ) by its algebraic equation of motion

$$\Sigma = -\frac{\mu}{2}\frac{\phi^\dagger \sigma_3 \phi + \chi^\dagger \sigma_3 \chi}{|\phi|^2 + |\chi|^2}. \tag{18}$$

In Eq.(15) the left hand sides constitute the triplet of the moment map (Killing potential) for the $U(1)$ gauge symmetry. Hence we see that these values are fixed to the FI parameters by integrating the auxiliary fields D and F_Σ, and that the hyper-Kähler quotient is obtained together with the $U(1)$ quotient (13). In the limit of $b = c = 0$ the singularity appears and the manifold becomes the orbifold $\mathbb{C}^2/\mathbb{Z}_2$, whereas the non-zero values of b and c resolve the orbifold singularity through the deformation of the complex structure and the blow up, respectively.

4. Eguchi–Hanson nonlinear sigma model after integrating vector multiplet in the $U(1)$ basis

Instead of taking the Wess–Zumino gauge in the component level, we can eliminate the auxiliary superfields V and Σ directly in the superfield formalism. Their equations of motion read from Eq.(6) as

$$\frac{\partial \mathcal{L}}{\partial V} = e^V |\phi|^2 - e^{-V} |\chi|^2 - c = 0 \, , \tag{19}$$

$$\frac{\partial \mathcal{L}}{\partial \Sigma} = \chi^T \phi - b = 0 \, , \tag{20}$$

in which V can be solved immediately to give

$$e^V = (c \pm \sqrt{c^2 + 4|\phi|^2|\chi|^2})/2|\phi|^2.$$

We thus obtain the Kähler potential

$$K = c\sqrt{1 + \frac{4}{c^2}|\phi|^2|\chi|^2} - c\log\left(1 + \sqrt{1 + \frac{4}{c^2}|\phi|^2|\chi|^2}\right) + c\log|\phi|^2 \, , \tag{21}$$

where we have chosen the plus sign of the solution for the positivity of the metric.

Fixing the complexified $U(1)$ gauge symmetry and solving (20), we can obtain the Lagrangian of the nonlinear sigma model in terms of independent superfields. We have presented some gauge fixing[1] applied to $T^*\mathbf{C}P^n$ for general n. Here we give a more symmetric gauge for $T^*\mathbf{C}P^1$ in the case of $b = 0$. In this case, we can fix the gauge as $\chi_1/\phi_2 = 1$, and (20) can be solved as $\phi^T = (x, y)$ and $\chi^T = (y, -x)$ (and hence $|\phi|^2 = |\chi|^2 = |x|^2 + |y|^2$). The Kähler potential becomes[46]

$$K = c\sqrt{1 + \frac{4}{c^2}|\phi|^4} - c\log\left(\frac{1 + \sqrt{1 + \frac{4}{c^2}|\phi|^4}}{|\phi|^2}\right) \, , \tag{22}$$

and the superpotential is given by

$$W = \mu xy \, . \tag{23}$$

The metric and its inverse can be calculated to give

$$g_{ij^*} = \frac{c}{|\phi|^4\sqrt{1 + \frac{4}{c^2}|\phi|^4}} \begin{pmatrix} |y|^2 + \frac{4}{c^2}|\phi|^6 & -x^*y \\ -y^*x & |x|^2 + \frac{4}{c^2}|\phi|^6 \end{pmatrix} \, ,$$

$$g^{ij^*} = \frac{c}{4|\phi|^4\sqrt{1 + \frac{4}{c^2}|\phi|^4}} \begin{pmatrix} |x|^2 + \frac{4}{c^2}|\phi|^6 & y^*x \\ x^*y & |y|^2 + \frac{4}{c^2}|\phi|^6 \end{pmatrix} \, , \tag{24}$$

where $g_{ij^*} = \partial^2 K/\partial\phi^i\partial\phi^{j*}$ and $\phi^i = (x, y)$. The scalar potential can be calculated as

$$V = g^{ij^*}\partial_i W \partial_{j^*} W^* = \frac{c|\mu|^2}{|\phi|^4\sqrt{1 + \frac{4}{c^2}|\phi|^4}}\left(\frac{1}{c^2}|\phi|^8 + |x|^2|y|^2\right) \, . \tag{25}$$

The manifold admits the tri-holomorphic isometry $SU(2)$, defined in Eq.(7).[c] The Killing vectors for this action $k_A^i - \frac{1}{\epsilon}\delta_\epsilon \phi^i = \frac{i}{2}(\sigma_A)^i{}_j \phi^j$ $(A = 1, 2, 3)$ are given as

$$(k_1, k_2, k_3) = \frac{i}{2}\left(\begin{pmatrix} y \\ x \end{pmatrix}, \begin{pmatrix} -iy \\ ix \end{pmatrix}, \begin{pmatrix} x \\ -y \end{pmatrix} \right). \tag{26}$$

The Killing potential $D_A(\phi, \phi^*)$ for these vectors, defined by $k_A^i = ig^{ij^*}\partial_{j^*}D_A$, are given as[36]

$$(D_1, D_2, D_3) = \frac{c}{2} \frac{\sqrt{1 + \frac{4}{c^2}|\phi|^4}}{|\phi|^2} \left(xy^* + yx^*, i(xy^* - yx^*), |x|^2 - |y|^2 \right). \tag{27}$$

Using these geometric quantities, the scalar potential can be rewritten by the square of the Killing vector

$$V = |\mu|^2 g_{ij^*} k_3^i k_3^{*j} = |\mu|^2 g^{ij^*}\partial_i D_3 \partial_{j^*}D_3 \tag{28}$$

as was shown in Eq.(5). It is now manifest that only the action of k_3 among three Killing vectors preserves the potential term and therefore is the symmetry of the whole Lagrangian as expected from the mass term in (6). The vacua are fixed points of the Killing vectors k_3 or the critical points of the Killing potential D_3.

5. Eguchi–Hanson nonlinear sigma model in $\mathcal{N} = 1$ superfields in the $O(2)$ basis

It is also useful to rewrite the model in terms of $O(2)$ gauge group instead of $U(1)$, since $O(2)$ basis is most frequently employed in the harmonic superspace formalism as given in the next section. Introducing $O(2)$ doublets $\tilde{\phi}^a$ as a column vector and $\tilde{\chi}_a$ as a row vector, superspace action in the $O(2)$ basis is given by

$$\mathcal{L}^{O(2)} = \left[\tilde{\phi}_a^\dagger (e^{VT})^a{}_b \tilde{\phi}^b + \tilde{\chi}_a (e^{-VT})^a{}_b \tilde{\chi}^{b\dagger} - cV \right]_{\theta^2 \bar{\theta}^2}$$
$$+ \left(\left[\Sigma \left(\tilde{\chi}_a T^a{}_b \tilde{\phi}^b - b \right) + \frac{\mu}{2}\tilde{\chi}_a \tilde{\phi}^a \right]_{\theta^2} + \text{c.c.} \right), \tag{29}$$

---

[c]The $SU(2)$ transformation law of ϕ^i as the coordinates of the quotient target manifold is unchanged from the one in Eq. (7) and hence is still *linear*, because the gauge fixing condition is invariant under the $SU(2)$ action. This is the advantage of our gauge fixing condition. To define the $SU(2)$ action in the cases of the other gauge condition,[1] we need an appropriate $U(1)$ gauge action to compensate the variation of the gauge condition, which makes the transformation law of ϕ^i *nonlinear*.

The diagonal $U(1)_D$ isometry defined in Eq.(8) of $U(2) \times U(1)_D = SU(2) \times U(1)_A \times U(1)_D$ isometry is holomorphic but not tri-holomorphic.

where the hermitian $O(2)$ generator is given from a diagonal generator σ_3 by

$$T \equiv \sigma_2 = e^{i\frac{\pi\sigma_1}{4}}\sigma_3 e^{-i\frac{\pi\sigma_1}{4}} = \frac{1+i\sigma_1}{\sqrt{2}}\sigma_3\frac{1-i\sigma_1}{\sqrt{2}}. \tag{30}$$

In order to establish a relation between superspace action in the $U(1)$ and $O(2)$ bases, it is convenient to define fields $\tilde{\phi}', \tilde{\chi}'$ with definite $U(1)$ charge by means of a rotation from the $O(2)$ basis

$$\tilde{\phi}' \equiv \frac{1-i\sigma_1}{\sqrt{2}}\tilde{\phi}, \qquad \tilde{\chi}' \equiv \tilde{\chi}\frac{1-i\sigma_1}{\sqrt{2}}. \tag{31}$$

In terms of these superfields, the action becomes

$$\mathcal{L}^{O(2)} = \left[\tilde{\phi}_a'^{\dagger}(e^{V\sigma_3})^a{}_b\tilde{\phi}'^b + \tilde{\chi}_a'(e^{V\sigma_3})^a{}_b\tilde{\chi}'^{b\dagger} - cV\right]_{\theta^2\bar{\theta}^2}$$
$$+ \left(\left[\Sigma\left(\tilde{\chi}_a'(\sigma_2)^a{}_b\tilde{\phi}'^b - b\right) + \frac{\mu}{2}\tilde{\chi}_a'(i\sigma_1)^a{}_b\tilde{\phi}'^a\right]_{\theta^2} + \text{c.c.}\right). \tag{32}$$

By identifying the $O(2)$ and $U(1)$ gauge fields, we find that the superfields ϕ_i, χ_i in the $U(1)$ basis should be related to the superfields $\tilde{\phi}'^i, \tilde{\chi}'_i$ in the $O(2)$ basis as

$$|\tilde{\phi}'^1|^2 + |\tilde{\chi}'_1|^2 = |\phi_1|^2 + |\phi_2|^2, \tag{33}$$
$$|\tilde{\phi}'^2|^2 + |\tilde{\chi}'_2|^2 = |\chi_1|^2 + |\chi_2|^2, \tag{34}$$
$$i\tilde{\chi}'_2\tilde{\phi}'^1 - i\tilde{\chi}'_1\tilde{\phi}'^2 = \chi_1\phi_1 + \chi_2\phi_2, \tag{35}$$
$$i\tilde{\chi}'_2\tilde{\phi}'^1 + i\tilde{\chi}'_1\tilde{\phi}'^2 = \chi_1\phi_1 - \chi_2\phi_2. \tag{36}$$

The most general solution of the conditions (33) and (34) is given in terms of two unitary matrices U, V

$$\begin{pmatrix} \tilde{\phi}'^1 \\ \tilde{\chi}'_1 \end{pmatrix} = U\begin{pmatrix} \phi_1 \\ \phi_2 \end{pmatrix}, \qquad U^{\dagger}U = 1, \tag{37}$$

$$\begin{pmatrix} \tilde{\phi}'^2 \\ \tilde{\chi}'_2 \end{pmatrix} = V\begin{pmatrix} \chi_1 \\ \chi_2 \end{pmatrix}, \qquad V^{\dagger}V = 1. \tag{38}$$

The constraints (35) and (36) give conditions

$$V^T\sigma_2 U = 1, \qquad V^T i\sigma_1 U = \sigma_3, \tag{39}$$

respectively. The first condition gives $V^T = U^{\dagger}\sigma_2$. By substituting it to the second condition we obtain

$$U^{\dagger}\sigma_3 U = \sigma_3. \tag{40}$$

The most general solution of these conditions is now given in terms of two arbitrary angle parameters α, β by

$$U = \begin{pmatrix} e^{i\alpha} & 0 \\ 0 & e^{i\beta} \end{pmatrix}, \qquad V = -\sigma_2 \begin{pmatrix} e^{-i\alpha} & 0 \\ 0 & e^{-i\beta} \end{pmatrix}. \tag{41}$$

These angles α, β clearly represent the $U(1) \times U(1)$ symmetry of our Lagrangian. Therefore a general solution for the identification of superfields ϕ_i, χ_i in the $U(1)$ basis and superfields $\tilde{\phi}^a, \tilde{\chi}_a$ in the $O(2)$ superfields is given by

$$\phi = \begin{pmatrix} \phi_1 \\ \phi_2 \end{pmatrix} = \begin{pmatrix} e^{-i\alpha}\tilde{\phi}'^1 \\ e^{-i\beta}\tilde{\chi}'_1 \end{pmatrix} = \frac{1}{\sqrt{2}} \begin{pmatrix} e^{-i\alpha}(\tilde{\phi}^1 - i\tilde{\phi}^2) \\ e^{-i\beta}(\tilde{\chi}_1 - i\tilde{\chi}_2) \end{pmatrix}, \tag{42}$$

$$\chi = \begin{pmatrix} \chi_1 \\ \chi_2 \end{pmatrix} = \begin{pmatrix} ie^{i\alpha}\tilde{\chi}'_2 \\ -ie^{i\beta}\tilde{\phi}'^2 \end{pmatrix} = \frac{1}{\sqrt{2}} \begin{pmatrix} e^{i\alpha}(\tilde{\chi}_1 + i\tilde{\chi}_2) \\ e^{i\beta}(-\tilde{\phi}^1 - i\tilde{\phi}^2) \end{pmatrix}. \tag{43}$$

From now on, we shall take $\alpha = \beta = 0$ case as a representative choice which is given by

$$\begin{pmatrix} \phi_1 \\ \phi_2 \end{pmatrix} = \frac{1}{\sqrt{2}} \begin{pmatrix} \tilde{\phi}^1 - i\tilde{\phi}^2 \\ \tilde{\chi}_1 - i\tilde{\chi}_2 \end{pmatrix}, \qquad \begin{pmatrix} \chi_1 \\ \chi_2 \end{pmatrix} = \frac{1}{\sqrt{2}} \begin{pmatrix} \tilde{\chi}_1 + i\tilde{\chi}_2 \\ -\tilde{\phi}^1 - i\tilde{\phi}^2 \end{pmatrix}. \tag{44}$$

Bosonic part of the Lagrangian in the $O(2)$ basis (29) is given in the Wess–Zumino gauge

$$\mathcal{L}_{\text{boson}}^{O(2)} = \mathcal{L}_{\text{kin}}^{O(2)} + \mathcal{L}_{\text{constr}}^{O(2)} + \mathcal{L}_{\text{pot}}^{O(2)}, \tag{45}$$

$$\begin{aligned}
\mathcal{L}_{\text{kin}}^{O(2)} &= -\left|\left(\partial_\mu + \frac{i}{2}v_\mu T\right)\tilde{\phi}\right|^2 - \left|\left(\partial_\mu + \frac{i}{2}v_\mu T\right)\tilde{\chi}^T\right|^2 \\
&= -|\partial_\mu\tilde{\phi}|^2 - |\partial_\mu\tilde{\chi}|^2 + \frac{i}{2}v^\mu\left(\tilde{\phi}^\dagger T \overset{\leftrightarrow}{\partial}_\mu \tilde{\phi} + \tilde{\chi}T\overset{\leftrightarrow}{\partial}_\mu\tilde{\chi}^\dagger\right) \\
&\quad - \frac{1}{4}v^\mu v_\mu\left(\tilde{\phi}^\dagger\tilde{\phi} + \tilde{\chi}\tilde{\chi}^\dagger\right),
\end{aligned} \tag{46}$$

$$\mathcal{L}_{\text{constr}}^{O(2)} = \frac{D}{2}\left(\tilde{\phi}^\dagger T\tilde{\phi} - \tilde{\chi}T\tilde{\chi}^\dagger - c\right) + \left(F_\Sigma\left(\tilde{\chi}T\tilde{\phi} - b\right) + \text{c.c.}\right), \tag{47}$$

$$\begin{aligned}
\mathcal{L}_{\text{pot}}^{O(2)} &= F_{\tilde{\phi}}^\dagger F_{\tilde{\phi}} + F_{\tilde{\chi}}F_{\tilde{\chi}}^\dagger \\
&\quad + \left(F_{\tilde{\chi}}\left(\Sigma T + \frac{\mu}{2}\mathbf{1}\right)\tilde{\phi} + \tilde{\chi}\left(\Sigma T + \frac{\mu}{2}\mathbf{1}\right)F_{\tilde{\phi}} + \text{c.c.}\right) \\
&\equiv -V(\phi, \chi, \Sigma).
\end{aligned} \tag{48}$$

The equation of motion for gauge field v_μ is given by

$$v_\mu = \frac{i\left(\tilde{\phi}^\dagger T \overset{\leftrightarrow}{\partial}_\mu \tilde{\phi} + \tilde{\chi} T \overset{\leftrightarrow}{\partial}_\mu \tilde{\chi}^\dagger\right)}{\tilde{\phi}^\dagger \tilde{\phi} + \tilde{\chi}\tilde{\chi}^\dagger}. \tag{49}$$

Eliminating the gauge field by this algebraic equation of motion, we obtain the kinetic term for hypermultiplets as

$$\mathcal{L}_{\text{kin}} = -|\partial_\mu \tilde{\phi}|^2 - |\partial_\mu \tilde{\chi}|^2 + \frac{\left|i\left(\tilde{\phi}^\dagger T \overset{\leftrightarrow}{\partial}_\mu \tilde{\phi} + \tilde{\chi} T \overset{\leftrightarrow}{\partial}_\mu \tilde{\chi}^\dagger\right)\right|^2}{4(\tilde{\phi}^\dagger \phi + \tilde{\chi}^\dagger \chi)}. \tag{50}$$

Integrating the Lagrange multiplier fields D and F_Σ in (47), we obtain two constraints

$$\tilde{\phi}^\dagger T \tilde{\phi} - \tilde{\chi} T \tilde{\chi}^\dagger - c = 0, \qquad \tilde{\chi} T \tilde{\phi} - b = 0. \tag{51}$$

Eliminating the algebraic equations of motion for the auxiliary fields F_ϕ, F_χ

$$F_\phi^\dagger = -\tilde{\chi}\left(\Sigma T + \frac{\mu}{2}\mathbf{1}\right), \qquad F_\chi^\dagger = -\left(\Sigma T + \frac{\mu}{2}\mathbf{1}\right)\tilde{\phi}. \tag{52}$$

After eliminating the auxiliary fields F_ϕ, F_χ by these algebraic equations of motion, we obtain the potential term

$$\begin{aligned}
V(\tilde{\phi}, \tilde{\chi}, \Sigma) \\
&= |F_{\tilde{\phi}}|^2 + |F_{\tilde{\chi}}|^2 \\
&= \left|\left(\Sigma T + \frac{\mu}{2}\mathbf{1}\right)\tilde{\phi}\right|^2 + \left|\tilde{\chi}\left(\Sigma T + \frac{\mu}{2}\mathbf{1}\right)\right|^2 \\
&= \left(\left|\frac{\mu}{2}\right|^2 + |\Sigma|^2\right)\left(|\tilde{\phi}|^2 + |\tilde{\chi}|^2\right) + \left(\frac{\mu^*}{2}\Sigma + \frac{\mu}{2}\Sigma^*\right)\left(\tilde{\phi}^\dagger T \tilde{\phi} + \tilde{\chi} T \tilde{\chi}^\dagger\right) \\
&= \left|\frac{\mu}{2}\right|^2 \left[|\tilde{\phi}|^2 + |\tilde{\chi}|^2 - \frac{\left(\tilde{\phi}^\dagger T \tilde{\phi} + \tilde{\chi} T \tilde{\chi}^\dagger\right)^2}{|\tilde{\phi}|^2 + |\tilde{\chi}|^2}\right].
\end{aligned} \tag{53}$$

In the last line, we have eliminated the scalar field Σ in the $\mathcal{N} = 2$ vector multiplet (V, Σ) by its algebraic equation of motion

$$\Sigma = -\frac{\mu}{2}\frac{\tilde{\phi}^\dagger T \tilde{\phi} + \tilde{\chi} T \tilde{\chi}^\dagger}{|\tilde{\phi}|^2 + |\tilde{\chi}|^2}. \tag{54}$$

6. A brief survey of harmonic superspace formalism

Harmonic superspace is defined as $(x^\mu,\ \theta_i,\ \bar\theta^i, u_i^\pm)$ which is called the central basis. The $u_i^\pm$ are called the harmonic variables which parameterize the coset $SU(2)_R/U(1)_r \sim S^2$, where $i = 1, 2$ is $SU(2)_R$ index and $\pm$ denotes $U(1)_r$ charge. The superfield in the Harmonic Superspace Formalism (HSF) is not defined in the central basis but in the subspace which is called the analytic subspace

$$\{\zeta_A, u_i^\pm | x_A^\mu = x^\mu - 2i\theta^{(i}\sigma^\mu\bar\theta^{j)}u_{(i}^+u_{j)}^-,\ \theta^+ = \theta^i u_i^+,\ \bar\theta^+ = \bar\theta^i u_i^+,\ u_i^\pm\}, \quad (55)$$

where parentheses for indices i, j mean symmetrization, for instance,

$$u_{(i}^+u_{j)}^- = (u_i^+u_j^- + u_j^+u_i^-)/2. \quad (56)$$

Hypermultiplet and vector multiplet superfields are defined as the function in the analytic subspace as $q^+(\zeta_A, u)$ and $V^{++}(\zeta_A, u)$, respectively, which are called the analytic superfields.

To describe the real action in terms of the analytic superfield, the star conjugation must be introduced in addition to the usual complex conjugation. The complex conjugation rules for the coefficients in the harmonic expansions $f^{i_1\cdots i_n}$ (see (71)), the Grassmann variables $\theta_{i\alpha}$ and the harmonic variables $u_i^\pm$ are defined as

$$\overline{f^{i_1\cdots i_n}} \equiv \bar f_{i_1\cdots i_n}, \qquad \overline{f_{i_1\cdots i_n}} = (-1)^n\bar f^{i_1\cdots i_n}, \quad (57)$$

$$\overline{\theta_{i\alpha}} = \bar\theta_{\dot\alpha}^i, \qquad \overline{\theta_\alpha^i} = -\bar\theta_{\dot\alpha i}, \quad (58)$$

$$\overline{u^{+i}} = u_i^-, \qquad \overline{u_i^+} = -u^{-i}, \quad (59)$$

respectively. The star conjugation rules are defined as

$$(f^{i_1\cdots i_n})^* = f^{i_1\cdots i_n}, \quad (60)$$

$$(\theta_\alpha^i)^* = \theta_\alpha^i, \quad (61)$$

$$(u^{+i})^* = u^{-i},\ \ (u_i^+)^* = u_i^-,\ \ (u^{-i})^* = -u^{+i},\ \ (u_i^-)^* = -u_i^+, \quad (62)$$

$$(u_i^\pm)^{**} = -u_i^\pm. \quad (63)$$

Note that the star conjugate acts only on the quantity having $U(1)_r$ charge. We write the combination of the complex and the star conjugation as

$$\overline{(q^+(\zeta_A, u))^*} \equiv \widetilde{q^+}(\zeta_A, u). \quad (64)$$

The combined conjugation rules are defined by

$$\widetilde{f^{i_1\cdots i_n}} = \overline{f^{i_1\cdots i_n}} \equiv \bar f_{i_1\cdots i_n}, \quad (65)$$

$$\widetilde{\theta^+} = \bar\theta^+,\quad \widetilde{\theta^-} = \bar\theta^-,\quad \widetilde{\bar\theta^+} = -\theta^+,\quad \widetilde{\bar\theta^-} = -\theta^-, \quad (66)$$

$$\widetilde{(u_i^\pm)} = u^{\pm i},\quad \widetilde{(u^{\pm i})} = -u_i^\pm. \quad (67)$$

The simple example of the real action is the free massless action of the Fayet–Sohnius hypermultiplet;

$$S = - \int d\zeta_A^{(-4)} du \, \widetilde{\phi^+} D^{++} \phi^+ \tag{68}$$

where D^{++} is defined by using the central charge Z

$$D^{++} = \partial^{++} - 2i\theta^+ \sigma^\mu \bar\theta^+ \partial_\mu^A - (\theta^{+2}\bar{Z} - \bar\theta^{+2}Z), \tag{69}$$

with $Z = 0$ for a massless hypermultiplet. Suffix A in the spacetime derivative ∂_μ^A denotes the variable appropriate in the analytic superspace,[41] ∂^{++} is the harmonic differential defined by $\partial^{++} = u_i^+ \frac{\partial}{\partial u_i^-}$. For details of notation in HSF, we refer to our paper[1] or a textbook.[41]

The action is real in the sense of ordinary complex conjugation $\bar{S} = S$. This property follows from the fact that $\widetilde{\widetilde{q^+}} = -q^+$.

Analytic superfields for the hypermultiplet $q_a^+(x_A, \theta^\pm, u)$ can be expanded in powers of Grassmann numbers θ as

$$\begin{aligned}
q_a^+(x_A, \theta^\pm, u) &= F_a^+(x_A, u) + \sqrt{2}\theta^+ \psi_a(x_A, u) + \sqrt{2}\bar\theta^+ \bar\varphi_a(x_A, u) \\
&+ i\theta^+ \sigma^\mu \bar\theta^+ A_{a\mu}^-(x_A, u) + \theta^+\theta^+ M_a^-(x_A, u) + \bar\theta^+ \bar\theta^+ N_a^-(x_A, u) \\
&+ \sqrt{2}\theta^+\theta^+\bar\theta^+ \chi_a^{--}(x_A, u) + \sqrt{2}\bar\theta^+\theta^+\theta^+ \xi_a^{--}(x_A, u) \\
&+ \theta^+\theta^+\bar\theta^+\bar\theta^+ D_a^{---}(x_A, u),
\end{aligned} \tag{70}$$

where a is a flavor index. Note that each component in the hypermultiplet analytic superfield (70) is a function of x_A, and the harmonic variables $u_i^\pm$. Therefore it includes infinite series of functions of x_A when expanded by the harmonic variables $u_i^\pm$ (harmonic expansions), for instance,

$$F_a^+(x_A, u) = \sum_{n=0}^\infty f^{(i_1 \cdots i_{n+1} j_1 \cdots j_n)}(x_A) u_{(i_1}^+ \cdots u_{i_{n+1}}^+ u_{j_1}^- \cdots u_{j_n)}^-. \tag{71}$$

Thus, the hypermultiplet includes infinitely many auxiliary fields in addition to physical fields.

We also use the convention to raise and lower the $SU(2)$ indices by means of ϵ_{ij} and ϵ^{ij},

$$\epsilon_{21} = \epsilon^{12} = 1, \qquad \epsilon_{12} = \epsilon^{21} = -1. \tag{72}$$

For instance the scalar fields for hypermultiplet $f_a^i, i = 1, 2$ have the property :

$$f_a^i = \epsilon^{ij} f_{aj}, \qquad f_{ai} = \epsilon_{ij} f_a^j, \tag{73}$$

$$\bar{f}_{ai} = \epsilon_{ij} \bar{f}_a^j, \qquad \bar{f}_a^i = \epsilon^{ij} \bar{f}_{aj}, \tag{74}$$

where lower index a denotes fundamental representation in flavor symmetry group. Therefore the scalar fields for hypermultiplet has the following reality property in conformity with our convention of complex conjugate in HSF (57) :

$$\left(f_a^i\right)^* = \overline{f_a^i} \equiv \bar{f}_{ai}.$$ (75)

Namely we have $\bar{f}_{a1} = -\bar{f}_a^2$, $\bar{f}_{a2} = \bar{f}_a^1$. We shall use f_a^i and its complex conjugate field $\bar{f}_{ai}$ as much as possible instead of $\bar{f}_a^i = \epsilon^{ij}\bar{f}_{aj}$.

7. The Eguchi–Hanson nonlinear sigma model in HSF

The massive HK sigma model on Eguchi–Hanson manifold[43] $(T^*\mathbf{C}P^1)$ is described in terms of harmonic superfields[41] integrated over the analytic subspace $d\zeta_A^{(-4)}du$

$$S = -\int d\zeta_A^{(-4)}du \left(\widetilde{q_1^+}D^{++}q_1^+ + \widetilde{q_2^+}D^{++}q_2^+ + V^{++}(\widetilde{q_1^+}q_2^+ - \widetilde{q_2^+}q_1^+ + \xi^{++})\right)$$ (76)

where the covariant derivative D^{++} defined in (69) contains the central charge Z satisfying the following eigenvalue equation

$$Zq_a^+(\zeta_A, u) = \frac{\mu}{2}q_a^+(\zeta_A, u).$$ (77)

This mass parameter can be attributed to the Sherk–Schwarz reduction from six dimensions[44] : $Z = -i(\partial_5 + i\partial_6)$.

The Lagrangian (76) is invariant under $O(2)$ gauge transformation

$$\delta q_1^+(\zeta_A, u) = -\lambda(\zeta_A, u)q_2^+(\zeta_A, u),$$ (78)

$$\delta q_2^+(\zeta_A, u) = \lambda(\zeta_A, u)q_1^+(\zeta_A, u),$$ (79)

$$\delta V^{++}(\zeta_A, u) = D^{++}\lambda(\zeta_A, u).$$ (80)

Similarly to the Grassmann expansion of hypermultiplets (70), the vector multiplet $V(\zeta_A, u)$ can also be expanded into infinitely many components when expanded in powers of Grassmann numbers θ. These components can then be expanded into power series in harmonic variables $u_i^\pm$. However, we can exploit the gauge transformation (80) to eliminate most of the auxiliary components in powers of Grassmann number θ and also in powers of harmonic variables $u_i^\pm$ in the vector multiplet. After eliminating infinitely many auxiliary fields by the gauge transformations, we obtain a gauge fixing

$$V_{\text{WZ}}^{++}(x_A,\theta^\pm,u) = \theta^+\theta^+\,\bar{M}_v(x_A) + \bar{\theta}^+\bar{\theta}^+ M_v(x_A) - 2i\theta^+\sigma^\mu\bar{\theta}^+ V_\mu(x_A)$$
$$+ \sqrt{2}\theta^+\theta^+\bar{\theta}^+\bar{\lambda}^i(x_A)u_i^- + \sqrt{2}\bar{\theta}^+\bar{\theta}^+\theta^+\lambda^i(x_A)u_i^-$$
$$+ \theta^+\theta^+\bar{\theta}^+\bar{\theta}^+ D_v^{(ij)}(x_A)u_{(i}^- u_{j)}^-, \tag{81}$$

which is called the Wess–Zumino gauge and is denoted by the suffix WZ. As a result, the remaining fields $M_v(x_A)$, $V_\mu(x_A)$, $\lambda^i(x_A)$ in (81) are physical fields except $D_v(x_A)^{(ij)}$, if there is a kinetic term for vector multiplet. The field $D_v(x_A)^{(ij)}$ is the usual SUSY auxiliary field. However, we will use here a vector multiplet with no kinetic term. Therefore we will eventually eliminate all these component fields in the vector multiplet giving rise to constraints for hypermultiplets.

After integrating Grassmann variables and the harmonic variables, and eliminating infinitely many auxiliary fields of the hypermultiplet expanded in powers of harmonic variables $u_i^\pm$, and taking the Wess–Zumino gauge for the vector multiplet in HSF, we obtain the bosonic part of the action as

$$\mathcal{L}_{\text{boson}}^{\text{HSF}}$$
$$= -\left(\partial_A^\mu f_1^i + V^\mu f_2^i\right)\left(\partial_\mu^A \bar{f}_{1i} + V_\mu \bar{f}_{2i}\right) - \left(\partial_A^\mu f_2^i - V^\mu f_1^i\right)\left(\partial_\mu^A \bar{f}_{2i} - V_\mu \bar{f}_{1i}\right)$$
$$-\frac{1}{2}\left(\bar{M}_v \bar{f}_{1i} - \frac{\bar{\mu}}{2}\bar{f}_{2i}\right)\left(M_v f_1^i - \frac{\mu}{2}f_2^i\right) - \frac{1}{2}\left(\bar{M}_v \bar{f}_{2i} + \frac{\bar{\mu}}{2}\bar{f}_{1i}\right)\left(M_v f_2^i + \frac{\mu}{2}f_1^i\right)$$
$$-\frac{1}{2}\left(M_v \bar{f}_{1i} + \frac{\mu}{2}\bar{f}_{2i}\right)\left(\bar{M}_v f_1^i + \frac{\bar{\mu}}{2}f_2^i\right) - \frac{1}{2}\left(M_v \bar{f}_{2i} - \frac{\mu}{2}\bar{f}_{1i}\right)\left(\bar{M}_v f_2^i - \frac{\bar{\mu}}{2}f_1^i\right)$$
$$-\frac{1}{3}D_{v(ij)}(-\bar{f}_1^{(i} f_2^{j)} + \bar{f}_2^{(i} f_1^{j)} + \xi^{(ij)})$$
$$= \mathcal{L}_{\text{kin}}^{\text{HSF}} + \mathcal{L}_{\text{constr}}^{\text{HSF}} + \mathcal{L}_{\text{pot}}^{\text{HSF}}, \tag{82}$$

$$\mathcal{L}_{\text{kin}}^{\text{HSF}} = -\partial_\mu^A f_a^i \partial_A^\mu \bar{f}_{ai} + \partial_\mu^A f_a^i \epsilon_{ab} V^\mu \bar{f}_{bi} - \epsilon_{ab} V^\mu f_a^i \partial_\mu^A \bar{f}_{bi}$$
$$- V^\mu V_\mu f_a^i \bar{f}_{ai}, \tag{83}$$

$$\mathcal{L}_{\text{constr}}^{\text{HSF}} = -\frac{1}{3}D_{v(ij)}(-\bar{f}_1^{(i} f_2^{j)} + \bar{f}_2^{(i} f_1^{j)} + \xi^{(ij)}), \tag{84}$$

where $a = 1,2$ denotes fundamental representation in $O(2)$ gauge group. The scalar potential $V(f,\bar{f})$ is given by

$$-\mathcal{L}_{\text{pot}}^{\text{HSF}} = V(f,\bar{f})$$
$$= \left(\left|\frac{\mu}{2}\right|^2 + |M_v|^2\right)\left(f_1^i \bar{f}_{1i} + f_2^i \bar{f}_{2i}\right)$$
$$+ \left(\frac{\mu}{2}\bar{M}_v - \frac{\bar{\mu}}{2}M_v\right)\left(f_1^i \bar{f}_{2i} - f_2^i \bar{f}_{1i}\right). \tag{85}$$

Let us stress once again that we adopt a convention for complex conjugation of complex scalar fields $\left(f_a^i\right)^* \equiv \bar{f}_{ai} = \epsilon_{ij}\bar{f}_a^j$, and use f_a^i and $\bar{f}_{ai}$ to denote a complex conjugate pair.

There are still auxiliary fields M_v and V^μ and $D_{v(ij)}$ of the vector multiplet. By changing variables, we can also introduce the most frequently used parameterization given by Curtright and Freedman[32]: four complex fields $\phi_i^\alpha, \alpha = 1, 2, \ i = 1, 2$

$$\phi_1^\alpha = \frac{1}{\sqrt{2}}(f_1^{2,\alpha} + if_2^{2,\alpha}), \qquad \phi_2^\alpha = \frac{1}{\sqrt{2}}(f_1^{1,\alpha} + if_2^{1,\alpha}), \tag{86}$$

where $f_a^{i,1} = f_a^i$ and $f_a^{i,2} = \bar{f}_a^i$.

The Lagrangian in the HSF (82)–(85) can be related to the component Lagrangian (45)–(48) in terms of $\mathcal{N} = 1$ superfields in the $O(2)$ basis in the Wess–Zumino gauge by the following identification :

$$M_v = i\Sigma, \qquad V_\mu = \frac{v_\mu}{2}, \tag{87}$$

and fields f_a^i of HSF can be identified with $\mathcal{N} = 1$ fields $\tilde{\phi}^a$, $\tilde{\chi}_a$ in the $O(2)$ basis (45)–(48) as

$$f_a^1 = \left(\tilde{\phi}^a\right)^*, \qquad f_a^2 = \tilde{\chi}_a. \tag{88}$$

The Fayet–Iliopoulos parameters $\xi^{(ij)}$ in HSF are identified with Fayet–Iliopoulos parameters c, b, b^* in the $\mathcal{N} = 1$ superfield formalism as

$$\xi^{11} = -ib^*, \qquad \xi^{22} = ib, \qquad \xi^{12} = \xi^{21} = \frac{ic}{2}. \tag{89}$$

The auxiliary fields $D_{v(ij)}$ in HSF are identified with the auxiliary fields D, F_Σ, F_Σ^* in the $\mathcal{N} = 1$ superfield formalism as

$$D_{v(11)} = 3iF_\Sigma^*, \qquad D_{v(22)} = -3iF_\Sigma, \qquad D_{v(12)} = -\frac{3iD}{2}. \tag{90}$$

These results are in conformity with the reality property of the Fayet–Iliopoulos parameters $\xi^{(ij)}$ and the auxiliary fields $D_{v(ij)}$ in HSF and those in $\mathcal{N} = 1$ superfield formalism

$$\xi^{(ij)} = \epsilon^{ik}\epsilon^{jl}\left(\xi^{(kl)}\right)^*, \qquad D_{v(ij)} = \epsilon_{ik}\epsilon_{jl}\left(D_{v(kl)}\right)^*. \tag{91}$$

$$b \in \mathbb{C}, \quad c \in \mathbb{R}, \qquad F_\Sigma \in \mathbb{C}, \quad D \in \mathbb{R}. \tag{92}$$

The identification (88) implies that the complex scalar fields f_a^1 belong to anti-chiral scalar superfields, and f_a^2 to chiral scalar superfields. The suffix a denotes fundamental representation of the gauge group $O(2)$.

The complex fields ϕ_i^α in the Curtright–Freedman basis (86) are more directly related to the complex scalar fields of the $\mathcal{N}=1$ superfields ϕ_i, χ_i in the $U(1)$ basis in (9)–(12) as

$$\phi_1{}^1 = \frac{1}{\sqrt{2}} \left(f_1^2 + if_2^2\right) = \frac{1}{\sqrt{2}} \left(\tilde{\chi}_1 + i\tilde{\chi}_2\right) = \chi_1, \tag{93}$$

$$\phi_2{}^1 = \frac{1}{\sqrt{2}} \left(f_1^1 + if_2^1\right) = \frac{1}{\sqrt{2}} \left(\tilde{\phi}_1^* + i\tilde{\phi}_2^*\right) = (\phi_1)^*, \tag{94}$$

$$\phi_1{}^2 = \frac{1}{\sqrt{2}} \left(\bar{f}_1^2 + i\bar{f}_2^2\right) = \frac{1}{\sqrt{2}} \left(-\bar{f}_{11} - i\bar{f}_{21}\right) = \frac{1}{\sqrt{2}} \left(-\tilde{\phi}^1 - i\tilde{\phi}^2\right) = \chi_2, \tag{95}$$

$$\phi_2{}^2 = \frac{1}{\sqrt{2}} \left(\bar{f}_1^1 + i\bar{f}_2^1\right) = \frac{1}{\sqrt{2}} \left(\bar{f}_{12} + i\bar{f}_{22}\right) = \frac{1}{\sqrt{2}} \left(\tilde{\chi}_1^* + i\tilde{\chi}_2^*\right) = (\phi_2)^*. \tag{96}$$

Therefore the complex scalar fields ϕ_1^i are identified as those of chiral scalar superfield, and ϕ_2^i are identified as those of anti-chiral scalar superfield. We also notice that all the complex fields $\phi_i^\alpha, i=1,2, \alpha=1,2$ in the Curtright-Freedman basis have $U(1)$ charge -1 in conformity with the charge assignment obtained in the model constructed by the tensor calculus for supergravity.[17] This supergravity model shows that our model can be embedded into supergravity. Moreover it explicitly demonstrates that our model can be extended to a model in five dimensions.

The equation of motion for the gauge field V_μ gives

$$V_\mu = \frac{\epsilon_{ab} \left(\partial_\mu^A f_a^i \bar{f}_{bi} - f_a^i \partial_\mu^A \bar{f}_{bi}\right)}{2 f_a^i \bar{f}_{ai}}. \tag{97}$$

After eliminating the vector field V^μ, we obtain the kinetic term for the scalar fields f_a^i in the hypermultiplets

$$\mathcal{L}_{\text{kin}}^{\text{HSF}} = -\partial_A^\mu f_1^i \partial_\mu^A \bar{f}_{1i} - \partial_A^\mu f_2^i \partial_\mu^A \bar{f}_{2i} + \frac{(f_2^i \overset{\leftrightarrow}{\partial_A^\mu} \bar{f}_{1i} - f_1^i \overset{\leftrightarrow}{\partial_A^\mu} \bar{f}_{2i})^2}{4(f_1^i \bar{f}_{1i} + f_2^i \bar{f}_{2i})}. \tag{98}$$

Integrating over scalar M_v and the auxiliary fields $D_{v(ij)}$ in the vector multiplet, we obtain constraints

$$-\bar{f}_1^{(i} f_2^{j)} + \bar{f}_2^{(i} f_1^{j)} + \xi^{(ij)} = 0. \tag{99}$$

This constraint makes the target space of the massive nonlinear sigma model into the Eguchi–Hanson manifold.[1] In the case of massless (without potential) model, the target metric for the four independent real bosonic fields has been shown to be just the Eguchi–Hanson metric.[32,40,45]

The equation of motion for scalar field M_v gives

$$M_v = \frac{\mu}{2} \frac{\epsilon_{ab} f_a^i \bar{f}_{bi}}{f_c^i \bar{f}_{ci}}, \tag{100}$$

where the flavor indices are summed. Integrating over M_v gives the potential term as

$$V(f_1, f_2) = \left|\frac{\mu}{2}\right|^2 \frac{1}{f_1^i \bar{f}_{1i} + f_2^i \bar{f}_{2i}} \left\{ -|f_1^i \bar{f}_{2i} - f_2^i \bar{f}_{1i}|^2 + (f_1^i \bar{f}_{1i} + f_2^i \bar{f}_{2i})^2 \right\}. \tag{101}$$

The parameters $\xi^{(ij)}$ have mass dimensions two and represent the scale of the curvature of the target manifold.

The bosonic action becomes in the Curtright–Freedman basis as

$$\mathcal{L}_{\text{boson}} = -\partial_A^\mu \phi_1 \partial_\mu^A \bar{\phi}_1 - \partial_A^\mu \phi_2 \partial_\mu^A \bar{\phi}_2 - \frac{(\phi_1 \overset{\leftrightarrow}{\partial_A^\mu} \bar{\phi}_1 + \phi_2 \overset{\leftrightarrow}{\partial_A^\mu} \bar{\phi}_2)^2}{4(|\phi_1|^2 + |\phi_2|^2)}$$
$$- \frac{\mu^2}{4(|\phi_1|^2 + |\phi_2|^2)} \left(-(\phi_1 \sigma^3 \bar{\phi}_1 + \phi_2 \sigma^3 \bar{\phi}_2)^2 + (|\phi_1|^2 + |\phi_2|^2)^2 \right). \tag{102}$$

8. Nonlinear sigma model in independent fields : Spherical coordinates and Gibbons–Hawking parameterization

Here we shall describe the model in terms of independent fields in several parameterizations by solving the constraints (99). In the following we shall take

$$\xi^{(12)} \equiv -i\xi, \qquad \xi^{(11)} = \xi^{(22)} = 0 \tag{103}$$

for simplicity. Then the constraints (99) become

$$|\phi_1|^2 - |\phi_2|^2 = 2\xi, \qquad \phi_1^* \phi_2 = \phi_2^* \phi_1 = 0. \tag{104}$$

It is convenient to introduce independent fields z^α, $\bar{z}^\alpha$, $\alpha = 1, 2$ through the following Ansatz[22,45]

$$\phi_1^\alpha = g(r)\frac{z^\alpha}{\sqrt{r}}, \qquad \phi_2^\alpha = f(r)i\sigma_2^{\alpha\beta}\frac{\bar{z}^\beta}{\sqrt{r}}, \tag{105}$$

$$r = z^1 \bar{z}^1 + z^2 \bar{z}^2, \tag{106}$$

where z^α are complex fields satisfying

$$\phi_1^1 \phi_2^2 - \phi_1^2 \phi_2^1 = -z^1 \bar{z}^1 - z^2 \bar{z}^2. \tag{107}$$

The real functions $f(r)$ and $g(r)$ are uniquely determined by the constraints (99) and (107) as

$$f(r)^2 = -\xi + \sqrt{r^2 + \xi^2}, \qquad g(r)^2 = \xi + \sqrt{r^2 + \xi^2}. \tag{108}$$

The action can be described without constraint by the independent complex fields z^α. These fields z^α are invariant under the $O(2)(U(1))$ gauge transformations in (80) which is used to take the quotient of the target manifold.

Another useful parameterization of the model is given by the spherical coordinates which are invariant under the $U(1)$ gauge transformations

$$z^1 = \sqrt{r}\cos\tfrac{\Theta}{2}\exp\tfrac{i}{2}(\Psi+\Phi), \tag{109}$$

$$z^2 = \sqrt{r}\sin\tfrac{\Theta}{2}\exp\tfrac{i}{2}(\Psi-\Phi), \tag{110}$$

$$0 \le r \le \infty, \quad 0 \le \Theta \le \pi, \quad 0 \le \Phi \le 2\pi, \quad 0 \le \Psi \le 2\pi, \tag{111}$$

$$f_1^1 = \frac{\phi_2^1 - \bar\phi_1^2}{\sqrt{2}} = -\frac{g(r)-f(r)}{\sqrt{2r}}\bar z^2 = -\frac{g(r)-f(r)}{\sqrt{2}}\sin(\frac{\Theta}{2})e^{\frac{i}{2}(-\Psi+\Phi)}, \tag{112}$$

$$f_2^1 = \frac{\phi_2^1 + \bar\phi_1^2}{i\sqrt{2}} = \frac{g(r)+f(r)}{i\sqrt{2r}}\bar z^2 = \frac{g(r)+f(r)}{i\sqrt{2}}\sin(\frac{\Theta}{2})e^{\frac{i}{2}(-\Psi+\Phi)}, \tag{113}$$

$$f_1^2 = \frac{\phi_1^1 + \bar\phi_2^2}{\sqrt{2}} = \frac{g(r)-f(r)}{\sqrt{2r}}z^1 = \frac{g(r)-f(r)}{\sqrt{2}}\cos(\frac{\Theta}{2})e^{\frac{i}{2}(\Psi+\Phi)}, \tag{114}$$

$$f_2^2 = \frac{\phi_1^1 - \bar\phi_2^2}{i\sqrt{2}} = \frac{g(r)+f(r)}{i\sqrt{2r}}z^1 = \frac{g(r)+f(r)}{i\sqrt{2}}\cos(\frac{\Theta}{2})e^{\frac{i}{2}(\Psi+\Phi)}, \tag{115}$$

$$\phi_1^1 = g(r)\cos(\frac{\Theta}{2})\exp(\frac{i}{2}(\Psi+\Phi)), \tag{116}$$

$$\phi_1^2 = g(r)\sin(\frac{\Theta}{2})\exp(\frac{i}{2}(\Psi-\Phi)), \tag{117}$$

$$\phi_2^1 = f(r)\sin(\frac{\Theta}{2})\exp(-\frac{i}{2}(\Psi-\Phi)), \tag{118}$$

$$\phi_2^2 = -f(r)\cos(\frac{\Theta}{2})\exp(-\frac{i}{2}(\Psi+\Phi)). \tag{119}$$

The bosonic action becomes in the spherical coordinates as

$$\mathcal{L}_{\text{boson}} = \frac{1}{2\sqrt{\xi^2+r^2}}\left[-\partial_\mu^A r\partial_A^\mu r - \left(r^2+\xi^2\right)\partial_\mu^A\Theta\partial_A^\mu\Theta \right.$$
$$- \left(r^2+\xi^2\sin^2\Theta\right)\partial_\mu^A\Phi\partial_A^\mu\Phi - r^2\partial_\mu^A\Psi\partial_A^\mu\Psi - 2r^2\cos\Theta\partial_\mu^A\Phi\partial_A^\mu\Psi$$
$$\left. -|\mu|^2\left(r^2+\xi^2\sin^2\Theta\right)\right]. \tag{120}$$

It is also useful to change variables into the following parameterization appropriate to describe the Gibbons–Hawking multi-center metric[47]

$$X^1 = r \sin \Theta \cos \Psi, \tag{121}$$

$$X^2 = r \sin \Theta \sin \Psi, \tag{122}$$

$$X^3 = \sqrt{r^2 + \xi^2} \cos \Theta, \tag{123}$$

$$\varphi = \Phi + \Psi. \tag{124}$$

By using the parameterisation (86)–(110) and (121)–(124), the bosonic part of the action (98) can be rewritten as

$$\mathcal{L} = -\frac{1}{2} \left\{ U \partial_\mu \mathbf{X} \cdot \partial^\mu \mathbf{X} + U^{-1} \mathcal{D}_\mu \varphi \mathcal{D}^\mu \varphi + \mu^2 U^{-1} \right\}, \tag{125}$$

where $\mathcal{D}_\mu \varphi = \partial_\mu \varphi + \mathbf{A} \cdot \partial_\mu \mathbf{X}$ and

$$\boldsymbol{\nabla} \times \mathbf{A} = \boldsymbol{\nabla} U. \tag{126}$$

The harmonic function U can be described

$$U = \frac{1}{2} \left[\frac{1}{|\mathbf{X} - \xi \mathbf{n}|} + \frac{1}{|\mathbf{X} + \xi \mathbf{n}|} \right], \tag{127}$$

where $\mathbf{n}$ is a unit three vector, which is given by $\mathbf{n} = (0,0,1)$. $\mathbf{A}$ is a potential whose solution is given as

$$A_1 = \frac{1}{2} \left\{ \frac{X^2}{|\mathbf{X} - \xi \mathbf{n}|(X^3 - \xi + |\mathbf{X} - \xi \mathbf{n}|)} + \frac{X^2}{|\mathbf{X} + \xi \mathbf{n}|(X^3 + \xi + |\mathbf{X} - \xi \mathbf{n}|)} \right\},$$

$$A_2 = \frac{1}{2} \left\{ \frac{-X^1}{|\mathbf{X} - \xi \mathbf{n}|(X^3 - \xi + |\mathbf{X} - \xi \mathbf{n}|)} + \frac{-X^1}{|\mathbf{X} + \xi \mathbf{n}|(X^3 + \xi + |\mathbf{X} + \xi \mathbf{n}|)} \right\},$$

$$A_3 = 0. \tag{128}$$

It is found that the target metric of the action (125) is just the Eguchi–Hanson metric.[32,40,45]

9. BPS equation and domain wall solution

In this section, we give the BPS domain wall solution in our model. In the following the complex mass parameter μ is taken to be real for simplicity. By requiring that the fermions conserve half of SUSY we obtain the BPS equations in HSF (for a detailed derivation, see our Ref.[1])

$$(M_v + V_2)f_2^1 + \left(\frac{\mu}{2} + \partial_2^A \right) f_1^1 = 0, \tag{129}$$

$$(M_v - V_2)f_2^2 + \left(\frac{\mu}{2} - \partial_2^A \right) f_1^2 = 0, \tag{130}$$

$$-(M_v + V_2)f_1^1 + \left(\frac{\mu}{2} + \partial_2^A \right) f_2^1 = 0, \tag{131}$$

$$-(M_v - V_2)f_1^2 + \left(\frac{\mu}{2} - \partial_2^A \right) f_2^2 = 0. \tag{132}$$

BPS wall solution should approach the supersymmetric discrete vacua as $y \to \pm\infty$. From the trivial solution of BPS equation (translational invariant solution), we find[1] that there are only two vacua : $(r, \Theta) = (0,0), (0,\pi)$ in terms of the spherical coordinates (109), (110). Another way of understanding these vacua is to observe from Eq.(120) that these two points are the minima of the scalar potential

$$V(r,\Theta,\Phi,\Psi) = \frac{|\mu|^2 \left(r^2 + \xi^2 \sin^2 \Theta\right)}{2\sqrt{\xi^2 + r^2}} \tag{133}$$

with vanishing vacuum energy

$$V(r = 0, \Theta = 0) = V(r = 0, \Theta = \pi) = 0. \tag{134}$$

Therefore we consider the domain wall solution connects these vacua, and we can expect that Θ has nontrivial configuration. After some algebra, we obtain four independent differential equations in terms of the spherical coordinates[1]

$$r' = \mu \cos\Theta \cdot r, \qquad r \cdot \Psi' = 0, \tag{135}$$

$$\Theta' = -\mu \sin\Theta, \qquad \sin\Theta \cdot \Phi' = 0. \tag{136}$$

The boundary condition of $r = 0$ at $y = -\infty$ dictates the solution of (135) to be $r = 0$ and $\Psi = 0$. The other two equations in (136) gives a nontrivial dependence in y resulting in the following BPS solutions

$$\Theta = \arccos[\tanh\mu(y + y_0)], \qquad \Phi = \varphi_0, \tag{137}$$

where y_0 and φ_0 are real constants: y_0 determines the position of the domain wall along y direction and φ_0 corresponds to the Nambu–Goldstone (NG) mode of $U(1)$ isometry of target space. The BPS wall solution is illustrated in Fig.2.

Our BPS wall solution is obtained from $\mathcal{N} = 2$ SUSY theory in four dimensions. However, subsequent study[17] revealed that our model can be extended to an $\mathcal{N} = 2$ SUSY theory in five dimensions. In fact we obtain precisely the same BPS solution as in (137) by taking the limit of vanishing gravitational coupling in the BPS wall solution in five-dimensional supergravity.[17] Therefore we can now use our BPS wall solution as a starting point for an interesting phenomenology for a unified model : our four-dimensional spacetime being a wall in higher dimensional spacetime following the brane-world scenario.

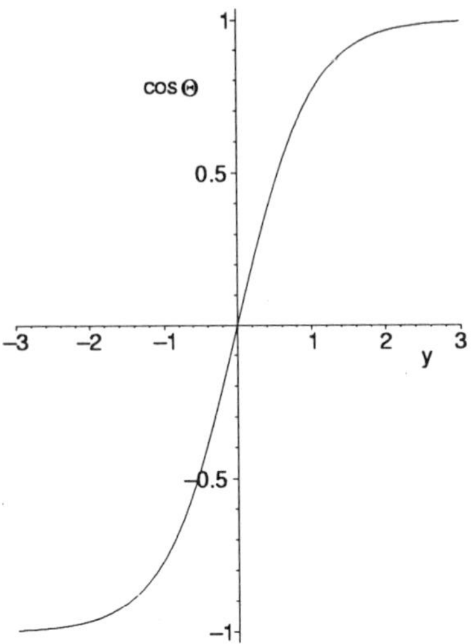

Fig. 2. BPS domain wall solution of field $\cos\Theta$ as a function of y with $y_0 = 0$.

In terms of harmonic superfields (76) and their bosonic components (98), the BPS solution is given by

$$q_1^+ = f_1^i u_i^+ = \sqrt{\frac{\xi}{2}}\,e^{\frac{i}{2}\varphi_0}\left(\begin{array}{c}-\sqrt{1-\tanh(\mu(y+y_0))}u_1^+\\ \sqrt{1+\tanh(\mu(y+y_0))}u_2^+\end{array}\right), \tag{138}$$

$$q_2^+ = f_2^i u_i^+ = -i\sqrt{\frac{\xi}{2}}\,e^{\frac{i}{2}\varphi_0}\left(\begin{array}{c}\sqrt{1-\tanh(\mu(y+y_0))}u_1^+\\ \sqrt{1+\tanh(\mu(y+y_0))}u_2^+\end{array}\right). \tag{139}$$

In terms of the fields in the Curtright-Freedman basis (102), the BPS wall solution is given by

$$\phi_1^1 = \sqrt{\xi(1+\tanh\mu(y+y_0))}\,e^{\frac{i}{2}\varphi_0}, \tag{140}$$

$$\phi_1^2 = \sqrt{\xi(1-\tanh\mu(y+y_0))}\,e^{-\frac{i}{2}\varphi_0}, \tag{141}$$

$$\phi_2^1 = \phi_2^2 = 0. \tag{142}$$

The BPS wall solution in the Gibbons–Hawking multi-center metric parameterization (125) is given by

$$X^1 = X^2 = 0, \tag{143}$$

$$X^3 = \xi\tanh\mu(y+y_0), \tag{144}$$

$$\varphi = \varphi_0. \tag{145}$$

Acknowledgements

One of the authors (NS) thanks Yoshiaki Tanii for useful discussion on hypermultiplets in six dimensions. This work is supported in part by Grant-in-Aid for Scientific Research from the Japan Ministry of Education, Science and Culture 13640269. The work of M. Naganuma is supported by JSPS Fellowship. The work of M. Nitta is supported by the U. S. Department of Energy under grant DE-FG02-91ER40681 (Task B).

References

1. M. Arai, M. Naganuma, M. Nitta, and N. Sakai, "Manifest Supersymmetry for BPS Walls in $\mathcal{N} = 2$ Nonlinear Sigma Models" *Nucl. Phys.* **B 652** (2003) 35 [hep-th/0211103].

2. S. Dimopoulos and H. Georgi, *Nucl. Phys.* **B193** (1981) 150; N. Sakai, Z. f. Phys. **C11** (1981) 153; E. Witten, *Nucl. Phys.* **B188** (1981) 513; S. Dimopoulos, S. Raby, and F. Wilczek, *Phys. Rev.* **D24** (1981) 1681.

3. N. Arkani-Hamed, S. Dimopoulos, and G. Dvali, *Phys. Lett.* **429B** (1998) 263 [hep-ph/9803315]; I. Antoniadis, N. Arkani–Hamed, S. Dimopoulos, and G. Dvali, *Phys. Lett.* **436B** (1998) 257 [hep-ph/9804398].

4. L. Randall and R. Sundrum, *Phys. Rev. Lett.* **83** (1999) 3370 [hep-ph/9905221]; *Phys. Rev. Lett.* **83** (1999) 4690 [hep-th/9906064].

5. E. Bogomol'nyi, *Sov. J. Nucl. Phys.* **B24** (1976) 449; M. K. Prasad and C. H. Sommerfield, *Phys. Rev. Lett.* **35** (1975) 760.

6. E. Witten and D. Olive, *Phys. Lett.* **B78** (1978) 97.

7. N. Seiberg and E. Witten, *Nucl. Phys.* **B426** (1994) 19 [hep-th/9407087]; *Nucl. Phys.* **B431** (1994) 484 [hep-th/9408099].

8. M. Cvetic, F. Quevedo, and S. Rey, *Phys. Rev. Lett.* **67** (1991) 1836; M. Cvetic, S. Griffies, and S. Rey, *Nucl. Phys.* **B381** (1992) 301 [hep-th/9201007]; G. Dvali and M. Shifman, *Nucl. Phys.* **B504** (1997) 127 [hep-th/9611213]; *Phys. Lett.* **396B** (1997) 64 [hep-th/9612128]; A. Kovner, M. Shifman, and A. Smilga, *Phys. Rev.* **D56** (1997) 7978 [hep-th/9706089]; B. Chibisov and M. Shifman, *Phys. Rev.* **D56** (1997) 7990 [hep-th/9706141]; A. Smilga and A. Veselov, *Phys. Rev. Lett.* **79** (1997) 4529 [hep-th/9706217]; J. Edelstein, M.L. Trobo, F. Brito, and D. Bazeia, *Phys. Rev.* **D57** (1998) 7561 [hep-th/9707016]; V. Kaplunovsky, J. Sonnenschein, and S. Yankielowicz, *Nucl. Phys.* **B552** (1999) 209 [hep-th/9811195]; G. Dvali, G. Gabadadze, and Z. Kakushadze, *Nucl. Phys.* **B562** (1999) 158 [hep-th/9901032]; B. de Carlos and J. M. Moreno, *Phys. Rev. Lett.* **83** (1999) 2120 [hep-th/9905165]; M. Naganuma and M. Nitta, *Prog. Theor. Phys.* **105** (2001) 501 [hep-th/0007184]; D. Binosi and T. ter Veldhuis, *Phys. Rev.* **D63** (2001) 085016, [hep-th/0011113].

9. N. Maru, N. Sakai, Y. Sakamura, and R. Sugisaka, *Phys. Lett.* **B496** (2000) 98, [hep-th/0009023].

10. N. Maru, N. Sakai, Y. Sakamura, and R. Sugisaka, *Nucl. Phys.* **B616** (2001)

47 [hep-th/0107204]; the Proceedings of the 10th Tohwa international symposium on string theory, American Institute of Physics, 607, pages 209–215, (2002) [hep-th/0109087]; "SUSY Breaking by stable non-BPS configurations", to appear in the Proceedings of the Corfu Summer Institute on Elementary particle Physics, Corfu, September 2001 [hep-th/0112244].

11. N. Sakai and R. Sugisaka, *Int. J. Mod. Phys.* **A17** (2002) 4697 [hep-th/0204214].

12. M. Eto, N. Maru, N. Sakai, and T. Sakata, *Phys. Lett.* **B553** (2003) 87 [hep-th/0208127].

13. E. R. C. Abraham and P. K. Townsend, *Nucl. Phys.* **B351** (1991) 313; G. Gibbons and P. Townsend, *Phys. Rev. Lett.* **83** (1999) 1727 [hep-th/9905196]; S.M. Carroll, S. Hellerman, and M. Trodden, *Phys. Rev.* **D61** (2000) 065001 [hep-th/9905217]; P.M. Saffin, *Phys. Rev. Lett.* **83** (1999) 4249–4252 [hep-th/9907066].

14. H. Oda, K. Ito, M. Naganuma, and N. Sakai, *Phys. Lett.* **B471** (1999) 140 [hep-th/9910095]; K. Ito, M. Naganuma, H. Oda, and N. Sakai, *Nucl. Phys.* **B586** (2000) 231 [hep-th/0004188]; *Nucl. Phys. Proc. Suppl.* 101 (2001) 304 [hep-th/0012182].

15. M. Naganuma, M. Nitta, and N. Sakai, *Phys. Rev.* **D65** (2001) 045016 [hep-th/0108179]; "BPS Walls and Junctions in $\mathcal{N} = 1$ SUSY Nonlinear Sigma Models" to appear in the the Proceedings of the third international Sakharov conference on Physics, [hep-th/0210205].

16. A. Gorsky and M. Shifman, *Phys. Rev.* **D61** (2000) 085001 [hep-th/9909015]; *Nucl. Phys.* **B586** (2000) 231 [hep-th/0004188]; D. Binosi and T. ter Veldhuis, *Phys. Lett.* **B476** (2000) 124 [hep-th/9912081]; S. Nam and K. Olsen, *JHEP* **0008** (2000) 001 [hep-th/0002176].

17. M. Arai, S. Fujita, M. Naganuma, and N. Sakai, "Wall Solution with Weak Gravity Limit in Five Dimensional Supergravity" [hep-th/0212175], to appear in *Phys. Lett.* **B**.

18. P. Fayet, *Nucl. Phys.* **B113** (1976) 135.

19. J. Wess and J. Bagger, "Supersymmetry and Supergravity", (1991), Princeton University Press.

20. L. Alvarez–Gaumé and D. Z. Freedman, *Comm. Math. Phys.* **80** (1981) 443.

21. B. Zumino, *Phys. Lett.* **87B** (1979) 203.

22. L. Alvarez–Gaumé and D. Z. Freedman, *Comm. Math. Phys.* **91** (1983) 87.

23. J. Sherk and J.H. Schwarz, *Phys. Lett.* **B82** (1979) 60.

24. G. Sierra and P. K. Townsend, *Nucl. Phys.* B 233 (1984) 289.

25. E. Abraham and P. K. Townsend, *Phys. Lett.* B 291 (1992) 85.

26. J. P. Gauntlett, D. Tong, and P.K. Townsend, *Phys. Rev.* **D63** (2001) 085001 [hep-th/0007124].

27. J. P. Gauntlett, R. Portugues, D. Tong, and P.K. Townsend, *Phys. Rev.* **D63** (2001) 085002 [hep-th/0008221].

28. J. P. Gauntlett, D. Tong, and P. K. Townsend, *Phys. Rev.* **D64** (2001) 025010 [hep-th/0012178].

29. D. Tong, *Phys. Rev.* **D66** (2002) 025013 [hep-th/0202012].

30. M. Naganuma, M. Nitta, and N. Sakai, *Grav. Cosmol.* **8** (2002) 129, [hep-

th/0108133].

31. R. Portugues and P. K. Townsend, *JHEP* **0204** (2002) 039 [hep-th/0203181].
32. T. L. Curtright and D. Z. Freedman, *Phys. Lett.* **90B** (1980) 71.
33. L. Alvarez–Gaumé and D. Z. Freedman, *Phys. Lett.* **94B** (1980) 171.
34. M. Roček and P. K. Townsend, *Phys. Lett.* **96B** (1980) 72.
35. U. Lindström and M. Roček, *Nucl. Phys.* **B222** (1983) 285; N. J. Hitchin, A. Karlhede, U. Lindström, and M. Roček, *Comm. Math. Phys.* **108** (1987) 535.
36. C. M. Hull, A. Karlhede, U. Lindström and M. Roček, *Nucl. Phys.* **B266** (1986) 1.
37. Y. Sakamura, "Superfield Description of Effectiver Theories on BPS Domain Walls", [hep-th/0207159].
38. A. Galperin, E. Ivanov, S. Kalitzin, V. Ogievetsky, and E. Sokatchev, *Class. Quantum Grav.* **1** (1984) 469; A. Galperin, E. Ivanov, V. Ogievetsky, and E. Sokatchev, *Class. Quantum Grav.* **2** (1985) 601; 617.
39. B. M. Zupnik, *Theor. Math. Phys.* **130** (2002) 213 [hep-th/0107012].
40. A. Galperin, E. Ivanov, V. Ogievetsky, and P. K. Townsend, *Class. Quantum Grav.* **3** (1986) 625.
41. A. Galperin, E. Ivanov, V. Ogievetsky, and E. Sokatchev, "Harmonic Superspace", Cambridge University Press, Cambridge (2001).
42. S. Ketov and C. Unkmeir, *Phys. Lett.* **422B** (1998) 179 [hep-th/9710185].
43. T. Eguchi and A. J. Hanson, *Phys. Lett.* **74B** (1978) 249; *Ann. Phys.* **120** (1979) 82.
44. B. Zupnik, *Sov. J. Nucl. Phys.* **44** (1986) 512; E. Ivanov, S. Ketov, and B. Zupnik, *Nucl. Phys.* **509** (1998) 53.
45. G. W. Gibbons, D. Olivier, P. J. Ruback, and G. Valent, *Nucl. Phys.* **B296** (1988) 679.
46. G. W. Gibbons and C. N. Pope, *Commun. Math. Phys.* **66** (1979) 267.
47. G. W. Gibbons and S. W. Hawking, *Phys. Lett.* **78B** (1978) 249.
48. M. Shifman and A. Yung, "Domain Walls and Flux Tubes in $\mathcal{N} = 2$ SQCD : D-brane Prototypes", [hep-th/0212293].

Current Algebra Approach to String Theory

Machiko Hatsuda

Theory Division, High Energy Accelerator Research Organization (KEK),
Tsukuba, Ibaraki, 305-0801, Japan
E-mail:mhatsuda@post.kek.jp

Warren Siegel

C.N. Yang Institute for Theoretical Physics (CNYITP),
State University of New York, Stony Brook, NY 11794-3840, USA
E-mail: siegel@insti.physics.sunysb.edu

We examine the BRST operators written by the current with an arbitrary representation. We find all irreducible representations of the current algebra constrained by the BRST nilpotency, resulting exactly the same set expected as the superstring theory.

1. Introduction

Superstring theories in curved backgrounds are widely examined recently[1-5] motivated by AdS/CFT correspondence.[6] Classical superstring mechanics describes symmetries depending on its backgrounds leading to the same number of degrees of freedom.[7,3] On the other hand quantum information of superstrings in curved backgrounds are hardly obtained except for a flat case; canonical quantization is difficult for a AdS superstring. Even if it is possible for a pp-wave superstring in the lightcone gauge, the critical dimension cannot be obtained by the absence of boost generator in this background.

It may be interesting to focus on a background independent feature and explore its quantum aspects.

It is shown[3] that the local symmetry algebra for a AdS superstring is the same as the flat superstring,[7] namely $\mathcal{ABCD}$ constraints algebra.[8,9] Since global symmetry of the pp-wave backgound is obtained by the Penrose limit from the AdS background without any sigular treatment,[10] it is plausible to expect the same $\mathcal{ABCD}$ constraints algebra as its local symmetry algebra.[5] The same local symmetry algebra leads to the same BRST charge,[9] while the constraints are written in terms of model dependent coordinates in general. However usual coordinates may not be neccessary especially in curved backgrounds, instead nonabelian currents give simpler description.[2] The quantum physical states are determined by the nilpotent BRST charge written by the current algebra language, and its consistency leads to its representation and the critical dimension.

In this paper we present the procedure for a flat superstring case without imposing the kappa symmetry (*i.e.* no $\mathcal{BC}$ constraints) toward general background applications. We will write the BRST operator for $\mathcal{AD}$ constraints in terms of the $\mathrm{OSp}(d|2)$ current by adding two anticommuting directions. This covariantized lightcone space $\mathrm{OSp}(d|2)$ is analyzed in $\mathrm{SO}(D)$, $D = d-2$ avoiding irrelevant complication caused by ghosts.[8] Then physical states specified by the BRST operator are the irreducible representation of the current algebra keeping both the Lorentz covariance and the conformal covariance manifestly.

In the next section we will review the BRST description for a particle with an arbitrary spin where degrees of freedom of spin are described by OSp currents. Since the BRST charge for this case is nilpotent automatically, all states with arbitrary spins are allowed. In the section 3, we examine the case for a spinning particle coupled to the Yang-Mills background whose mechanics are similar to string theories.[11] In this case the nilpotency of the BRST charge gives conditions on the representation of the OSp group. We will solve this nilpotency condition analogously to Ref.[12] In section 4 a spinning string is examined. The BRST nilpotency conditions are solved in terms of the highest weight state for the $\mathrm{OSp}(d|2)$ current algebra rather than original Virasoro algebra which is shifted by the $\mathrm{Sp}(2)$ components of the $\mathrm{OSp}(d|2)$ current.

2. Spinning particle

A massless particle with an arbitrary spin is governed by the set of constraints[8,11,13,14]

$$A = p^2 = 0,$$

$$D^a = [K^a, A] = \frac{1}{2}\left\{J^{ab}, p_b\right\} + \frac{1}{2}\left\{\triangle, p^a\right\} = s^{ab}p_b + sp^a = 0, \qquad (1)$$

where p^a, K^a, J^{ab}, $\triangle$ are the momentum, the conformal boost generators, the Lorentz generators and the dilatation generator making the conformal group. The spin variable, s^{ab}, and spin, s, are in an arbitrary representation. They take following expressions for a scalar case $s^{ab} = 0$ and $s = 0$, for a spinor case $s^{ab} = \frac{1}{4}\gamma^{[a}\gamma^{b]}$ and $s = 1/2$, for a vector case $s^{ab} = |^{[a}\rangle\langle^{b]}|$ and $s = 1$, and so on. It is crucial that solutions of the constraints $D^a\Psi = 0$ are field strengths rather than gauge fields. The fact that $D^a\Psi = 0$ is satisfied without any gauge fixing condition leading to that Ψ is a gauge invariant quantity rather than a gauge field.

The study of string theory gives the guiding principle to construct a second quantized field theory from a first quantized BRST charge Q_B as $\int \Psi Q_B \Psi$, where Q_B is a linear combination of the set of first class constraints multiplied by ghosts and terms are added to satisfy $Q_B^2 = 0$. The reparametrization invariance constraint $A = 0$ becomes the kinetic operator in the field theory action in the Fermi-Feynman gauge. Naively the set of constraints (1) are replaced by the BRST charge as $Q_B = cA + \gamma_a D^a + \cdots$ with ghosts c, γ^a, but $\int \Psi Q_B \Psi$ action does contain scalar and spinor fields but no usual gauge theory action as explained in detail.[15]

In order to get the field theory actions by the BRST procedure it is necessary to find the BRST charge in such a way that its cohomology contains gauge fields instead of field strengths.

The way to resolve it is to reconstruct the BRST charge in the lightcone gauge where the gauge field can be expressed in terms of the field strength as $A_i = (p_+)^{-1}F_{+i}$ with $A_+ = 0$, $i = 1, \cdots, d-2$.[8,14] The lightcone symmetry can be covariantized to the $\mathrm{OSp}(d|2)$ symmetry by adding two commuting and two anticommuting degrees of freedom. The spin variables s^{ij}, $i, j \in \{0, 1, \cdots, d-1, \oplus, \ominus\}$ are $\mathrm{OSp}(d|2)$ generators satisfying

$$[s^{ij}, s^{kl}\} = \eta^{[l|[i}s^{j)|k)}, \tag{2}$$

where $[A, B\}$ denotes anticommutator for Grassman odd A, B and commutator otherwise, and $[ij)$ denotes symmetrization for $i, j = \oplus$ or $\ominus$ and antisymmetrization otherwise. We begin with the BRST charge which has analogous form of the one obtained as one of the $\mathrm{OSp}(d \mid 2)$ generator

$$\tilde{Q}_B = \frac{1}{p_+}[cA + D^\oplus - s^{\oplus\oplus}b], \quad D^\oplus = s^{\oplus a}p_a \tag{3}$$

ignoring the non-minimal term. This BRST charge (3) is nilpotent. The

BRST cohomology is solved in the gauge

$$b\Psi = 0, \tag{4}$$

as

$$A \, \frac{1}{p_+} \Psi = 0,$$

$$D^\oplus \, \frac{1}{p_+} \Psi = 0 \quad \rightarrow \quad D^i \, \frac{1}{p_+} \Psi = 0 \tag{5}$$

from the $OSp(d \mid 2)$ covariance. The combination $(p_+)^{-1}\Psi$ is nothing but the gauge field, so the BRST field theory actions $\int \Psi Q_B \Psi$ contain usual gauge field actions.

The BRST charge (3) reduces into the following BRST charge by the similarity transformation

$$Q_B = U \tilde{Q}_B U^{-1} = cA + D^\oplus - s^{\oplus\oplus}b, \quad \ln U = \ln p_+ (\frac{1}{2}[c,b] - s^{\oplus\ominus}). \tag{6}$$

Q_B is also nilpotent without any restriction on the representation

$$\tilde{Q}_B^2 = 0 = Q_B^2, \tag{7}$$

allowing arbitrary spin representations. We take the form of the BRST charge (6) as a starting point to extend general interacting theories and string theories.

3. Spinning particle in the Yang–Mills background

3.1. *BRST nilpotency conditions*

We consider the coupling to the background YM field. The momentum variable is replaced by the covariant derivative; $p_a \rightarrow \nabla_a = p_a + \Gamma_a$ satisfying

$$[\nabla_a, \nabla_b] = iF_{ab}, a = \{0, \cdots, d-1\}. \tag{8}$$

The closure of the constraint algebras requires the following modification of (1)

$$\mathcal{A} = \frac{1}{2}(\nabla^2 + iF_{ab}s^{ab}) = 0, \quad \mathcal{D}^a = s^{ab}\nabla_b + s\nabla^a = 0, \tag{9}$$

and the field equation for the YM background when coupled to spin 1

$$[\nabla^a, F_{ab}] = 0 \tag{10}$$

together with the Bianchi identity $\nabla_{[a} [\nabla_b, \nabla_{c]}] = 0.$

We construct the BRST charge for a massless spinning particle coupled to the Yang-Mills background analogously to (6) as

$$Q_B = Q_0 + Q_1 + Q_2, \tag{11}$$
$$Q_0 = c\mathcal{A},$$
$$Q_1 = \mathcal{D}^{\oplus} = s^{\oplus a}\nabla_a,$$
$$Q_2 = -s^{\oplus\oplus}b.$$

Square of BRST charge is

$$Q_B^2 = -\frac{i}{2}F_{ab}\phi^{\oplus\oplus ab} + \frac{i}{4}c\psi^{abc}\left[F_{bc}, \nabla_a\right], \tag{12}$$

$$\phi^{\oplus\oplus ab} = s^{\oplus\oplus}s^{ab} - \frac{1}{2}\left[s^{\oplus a}, s^{\oplus b}\right], \tag{13}$$

$$\psi^{\oplus abc} = \left\{s^{\oplus a}, s^{bc}\right\} - \frac{2}{3!}s^{\oplus[a}s^{bc]}. \tag{14}$$

The nilpotency of Q_B requires

$$Q_B^2 = 0 \Leftrightarrow \phi^{\oplus\oplus ab} = 0 = \psi^{\oplus abc}. \tag{15}$$

The constraint $\phi^{\oplus\oplus ab} = 0$ and $\psi^{\oplus abc} = 0$ are brought into $\Phi^{ijkl} = 0$ by OSp rotations with antisymmetrized indices $i \leftrightarrow j$ and $k \leftrightarrow l$ and symmetrized indices $i \leftrightarrow k$ and $j \leftrightarrow l$, $\begin{array}{|c|c|}\hline i & k \\\hline j & l \\\hline\end{array} = 0$. For the simplicity the following computations are performed for the SO(D) case, while D is related to the OSp($d, 2$) as $D = d-2$. The irreducible part of the constraints (13) and (14) is given by

$$\Phi^{ijkl} = -\frac{1}{4}s^{(i|(j}s^{l)|k)} + \frac{1}{D-2}\eta^{[k|[i}(s^2)^{j]|l]}_{\text{sym}-\text{trls}}$$

$$+\frac{1}{D(D-1)}\eta^{i[k}\eta^{l]j}\text{tr}s^2 - \eta^{ij}\eta^{kl}c_{\text{rep.}}, \tag{16}$$

$$\text{for } D \neq 2$$

with $(s^2)^{\text{sym}-\text{trls}}_{ij} = \frac{1}{2}(s_i{}^k s_{kj}+s_j{}^k s_{ki})-\eta_{ij}\frac{1}{D}\text{tr}s^2$, $\text{tr}s^2 = s^{ij}s_{ji}$ and $c_{\text{rep.}}$ is a double-trace part arisen on highest weight states;[d] $c_{\text{scalar}} = 0$, $c_{\text{spinor}} = 1/8$

---

[d]Double trace part is given by

$$c_{\text{rep.}} = \begin{cases} h_{+I-I+J-J} = h_{-I+I-J+J} = \frac{1}{4}(h_I - h_J) + \frac{1}{2}h_I h_J \\ h_{+I-I-J+J} = h_{-I+I+J-J} = \frac{1}{4}(h_I + h_J) - \frac{1}{2}h_I h_J \end{cases}$$
$$, \quad I<J \quad , \tag{17}$$

in the notation $i = \pm I$, $I = 1, \cdots, r$ with rank r.

and $c_{\text{vector}} = 1/4$. For the case $D = d - 2 = 2$ this condition should be considered in $SO(d)$ space

$$\Phi^{abcd} = -\frac{1}{4}s^{(a|(b}s^{d)|c)} + \frac{1}{d(d-1)}\eta^{a[c}\eta^{d]b}\text{tr}s^2,$$

$$\text{for } D = d - 2 = 2. \tag{18}$$

3.2. *Solution to the constraint*

We next solve constraints (17) and (18). Highest weight states $|h\rangle$ of $SO(D)$ (*e.g.*, Ref.[19]) are defined by

$$H^I|h\rangle = s^{+I-I}|h\rangle = h_I|h\rangle$$

$$E^+|h\rangle = \begin{cases} s^{+I+J}|h\rangle \\ s^{+I-J}|h\rangle, & I<J \\ (\, s^{+I0}|h\rangle, \text{ for odd } D\,) \end{cases} = 0,$$

where $i = \pm I$, $I = 1, \cdots, r$ with rank $r = D/2$ for even D and $i = 0$, $\pm I$, $I = 1, \cdots, r$ with rank $r = (D-1)/2$ for odd D.

For even D the constraint (17) on a highest state for $i = +I, j = +J, k = -I, l = -J$ with $I < J$ is written as

$$\Phi^{+I+J-I-J}|h\rangle$$

$$= \begin{pmatrix} -\dfrac{1}{2}h_I \\[4pt] +\dfrac{1}{D-2}\left\{ \begin{aligned} &\sum_{K=1}^{I-1} h_K + (h_I - I + r)h_I \\ &+\sum_{K=1}^{J-1} h_K + (h_J - J + r)h_J \end{aligned} \right\} \\[10pt] -\dfrac{2}{(D-1)(D-2)}\sum_{K=1}^{r}\{\sum_{K'=1}^{K-1} h_{K'} + (h_K - K + r)h_K\} \end{pmatrix} |h\rangle = 0 \,. \tag{19}$$

The differences between two consecutive I, say $I = n$ and $I = n + 1$ lead to the constraints among the highest weights

$$\Phi^{+I+J-I-J}\big|_{I=n+1} - \Phi^{+I+J-I-J}\big|_{I=n}$$

$$= \frac{1}{D-2}(h_{n+1} - h_n)(h_{n+1} + h_n - n) = 0$$

$$\Leftrightarrow h_{n+1} = h_n \text{ or } -h_n + n. \tag{20}$$

If we take $I = 1,\ J = 2$, then (19) gives

$$\Phi^{+I+J-I-J} = -\frac{1}{2} + \frac{1}{D-2}\left\{\begin{array}{l}(h_1 - 1 + r)h_1 \\ +h_1 + (h_2 - 2 + r)h_2\end{array}\right\}$$

$$= -\frac{2}{(D-1)(D-2)}\left\{\begin{array}{l}(h_1 - 1 + r)h_1 \\ +h_1 + (h_2 - 2 + r)h_2 \\ +(h_1 + h_2) + (h_3 - 3 + r)h_3 \\ +\cdots\end{array}\right\}$$

$$= \frac{D-3}{(D-1)(D-2)}\left((h_1)^2 + (h_2)^2 - h_1 + \frac{(D-4)(D-5)}{2(D-3)}h_2\right)$$

$$-\frac{2}{(D-1)(D-2)}\sum_{K=3}^{r}\left(\sum_{K'=3}^{K-1} h_{K'} + (h_K - K + r)h_K\right) = 0$$

$$(21)$$

which leads to

$$\left(h_1 - \frac{1}{2}\right)^2 + \left(h_2 + \frac{(D-4)(D-5)}{4(D-3)}\right)^2$$

$$= \frac{1}{4} + \left(\frac{(D-4)(D-5)}{4(D-3)}\right)^2$$

$$+ \frac{2}{D-3}\sum_{K=3}^{r}\left(\sum_{K'=3}^{K-1} h_{K'} + (h_K - K + r)h_K\right) . \qquad (22)$$

This condition draws a circle in the (h_1, h_2) plane with radius greater than $1/2$. The value of highest weight h_I must be half-integer, and $h_i \geq h_j$ and $|h_i| \geq |h_j|$ for $i > j$. Then sets of the first two values are determined to be $(h_1, h_2) = \{(0,0), (1/2, 1/2), (1/2, -1/2), (1,0)\}$. The recursive relation (20) gives the set of highest weights

$$(h_1, h_2, \cdots, h_r) = \{(0, 0, \cdots, 0)\ , \ (\frac{1}{2}, \frac{1}{2}, \cdots, \pm\frac{1}{2}), \ (1, 0, \cdots, 0)\}. \quad (23)$$

For the case $D = d - 2 = 2$ the condition (18) for $a = +I, b = +J, c = -I, d = -J$ with $I = 1,\ J = 2$ on a highest weight state is written as

$$\Phi^{+I+J-I-J}|h\rangle = \frac{1}{6}\left((h_1 - \frac{1}{2})^2 + (h_2)^2 - \frac{1}{4}\right)|h\rangle = 0 . \qquad (24)$$

Again this condition draws a circle with a center $(h_1, h_2) = (1/2, 0)$ and a radius $1/2$, then solutions in $D = 2\ (d = 4)$ are

$$(h_1, h_2) = \{(0,0),\ (1/2, \pm 1/2),\ (1,0)\} \qquad (25)$$

corresponding to scalar, spinor, vector respectively.

For odd D the constraint (17) on a highest state for $i = 0, j = +I, k = 0, l = -I$ is written as

$$
\begin{aligned}
&\Phi^{0+I\ 0-I}|h\rangle \\
&= \begin{pmatrix} -\dfrac{1}{2}h_I \\[2mm] +\dfrac{1}{D-2}\left\{ \displaystyle\sum_{K=1}^{I-1} h_K + (h_I - I + r + \dfrac{1}{2})h_I \right\} \\[4mm] -\dfrac{2}{(D-1)(D-2)} \displaystyle\sum_{K=1}^{r}\left\{ \sum_{K'=1}^{K-1} h_{K'} + (h_K - K + 1)h_K \right\} \end{pmatrix} |h\rangle = 0 \ .
\end{aligned}
\tag{26}
$$

The differences between two consecutive I, say $I = n$ and $I = n+1$, lead to the same recursive relation (20). Three solutions (23) satisfy the odd D condition (26) also. In all dimension D scalar, spinor and vector representations, but no more higher rank tensor representations, are allowed as the BRST description of spinning particle in a YM background.

4. Spinning string

4.1. *BRST nilpotency conditions*

We now consider the string first quantized BRST charge. The covariant derivative of (8) is replaced by the holomorphic part of the string extended momenta; $\nabla_a \to \hat{P}_a(z)$ satisfying

$$
\hat{P}^a(z)\hat{P}^b(w) \sim \frac{\eta^{ab}}{(z-w)^2} \ .
\tag{27}
$$

Ghost and antighost $c(z)$ and $b(z)$ and the OSp($d|2$) current $s^{ij}(z)$ satisfy following OPE

$$
c(z)b(w) \sim \frac{1}{z-w} \sim b(z)c(w)
$$

$$
s^{ij}(z)s^{kl}(w) \sim \frac{1}{z-w}\eta^{[l|[i}s^{j)|k)}(w) - \frac{\tilde{k}\eta^{k[i}\eta^{j)l}}{(z-w)^2} \ .
\tag{28}
$$

The BRST charge for a spinning string has the following form by stringy extension of (11)

$$Q_B = Q_0 + Q_1 + Q_2 \, , \tag{29}$$

$$Q_0 = \oint \frac{dz}{2\pi i}[c\frac{\hat{P}^2}{2} + cc'b + \frac{3}{2}c'' + cT_s](z) \, ,$$

$$Q_1 = \oint \frac{dz}{2\pi i} s^{\oplus a}\hat{P}_a(z) \, ,$$

$$Q_2 = \oint \frac{dz}{2\pi i}[-s^{\oplus\oplus}b](z) \, ,$$

where T_s is a spin part of stress-energy tensor satisfying

$$T_s(z)T_s(w) \sim \frac{c_s}{2(z-w)^4} + \frac{2T_s(w)}{(z-w)^2} + \frac{\partial T_s(w)}{z-w} \, . \tag{30}$$

Square of the BRST charge is

$$Q_B^2 = \oint \frac{dw}{2\pi i}[\frac{-d+26-c_s}{12}cc''' + \Psi^{\oplus} - \chi^{\oplus\oplus} + \Phi^{\oplus\oplus}](w) \, ,$$

$$\Psi^{\oplus}(w) = \oint_{C_w} \frac{dz}{2\pi i}[c(z)(T_s(z)s^{\oplus a}(w))\hat{P}_a(w) - z \leftrightarrow w]$$

$$\qquad\qquad - 2c\hat{P}_a(s^{\oplus a})'(w) \, , \tag{31}$$

$$\chi^{\oplus\oplus}(w) = \oint_{C_w} \frac{dz}{2\pi i}[c(z)(T_s(z)s^{\oplus\oplus}(w))b(w) - z \leftrightarrow w]$$

$$\qquad\qquad + 2(cb' + 2c'b)s^{\oplus\oplus}(w) \, , \tag{32}$$

$$\Phi^{\oplus\oplus}(w) = \oint_{C_w} \frac{dz}{2\pi i}[\frac{1}{z-w}(T_s(z)s^{\oplus\oplus}(w))|_{\text{singular part}} - z \leftrightarrow w]$$

$$\qquad\qquad - : (s^{\oplus a})'s^{\oplus}{}_a : +2 : T_s s^{\oplus\oplus} : (w) \, . \tag{33}$$

Nilpotency of BRST charge gives the following constraints;

$$Q_B^2 = 0 \Leftrightarrow -d + 26 - c_s = 0 = \oint \Psi^{\oplus} = \oint \chi^{\oplus\oplus} = \oint \Phi^{\oplus\oplus} \, , \tag{34}$$

$$\oint \Psi^{\oplus} = 0 \Leftrightarrow T_s(z)s^{\oplus a}(w) \sim \frac{1}{z-w}(s^{\oplus a})'(w) \, , \tag{35}$$

$$\oint \chi^{\oplus\oplus} = 0 \Leftrightarrow T_s(z)s^{\oplus\oplus}(w)$$

$$\qquad\qquad \sim -\frac{1}{(z-w)^2}s^{\oplus\oplus}(w) + \frac{1}{z-w}(s^{\oplus\oplus})'(w) \, , \tag{36}$$

$$\oint \Phi^{\oplus\oplus} = 0 \Leftrightarrow \oint[: T_s s^{\oplus\oplus} : -\frac{1}{2} : (s^{\oplus a})'s^{\oplus}{}_a :] = 0 \, . \tag{37}$$

Constraints (35) and (36) give OPE of T_s and s's telling that $s^{\oplus a}$ and $s^{\oplus\oplus}$ are conformal fields with weight 0 and -1 respectively. The third constraint (37) contains $T_s(z)$ which is not OSp covariant operator, so we need to rewrite it to be OSp covariant way.

4.2. *Sugawara–Sommerfeld construction*

In order to make the constraint (37) to be the OSp irreducible piece, we introduce OSp covariant stress-energy tensor which gives the same conformal dimension to all component of OSp currents, s^{ij}. This stress-energy tensor is given by the Sugawara–Sommerfeld construction (*e.g.*, Ref.[16]). Given the current algebra (28), the Virasoro operator is constructed by [e]

$$\mathcal{T} = \frac{1}{2\tilde{k} + C_{\text{adj}}} \frac{1}{2} : s^{ij} s_{ji} : \tag{42}$$

---

[e]The normalization of the Sugawara–Sommerfeld–Virasoro operator is determined by $s^{ij}(z)$ to be a conformal field with weight 1 with respect to $\mathcal{T}(z)$ (44): Set the normalization of it as $1/N$

$$\mathcal{T} = \frac{1}{N} : s^{ij} s_{ji} :, \tag{38}$$

and consider a state $\mathcal{T}(z) \mid h\rangle$ with the highest weight states of the affine Kac–Moody algebra $\mid h\rangle$ defined as (51)

$$\mathcal{T}(z)|h\rangle = \frac{1}{N}\, s^{ij}(z)\frac{t_{ji}}{z}|h\rangle \ . \tag{39}$$

Then act $s^{kl}(z')$ on it

$$\text{l.h.s.} = s^{kl}(z')\mathcal{T}(z)|h\rangle \sim \frac{1}{z^3} t^{kl}|h\rangle \ ,$$

$$\text{r.h.s.} = \frac{1}{N} s^{kl}(z') s^{ij}(z)\frac{t_{ji}}{z}|h\rangle \sim \frac{1}{N}\frac{1}{z^3}\left(2(D-2) + 2\tilde{k}\right) t^{kl}|h\rangle \ ,$$

where $2(D-2) = C_{\text{adj}}$ is the quadratic Casimir of the adjoint representation as

$$f^{ACD} f^{BCD} = \delta^{AB} C_{\text{adj}} \ , \tag{40}$$

we get

$$N = 2\tilde{k} + C_{\text{adj}} \ . \tag{41}$$

satisfying following OPE

$$\mathcal{T}(z)\mathcal{T}(w) \sim \frac{c_{ss}}{2(z-w)^4} + \frac{2\mathcal{T}(w)}{(z-w)^2} + \frac{\partial\mathcal{T}(w)}{z-w} \ ,$$

$$c_{ss} = \frac{2\tilde{k}|G|}{2\tilde{k} + C_{\mathrm{adj}}} \ , \tag{43}$$

$$\mathcal{T}(z)s^{ij}(w) \sim \frac{s^{ij}(w)}{(z-w)^2} + \frac{\partial s^{ij}(w)}{z-w} \ , \tag{44}$$

where $\tilde{k}$ is introduced in (28), C_{adj} is the quadratic Casimir for the adjoint representation $f^{ACD}f^{BCD} = C_{\mathrm{adj}}\delta^{AB}$ and $|G|$ is the dimension of G. The level of the current algebra is given by $k = 2\tilde{k}/\lambda$ where λ is a normalization of a representation of the group as $\mathrm{tr}\, T^A T^B = \lambda\delta^{AB}$.

The OPE (44) shows that all component of OSp currents s^{ij} have dimension 1 with respect to this Virasoro operator $\mathcal{T}(z)$. The relation between $T_s(z)$ in (29) and $\mathcal{T}(z)$ is determined from (35), (36) and (44) as

$$T_s(z) = \mathcal{T}(z) + \partial J_s(z) \ , \tag{45}$$

where $J_s(z)$ is the spin part of the ghost number current

$$J_s(z) = -s^{\oplus\ominus} \ . \tag{46}$$

The central charge c_s in (30) is now related to c_{ss} in (43) as

$$c_s = c_{ss} + 12\tilde{k} \ . \tag{47}$$

The constraint for the nilpotency of the Q_B (37) is now written as

$$\oint \Phi^{\oplus\oplus} = \oint [: \mathcal{T}s^{\oplus\oplus} : -\frac{1}{2} : (s^{\oplus k})'s^{\oplus}{}_k :] = 0 \ . \tag{48}$$

This constraint is brought into second rank anti-symmetric tensor piece $\boxed{\begin{array}{c} i \\ \hline j \end{array}}$

$= 0$ by OSp rotations.

4.3. *Solution of the constraint*

The primary field $\phi(w)$ corresponding to the highest weight state satisfies the following OPE's

$$\mathcal{T}(z)\phi(w) \sim \frac{h\phi(w)}{(z-w)^2} + \frac{\partial\phi(w)}{z-w} \ , \tag{49}$$

$$s^{ij}(z)\phi(w) \sim \frac{t^{ij}}{z-w}\phi(w) \ , \tag{50}$$

with the matrix representation of t^{ij}. The highest-weight state $|h\rangle = \phi(0)|0\rangle$ is determined by

$$H^I(z)|h\rangle = s^{+I-I}(z)|h\rangle = \frac{h_I}{z}|h\rangle \, , \tag{51}$$

$$E^+(z)|h\rangle = \begin{cases} s^{+I+J}(z)|h\rangle \\ s^{+I-J}(z)|h\rangle \, , \ \ I<J \\ (\ s^{+I0}(z)|h\rangle \, , \ \text{odd } D\ ) \end{cases} = 0 \, ,$$

with $i = \pm I$ and $I = 1, \cdots, r$ with rank $r = D/2$ for even D and $i = 0, \pm I$ and $I = 1, \cdots, r$ with rank $r = (D-1)/2$ for odd D. The Sugawara–Sommerfeld Virasoro operators act as

$$\mathcal{T}(z)|h\rangle = \frac{h}{z^2}|h\rangle \tag{52}$$

where h is given by

$$h = \frac{C}{2\tilde{k} + C_{\text{adj}}} \, , \quad C = \frac{1}{2}\text{tr}(s_0^{ij})^2 = \sum_I \left((h_I)^2 + h_I(r - I) + \sum_{J=1}^{I-1} h_J \right) \tag{53}$$

with C is the quadratic Casimir for the representation $|h\rangle$.

The constraint (48) is evaluated on a highest weight state

$$\oint \Phi^{\oplus\oplus}|h\rangle = \oint [: \mathcal{T}s^{\oplus\oplus} : -\frac{1}{2} : (s^{\oplus k})'s^{\oplus}{}_k :]|h\rangle = 0 \, . \tag{54}$$

Choosing $i = +I, j = -I$ of Φ^{ij} by the OSp rotation in the constraint (54) gives

$$\Phi^{+I-I}(z)|h\rangle = \left(\frac{h_I h}{z^3} - \frac{2h_I}{z^3} \right)|h\rangle + \partial\left(\frac{h_I}{(z-w)^2}|h\rangle \right) + \frac{1}{2}\frac{h_I(r-1)}{z^3}|h\rangle$$

$$= \frac{h_I(h - 2 + \frac{1}{2}(r-1))}{z^3}|h\rangle + \partial\left(\frac{h_I}{(z-w)^2}|h\rangle \right) \, , \tag{55}$$

and the condition $\oint \Phi^{+I-I}|h\rangle = 0$ becomes

$$h_I(h - \frac{5-r}{2}) = 0 \tag{56}$$

leading to the following recursion relation

$$(h_{n+1} - h_n)(h - \frac{5-r}{2}) = 0 \, . \tag{57}$$

One solution is $h_1 = 0$ by (56) and $h_1 = h_2 = \cdots = 0$ by (57)

$$(h_1, \cdots, h_r) = (0, \cdots, 0) \, , \quad h = 0 \, , \tag{58}$$

which is independent of the value of $\tilde{k}$. This is a scalar state. The anomaly cancellation condition (34) using with (47) restricts to

$$\tilde{k} = 0, \; c_{ss} = 0, \; c_s = 0 \;\rightarrow\; d = 26 \;. \tag{59}$$

This is nothing but a bosonic string ground state.

Other solutions satisfy $h - (5-r)/2 = 0$ of (56) and (57). Solutions are classified by values of the rank r as a positive integer, and the highest weight of the SS Virasoro (52) h is determined by $h - (5-r)/2 = 0$ of (56) and (57). The possible value of the level of the current algebra $k = \tilde{k}$ ($\lambda = 2$) is restricted by (53)

$$\tilde{k} = 2 - D + \frac{C_{\text{rep.}}}{2h} \;. \tag{60}$$

depending on the representations, while the quadratic Casimir takes the following values

$$(1/2, 1/2, \cdots, \pm 1/2) \;,\; C_{\text{spinor}} = \frac{1}{4} r(2r-1) \;,$$
$$(1, 0, \cdots) \;,\; C_{\text{vector}} = 2r - 1 \;,$$
$$(\overbrace{1, \cdots, 1}^{n}, 0 \cdots, 0) \;,\; C_{n-\text{th tensor}} = 2nr - n^2 \;,$$
$$(m, 0, \cdots, 0) \;,\; C_m = 2mr + (m^2 - 2m) \;,$$
$$etc. \tag{61}$$

The central charge for the SS Virasoro and for the original Virasoro are determined by the level of the current algebra by (43) and (47). The anomaly cancellation condition (34) is strong enough to suppress representations, and the remaining representations are spinor and vector states for the case

$$r = 4, \; h = \frac{1}{2}, \; C_{\text{spinor}} = C_{\text{vector}} = 7, \; \tilde{k} = 1, \; c_{ss} = 4, \; c_s = 16$$
$$\rightarrow \; d = 10 \;. \tag{62}$$

This is nothing but the case for the superstring. This solution is for the holomorphic part, and the antiholomorphic part has the same solution. As a result solutions of the whole string becomes the product of both parts

$$(\text{spinor} \oplus \text{vector}) \otimes (\text{spinor}' \oplus \text{vector}) \tag{63}$$

which are the expected sets of fields as the superstring theory. Chirality of the second spinor representation is the opposite to the first one for the type IIA and the same for the type IIB.

340   *M. Hatsuda and W. Siegel*

5. Discussions

We have presented current algebra approach to BRST field theories and physical states are obtained as the irreducible $OSp(d|2)$ representation. Solutions are space-time spinor, co-spinor and vector states for type I and their product for type II theories, only if the condition (62) is satisfied.

The Neveu–Schwarz–Ramond model[17] has the expression of the current (28) as

$$s^{ij}(z) = \frac{1}{2} : \psi^{[i}\psi^{j]} : (z) \ , \quad \psi^i(z)\psi^j(w) \sim \frac{\delta^{ij}}{z-w}, \quad \tilde{k} = 1 \qquad (64)$$

for the anticommuting vector ψ^i, $i = 1, \cdots, 8$. This is consistent case as a free fermion representation. The Green–Schwarz model[18] is represented as

$$s^{ij}(z) = \frac{1}{2} : \zeta\Gamma^{ij}\theta : (z) \ , \quad \theta^\alpha(z)\zeta_\beta(w) \sim \frac{\delta^\alpha_{\ \beta}}{z-w}, \quad \tilde{k} = 2 \qquad (65)$$

for the anticommuting Majorana–Weyl spinor θ^α, $\alpha = 1, \cdots, 8$ and co-spinor ζ_β. Imposing 8 second class constraints reduces the level into $\tilde{k} = 1$ which is the same with the NSR case.

The twistor string model[20] may correspond to the expression for the pure spinor

$$s^{ij}(z) = \frac{1}{2} : \omega\Gamma^{ij}\lambda : (z), \quad \lambda^\alpha(z)\omega_\beta(w) \sim \frac{\delta^\alpha_{\ \beta}}{z-w}, \quad \tilde{k} = -\frac{7}{4} \qquad (66)$$

for the commuting Majorana–Weyl spinor λ^α and co-spinor ω_β satisfying "the pure spinor condition" in 8-dimensions. [f]

$$\lambda(C\Gamma^{ijkl})\lambda = 0 = \omega(\Gamma^{ijkl}C^{-1})\omega \ , \qquad (70)$$

giving 7-pairs degrees of freedom for $D = 8$. The pure spinor has the geometrical meaning that any pure spinor determines a projective $\frac{1}{2}(D-3)$-plane. The $SO(8)$ triality brings this pure spinor into the vector which is

---

[f]The pure spinor condition in D-dimensions is given by[21]

$$\lambda(C\Gamma^{[N]})\lambda = 0 = \omega(\Gamma^{[N]}C^{-1})\omega \qquad (67)$$

with $\Gamma^{[N]} = \Gamma^{[i_1 i_2 \cdots i_N]}/N!$ satisfying

$$D - 2N = 0, 1, 7 \pmod 8 \ . \qquad (68)$$

The number of the pure spinor is given by[21]

$$\frac{1}{8}D(D-2) + 1 \ \text{ even } D, \quad \frac{1}{8}(D^2 - 1) + 1 \ \text{ odd } D \ . \qquad (69)$$

null resulting 7 independent components. It can be seen that the current algebra (66) and (28) is not affected by the pure constraint by examining "the Dirac bracket". The Dirac bracket is defined for the strongly realizing second class constraint $\chi_N = 0$ as

$$\{A, B\}_D \equiv \{A, B\} - \sum_{N,M} \{A, \chi_N\} \left(\{\chi, \chi\}\right)^{-1}_{NM} \{\chi_M, B\} \ . \tag{71}$$

We replace a Poisson bracket by a single pole part of OPE. The pure spinor condition (70) are second class

$$\chi_1 = \lambda(C\Gamma^{ijkl})\lambda = 0, \quad \chi_2 = \omega(\Gamma^{ijkl}C^{-1})\omega,$$
$$\left\{(\chi_1)^{ijkl}, (\chi_2)^{i'j'k'l'}\right\} = -\left\{(\chi_2)^{ijkl}, (\chi_1)^{i'j'k'l'}\right\}$$
$$\approx 8\{(\delta^{ii'}\delta^{jj'}\delta^{kk'}\delta^{ll'} + [ijkl]) + \epsilon^{ijkli'j'k'l'}\}\omega\lambda \tag{72}$$

as long as $\omega\lambda \neq 0$, where $\approx$ means that the Lorentz currents and the pure spinor constraints (70) are set to be zero. By the fact that the Lorentz current commute with these second class constraints

$$\left\{s^{ij}, \chi_N\right\} \approx 0 \ , \tag{73}$$

where we used the antisymmetry of the $C\Gamma^{k'l'}$ [g] and the pure spinor constraint (70), the Lorentz current algebra using with the Dirac bracket is the same with the one calculated with the usual commutator $\left\{s^{ij}, s^{kl}\right\}_D = \left\{s^{ij}, s^{kl}\right\}$. Therefore the pure spinor constraints (70) eliminates degrees of freedom without changing the Lorentz algebra (66). If the suitable combination of the Greesn-Schwarz variables and the pure spinors could gives the current algebra level to be $\tilde{k} = 1$, that may give new string model.

Acknowledgments

M.H. would like to express my deep gratitude to Professor Ezawa for stimulating courses at Gakushuin University and his encouragement. In particular the sweet TAIYAKI (Japanese fish-shaped cakes), which he kindly served to his students studying late in the evening, were unforgettable!

    She also thanks C.N.YITP for its kind hospitality and support.

---

[g]In Euclidean 8-dimension the Majorana–Weyl spinor exists with $C^T_{(-)} = C_{(-)}$, $(C\Gamma^{ij})^T = -C\Gamma^{ij}$.

References

1. R. R. Metsaev and A. A. Tseytlin, *Nucl. Phys.* **B533** (1998) 109, hep-th/9805028. R. Kallosh and J. Rahmfeld, *Phys. Lett.* **B443** (1998) 143, hep-th/9808038;
   J. Rahmfeld and A. Rajaraman, *Phys. Rev.* **D60** (1999) 64014, hep-th/9809164;
   I. Pesando, *J. High Energy Phys.* 11 (1998) 002, hep-th/9808020; *J. High Energy Phys.* 02 (1999) 007, hep-th/9809145;
   J. Park and S-J. Rey, *J. High Energy Phys.* 01 (1999) 001, hep-th/9812062;
   J-G. Zhou, *Nucl. Phys.* **B559** (1999) 92, hep-th/9906013;
   N. Berkovits, M. Bershadsky, T. Hauer, S. Zhukov and B. Zwiebach, *Nucl. Phys.* **B567** (2000) 61, hep-th/9907200;
   M. Hatsuda and M. Sakaguchi, *Phys. Rev.* **D66** (2002) 045020, hep-th/0205092.

2. R. Roiban and W. Siegel, *J. High Energy Phys.* 0011 (2000) 024, hep-th/0010104.

3. M. Hatsuda and K. Kamimura, *Nucl. Phys.* **B611** (2001) 77, hep-th/0106202.

4. R. R. Metsaev, *Nucl. Phys.* **B625** (2002) 70, hep-th/0112044;
   R. R. Metsaev and A. A. Tseytlin, *Phys. Rev.* **D65** (2002) 126004, hep-th/0202109.

5. M. Hatsuda, K. Kamimura and M. Sakaguchi,
   Nucl. Phys. **B644** (2002) 40, hep-th/0207157.

6. L. Maldacena, *Adv. Theor. Math. Phys.* **2** (1998) 231, hep-th/9711200;
   S. S. Gubser, I. R. Klebanov and A. M. Polyakov, *Phys. Lett.* **B248** (1998) 105, hep-th/9802109;
   E. Witten, *Adv. Theor. Math. Phys.* **2** (1998) 253, hep-th/9802150.

7. W. Siegel, *Nucl. Phys.* **B263** (1985) 93.

8. W. Siegel, *"Introduction to string field theory"* (World Scientific Publishing Co. Pte. Ltd., Singapore, 1988), hep-th/0107094; *"Fields"*, (1999), hep-th/9912205.

9. F. Essler, E. Laenen, W. Siegel and J. P. Yamron, *Phys. Lett.* **B254** (1991) 411;
   F. Essler, M. Hatsuda, E. Laenen, W. Siegel and J. P. Yamron, T. Kimura and A. Mikovic, *Nucl. Phys.* **B364** (1991) 67;
   W. Siegel, *Nucl. Phys.* **B263** (1985) 93.

10. M. Hatsuda, K. Kamimura and M. Sakaguchi, *Nucl. Phys.* **B632** (2002) 114, hep-th/0202190; *Nucl. Phys.* **B637** (2002) 168, hep-th/0204002.

11. W. Siegel, *Nucl. Phys.* **B263** (1986) 93.

12. W. Siegel, *Int. J. Mod. Phys.* **A4** (1989) 2015.

13. A. J. Bracken, *Lett. Nuovo. Cim.* **2** (1971) 574;
    A. J. Bracken and B. Jessup, *J. Math. Phys.* **23** (1982) 1925.

14. W. Siegel and B. Zwiebach, *Nucl. Phys.* **B282** (1987) 125.

15. M. Hatsuda, *Int. J. Mod. Phys.* **A7** (1992) 1187.

16. S. V. Ketov, *"Conformal Field Theory"* (1995), World Scientific Publishing Co. Pte. Ltd. Singapore.

17. A. Neveu and J. H. Schwarz, *Nucl. Phys.* **B31** (1971) 86;
    P. Ramond, *Phys.Rev.* **D3** (1971) 2415.
18. M. B. Green and J. H. Schwarz, *Phys. Lett.* **B136** (1984) 367; *Nucl. Phys.* **B243** (1984) 285.
19. H. Georgi, *"Lie Algebra in Particle Physics"* (1982), Addison–Wesley Publishing Company.
20. N. Berkovits, *Phys. Lett.* **B232** (1989) 184; **B241** (1990) 497.
21. R. Penrose and W. Rindler, *"Spinors and spce-time"* vol.II (1986), p.453 Cambridge Univ. Press, Cambridge.

Statistical Mechanics of Thermodynamic Processes

Research group in mathematical physics

Collaboration of J. Fröhlich, M. Merkli, S. Schwarz, D. Ueltschi

Department of Physics, ETH-Hönggerberg
CH-8093 Zürich
E-mails: juerg@itp.phys.ethz.ch, merkli@itp.phys.ethz.ch,
sschwarz@itp.phys.ethz.ch, ueltschi@math.ucdavis.edu

This note is dedicated to H. Ezawa on the occasion of his 70th birthday, with respect and affection.

1. Time-dependent thermodynamic processes

In this note we describe some results concerning non-relativistic quantum systems at positive temperature and density confined to macroscopically large regions, Λ, of physical space $\mathbb{R}^3$ which are under the influence of some local, time-dependent external forces. We are interested in asymptotic properties of such systems, as Λ increases to all of $\mathbb{R}^3$. It might thus appear natural to directly study such systems in the thermodynamic limit, $\Lambda \nearrow \mathbb{R}^3$. But for reasons of technical simplicity and ease of exposition we prefer to first consider finite systems and then extend our results to the thermodynamic limit. An important reference is Ref.[3] Details of our results appear in Refs.[1,5,4]

The Hilbert space of pure state vectors of a system confined to Λ is denoted by $\mathcal{H}^\Lambda$, and its dynamics is generated by a *time-dependent* Hamiltonian H_t^Λ with the properties that H_t^Λ is a selfadjoint operator on $\mathcal{H}^\Lambda$, for each time t, that its domain of definition is time-independent, and that $\dot{H}_t^\Lambda = \frac{d}{dt} H_t^\Lambda$ is bounded by H_s^Λ, *e.g.* in the sense of Kato-Rellich,[8] for arbitrary times t and s. If the sytem is in a state corresponding to a vector

$\psi_s \in \mathcal{H}^\Lambda$, at time s, then its state vector, ψ_t, at time t is given by

$$\psi_t = U^\Lambda(t, s)\psi_s, \tag{1}$$

where $U^\Lambda(t, s)$ denotes the *unitary propagator* on $\mathcal{H}^\Lambda$. This operator is the solution of the equation

$$\frac{\partial}{\partial t}U^\Lambda(t, s) = -iH_t^\Lambda U^\Lambda(t, s) \tag{2}$$

with

$$U^\Lambda(s, s) = \mathbb{1} \tag{3}$$

and has the property that

$$U^\Lambda(t, s) = U^\Lambda(t, u)U^\Lambda(u, s), \tag{4}$$

for arbitrary pairs $(t, s), (t, u)$ and (u, s) of times. (We are using units such that Planck's constant $\hbar = 1$.)

The kinematics of the system is encoded in an algebra $\mathcal{F}^\Lambda$ of bounded operators on $\mathcal{H}^\Lambda$ with the properties that

$$\mathcal{F}^\Lambda \subseteq \mathcal{B}(\mathcal{H}^\Lambda), \tag{5}$$

and

$$U^\Lambda(s, t)AU^\Lambda(t, s) \in \mathcal{F}^\Lambda, \tag{6}$$

for all $A \in \mathcal{F}^\Lambda$ and all times s, t.

The time evolution of a time-dependent family of operators

$$\{A_t\}_{t \in \mathbb{R}} \subset \mathcal{F}^\Lambda$$

in the *Heisenberg picture* is given by

$$A(t) := U^\Lambda(t_0, t)A_tU^\Lambda(t, t_0) \tag{7}$$

where t_0 denotes the "initial time"; (*e.g.*, the time when an experiment involving the system is started). We also denote the r.h.s. of (7) by

$$\alpha_{t_0, t}^\Lambda(A_t). \tag{8}$$

Then

$$A(t) = \alpha_{t_0, t}^\Lambda(A_t) = \alpha_{t_0, t_1}^\Lambda \circ \alpha_{t_1, t}^\Lambda(A_t), \tag{9}$$

for an arbitrary time t_1. One easily verifies that

$$\frac{d}{dt}A(t) = \alpha_{t_0, t}^\Lambda\left(\frac{DA_t}{Dt}\right), \tag{10}$$

where the *Heisenberg derivative*, DA_t/Dt, is defined by

$$\frac{DA_t}{Dt} = i[H_t^\Lambda, A_t] + \dot{A}_t. \tag{11}$$

We assume that the Hamiltonians H_t^Λ are of the form

$$H_t^\Lambda = H_0^\Lambda + W_t, \tag{12}$$

where the term W_t describes a time-dependent perturbation of the system. When this perturbation is turned off the propagator is given by the unitary group $\{e^{itH_0^\Lambda}\}_{t\in\mathbb{R}}$ on $\mathcal{H}^\Lambda$ implementing the Heisenberg time evolution

$$\alpha_t^{0,\Lambda}(A_t) := e^{itH_0^\Lambda} A_t e^{-itH_0^\Lambda}. \tag{13}$$

The unperturbed system may exhibit a group of dynamical internal symmetries unitarily represented on $\mathcal{H}^\Lambda$. For reasons of simplicity of our exposition, we assume that the symmetry group is a connected compact Lie group $\mathcal{G}$. Let $\mathcal{Z}$ denote an n-dimensional continuous connected subgroup contained in or equal to the centre of the group $\mathcal{G}$, and let $Q_1^\Lambda, \dots, Q_n^\Lambda$ denote the generators of the unitray representation of $\mathcal{Z}$ on $\mathcal{H}^\Lambda$. The operators $Q_1^\Lambda, \dots Q_n^\Lambda$ are selfadjoint operators on $\mathcal{H}^\Lambda$ with

$$[Q_i^\Lambda, Q_j^\Lambda] = 0, \quad \text{for all } i, j = 1, \dots, n, \tag{14}$$

(in the sense that their spectral projections commute), and, since $\mathcal{G}$ has been assumed to be a group of *dynamical* symmetries,

$$e^{i\boldsymbol{\tau}\cdot\boldsymbol{Q}^\Lambda} e^{itH_0^\Lambda} = e^{itH_0^\Lambda} e^{i\boldsymbol{\tau}\cdot\boldsymbol{Q}^\Lambda}, \tag{15}$$

for arbitrary $\boldsymbol{\tau} = (\tau_1, \dots, \tau_n)$ and arbitrary t; (here $\boldsymbol{\tau}\cdot\boldsymbol{Q}^\Lambda := \sum_{j=1}^n \tau_j Q_j^\Lambda$). We define *gauge transformations of the first kind* by

$$\phi_{\boldsymbol{\tau}}^\Lambda(A) := e^{i\boldsymbol{\tau}\cdot\boldsymbol{Q}^\Lambda} A e^{-i\boldsymbol{\tau}\cdot\boldsymbol{Q}^\Lambda}, \tag{16}$$

for arbitrary $A \in \mathcal{F}^\Lambda$. It is assumed that $\phi_{\boldsymbol{\tau}}^\Lambda$ are $*$automorphisms of $\mathcal{F}^\Lambda$. We define the C^*-algebra $\mathcal{A}^\Lambda$ of "observables" to be the fixed-point subalgebra of the algebra $\mathcal{F}^\Lambda$ with respect to the automorphism group $\{\phi_{\boldsymbol{\tau}}^\Lambda\}_{\boldsymbol{\tau}\in\mathbb{R}^n}$; *i.e.*,

$$\mathcal{A}^\Lambda := \{A \in \mathcal{F}^\Lambda \mid \phi_{\boldsymbol{\tau}}^\Lambda(A) = A, \forall\boldsymbol{\tau}\}. \tag{17}$$

Mixed states of the system are described by *density matrices*, ϱ, on $\mathcal{H}^\Lambda$, (*i.e.*, by non-negative trace-class operators with $\text{tr}\varrho = 1$). If the perturbation W_t of the system vanishes, *i.e.*,

$$H_t^\Lambda = H_0^\Lambda, \quad \text{for all times } t, \tag{18}$$

then the notion of *thermal equilibrium* of the system is meaningful. At *inverse temperature* β and *chemical potentials* $\mu_1, \ldots, \mu_n$, the equilibrium state is given by the density matrix

$$\varrho_{\beta,\mu} := \left(\Xi_{\beta,\mu}^\Lambda\right)^{-1} \exp -\beta[H_0^\Lambda - \mu \cdot Q^\Lambda]. \tag{19}$$

It is assumed, here, that $\exp -\beta[H_0^\Lambda - \mu \cdot Q^\Lambda]$ is trace-class, for arbitrary $\beta > 0$, $\mu \in \mathbb{R}^n$; the normalization factor $\Xi_{\beta,\mu}^\Lambda$, the so-called *grand partition function*, is chosen such that $\operatorname{tr}\varrho_{\beta,\mu} = 1$, and one commonly assumes that the system is *thermodynamically stable*, in the sense that the thermodynamic potential, G^Λ, given by

$$\beta G^\Lambda(\beta, \mu) := -\ln \Xi_{\beta,\mu}^\Lambda \tag{20}$$

is *extensive*, *i.e.*, bounded in absolute value by a constant times the volume of Λ, for arbitrary $\beta > 0, \mu \in \mathbb{R}^n$, and $\Lambda \nearrow \mathbb{R}^3$.

If, at time t_0, the system is in a mixed state $\varrho(t_0)$ then its state at time t is given by the density matrix

$$\varrho(t) = U^\Lambda(t, t_0)\varrho(t_0)U^\Lambda(t_0, t) = \alpha_{t,t_0}^\Lambda(\varrho(t_0)). \tag{21}$$

Then, using equations (7) and (21), we find that

$$\langle A_t\rangle_{\varrho(t)} := \operatorname{tr}\left(\varrho(t)A_t\right) = \operatorname{tr}\left(\varrho(t_0)A(t)\right) =: \langle A(t)\rangle_{\varrho(t_0)}, \tag{22}$$

as expected.

The *entropy* of a state given by a density matrix ϱ is defined by

$$S(\varrho) = -\operatorname{tr}\left(\varrho \ln \varrho\right). \tag{23}$$

(We use units such that Boltzmann's constant $k_B = 1$.)

Since $U^\Lambda(t, t_0)$ is unitary, for arbitrary t, t_0, it follows from (21) and the cyclicity of the trace that

$$S(\varrho(t)) = S(\varrho(t_0)), \tag{24}$$

for arbitrary t, t_0.

Next, we introduce the notion of a (time-dependent) *thermodynamic process*. We imagine that, for all times $t \leq t_0$, the Hamiltonian $H_t^\Lambda = H_{t_0}^\Lambda =: H_0^\Lambda$ is *independent* of time t, and that the initial state at time t_0 of the system is given by an *equilibrium state* $\varrho_{\beta,\mu} =: \varrho_{\beta,\mu}(t_0)$, as defined in equation (19), for some inverse temperature β and chemical potentials μ. We are interested in studying the effects of *local, external perturbations* acting on the system. In order to make more precise what we are talking about, we assume that the systems considered in this note have a *local structure*:

if Λ_0 is an arbitrary convex subset of the convex region Λ containing the system, and $\Lambda\backslash\Lambda_0$ denotes its complement then the Hilbert space $\mathcal{H}^\Lambda$ of the system can be factorized into

$$\mathcal{H}^\Lambda = \mathcal{H}^{\Lambda_0} \otimes \mathcal{H}^{\Lambda\backslash\Lambda_0},$$

where $\mathcal{H}^{\Lambda_0}$ can be interpreted as the Hilbert space of pure state vectors of the degrees of freedom localized in Λ_0. Let $\mathcal{F}^{\Lambda_0} \subseteq \mathcal{B}(\mathcal{H}^{\Lambda_0})$ be the kinematical algebra associated to the region Λ_0, see (5). Then the subalgebra

$$\mathcal{F}^{\Lambda_0} \otimes \mathbb{1}\big|_{\mathcal{H}^{\Lambda\backslash\Lambda_0}} \subset \mathcal{F}^\Lambda \tag{25}$$

is naturally identified with $\mathcal{F}^{\Lambda_0}$. Without any essential loss of generality, we may assume that the gauge transformations ϕ_τ^Λ introduced in equation (16) leave the subalgebra $\mathcal{F}^{\Lambda_0}$ of $\mathcal{F}^\Lambda$ invariant, and when restricted to $\mathcal{F}^{\Lambda_0}$ coincide with $\phi_\tau^{\Lambda_0} \otimes \mathrm{id}\big|_{\mathrm{Aut}(\mathcal{F}^{\Lambda\backslash\Lambda_0})}$, where $\mathrm{Aut}(\mathcal{F}^{\Lambda\backslash\Lambda_0})$ is the set of all automorphisms of $\mathcal{F}^{\Lambda\backslash\Lambda_0}$. Then the algebra $\mathcal{A}^{\Lambda_0}$ can be identified with

$$\{A \in \mathcal{F}^{\Lambda_0} \mid \phi_\tau^\Lambda(A) = A\} \tag{26}$$

and will be viewed as a subalgebra of $\mathcal{A}^\Lambda$, for arbitrary $\Lambda_0 \subset \Lambda$.

In the following, we shall keep $\Lambda_0 \subset \Lambda$ fixed and view the degrees of freedom localized in Λ_0 as a finite subsystem of the entire system, while the regions Λ will be let to increase to $\mathbb{R}^3$, eventually.

The interaction term W_t, given in (12), describes the dynamical effects of an external perturbation acting on the system and is assumed to have the following properties:

(i) $W_t = 0$, for $t < t_0$; and
(ii) W_t is *local* in the sense that $W_t \in \mathcal{F}^{\Lambda_0}$, for all times t, where Λ_0 is an arbitrary, but fixed bounded, convex subset of $\mathbb{R}^3$ (*independent* of t).

A thermodynamic process is *charge-conserving* iff W_t is *gauge-invariant*, i.e., $W_t \in \mathcal{A}^{\Lambda_0}$, for all times t.

Later, we shall also assume that W_t is *small* in the sense that a suitable norm of W_t is assumed to be small, *uniformly* in t.

The perturbation W_t may describe, for example, the effects of shining a focussed beam of light into the system, or of local, time-dependent variations of an external magnetic field applied to the system, or of the motion of a piston confining particles to a time-dependent subset, Λ_t, of Λ, with $\Lambda\backslash\Lambda_t \subseteq \Lambda_0$. Thus, W_t is typically of the form

$$W_t = W(\boldsymbol{\lambda}(t)), \tag{27}$$

where $\boldsymbol{\lambda} = (\lambda_1, \ldots, \lambda_k)$ is a finite set of *external control parameters*, and the time-dependence of W_t is entirely due to a possible time-dependence of the control parameters $\boldsymbol{\lambda}$.

If the ratio volume(Λ_0):volume(Λ) is very small the subsystem in the region $\Lambda \backslash \Lambda_0$ can be interpreted as a *thermostat* for the small subsystem in Λ_0, keeping the values of the temperature and the chemical potentials constant throughout a thermodynamic process. Since we have assumed that the initial state, $\varrho(t_0) = \varrho_{t_0}$, of the entire system at time t_0 is an equilibrium state,

$$\varrho(t_0) = \varrho_{t_0} = \varrho_{\beta,\boldsymbol{\mu}}, \tag{28}$$

as defined in equation (19), with $H_{t_0}^\Lambda = H_0^\Lambda$, we are studying thermodynamic processes at constant temperature β^{-1} and constant chemical potentials $\boldsymbol{\mu}$; (at least after passing to the thermodynamic limit $\Lambda \nearrow \mathbb{R}^3$, with Λ_0 kept fixed).

The *true state* of the system at time t is given by

$$\varrho(t) = \alpha_{t,t_0}^\Lambda(\varrho(t_0)), \tag{29}$$

see equation (21). If the time-dependence of the perturbation $W_t = W(\boldsymbol{\lambda}(t))$ is *slow* it is of interest to compare the true state $\varrho(t)$ of the system with a reference state, ϱ_t, given by

$$\varrho_t := e^{\beta G^\Lambda(\beta,\boldsymbol{\mu};\boldsymbol{\lambda}(t))} \exp -\beta \left[H_t^\Lambda - \boldsymbol{\mu} \cdot \boldsymbol{Q}^\Lambda \right], \tag{30}$$

where

$$\beta G^\Lambda(\beta, \boldsymbol{\mu}; \boldsymbol{\lambda}(t)) = -\ln \operatorname{tr} \left(\exp -\beta \left[H_t^\Lambda - \boldsymbol{\mu} \cdot \boldsymbol{Q}^\Lambda \right] \right), \tag{31}$$

and $\boldsymbol{\lambda}(t)$ are the time-dependent control parameters that give rise to the time-dependence of $H_t^\Lambda = H_0^\Lambda + W(\boldsymbol{\lambda}(t))$. We shall call the state ϱ_t in equation (30) the *reference state* at time t.

An important quantity in the characterization of thermodynamic processes is the *relative entropy* of the reference state ϱ_t with respect to the true state $\varrho(t)$ of the system, which is given by

$$S(\varrho_t|\varrho(t)) := -\operatorname{tr}\left(\varrho(t)[\ln \varrho_t - \ln \varrho(t)] \right) = -\operatorname{tr}\left(\varrho(t) \ln \varrho_t \right) - S(\varrho(t)), \tag{32}$$

where the entropy $S(\varrho(t))$ of $\varrho(t)$ has been defined in (23). If A is a nonnegative trace-class operator and B is a strictly positive trace-class operator then

$$-\operatorname{tr}\left(A \ln B - A \ln A \right) \geq \operatorname{tr}\left(A - B \right);$$

see *e.g.* Lemma 6.2.21 of Ref.[3] Setting $A = \varrho(t)$ and $B = \varrho_t$, and using that $\operatorname{tr} \varrho(t) = \operatorname{tr} \varrho_t = 1$, we conclude that

$$S(\varrho_t|\varrho(t)) \geq 0, \tag{33}$$

for all times t. By equation (24),

$$S(\varrho(t)) = S(\varrho(t_0)) =: S(t_0). \tag{34}$$

It then follows from (32) and (33) that

$$S(t) := -\operatorname{tr}\big(\varrho(t) \ln \varrho_t\big) \geq S(t_0), \tag{35}$$

for all times $t \geq t_0$. Next, we note that, by equation (30),

$$S(t) = \beta \left[\left\langle H_t^\Lambda - \boldsymbol{\mu} \cdot \boldsymbol{Q}^\Lambda \right\rangle_{\varrho(t)} - G(\beta, \boldsymbol{\mu}; \boldsymbol{\lambda}(t)) \right]. \tag{36}$$

It is natural to define the *internal energy*, $U^\Lambda(t)$, of the sytem at time t by

$$U^\Lambda(t) := \left\langle H_t^\Lambda \right\rangle_{\varrho(t)}, \tag{37}$$

and the various *charge densities* by

$$q_j^\Lambda(t) := \left\langle Q_j^\Lambda \right\rangle_{\varrho(t)}. \tag{38}$$

Dropping superscripts Λ, it follows that

$$G = U - \boldsymbol{\mu} \cdot \boldsymbol{q} - TS, \tag{39}$$

which is the usual relation between the Gibbs potential G and the internal energy, charge densities and the entropy. All these quantities are *extensive* and, hence, do not have a limit, as Λ increases to $\mathbb{R}^3$. It is more useful to consider their *time derivatives*. Taking the time derivative of equation (36) it follows that

$$\dot{S}(t) = \beta \dot{U}(t) - \beta \boldsymbol{\mu} \cdot \dot{\boldsymbol{q}}(t) - \beta \frac{\partial G}{\partial \boldsymbol{\lambda}} \cdot \dot{\boldsymbol{\lambda}}(t). \tag{40}$$

The combination

$$ðA := -\boldsymbol{\mu} \cdot d\boldsymbol{q} - \frac{\partial G}{\partial \boldsymbol{\lambda}} \cdot d\boldsymbol{\lambda} \tag{41}$$

is commonly interpreted as the *work* done by the system during a change of state. Hence we conclude that

$$\dot{U} = T\dot{S} - \frac{ðA}{dt}, \tag{42}$$

which summarizes the first and second law of thermodynamics (for *reversible processes*).

It is important to notice that, under our assumptions on the perturbation operator $W_t = W(\boldsymbol{\lambda}(t))$, the quantities

$$\dot{U}^\Lambda(t), \dot{\boldsymbol{q}}^\Lambda(t) \ \text{ and } \ (\partial G^\Lambda/\partial\boldsymbol{\lambda}) \cdot \dot{\boldsymbol{\lambda}}(t)$$

have finite thermodynamic limits. By equations (29) and (37), and because

$$\frac{DH_t^\Lambda}{Dt} = \dot{H}_t^\Lambda = \dot{W}_t = \frac{\partial W(\boldsymbol{\lambda}(t))}{\partial\boldsymbol{\lambda}} \cdot \dot{\boldsymbol{\lambda}}(t)$$

it follows that

$$\dot{U}^\Lambda(t) = \left\langle \frac{\partial W(\boldsymbol{\lambda}(t))}{\partial\boldsymbol{\lambda}} \right\rangle_{\varrho(t)} \cdot \dot{\boldsymbol{\lambda}}(t). \tag{43}$$

Similarly,

$$\dot{\boldsymbol{q}}^\Lambda(t) = \left\langle \frac{D\boldsymbol{Q}^\Lambda}{Dt} \right\rangle_{\varrho(t)}$$

$$= i\left\langle [W(\boldsymbol{\lambda}(t)), \boldsymbol{Q}^\Lambda] \right\rangle_{\varrho(t)} = -\frac{\partial}{\partial\boldsymbol{\tau}} \left\langle \phi_{\boldsymbol{\tau}}^\Lambda \big(W(\boldsymbol{\lambda}(t))\big) \right\rangle_{\varrho(t)} \Big|_{\boldsymbol{\tau}=0}, \tag{44}$$

where the gauge transformations $\phi_{\boldsymbol{\tau}}^\Lambda$ have been defined in equation (16). Under our hypotheses on $W(\boldsymbol{\lambda}(t))$, the operators $\frac{\partial W(\boldsymbol{\lambda}(t))}{\partial\boldsymbol{\lambda}}$ and $\frac{\partial\phi_{\boldsymbol{\tau}}^\Lambda(W(\boldsymbol{\lambda}(t)))}{\partial\boldsymbol{\tau}}$ are strictly *local*, in the sense that they are elements of the algebra $\mathcal{F}^{\Lambda_0} \subset \mathcal{F}^\Lambda$, see (25), and *independent* of Λ. In the next section, we shall see that the propagators in the *interaction picture*

$$U^{I,\Lambda}(t,s) := e^{itH_0^\Lambda} U^\Lambda(t,s) e^{-isH_0^\Lambda}$$

have thermodynamic limits, as $\Lambda \nearrow \mathbb{R}^3$, under standard assumptions on the unperturbed evolution generated by H_0^Λ. If expectations of local, bounded operators in the initial state, $\varrho_{t_0} = \varrho_{\beta,\mu}$, of the system have thermodynamic limits, as $\Lambda \nearrow \mathbb{R}^3$, which is a standard assumption (or result — see Ref.[3] for examples —) then it follows that the thermodynamic limit of the quantities on the r.h.s. of equations (43) and (44) exist.

Finally, from the definition of G^Λ, equation (31), and using the cyclicity of the trace, we find that

$$\frac{\partial G^\Lambda}{\partial\boldsymbol{\lambda}}(\beta, \boldsymbol{\mu}; \boldsymbol{\lambda}(t)) \cdot \dot{\boldsymbol{\lambda}}(t) = \left\langle \dot{H}_t^\Lambda \right\rangle_{\varrho_t} = \left\langle \frac{\partial W(\boldsymbol{\lambda}(t))}{\partial\boldsymbol{\lambda}} \right\rangle_{\varrho_t} \cdot \dot{\boldsymbol{\lambda}}(t),$$

hence

$$\frac{\partial G}{\partial\boldsymbol{\lambda}}(\beta, \boldsymbol{\mu}; \boldsymbol{\lambda}(t)) = \left\langle \frac{\partial W(\boldsymbol{\lambda}(t))}{\partial\boldsymbol{\lambda}} \right\rangle_{\varrho_t}, \tag{45}$$

which, under the same standard assumptions, has a well defined thermodynamic limit. We have thus proven the equation

$$\dot{S}(t) = \beta \left[\left\langle \frac{\partial W(\boldsymbol{\lambda}(t))}{\partial \boldsymbol{\lambda}} \right\rangle_{\varrho(t)} - \left\langle \frac{\partial W(\boldsymbol{\lambda}(t))}{\partial \boldsymbol{\lambda}} \right\rangle_{\varrho_t} \right] \cdot \dot{\boldsymbol{\lambda}}(t)$$
$$+ \beta\boldsymbol{\mu} \cdot \frac{\partial}{\partial \boldsymbol{\tau}} \left\langle \phi_{\boldsymbol{\tau}}^{\Lambda}\big(W(\boldsymbol{\lambda}(t)) \big) \right\rangle_{\varrho(t)} \Big|_{\boldsymbol{\tau}=0} \tag{46}$$

for the rate of change in time of the entropy $S(t)$, and we have convinced ourselves that all three terms on the r.h.s. of equation (46) have well defined thermodynamic limits.

Returning to equations (32), (33) and (35), one may ask under what conditions the inequalities in (33) and (35) are saturated. The answer is given in Lemma 6.2.21 of Ref.[3]:

$$S(\varrho_t | \varrho(t)) = 0 \quad \text{iff} \quad \varrho_t = \varrho(t), \tag{47}$$

i.e., iff the reference state ϱ_t coincides with the true state $\varrho(t)$. In view of equation (46) one may expect that a similar result holds in the thermodynamic limit. Equation (46) proves that the thermodynamic limit of

$$\Delta S(t) := S(t) - S(t_0) = \int_{t_0}^{t} \dot{S}(t')dt' \tag{48}$$

exists, and one expects that if $\Delta S(t) = 0$ then the true state of the system at time t is given by the thermodynamic limit of the reference states ϱ_t; (see Ref.[5]). Furthermore, if, at time t, the restrictions of the true and the reference state to the subalgebra $\mathcal{F}^{\Lambda_0}$ coincide in the thermodynamic limit then $\Delta\dot{S}(t) = \dot{S}(t) = 0$, as follows from equation (46).

Of course, if W_t depends non-trivially on time t, for $t > t_0$, there is no reason why the true and the reference state should ever coincide at times $t > t_0$. Then a relevant and interesting problem is to study the rate of change of the entropy under the assumption that the perturbation W_t depends slowly on time,

$$W_t = W(\boldsymbol{\lambda}(t/T)),$$

for some large T. In this situation one would like to prove an *adiabatic theorem* yielding sharp estimates on the rate at which $\dot{S}(t)$ tends to 0, as $T \to \infty$. More interestingly, such a theorem would tell us at which rate the differences of expectation values of local operators in the true state and in the reference state tend to 0, as $T \to \infty$. In this note we shall not address this problem.

The essential features of our definition of the entropy $S(t)$ can be summarized as follows:

(1) It is compatible with the first and second law of thermodynamics; see equations (40)-(42).

(2) $S(t) \geq S(t_0)$, for all $t \geq t_0$, *i.e.*, entropy tends to increase.

(3) The thermodynamic limit of $\dot{S}(t)$, and hence of $\Delta S(t) = S(t) - S(t_0)$ exists, for all times t.

(4) If the time evolution is *adiabatic*, in the sense that the norm of the difference of the restrictions of the true and of the reference state to the algebra $\mathcal{F}^{\Lambda_0}$ is bounded by some small, positive number $\epsilon > 0$, for all times $t \leq t_1$, then $|\dot{S}(t)| < O(\epsilon)$ and $\Delta S(t) < O(\epsilon)$, for $t \leq t_1$, *i.e.*, the entropy remains approximately constant. For more detail, see Ref.[5]

In the following, we consider two typical examples of time-dependent thermodynamic processes.

Process (I). The perturbation W_t converges to a limiting operator $W_\infty \neq 0$, as $t \to \infty$, with

$$\int^\infty \|W_t - W_\infty\| dt < \infty. \tag{49}$$

For such perturbations we study the phenomenon of *return to equilibrium*: Under suitable assumptions on the unperturbed dynamics in the thermodynamic limit ("dispersiveness") and assuming that a suitable norm of W_∞ is small enough, we show that, in the thermodynamic limit, the true state of the system converges to an equilibrium state w.r.t. the dynamics determined by $H_\infty^\Lambda = H_0^\Lambda + W_\infty$ at temperature β^{-1}, as $t \to \infty$, if the initial state is an equilibrium state of the unperturbed dynamics at temperature β^{-1} or a local perturbation thereof. For earlier results, see Refs.[11,7,1] Our analysis is an extension of results in Refs.[6,11,2] and is based on methods developed in Ref.[4]

The result described here is a kind of adiabatic theorem and shows that thermodynamic processes with perturbations W_t as specified above are *reversible*. It implies that, in the thermodynamic limit,

$$\lim_{t \to \infty} \dot{S}(t) = 0, \quad \text{and hence} \quad \lim_{t \to \infty} \lim_{\Lambda \nearrow \mathbb{R}^3} \dot{\boldsymbol{q}}^\Lambda(t) = 0, \tag{50}$$

i.e., the entropy production rate and the rate of change of the charges $\boldsymbol{q}^\Lambda(t)$ vanish in the thermodynamic limit, as time tends to infinity.

Process (II). For $t \geq t_0$, the perturbation W_t depends *periodically* on time t, with period T, see Ref.[9], and Refs.[10,12] for recent experiments involving time-periodic perturbations. Under the same assumptions on the unperturbed dynamics in the thermodynamic limit as in (I) and if a suitable norm of W_t is small enough, for all $t \in [t_0, t_0 + T]$, we prove that, in the thermodynamic limit, the true state of the system converges to a *time-periodic state* of period T, as time t tends to infinity.

It is not hard to generalize this to perturbations W_t with the property that

$$\int^{\infty} \|W_t - W_t^{\infty}\| dt < \infty, \tag{51}$$

where W_t^{∞} is periodic in t with some period T.

2. Processes (I) and (II) in the thermodynamic limit

In this section we study the thermodynamic limit, $\Lambda \nearrow \mathbb{R}^3$, of thermodynamic processes, in particular of processes (I) and (II) described at the end of Section 1. It is convenient to introduce a C^*-algebra, $\mathcal{F}$, of operators for the infinite system. We define

$$\overset{\circ}{\mathcal{F}} = \bigvee_{\Lambda \nearrow \mathbb{R}^3} \mathcal{F}^{\Lambda} \tag{52}$$

to be the algebra generated by all the algebras $\mathcal{F}^{\Lambda}$, for an increasing sequence of bounded convex regions $\Lambda \nearrow \mathbb{R}^3$. The C^*-algebra $\mathcal{F}$ is defined as the closure of $\overset{\circ}{\mathcal{F}}$ in the operator norm. For an operator $A \in \mathcal{F}^{\Lambda}$ and a set $\Lambda' \supseteq \Lambda$, one can define

$$\alpha_t^{0,\Lambda'}(A) := e^{itH_0^{\Lambda'}} A e^{-itH_0^{\Lambda'}}. \tag{53}$$

It is a standard assumption (that can be verified in physically relevant examples – see Section 3) that the norm-limit

$$n - \lim_{\Lambda' \nearrow \mathbb{R}^3} \alpha_t^{0,\Lambda'}(A) =: \alpha_t^0(A) \tag{54}$$

exists, for all $A \in \mathcal{F}^{\Lambda}$, for an arbitrary bounded convex set $\Lambda \subset \mathbb{R}^3$. Since $\alpha_t^{0,\Lambda'}(A)$ belongs to $\mathcal{F}^{\Lambda'}$, it follows that $\alpha_t^0(A) \in \mathcal{F}$. By continuity, α_t^0 can be extended to a $*$automorphism group of the C^*-algebra $\mathcal{F}$.

For a finite system confined to a region $\Lambda \supset \Lambda_0$, we define the *propagator in the interaction picture* by

$$U^{I,\Lambda}(t,s) := e^{itH_0^{\Lambda}} U^{\Lambda}(t,s) e^{-isH_0^{\Lambda}}. \tag{55}$$

Assuming that the perturbation W_t is norm-continuous in t, we can expand $U^{I,\Lambda}(t,s)$ in a Dyson series that converges in norm, uniformly in Λ, and

with the property that all terms in the series have a thermodynamic limit.
Setting

$$W_t^{I,\Lambda} := \alpha_t^{0,\Lambda}(W_t),$$

and

$$W_t^I := \alpha_t^0(W_t), \tag{56}$$

we find that

$$U^{I,\Lambda}(t,s) = \mathbb{1} + \sum_{n \geq 1}(-i)^n \int_s^t dt_1 \cdots \int_s^{t_{n-1}} dt_n \, W_{t_1}^{I,\Lambda} \cdots W_{t_n}^{I,\Lambda}$$

and this operator converges in norm to

$$U^I(t,s) = \mathbb{1} + \sum_{n \geq 1}(-i)^n \int_s^t dt_1 \cdots \int_s^{t_{n-1}} dt_n \, W_{t_1}^I \cdots W_{t_n}^I, \tag{57}$$

as $\Lambda \nearrow \mathbb{R}^3$. The propagator $U^I(t,s)$ solves the differential equation

$$\begin{aligned}\tfrac{\partial}{\partial t}U^I(t,s) &= -iW_t^I U^I(t,s),\\ U^I(s,s) &= \mathbb{1}\end{aligned} \tag{58}$$

and $U^I(t,s) \in \mathcal{F}$, for all finite times t, s.

These remarks enable us to define the time evolution of an operator $A \in \mathcal{F}$ in the Heisenberg picture from time s to time t by

$$\alpha_{s,t}(A) = n - \lim_{\Lambda \nearrow \mathbb{R}^3} \alpha_{s,t}^\Lambda(A) = \alpha_{-s}^0\left(U^I(s,t)\alpha_t^0(A)U^I(t,s)\right); \tag{59}$$

see equations (55) and (8). Equation (59) shows that $\alpha_{s,t}$ is a $*$automorphism of the algebra $\mathcal{F}$, for arbitrary times s and t, with

$$\alpha_{s,t} = \alpha_{s,t'} \circ \alpha_{t',t}.$$

Next, we describe two key hypotheses enabling us to study thermodynamic processes, such as processes (I) and (II) described at the end of Section 1, in the thermodynamic limit.

Hypothesis (A) There is a class $\mathcal{W}$ of (time-dependent) interactions s.t. the $*$automorphisms α_t^0 and $\alpha_{s,t}$, defined in (54) and (59), with $W_t \in \mathcal{W}$, satisfy the following property: for arbitrary $A \in \mathcal{F}$,

$$n - \lim_{s \to \mp\infty} \alpha_s^0(\alpha_{s,0}(A)) =: \sigma_\pm(A) \tag{60}$$

exists and defines a $*$endomorphism of $\mathcal{F}$.

Hypothesis (B) For $H_{t_0}^\Lambda = H_0^\Lambda$, the thermodynamic limit of the equilibrium states $\varrho_{\beta,\mu}$ defined in equation (19) exists on the C^*-algebra $\mathcal{F}$ and satisfies the KMS condition (see *e.g.* Refs.[3,4]), for arbitrary $\beta > 0$, $\mu \in \mathbb{R}^n$.

These two hypotheses can be verified in some simple, but physically relevant examples; see Section 3 and Refs.[4,5] We now use them to discuss processes (I) and (II). Let

$$\omega^0 = \omega_{\beta,\mu} \tag{61}$$

denote the state of $\mathcal{F}$ obtained as the thermodynamic limit of the equilibrium states $\varrho_{\beta,\mu}$ of equation (19). We are interested in understanding the time dependence of the states

$$\omega_t(A) := \omega^0(\alpha_{t_0,t}(A)), \quad A \in \mathcal{F}. \tag{62}$$

We first study this problem for **process (I)**, with a limiting interaction $W_\infty \in \mathcal{W}$. Let $\alpha_{s,t}^\infty \equiv \alpha_{t-s}^\infty$ denote the $*$automorphism of $\mathcal{F}$ constructed in equation (59) in the example where $H_t^\Lambda = H_0^\Lambda + W_\infty$, for all t; (H_t^Λ is then time-independent, hence $\alpha_{s,t}^\infty \equiv \alpha_{t-s}^\infty$ only depends on time differences). We consider the operator

$$\mathcal{E}_A(t,s) := \alpha_{t_0,t}(\alpha_{s-t}^\infty(A)). \tag{63}$$

The fundamental theorem of calculus yields

$$\mathcal{E}_A(t,s) = \mathcal{E}_A(t_0,s) + \int_{t_0}^t \mathcal{E}_A{}'(u,s)\,du,$$

with

$$\mathcal{E}_A{}'(u,s) := \frac{\partial}{\partial u}\mathcal{E}_A(u,s) = i\alpha_{t_0,u}\left([W_u - W_\infty, \alpha_{s-u}^\infty(A)]\right). \tag{64}$$

Note that

$$\|\mathcal{E}_A{}'(u,s)\| \leq 2\|A\|\,\|W_u - W_\infty\|, \tag{65}$$

and the r.h.s. in (65) tends to 0, as $u \to \infty$, at an integrable rate, see (49). Hence

$$\mathcal{E}_A(t,s) = \alpha_{s-t_0}^\infty(A) + i\int_{t_0}^t du\,\alpha_{t_0,u}\left([W_u - W_\infty, \alpha_{s-u}^\infty(A)]\right). \tag{66}$$

Since ω^0 is invariant under the unperturbed time evolution α_t^0 (see equation (54), (61)), it follows that

$$\omega^0(\mathcal{E}_A(t,s)) = \omega^0(\alpha_{t_0-s}^0(\alpha_{s-t_0}^\infty(A)) + i\int_{t_0}^t du\,\omega_u\left([W_u - W_\infty, \alpha_{s-u}^\infty(A)]\right), \tag{67}$$

see (64), (62), and the integral on the r.h.s. of (67) converges uniformly in t.

By Hypothesis (A),

$$\lim_{s \to \infty} \omega^0(\alpha^0_{t_0-s}(\alpha^\infty_{s-t_0}(A))) = \omega^0(\sigma_+(A)) \tag{68}$$

exists, for arbitrary $A \in \mathcal{F}$. Furthermore, for arbitrary $u < \infty$,

$$\lim_{s \to \infty} \omega_u \left([W_u - W_\infty, \alpha^\infty_{s-u}(A)]\right) = \lim_{s \to \infty} \omega_u \left([W_u - W_\infty, \alpha^0_{s-u}(\sigma_+(A))]\right),$$

again by Hypothesis (A), and it follows from the property of return to equilibrium for the unperturbed time evolution, α^0_s, that

$$\omega_u \left([W_u - W_\infty, \alpha^0_{s-u}(\sigma_+(A))]\right) \to 0, \tag{69}$$

as $s \to \infty$; see Section 3 for an example where return to equilibrium holds for α^0_s, and Refs.[7,4]

In conclusion, the limit

$$\lim_{t \to \infty} \omega^0(\alpha_{t_0,t}(A)) = \lim_{t \to \infty} \omega^0(\mathcal{E}_A(t,t)) = \omega^0(\sigma_+(A)), \quad A \in \mathcal{F}, \tag{70}$$

exists, by equations (66), (68) and (69). It is known from Ref.[3] that the state $\omega^0(\sigma_+(\cdot))$ is an equilibrium (*i.e.* KMS) state for the asymptotic dynamics α^∞_t.

We note that equation (70) is valid for any initial state ω^0 which is invariant under α^0_t and has the property of return to equilibrium, see also Section 3. This completes our discussion of the thermodynamic process (I).

Next, we examine **process (II)**. Since we are interested in times $t \geq t_0$ we may consider the interaction W_t to be periodic with period T for *all* times, $W_{t+T} = W_t$, for $t \in \mathbb{R}$. It follows that

$$\alpha_{s,t}(A) = \alpha_{s+nT,t+nT}(A), \tag{71}$$

for arbitrary times s, t, $n \in \mathbb{Z}$ and $A \in \mathcal{F}$. Decomposing the time variable $t \in \mathbb{R}$ uniquely as $t = n(t)T + \tau(t)$, with $n(t) \in \mathbb{Z}$ and $\tau(t) \in [0,T)$ we obtain from (71) the equation

$$\alpha_{t_0,t}(A) = \alpha_{t_0-n(t)T,0}\left(\alpha_{0,\tau(t)}(A)\right).$$

The invariance of ω^0 under α^0_t and Hypothesis (A) then imply that

$$\lim_{t \to \infty} \left|\omega^0(\alpha_{t_0,t}(A)) - \omega^0\left(\sigma_+\left(\alpha_{0,\tau(t)}(A)\right)\right)\right| = 0, \tag{72}$$

for all $A \in \mathcal{F}$.

The state given by

$$\omega^P_t(A) := \omega^0\left(\sigma_+(\alpha_{0,\tau(t)}(A))\right), \quad A \in \mathcal{F},$$

is periodic in t with period T (because $t \mapsto \tau(t)$ is). This shows that ω_t approaches a *time-periodic* state as $t \to \infty$. Notice that (72) holds for an arbitrary α_t^0-invariant initial state ω^0.

Remark. The approach to the asymptotic state (which is stationary for Process (I) and time-periodic for Process (II)), for large times, holds for arbitrary initial states which are normal w.r.t. the state ω^0 given in (61). In other words, relations (70) and (72) hold if we replace $\omega^0(\alpha_{t_0,t}(A))$ by $\omega(\alpha_{t_0,t}(A))$, for any state ω on $\mathcal{F}$ which is normal w.r.t. ω^0. The proof can be found in Ref.[4]

3. Thermodynamic processes for a reservoir of non-relativistic non-interacting fermions

We consider an ideal quantum gas of fermionic particles, *e.g.* modelling non-interacting, non-relativistic electrons in a metal or a semi-conductor, subject to a time-dependent perturbation. For the purpose of exposition, we concentrate here on spinless fermions; a more general treatment can be found in Refs.[5,4]

The Hilbert space of pure states of the system confined to a bounded region $\Lambda \subset \mathbb{R}^3$ is given by the fermionic Fock space over $L^2(\Lambda, d^3x)$,

$$\mathcal{H}^\Lambda := F_-\left(L^2(\Lambda, d^3x)\right) = \bigoplus_{n \geq 0} P_-\left(L^2(\Lambda, d^3x)\right)^{\otimes n}, \qquad (73)$$

where P_- denotes the projection operator onto the subspace of antisymmetric functions, and where the subspace for $n = 0$ is $\mathbb{C}$.

The non-interacting Hamiltonian is given by

$$H_0^\Lambda := \bigoplus_{n \geq 0} h_n^\Lambda, \qquad (74)$$

where h_n^Λ acts on $P_-\left(L^2(\Lambda, d^3x)\right)^{\otimes n}$ as

$$h_n^\Lambda = \sum_{k=1}^{n} \mathbb{1} \otimes \cdots \otimes \mathbb{1} \otimes (-\Delta) \otimes \mathbb{1} \cdots \otimes \mathbb{1},$$

and $-\Delta$ is the Laplacian on $L^2(\Lambda, d^3x)$ with selfadjoint (*e.g.* Dirichlet-, Neumann-, or periodic) boundary conditions, acting on the k-th factor. We set $h_0^\Lambda = 0$.

We define the field algebra as the CAR algebra (CAR for "canonical anti-commutation relations")

$$\mathcal{F}^\Lambda := \mathrm{CAR}\left(L^2(\Lambda, d^3x)\right),$$

which is the C^*-algebra generated by creation- and annihilation operators,

$$\{a^\#(f) \mid f \in L^2(\Lambda, d^3x)\}.$$

The symbol $a^\#$ denotes either a or a^*; recall that the annihilation operator $a(f)$ acts on a wave-function $\psi \in F_-(L^2(\Lambda, d^3x))$ as

$$(a(f)\psi)_n(x_1,\dots,x_n) = \sqrt{n+1}\int \overline{f(x_{n+1})}\psi_{n+1}(x_1,\dots,x_n,x_{n+1})d^3x_{n+1},$$

where ψ_n is the projection of ψ onto the n-particle subspace of Fock space. The creation operators $a^*(f)$ (adjoint of $a(f)$) and annihilation operators satisfy the canonical anti-commutation relations

$$\{a(f), a^*(g)\} := a(f)a^*(g) + a^*(g)a(f) = \int \overline{f(x)}g(x)d^3x$$
$$\{a^*(f), a^*(g)\} = \{a(f), a(g)\} = 0,$$

for any $f, g \in L^2(\Lambda, d^3x)$. Notice that the C^*-algebra $\mathcal{F}^\Lambda$ is weakly dense in $\mathcal{B}(\mathcal{H}^\Lambda)$,

$$\left(\mathcal{F}^\Lambda\right)'' = \mathcal{B}(\mathcal{H}^\Lambda)$$

(double commutant). The non-interacting Hamiltonian (74) generates a $*$automorphism group $\alpha_t^{0,\Lambda}$ of $\mathcal{F}^\Lambda$, according to formula (13), given by

$$\alpha_t^{0,\Lambda}\left(a^\#(f)\right) = a^\#(e^{-it\Delta}f), \quad f \in L^2(\Lambda, d^3x).$$

In this paper, we limit our discussion to only one dynamical symmetry, namely the one corresponding to the charge operator

$$N^\Lambda := \bigoplus_{n \geq 0} n\mathbb{1}\Big|_{P_-(L^2(\Lambda, d^3x))^{\otimes n}}, \tag{75}$$

i.e., the particle number operator. We refer the reader to Refs.[5,4] for a discussion involving more general charges. It is obvious that the commutation relation (15) is satisfied for $\boldsymbol{Q}^\Lambda = N^\Lambda$ and H_0^Λ given by (75) and (74), and that the charge (75) generates a $*$automorphism group ϕ_τ^Λ on $\mathcal{F}^\Lambda$, according to (16). The observable algebra $\mathcal{A}^\Lambda$ defined in (26) corresponds to the C^*-algebra generated by monomials in creation- and annihilation operators (smeared out with functions in $L^2(\Lambda, d^3x)$) in which the number of creation operators equals the number of annihilation operators.

We take the initial state of the system, at some fixed time t_0, to be given by the density matrix

$$\varrho_{\beta,\mu} := \left(\Xi^\Lambda_{\beta,\mu}\right)^{-1} \exp -\beta \left[H^\Lambda_0 - \mu N^\Lambda\right]. \tag{76}$$

It is a standard result (see *e.g.* Section 5.2.4 of Ref.[3]) that the dynamics $\alpha_t^{0,\Lambda}$ has a thermodynamic limit, α_t^0, in the sense of equation (54) and moreover, that the equilibrium state (76) has a thermodynamic limit in the sense that

$$\lim_{\Lambda \nearrow \mathbb{R}^3} \left(\Xi^\Lambda_{\beta,\mu}\right)^{-1} \mathrm{tr}\, \left(\varrho_{\beta,\mu} A\right) =: \omega_{\beta,\mu}(A)$$

exists for all $A \in \mathrm{CAR}(L^2(\Lambda', d^3x))$ and any bounded $\Lambda' \subset \mathbb{R}^3$ and defines an equilibrium (KMS) state at inverse temperature β and chemical potential μ on the C^*-algebra

$$\mathcal{F} = \mathrm{CAR}(L^2(\mathbb{R}^3, d^3x)) = \overline{\bigvee_{\Lambda \nearrow \mathbb{R}^3} \mathrm{CAR}(L^2(\Lambda, d^3x))}. \tag{77}$$

This means that Hypothesis (B) is verified.

It is known that the property of return to equilibrium holds for the KMS state $\omega_{\beta,\mu}$ (relative to the dynamics α_t^0), see Refs.[7,1,4]; this means that

$$\lim_{t \to \pm\infty} \omega^0(B\alpha_t^0(A)C) = \omega^0(BC)\omega^0(A),$$

for all $A, B, C \in \mathcal{F}$.

So far, we have verified that our example is structurally compatible with the theory outlined in Sections 1,2, and that the α_t^0-invariant initial state ω^0 satisfies the property of return to equilibrium. We are left with the specification of a class $\mathcal{W}$ of interactions satisfying Hypothesis (A).

The description of such a class has been given in Ref.[4] for time-independent interactions, and we indicate here an extension to time-dependent ones. Set $x^{(N)} := (x_1, \dots, x_N)$, $x_j \in \mathbb{R}^3$ and similarly for $y^{(N)}$, and let

$$w^N(t, x^{(N)}, y^{(N)})$$

be a function which is bounded and continuously differentiable in $t \in \mathbb{R}$ and smooth and with support in a compact region Λ_0 in each variable $x_j, y_j \in \mathbb{R}^3$. We denote by

$$a^*(x^{(N)}) := a^*(x_1) \cdots a^*(x_N)$$

the product of creation operators $a^*(x_j)$ at positions $x_j \in \mathbb{R}^3$; $a(y^{(N)})$ is defined similarly. The operator

$$W_t^N := \int a^*(x^{(N)}) w^N(t, x^{(N)}, y^{(N)}) a(y^{(N)}) dx^{(N)} dy^{(N)}, \qquad (78)$$

where we integrate over all spatial variables x_j and y_j in Λ_0, defines an element of the C^*-algebra $\mathcal{F}$ introduced in (77). A norm-summable sequence of operators $\{W_t^N\}_{N \geq 1}$ of the form (78) determines an operator

$$W_t := \sum_{N \geq 1} W_t^N \in \mathcal{F}. \qquad (79)$$

The class $\mathcal{W}$ consists of interactions of the form (79) which satisfy the smallness condition

$$\|W_t\|_\infty' < \frac{1}{24\pi},$$

where we have introduced the norm

$$\|W_t\|_\infty' := \sum_{N \geq 1} 2^{5N} N \sup_{t \in \mathbb{R}} \|w^N(t, \cdot, \cdot)\|_{6N}',$$

with

$$\|f\|_M' := \frac{1}{2^{3M/2}} \left\langle f, \prod_{k=1}^{M} \left(-\frac{d^2}{dx_k^2} + x_k^2 + 1 \right)^3 f \right\rangle_{L^2(\mathbb{R}^M)}^{1/2},$$

for a function $f \in L^2(\mathbb{R}^M)$ and where $\langle \cdot, \cdot \rangle_{L^2(\mathbb{R}^M)}$ is the inner product in $L^2(\mathbb{R}^M)$.

It is shown in Ref.[4] that the limit (60) exists, for time-independent interactions in $\mathcal{W}$; the proof of convergence given there generalizes readily to the time-dependent case, hence Hypothesis (A) holds.

Remark. One can explicitly calculate relevant physical quantities, such as rates of change in time of internal energy, $\dot{U}(t)$, charge, $\dot{q}(t)$, or entropy, $\dot{S}(t)$ (see (43), (44), (46)) in the thermodynamic limit in a *perturbative way* by using a Dyson series expansion (which is norm-convergent uniformly in time); see equations (63)–(70). Furthermore, expectations in the reference state ϱ_t have a convergent perturbation expansion, as well. Thus, our methods are quantitative.

Acknowledgements

We thank Walter Kohn for attracting our interest to problems of statistical mechanics involving time-periodic Hamiltonians and for drawing our attention to Ref.[9]

References

1. V. Bach, J. Fröhlich, I. M. Sigal, *J. Math. Phys.* **41**, 3985 (2000).
2. D. D. Botvich, V. A. Malishev, *Comm. Math. Phys.* **91**, 301 (1983).
3. O. Bratteli, D. Robinson, *Operator Algebras and Quantum Statistical Mechanics I, II* (Texts and Monographs in Physics, Springer Verlag, 1987, 1997).
4. J. Fröhlich, M. Merkli, D. Ueltschi, submitted to Ann. H. Poincaré, math-ph/0212062.
5. J. Fröhlich, M. Merkli, S. Schwarz, D. Ueltschi, in preparation.
6. K. Hepp, *Solid State Communications* **8**, 2087 (1970).
7. V. Jakšić, C.-A. Pillet, *Comm. Math. Phys.* **178**, 627 (1996).
8. T. Kato, *Perturbation Theory for Linear Operators* (Springer Verlag, 1966).
9. W. Kohn, *J. Stat. Phys.* **103**, 417 (2001).
10. R. Mani, J. H. Smet, K. von Klitzing, V. Narayanamurti, W.B. Johnson, and V. Umansky, *Nature* **420**, 646 (2002).
11. D. Robinson, *Comm. Math. Phys.* **31**, 171 (1973).
12. M.A. Zudov, R.R. Du, L.N. Pfeiffer, and K.W. West, cond-mat/0210034.

How to Formulate Non-Equilibrium Local States in QFT?
— General Characterization and Extension to Curved Spacetime—

Izumi Ojima

Research Institute for Mathematical Sciences,
Kyoto University, Kyoto 606-8502
E-mail: ojima@kurims.kyoto-u.ac.jp

Dedicated to Professor Hiroshi Ezawa
on the occasion of his seventieth birthday

The essence of a general formulation to accommodate non-equilibrium local states in relativistic quantum field theory is explained from the viewpoint of comparison at a spacetime point between unknown generic states to be characterized as such states and the known family of statistical mixtures of equilibrium states. Taking advantage of the local nature of the problem, we extend the formalism to the general-relativistic context with curved spacetimes.

1. Introduction

It is a great honour for me to make a contribution to this volume to commemorate Professor Ezawa's 70th anniversary, especially because I have benefited very much from him and from what he wrote and said on physics in general and on quantum field theory in particular, at various stages of my research career. I have been impressed by his wide perspectives ranging from mathematical physics to any kind of physical aspects of nature, among which non-equilibrium statistical physics to be discussed in the following, is an important common subject between him and myself.

"Non-equilibrium" seems to be one of the characteristic features of domains or phenomena in which nature exhibits its most vivid essence. However, theoretical attempts of systematic approaches to it starting *from the first principles of microscopic quantum theory* seem to have been rather rare (aside from some attractive phenomenological theories in thermodynamic frameworks), in sharp contrast with equilibrium cases. For the latter, we know the existence of variety of achievements successfully attained in the clear-cut formulation based upon the notion of Gibbs ensembles or *Kubo–Martin–Schwinger (KMS) states* (as generalized version of the former applicable to infinitely extended systems in thermodynamic limit), ranging from detailed analyses of concrete models to abstract sophisticated mathematical treatments of general infinite systems. The reason for such a difference seems to be evident: pursuits for concise and universal characterization of non-equilibrium have been given up, for such reasons as

i) strong negative influence of *poor and ambiguous images* originating from negative ideas and pictures: in the word "*non*-equilibrium (states)" one sees only *simple negation* of equilibrium, missing positive contents.

ii) On the positive side, the experiences of being confronted with the huge *variety* exhibited by non-equilibrium domains can easily mislead one to a superstition that emphasis on *ample individual features at macroscopic levels* is equivalent to negating connections with their universal microscopic bases, which ends up with pessimism against *ab initio* discussions starting from the "first principles" of microscopic quantum theory.

We need to recall, however, that the great achievements in equilibrium statistical mechanics and solid state physics should be found in their *unified understanding of macroscopic variety on the universal basis of microscopic quantum theory*. At this point, we recall also many fruitful positive examples in the history of transitions to such domains with "non-", as ⟨⟨from Euclidean to *non*-Euclidean geometries⟩⟩, ⟨⟨from commutative classical world to *non*-commutative quantum one⟩⟩, or ⟨⟨from standard to *non*-standard logic⟩⟩, and so on.

Taking analogy to the basic idea of *manifolds* exhibiting the process ⟨⟨from Euclidean to *non*-Euclidean geometries⟩⟩, I try here to re-view the general conceptual essence of our recent work[1,2] towards a general framework for treating nonequilibrium in relativistic QFT, where nonequilibrium local states are specified by a concise *selection criterion* and their thermal *interpretations* are canonically fixed. (The emphasis here is on the conceptual aspects with technical details omitted.)

When we try to understand non-Euclidean geometry as the geometry of curved spaces M, *e.g.*, surface of the earth, in a precise way, the aim is never attained with such insistence that what is curved should be treated as it stands. According to the common wisdom, we start from a small neighbourhood U_i in M so that effects of curvature are negligible and try to draw a precise *map* of U_i *on a flat Euclidean space,* which is nothing but a usual local map in the case of surface of the earth. While the whole sphere cannot correctly be drawn on one sheet, a (geometrically) *"precise"* description of such a curved space as the earth sphere can be attained by an *atlas* as the totality of many local charts $\varphi_i : U_i \to \mathbb{R}^d$ covering $M = \cup_i U_i$.

This familiar discussion found at the beginning of any textbooks on manifolds tells us the following points: In our scientific attempts to describe something in nature, *e.g.,* "curved space" M, as

i) our *unknown target object* to be described,

we inevitably need to relate it with

ii) something familiar to serve as a *standard reference frame,*

(such as a flat Euclidean space $\mathbb{R}^d$ in the example), which is implemented by

iii) the processes of measuring unknown target objects so that *i) is mapped to ii)* [: as the case of φ_i above],

and then,

iv) the data obtained in iii) need be collected and *organized into a coherent interpretation* (through which a description is realized)   [: the atlas in the example].

These are just the minimum ingredients for our purpose. What corresponds to these in our discussion of non-equilibrium local states can be identified as follows:

i′) [*unknown target object*]=unknown quantum state ω to be identified as our non-equilibrium local state,

ii′) [*standard reference frame*]=[*family K of thermal reference states*] consisting of convex combinations of global equilibrium states ω_β at all the possible temperatures,

iii′) [processes of measuring to *map i′) to ii′)*]=[*local thermal observables $\mathcal{T}$*] which detect local thermal properties and whose measured values in an unknown state ω in i′) are to be compared with the corresponding values in known states belonging to K, iv$'$) [*to organize data obtained in iii$'$) into a coherent interpretation*]= [*criterion to select non-equilibrium local states + thermal interpretations of selected states*] in terms of measured values of thermal quantities obtained in iii$'$).

In the following sections, the actual contents of i$'$)–iv$'$) will be explained.

2. Thermal reference states

In relativistic QFT we identify global thermal equilibria with relativisitc **KMS states** given as follows. The KMS condition[3,4] is a mathematical characterization of a KMS state ω_β which generalizes familiar Gibbs states into a form applicable to infinitely extended systems by extracting characteristic relation

$$Tr(e^{-\beta H} AB(t)) = Tr(e^{-\beta H} A e^{iHt} B e^{-iHt}) = Tr(e^{-\beta H} B(t - i\beta)A). \quad (1)$$

In a relativistic version, a state ω_β as an *expectation functional* on the algebra $\mathcal{A}$ of observables ($\omega_\beta : \mathcal{A} \ni A \longmapsto \omega_\beta(A) \in \mathbb{C}$) is called a *relativistic KMS state* with an inverse temperature 4-vector $\beta = (\beta^\mu) \in V_+ (:= \{x \in \mathbb{R}^4; x^2 \equiv (x^0)^2 - (\vec{x})^2 > 0, x^0 > 0\})$, if it satisfies the following *relativistic KMS condition*[5]: for any pair $A, B \in \mathcal{A}$ there is a function $h = h_{A,B}$, analytic in $D_\beta := \mathbb{R}^4 + i\, (V_+ \cap (\beta - V_+))$, and continuous on $\overline{D_\beta}$ with the boundary conditions

$$h(a) = \omega_\beta(A\alpha_a(B)), \quad h(a + i\beta) = \omega_\beta(\alpha_a(B)A), \quad (2)$$

where $\mathbb{R}^4 \ni a \longmapsto \alpha_a \in Aut(\mathcal{A})$ [: *-automorphism group of $\mathcal{A}$] is a spacetime translation acting on $\mathcal{A}$. Then, ω_β can be seen to describe a global thermal equilibrium at a temperature $T = (k_B \sqrt{\beta^2})^{-1}$ in a *rest frame* determined by a timelike unit vector $e = \beta/\sqrt{\beta^2} \in V_+$.

The totality K_β of relativistic KMS state with $\beta \in V_+$ is known[4] to be a *simplex* admitting for each state $\in K_\beta$ a unique decomposition into a convex combination of extremal points, which can be identified with *thermodynamic pure phases* and can be parametrized uniquely by such thermodynamic parameters as β (in combination with some such additional ones as chemical potentials μ, if necessary).

A non-trivial Poincaré transformation $\lambda = (a, \Lambda) \in \mathcal{P}_+^\uparrow := \mathbb{R}^4 \rtimes L_+^\uparrow$ transforms $\omega_\beta \in K_\beta$ into another one $\omega_\beta \circ \alpha_\lambda^{-1} \in K_{\Lambda\beta}$ with inverse temperature $\Lambda\beta$, through which a temparature defined by $1/\sqrt{\beta^2} = 1/\sqrt{(\Lambda\beta)^2}$ is unchanged but the state $\omega_\beta \circ \alpha_\lambda^{-1} \neq \omega_\beta$ can be different from the original one ω_β, because of the change of reference frame $e = \beta^\mu/\sqrt{\beta^2} \to \Lambda e \neq e$: this

is the spontaneous breakdown of Lorentz invariance due to temperature.[6]
When convenient for simplification in the following, we introduce the assumption of the *absence of phase transitions* formulated as the uniqueness of KMS state at each temperature $\beta \in V_+$, which implies the relation $\omega_\beta \circ \alpha_\lambda^{-1} = \omega_{\Lambda\beta}$, and hence, the invariance of ω_β under spacetime translations as well as its isotropy in the rest frame.

As ii') [*family K of thermal reference states*] for local thermal description of an unknown state ω, we take all the possible statistical mixtures $\omega = \int_B d\rho(\beta)\,\omega_\beta =: \omega_\rho$ of KMS states ω_β in which temperature β is fluctuating over some compact subsets B in V_+ with such probability distributions $d\rho(\beta)$ that supp$(\rho) \subset B$. Namely, we adopt the definition

$$K := \bigcup_{B:\ \mathrm{cpt}\ \subset V_+} K_B \tag{3}$$

with

$$K_B :=$$
$$\{\omega_\rho = \int_B d\rho(\beta)\,\omega_\beta;\ \rho\text{: probability measure on }V_+\text{ with supp}(\rho) \subset B\}.$$

In more general situations, the requirement of compact supports of ρ may have to be removed and, when the assumption of no phase transition is invalidated, we also add other order-parameters like chemical potential μ, *etc.*, so as for all the relevant thermodynamic pure phases to be discriminated. Then we denote generically the spaces of all the relevant thermodynamic parameters (β, μ) and of (a suitable class of) probability measures $d\rho(\beta, \mu)$, respectively, by B_K and by $Th\ (= M_1(B_K)$ = the space of probability measures on B_K, for instance): $K \ni \omega_\rho := \int_{B_K} d\rho(\beta, \mu)\,\omega_{\beta,\mu}$, $(\beta, \mu) \in B_K$, $\rho \in Th$.

3. "Coordinatization" by local thermal observables

Next, we need "coordinatization map" iii') connecting i') to ii'). The basic idea for this is first to measure some thermal observables $\{\Phi_i\}$ in an unknown state ω under consideration and then to plot on "$\{\Phi_i\}$-space" the measured values $\{\omega(\Phi_i) = \Phi_i(\omega)\}$ regarding them as "Φ_i-coordinates" of ω. If we find a data set $\{\Phi_i(\omega_\beta)\}$ for a known equilibrium state $\omega_\beta \in K_\beta$ s.t. $\Phi_i(\omega) = \Phi_i(\omega_\beta)$, then our unknown ω can be identified with this known $\omega_\beta \in K$ *as far as thermal properties determined by quantities* $\{\Phi_i\}$ *are concerned*: $\omega \underset{\{\Phi_i\}}{\equiv} \omega_\beta$. (The inclusion of mixture $\omega_\rho = \int_B d\rho(\beta)\,\omega_\beta$ is just for the sake of wider range of data search to include temperature fluctuations.)

Now the problem is *how to find physical quantities* $\{\Phi_i\}$ *suitable for describing local thermal properties of non-equilibrium states in the framework of relativistic QFT?*

The aim of this section is to give an answer to this. While the most desirable form of the answer would be such that thermal properties of unknown states in a small spacetime region $\mathcal{O}$ are determined by local observables measurable within $\mathcal{O}$. In contrast to our motivating example of manifold, however, it is almost impossible to attain directly this goal starting from a finitely extended region.

Our strategy here is *first to concentrate on a spacetime point x* and *then to extend the obtained results to a finitely extended region.* However, we immediately encounter the well-known difficulty of ultraviolet (UV) divergences invalidating a quantum field at a point, which seems to make our desire hopeless! (Perhaps, this may be one of the reasons for which the programme to construct general framework of non-equilibrium starting from microscopic quantum theory has been discouraged for a long time.) It is too early, however, to give up here! *There is an escape through which quantum field $\hat{\phi}(x)$ at a point can be made meaningful.* Since the cause to invalidate $\hat{\phi}(x)$ is just the UV divergenges due to high-frequency modes in quantum fields, $\hat{\phi}(x)$ *makes sense in states to which high energy components make no significant contributions.*

This idea can be formulated in a mathematically meaningful form as follows. In many model examples in constructive field theory, the validity of *energy-bound inequality* has been checked[7]: for $\forall l > 0$ there exist $m > 0$ and a constant $c > 0$ s.t.

$$\|(1 + H)^{-m}\,\hat{\phi}(f)\,(1 + H)^{-m}\| \leq c \int dx\, |(1 - \Delta)^{-l} f(x)|, \qquad (4)$$

holds for $\forall f \in \mathcal{S}(\mathbb{R}^4)$. Here H is a Hamiltonian defined in the vacuum representation where operator norm $\|\cdot\|$ is defined and Δ is the Laplacian in $\mathbb{R}^4$. Taking a sequence δ_i of test functions s.t. $\delta_i \underset{i \to \infty}{\to} \delta_x$ (: Dirac measure on x), we have, for sufficiently large $m > 0$,

$$\lim_{i \to \infty} (1 + H)^{-m}\,\hat{\phi}(\delta_i)\,(1 + H)^{-m} =: (1 + H)^{-m}\,\hat{\phi}(x)\,(1 + H)^{-m} \qquad (5)$$

which justifies $\hat{\phi}(x)$ mathematically. Then, $\omega(\hat{\phi}(x))$ is meaningful for any state ω s.t. $\omega((1 + H)^{2m}) < \infty$.

By replacing H meaningful only in the vacuum represetation with a *local Hamiltonian $H_{\mathcal{O}}$* playing the role of H in a local region $\mathcal{O}$ *independently of* *representations*, we arrive at the condition

$$\omega((1 + H_{\mathcal{O}})^{2m}) < \infty, \tag{6}$$

to be imposed on states ω in question. Actually this is automatically satisfied by any states admitting local thermal interpretation which should have finite energy locally. We denote $E_{\mathcal{O}}$ the totally of states ω satisfying Eq.(6) with a suitable $m > 0$,

$$E_{\mathcal{O}} := \{\omega; \omega\text{: state of } \mathcal{A} \text{ and } \exists m > 0 \text{ s.t. } \omega((1 + H_{\mathcal{O}})^{2m}) < \infty\} \supset K, \tag{7}$$

whose pointlike limit (projective limit)

$$E_x (= \varprojlim_{\mathcal{O} \to x} E_{\mathcal{O}}) \hookleftarrow K \tag{8}$$

is given by the set of equivalence classes in $\cup_{\mathcal{O}} E_{\mathcal{O}}$ with respect to the equivalence relation $\sim$ defined by

$$\omega_1 \sim \omega_2 \overset{\text{def}}{\Longleftrightarrow} \exists \mathcal{O}\text{: neighbourhood of } x \text{ s.t. } \omega_1 \lceil_{\mathcal{O}} = \omega_2 \lceil_{\mathcal{O}}. \tag{9}$$

(The family $\mathcal{O} \longmapsto E_{\mathcal{O}}$ constitutes a presheaf of state germs[8] whose stalk at x is given by E_x.) States in E_x (or $E_{\mathcal{O}}$) are just what to be examined as i').

While the product structure of quantum fields is lost through this procedure, it can effectively be recovered by the notion of *normal products* in the operator-product expansion (OPE) reformulated recently by Ref.[9] in a mathematically rigorous form. Namely, linear spaces $\mathcal{N}(\hat{\phi}^2)_{q,x}$ consisting of normal products appearing in the expansion of $\hat{\phi}(x + \zeta)\hat{\phi}(x - \zeta)$ around $\zeta = 0$ (valid for sufficiently large $n \in \mathbb{N}$),

$$\|(1 + H_{\mathcal{O}})^{-n} \left[\hat{\phi}(x + \zeta)\hat{\phi}(x - \zeta) - \sum_{j=1}^{J(q)} c_j(\zeta)\, \hat{\Phi}_j(x) \right] (1 + H_{\mathcal{O}})^{-n}\| \leq c' |\zeta|^q, \tag{10}$$

as coefficients $\hat{\Phi}_j(x)$ of c-number singular functions $c_j(\zeta)$ in ζ are seen to serve as substitutes for the ill-defined $\hat{\phi}(x)^2$, and similarly $\mathcal{N}(\hat{\phi}^p)_{q,x}$ for higher power $\hat{\phi}(x)^p$.

Through the similar expansion of $\partial_\zeta \hat{\phi}(x + \zeta)\hat{\phi}(x - \zeta)$, the derivatives in the relative coordinates ζ (called "*balanced derivatives*" here),

$$\|(1 + H_{\mathcal{O}})^{-n} \left[\partial_\zeta \hat{\phi}(x + \zeta)\hat{\phi}(x - \zeta) - \sum_{j=1}^{J(q)} \partial_\zeta c_j(\zeta)\, \hat{\Phi}_j(x) \right] (1 + H_{\mathcal{O}})^{-n}\|$$

$$\leq c'' |\zeta|^r, \tag{11}$$

can be similarly made meaningful (for sufficiently large $n, q \in \mathbb{N}$), which describe internal structures of composite operators at the same spacetime point x. The expectation values of these normal products determine p-point correlation functions around x. While derivatives ∂_x in the centre of mass coordinates are sensitive to the spacetime inhomogeneity of an unknown state ω, the corresponding quantities to $\omega(\partial_x(\cdots))$ in thermal reference states are all vanishing $\omega_\beta(\partial_x(\cdots)) = 0$, owing to the translational invariance of ω_β. Since the comparison iii') between ω and ω_β is for the sake of clarifying thermal properties of ω instead of spacetime ones, this discrepancy indicates that local observables involving ∂_x should not be counted as *local thermal observables* suitable for detecting local thermal properties. With all such irrelevant observables excluded, a suitable choice of *local thermal observables* as "coordinatization map" in iii') amounts to the *linear space $\mathcal{T}_x$ of point-like fields*,

$$\mathcal{T}_x := \sum_{p,q} \mathcal{N}(\hat{\phi}_0^p)_{q,x}\,, \tag{12}$$

consisting of basic fields $\hat{\phi}_0(x)$ at x together with their normal products $\mathcal{N}(\hat{\phi}_0^p)_{q,x}$.

What is remarkable about $\mathcal{T}_x$ is its natural *hierarchical nesting structure* ordered by indices m, p, q related to energy bound and OPE, according to their increasing orders starting from scalar multiples of identity with basic fields $\hat{\phi}_0(x)$ coming next, and so on. Since p-point functions with larger p govern those with smaller p, this hierarchy has an operationally intrinsic meaning in such a form as "the larger p, the *finer resolution of thermal properties* is provided by $\mathcal{N}(\hat{\phi}_0^p)_{q,x}$, $q > 0$". In this context, macroscopic properties of thermal states are expected to be described by subspaces $\mathcal{N}(\hat{\phi}_0^p)_{q,x}$ with smaller p,q.

—Macroscopic interpretations of $\mathcal{T}_x$—

We now examine how these local thermal observables in $\mathcal{T}_x$ provide information about macroscopic thermal properties of states in K. This is materialized by *thermal functions as macroscopic observables* as follows. In thermodynamics, the physical contents of relevant thermodynamic quantities like internal energy, entropy, *etc.*, are specified by their dependence on temperature (together with other necessary thermodynamic parameters like pressure, chemical potentials, *etc.*). In parallel with this, all intensive thermal parameters associated with states in K can be represented here by functions $B_K \ni (\beta, \mu) \mapsto F(\beta, \mu)$, which we call *thermal functions*.

The relation between *quantum local thermal observables* and *classical macroscopic observables* is described by a map $\mathcal{C}$ associating a thermal function $\mathcal{C}(\hat{A})$ to each quantum observable $\hat{A} \in \mathcal{A}$ or $\mathcal{T}_x$ by

$$\mathcal{C} : \hat{A} \longmapsto \mathcal{C}(\hat{A}) := [(\beta, \mu) \longmapsto \omega_{\beta, \mu}(\hat{A})] \in C(B_K). \tag{13}$$

In the case where all the thermal reference states ω_β are translation invariant, the thermal function $\mathcal{C}(\hat{\Phi}(x)) = \Phi_x$ corresponding to $\hat{\Phi}(x) \in \mathcal{T}_x$ is x-independent, which is invalidated, for instance, by the crystalline structures to break the spatial homogeneity, though. In the special case of *no* phase transitions where we have $\omega_\beta \circ \alpha_{(\Lambda, a)}^{-1} = \omega_{\Lambda \beta}$, the thermal function $\mathcal{C}(\hat{\Phi}(x)) = \Phi$ even becomes an x-independent Lorentz tensor in β.

We see now that thermal interpretation of each $\hat{\Phi}(x) \in \mathcal{T}_x$ is given by thermal function $(\beta, \mu) \longmapsto \Phi_x(\beta, \mu) = \mathcal{C}(\hat{\Phi}(x))(\beta, \mu) = \omega_{\beta, \mu}(\hat{\Phi}(x))$ (which amounts to recording the mean values of a local thermal observable $\hat{\Phi}(x)$ in all equilibrium states $\omega_{\beta, \mu}$).

Since the map $\mathcal{C}$ is normalized and positive linear, $\mathcal{C}(\mathbf{1}) = 1, \mathcal{C}(\hat{A}^* \hat{A}) \geq 0$ taking values in a commutative algebra $C(B_K)$, it is a *completely positive* (CP) map characterized by the condition $\sum_{ij=1}^{n} \bar{f}_i \mathcal{C}(\hat{A}_i^* \hat{A}_j) f_j \geq 0$ for $\forall n \in \mathbb{N}, \forall f_1, \cdots, \forall f_n \in C(B_K)$ and $\forall \hat{A}_1, \cdots, \forall A_n \in \mathcal{A}$. The dual map $\mathcal{C}^*$ of CP map $\mathcal{C}$ defined on states by

$$\mathcal{C}^*(\rho)(\hat{A}) = \rho(\mathcal{C}(\hat{A})) = \int_{B_K} d\rho(\beta, \mu) \mathcal{C}(\hat{A})(\beta, \mu) = \int_{B_K} d\rho(\beta, \mu) \omega_{\beta, \mu}(\hat{A}),$$

$$\Longrightarrow \mathcal{C}^*(\rho) = \int_{B_K} d\rho(\beta, \mu) \omega_{\beta, \mu} = \omega_\rho \in K, \tag{14}$$

becomes a *classical-quantum (c→q) channel*[10] $\mathcal{C}^* : Th \ni \rho \longmapsto \mathcal{C}^*(\rho) \in K$, mapping classical probabilities ρ into quantum states $\mathcal{C}^*(\rho) \in K$. (Recall that $Th = M_1(B_K)$ is the space of classical thermal states identified with probability measures ρ on B_K describing the mean values of thermodynamic parameters (β, μ) together with their fluctuations.) Measuring a local thermal observable $\hat{\Phi}(x) \in \mathcal{T}_x$ in this thermal reference state $\mathcal{C}^*(\rho)$, we obtain

$$\mathcal{C}^*(\rho)(\hat{\Phi}(x))$$
$$= \int_{B_K} d\rho(\beta, \mu) \, \omega_{\beta, \mu}(\hat{\Phi}(x)) = \int_{B_K} d\rho(\beta, \mu) [\mathcal{C}(\hat{\Phi}(x))](\beta, \mu) = \rho(\Phi_x). \tag{15}$$

Thus the *thermal interpretation* of a quantum observable $\hat{\Phi}(x)$ in all thermal reference states of the form $\mathcal{C}^*(\rho) = \omega_\rho \in K$ is given by the corresponding macroscopic *thermal function* $\mathcal{C}(\hat{\Phi}(x))$ evaluated with the classical probability $d\rho(\beta,\mu)$ which describes the *fluctuations* of thermodynamic configurations (β,μ) in ω_ρ.

This applies to the case where ρ is already known. What we need in the actual situations is how to determine the *unknown* ρ from the given data list $\Phi \longmapsto \omega_\rho(\hat{\Phi}(x)) = \rho(\Phi_x)$ of expectation values of thermal functions $\Phi_x = \mathcal{C}(\hat{\Phi}(x))$ (which is the problem of state estimation): this problem can be solved if $\mathcal{T}_x$ has sufficiently many local thermal observables so that the image $\mathcal{C}(\mathcal{T}_x)$ of $\mathcal{T}_x$ is dense in $C(B_K)$ so as to approximate arbitrary continuous functions of (β,μ) (which need be checked in each concrete model). In this case ρ is given as the unique solution to a (generalized) "moment problem". Thus we see:

★ If the set $\mathcal{T}_x$ of local thermal observables is large enough to discriminate all the thermal reference states in K, any reference state $\in K$ can be written as $\mathcal{C}^*(\rho)$ in terms of a uniquely determined probability measure ρ on B_K describing the statistical fluctuations of thermal parameters in the state in question. Then local thermal observables $\hat{\Phi}(x) \in \mathcal{T}_x$ provide the same information on the thermal properties of states in K as that provided by the corresponding classical macroscopic thermal functions $\Phi = \mathcal{C}(\hat{\Phi})$ [*e.g.*, internal energy, entropy density, *etc.*]: $\omega_\rho(\hat{\Phi}) = \rho(\Phi)$.

In this situation, *any continuous function F on compact $B \subset V_+$ can be* approximated by thermal functions $\Phi_x = \mathcal{C}(\hat{\Phi}(x))$ with arbitrary precision, *even if F itself is not an image of $\mathcal{C}$*. In spite of the absence of quantum $\hat{s}(x) \in \mathcal{T}_x$ s.t. $\omega_\beta(\hat{s}(x)) = s(\beta)$, *entropy density $s(\beta)$ can be treated as an approximate* thermal function.

The above (★) ensures the existence of *inverse of $c\to q$ channel $\mathcal{C}^*$* on K:

$$K \ni \omega_\rho = \mathcal{C}^*(\rho) \longleftrightarrow (\mathcal{C}^*)^{-1}(\omega_\rho) = \rho \in Th, \tag{16}$$

and the thermal interpretation of thermal reference states $\in K$ is just given by this *$q \to c$ channel*[2] $(\mathcal{C}^*)^{-1} : K \ni \omega \longmapsto \rho \in Th$ s.t. $\omega = \mathcal{C}^*(\rho)($, which can be regarded as a simple adaptation and extension of the notions of classifying spaces and classifying maps to the context involving (quantum) probability theory). We express formally the essence of the above situation (★) by

$$K(\omega,\mathcal{C}^*(\rho))/\mathcal{T}_x \overset{q\rightleftarrows c}{\simeq} Th((\mathcal{C}^*)^{-1}(\omega),\rho)/\mathcal{C}(\mathcal{T}_x), \tag{17}$$

with a quantum state $\omega \in E_x$ and a probability measure $\rho \in Th$. (The precise meaning of this can be understood as a categorical adjunction between two functors given by $c \to q$ $(\mathcal{C}^*)$ and $q \to c$ $((\mathcal{C}^*)^{-1})$ channels which connect K and Th both regarded as groupoids[h] corresponding to the equivalence relations $\omega_1 \underset{\mathcal{T}_x}{\equiv} \omega_2$ and $\rho_1 \underset{\mathcal{C}(\mathcal{T}_x)}{\equiv} \rho_2$ defined, respectively, by $(\omega_1 - \omega_2)(\mathcal{T}_x) = \{0\}$ and $(\rho_1 - \rho_2)(\mathcal{C}(\mathcal{T}_x)) = \{0\}$; in the form (17), the essence of $(\bigstar)$ can be generalized to wider contexts as selection criteria to choose states of relevance[11].) From the conceptual viewpoint, what is important here is that *two different levels*, quatum statistical mechanics with family K of mixtures of KMS states and macroscopic thermodynamics described by Th of probability measures of fluctuating thermal parameters on the parameter space B_K, are so interrelated by the two channels, $c \to q$ $(\mathcal{C}^*)$ and $q \to c$ $((\mathcal{C}^*)^{-1})$, that the following two points are simultaneously attained:

a) characterization of thermal reference states K as image of $\mathcal{C}^*$, $\omega_\rho = \mathcal{C}^*(\rho)$: *selection criterion* for K,

b) *thermal interpretation* of selected states in K in terms of classical data, $\Phi_x = \mathcal{C}(\hat{\Phi}(x))$ and $\rho = (\mathcal{C}^*)^{-1}(\omega_\rho)$.

Then the problem is now boiled down into how to select suitable classes of *non-equilibrium states* $\omega \notin K$ in such a way that some thermal interpretations are still guaranteed. This is what to be answered in the next section.

4. Characterization of non-equilibrium local states by hierarchized zeroth law of local thermodynamics and their thermal interpretations

Hierarchized zeroth law of local thermodynamics[2]: to meet simultaneously the two requirements of *characterizing an unkown state* ω *as a non-equilibrium local state* and of *establishing its thermal interpretation* in a

---

[h]A groupoid Γ is, roughly speaking, a generalization of a group so that there are many unit elements constituting a set Γ_0. Each element $\gamma \in \Gamma$ has its source $s(\gamma)$ and target $r(\gamma)$ in Γ_0 and these points are thought to be connected by γ, $s(\gamma) \overset{\gamma}{\to} r(\gamma)$, in an invertible way: $r(\gamma) \overset{\gamma^{-1}}{\to} s(\gamma)$. Two elements $\gamma_1, \gamma_2 \in \Gamma$ are not always composable but $\gamma_1 \gamma_2$ is meaningful only when $r(\gamma_2) = s(\gamma_1)$. There is a one-to-one and onto correspondence between a groupoid Γ and an *equivalence relation* $\sim$ on a set Γ_0 through $[a \sim b$ for $a, b \in \Gamma_0] \iff [\exists \gamma \in \Gamma$ s.t. $a = r(\gamma)$ and $b = s(\gamma)]$. As a category, Γ is one with Γ_0 as the set of objects and with all its morphisms being invertible. In our case, they are defined by $\Gamma_0 := K$ or Th together with the equivalence relation $\underset{\mathcal{T}_x}{\equiv}$ or $\underset{\mathcal{C}(\mathcal{T}_x)}{\equiv}$, respectively.

similar way to the above a) and b), we compare ω with thermal reference states $\in K = \mathcal{C}^*(Th)$ by means of local thermal observables $\in \mathcal{T}_x$ at x whose physical meanings are exhibited by the associated thermal functions $\in \mathcal{C}(\mathcal{T}_x)$.

In view of the above conclusion [$q \to c$ channel $(\mathcal{C}^*)^{-1} =$ thermal interpretation of quantum states] and also of the hierarchy in $\mathcal{T}_x$, we relax the requirement for ω to agree with $\exists \omega_\rho := \mathcal{C}^*(\rho_x) \in K$ up to some suitable *sub*space $\mathcal{S}_x$ of $\mathcal{T}_x$. Then, we characterize ω as a non-equilibrium local state by the equalities

$$\omega(\hat{\Phi}(x)) = \omega_{\rho_x}(\hat{\Phi}(x)) = \mathcal{C}^*(\rho_x)(\hat{\Phi}(x)) \tag{18}$$

valid for $\forall \hat{\Phi}(x) \in \mathcal{S}_x$. Namely, the unknown ω should *look like* a thermal reference state ω_{ρ_x} *as far as* the thermal properties described by $\hat{\Phi}(x) \in \mathcal{S}_x$ are concerned. We denote this selection criterion by

$$\omega \underset{\mathcal{S}_x}{\equiv} \mathcal{C}^*(\rho_x), \tag{19}$$

and call such ω an $\mathcal{S}_x$-*thermal* state. In terms of thermal functions $\Phi_x := \mathcal{C}(\hat{\Phi}(x)) \in \mathcal{C}(\mathcal{S}_x)$, this can be rewritten as

$$\omega(\Phi)(x) := \omega(\hat{\Phi}(x)) = \rho_x(\Phi_x). \tag{20}$$

So, ω: $\mathcal{S}_x$-*thermal* implies that the selection criterion $\omega \underset{\mathcal{S}_x}{\equiv} \mathcal{C}^*(\rho_x)$ can be "solved" conditionally in favour of ρ_x as "$(\mathcal{C}^*)^{-1}$"$(\omega) \underset{\mathcal{C}(\mathcal{S}_x)}{\equiv} \rho_x$, which provides the local thermal interpretation of ω.[2] Physically this means the state ω looks like a thermal equilibirum $\mathcal{C}^*(\rho_x)$ *locally* at x to within a level controlled by a subset $\mathcal{S}_x$ of thermal observables.

To be precise mathematically, we need here to be careful about the meaning of such a heuristic expression as "$(\mathcal{C}^*)^{-1}$"(ω) for $\omega \notin K$ in relation to our observation above: $\omega \notin K = \mathcal{C}^*(Th)$. As we shall see below, "$(\mathcal{C}^*)^{-1}$" outside of K is certainly *not* a $q \to c$ channel preserving the positivity, whereas it can be seen to be still definable on the states ω selected out by the above criterion Eq.(19), by means of its equivalent reformulation given by:

Criterion *For a subspace $\mathcal{S}_x$ of $\mathcal{T}_x$ containing $\mathbf{1}$, a state $\omega \in E_x$ is $\mathcal{S}_x$-thermal iff there is a compact set $B \subset V_+$ s.t.*

$$|\omega(\hat{\Phi}(x))| \leq \tau_B(\hat{\Phi}(x)) := \sup_{(\beta,\mu) \in B_K, \beta \in B} |\omega_{\beta,\mu}(\hat{\Phi}(x))|$$

$$= \left\| \mathcal{C}(\hat{\Phi}(x)) \right\|_B, \quad \text{for } \hat{\Phi}(x) \in \mathcal{S}_x. \tag{21}$$

(The above semi-norm is well-defined under the condition that $B_K \ni (\beta, \mu) \longmapsto \omega_{\beta,\mu} \in K$ is (weakly) continuous, which requires singularities of critical points to be excluded from our considerations.)

While the requirement for "$(C^*)^{-1}$"(ω) to be a probability measure forces ω to be among the reference states belonging to K, the above inequality (21) combined with the Hahn-Banach extension theorem (under the assumption for τ_B to be a norm) allows us to extend $\mathcal{C}(\mathcal{S}_x) \ni \mathcal{C}(\hat{\Phi}(x)) \longmapsto \omega(\hat{\Phi}(x))$ as a *linear functional* defined on $\mathcal{C}(\mathcal{S}_x)$ to one ν defined on $\overline{\mathcal{C}(\mathcal{T}_x)} = C(B_K)$, which should *not* be a positive-definite measure but is a *signed* measure: $\nu = \nu_+ - \nu_-$, $0 \le \nu_\pm \in C(B_K)^*_+$, $\nu_- \ne 0$, $\nu_- \!\upharpoonright_{\mathcal{C}(\mathcal{S}_x)} = 0$, $C^*(\nu_+) \!\upharpoonright_{\mathcal{S}_x} = \omega \!\upharpoonright_{\mathcal{S}_x}$. The similar argument for this has already been used in Ref.[1] to ensure the existence of a genuine *non-equilibrium* local state ω which is $\mathcal{S}_x$-thermal with finite-dimensional subspace $\mathcal{S}_x$ at a lower level of hierarchy in $\mathcal{T}_x$, but which shows deviations from K for observables outside of $\mathcal{S}_x$. (See this discussion also for the case with τ_B being a *semi*-norm.) Thus, understanding the meaning of $(C^*)^{-1}(\omega)$ as the set of inverse images of ω under C^* in the space $C(B_K)^*$ of linear functionals,

$$(C^*)^{-1}(\omega) := \{\nu \in C(B_K)^*; \nu = \nu_+ - \nu_-, \nu_\pm \ge 0,$$
$$\nu_- \!\upharpoonright_{\mathcal{C}(\mathcal{S}_x)} = 0, C^*(\nu_+) \!\upharpoonright_{\mathcal{S}_x} = \omega \!\upharpoonright_{\mathcal{S}_x}\}, \tag{22}$$

we can put Eq.(19) into the similar form to Eq.(17) as

$$E_x(\omega, C^*(\rho_x))/\mathcal{S}_x \overset{q \rightleftarrows c}{\simeq} Th((C^*)^{-1}(\omega), [\rho_x])/\mathcal{C}(\mathcal{S}_x), \tag{23}$$

where $[\rho_x] := \{\sigma \in Th; \sigma \!\upharpoonright_{\mathcal{C}(\mathcal{S}_x)} = \rho_x \!\upharpoonright_{\mathcal{C}(\mathcal{S}_x)}\}$ enters here owing to the non-uniqueness of $\rho_x \in Th$ in Eq.(19). This relation can be viewed as a form of "*hierarchized zeroth law of local thermodynamics*"; the reason for mentioning the "zeroth law" here is due to the implicit relevance of measuring processes of local thermal observables validating the above equalities, which require the *contacts of two bodies*, measured object(s) and measuring device(s), in a local thermal equilibrium, conditional on the chosen $\mathcal{S}_x$. (The transitivity of this contact relation just corresponds to the localized and hierarchized version of the standard zeroth law of thermodynamics.)

It is interesting to note that, in view of the relation

$$\exists \nu = \nu_+ - \nu_- \in (C^*)^{-1}(\omega) \text{ with } \nu_- = 0 \Longleftrightarrow (C^*)^{-1}(\omega) = \{\nu\} \subset Th$$
$$\Longleftrightarrow \omega \in K \Longleftrightarrow [\text{maximal choice of } \mathcal{S}'_x \text{ s.t. } C^*(\nu_+) \!\upharpoonright_{\mathcal{S}'_x} = \omega \!\upharpoonright_{\mathcal{S}'_x}] = \mathcal{T}_x, \tag{24}$$

we can specify the extent to which a non-equilibrium $\mathcal{S}_x$-thermal ω deviates from equilibria belonging to K by the *failure of state positivity* $(\nu_- \ne 0)$

and can also measure it by the *maximal size* of $\mathcal{S}'_x$ within the hierarchy of subspaces $\mathcal{S}'_x$ in $\mathcal{T}_x$ such that $\mathcal{S}'_x \supset \mathcal{S}_x$, $\nu_- \lceil_{\mathcal{C}(\mathcal{S}'_x)} = 0$ with all the possible choices of $\nu \in (\mathcal{C}^*)^{-1}(\omega)$: owing to the presence of ν_-, ω ceases to be $\mathcal{S}'_x$-thermal when $\mathcal{S}'_x$ is so enlarged that $\nu_- \lceil_{\mathcal{C}(\mathcal{S}'_x)} = 0$ is invalidated, which shows that ω shares with reference states in K only gross thermal properties described by smaller $\mathcal{S}'_x$. In this sense, the hierarchy of $\mathcal{S}'_x$ in $\mathcal{T}_x$ should have a close relationship with the thermodynamic hierarchy at various scales appearing in the transitions between non-equilibrium and equilibrium controlled by certain family of *coarse graining* procedures. Thus, we see that our selection criterion can give a characterization of states identifiable as non-equilibrium ones and, at the same time, provide associated relevant physical interpretations of the selected states in a systematic way.

—Fluctuations of thermal quantities; temperature as a physical quantity—

The present framework allows one also to judge whether a thermal function Φ has locally a sharply specified value in a state ω or is statistically fluctuating, which can be implemented if $\mathcal{S}_x$ is large enough for the *mean value* of Φ_x together with its *fluctuations* to be determined within it. For instance, if $\mathcal{S}_x$ contains local observables $\hat{\Phi}_1(x)$ and $\hat{\Phi}_2(x)$ corresponding respectively to $\Phi_x = \mathcal{C}(\hat{\Phi}_1(x))$ and $\Phi_x^2 = \mathcal{C}(\hat{\Phi}_2(x))$, we have a thermal function $(\Phi_x - \kappa\,1)^2 = \mathcal{C}(\delta\hat{\Phi}_\kappa(x))$ with

$$\delta\hat{\Phi}_\kappa(x) := \hat{\Phi}_2(x) - 2\kappa\,\hat{\Phi}_1(x) + \kappa^2\,1, \quad \kappa \in \mathbb{R}. \tag{25}$$

Since this is non-negative in all thermal reference states $\in K$ and vanishes only in those states with Φ_x having a sharp value κ, we can conclude that Φ_x at x has the sharp value κ in such an $\mathcal{S}_x$-thermal state ω that $\omega(\delta\hat{\Phi}_\kappa(x)) = 0$.

In virtue of this scheme it is meaningful to treat the (inverse) temperature β as a real physical quantity to be determined a posteriori through its measurements, which is in sharp contrast to the standard idea in statistical mechanics of treating it as an *a priori* given *parameter*. Choosing suitable spaces $\mathcal{S}_x$ (which is finite dimensional in generic cases[1]), we can select a state ω having locally a *sharp temperature* vector β_x, i.e., $\omega \underset{\mathcal{S}_x}{\equiv} \omega_{\beta_x}$. In this case, all thermal functions Φ_x corresponding to $\hat{\Phi}(x) \in \mathcal{S}_x$ have locally definite values. Concerning the possible objections against the introduction of a temperature β_x (as well as any kind of thermal objects) *at a point* x (without any extension), it is important to recall here that any state $\omega \in E_x$ relevant to our present context is already *extended in spacetime* effectively, owing to the regularity condition (6) imposed on it. In this sense, the point x in β_x should not literally be understood to refer to a strictly microscopic spacetime point but to a macroscopic one with certain fuzzy extensions, which is taken into account on the side of chosen states.

—Space-time evolution of thermal properties—

Extending our formalism from a point x to a *(finitely extended) subregion* $\mathcal{O} \subset \mathbb{R}^4$, we can now incorporate *local states with thermal interpretation in $\mathcal{O}$*. For simplicity, we keep the set of thermal functions fixed in each region, by identifying the spaces $\mathcal{S}_x$, $x \in \mathcal{O}$ through translations:

$$\mathcal{S}_x := \alpha_x(\mathcal{S}_0), \quad x \in \mathcal{O}. \tag{26}$$

With this convention understood, we say that a state $\omega \in E_\mathcal{O}$ is $\mathcal{S}_\mathcal{O}$-*thermal in $\mathcal{O}$*, if there exists $\omega_{\rho(x)} \in K$ for each $x \in \mathcal{O}$, s.t. $\omega \underset{\mathcal{S}_x}{\equiv} \omega_{\rho(x)}$. The resulting functions $\mathcal{O} \ni x \mapsto \omega(\Phi)(x) := \omega(\hat{\Phi}(x))$ describe the space-time behaviour of mean values of thermal functions Φ. Hence they provide the *link between microscopic dynamics α_x and the evolution $x \mapsto \omega(\Phi)(x)$ of macroscopic thermal properties, i.e. thermo-dynamics* of states. We have reached the same level as the familiar local formulation of manifolds at the beginning.

On this setting, thermal functions in a state near equilibrium are generally shown to satisfy a linear evolution equations, which can be viewed as a generalization of low energy theorems to thermal situations. This is consistent with interpretation of perturbations to equilibrium in terms of quasi-particles.

The two goals of identifying non-equilibrium local states admitting local thermal interpretation and of describing their specific thermodynamic properties are solved simultaneously by the above selection criterion based upon a *localized and hierarchized form of the zeroth law* of thermodynamics. In this framework, we have identified at least three different kinds of sources of derivations of an $\mathcal{S}_x$-thermal non-equilibrium local state $\omega \in E_x$ from the genuine equilibrium states ω_β as

a) *spacetime dependence* of thermal parameters such as temperature distributions $x \longmapsto \beta(x)$,

b) *statistical fluctuations* of thermal parameters at x described by probability distributions $d\rho_x(\beta) \in Th$,

and

c) essential deviations of local states $\omega \in E_x$ from states in K expressed by the *positivity-violating* term $\nu_- \neq 0$ in $\nu = \nu_+ - \nu_- \in (\mathcal{C}^*)^{-1}(\omega) \subset C(B_K)^*$ with $\nu_- \upharpoonright_{C(\mathcal{S}_x)} = 0, \mathcal{C}^*(\nu_+) \upharpoonright_{\mathcal{S}_x} = \omega \upharpoonright_{\mathcal{S}_x}$.

For concrete examples to exhibit the basic features, see Ref.[1] Here discussions have been focused on the conceptual aspects developed in Ref.[2]

5. General-relativistic extension: global vs. local and flat vs. curved

—Global vs. local—

Looking back over what is done so far, we notice here some room for further improvements and generalizations in view of such restricted choices of *global* KMS states[12] on the side of thermal reference states. When compared with our motivating discussion of manifolds, it also looks strange that the whole theory is restricted only to within the *flat* Minkowski spacetime[13] in spite of the emphasis on the *local* aspects of states to be examined. Actually, in the comparison of an unknown state ω with a known reference state $\omega_\rho \in K$, the latter serves only to provide a reference data set $\omega_\rho(\hat{A})$ involving *local* thermal observables $\hat{A} \in \mathcal{T}_x$ or $\mathcal{S}_x$ near the focus point x, which requires only the *local restrictions* of states $\omega_\rho \in K$ onto small neighbourhoods of x (*i.e.*, local state germs in $K \hookrightarrow E_x$). Therefore, once the reference states ω_ρ can properly be specified *within small neighbourhoods* of spacetime, we expect the freedom to go across the barriers separating the flat and curved spacetimes, existing at the *global* level but irrelevant locally. Such a possibility can naturally be read off in our selection criterion formulated as a relation of *comparison* (more appropriately, a categorical adjunction), Eq.(23), which is *not* necessarily required to be a strict equality between ω and $C^*(\rho)$ but should be a well-defined and suitably controllable *relation*, as is common in many cases of reference to standard objects constituting a *model space*, such as local charts referring to $\mathbb{R}^n$ in manifolds. On the basis of this observation, a more flexible setting up is envisaged to emerge through such a possibility that unknown ω to be examined can be generalized to those states living in *curved* background spacetimes: the mathematical basis for this physical ideas can already be found in the notions of local definiteness and/or local normality (see Refs.[14,15]), which allow one to treat generic localized states of quantum fields in curved spacetime backgrounds in the same Hilbert space of the vacuum representation in the Minkowski spacetime.

—Flat vs. curved spacetimes—

From the above point of view, we try now to extend the previous scheme to the situation with quantum fields in a *curved* spacetime by restricting the original thermal reference states to *local* small regions in the *flat* Minkowski spacetime. Note here that, in sharp contrast to the examined unknown ω being allowed to be a state in a *curved* spacetime, the reference states ω_ρ are understood as the local restrictions of mixtures of global KMS states still living in the *flat* Minkowski spacetime; this is not only in harmony with the line of thought found in our starting discussion of manifolds taking the flat $\mathbb{R}^n$ as the model space, but also is a very important and inevitable choice necessitated by the possible *absence of appropriate vacua and/or KMS states* in generic curved spacetimes.

In spite of the *inherent slight delocalization* of our selected states ω due to the condition (6), we can here benefit from the expressions referring to one spacetime point x as follows. To a small neighbourhood of a point x in a curved spacetime M we can apply Einstein's basic idea of *equivalence principle* based upon the *free-falling frame* at x whose mathematical expression can be found in the notion of *normal coordinates*,[16] the coordinates along geodesic flows starting from x which are always definable in some neighbourhood $\mathcal{O}_0$ of the origin 0 of the tangent space $T_x(M)$ at x even for *incomplete* geodesic flows,

$$Exp_x : (T_x(M) \supset)\mathcal{O}_0 \to Exp_x(\mathcal{O}_0) = \mathcal{O}_x(\subset)M.$$

Corresponding to this local diffeomorphism Exp_x, we can define a mapping Ψ_x to transform locally a QFT defined in a flat $\mathcal{O}_0(\hookrightarrow \mathbb{R}^4)$ into the local restriction of a QFT in curved M on its small neighbourhood $\mathcal{O}_x$ of x (owing to the *functoriality* in the definition of quantum fields on curved spacetimes[17,18]):

$$\Psi_x : \mathcal{A}_0(\mathcal{O}_0) \to \mathcal{A}_M(\mathcal{O}_x),$$

where $\mathcal{O} \longmapsto \mathcal{A}_0(\mathcal{O})$ and $\mathcal{O} \longmapsto \mathcal{A}_M(\mathcal{O})$ denote the corresponding local nets, respectively, on the flat Minkowski spacetime and on a curved spacetime M. Then what we need is simply to modify our selection criterion Eq.(23) into

$$E_{M,x}(\omega, \mathcal{C}_x^*(\rho))/\Psi_x(\mathcal{S}_0) := E_0(\omega \circ \Psi_x, \mathcal{C}^*(\rho))/\mathcal{S}_0$$
$$\overset{q \rightleftarrows c}{\simeq} Th((\mathcal{C}_x^*)^{-1}(\omega), [\rho])/\mathcal{C}(\mathcal{S}_0),$$

where $E_{M,x}$ and $\Psi_x(\mathcal{S}_0)$ are the sets of local thermal states and of local thermal observables, respectively, at x in the curved spacetime M and $\mathcal{C}_x = \mathcal{C} \circ \Psi_x^{-1}$. In this way, our formulation of non-equilibrium local states can safely be extended to the general-relativistic context by simple restriction of reference states onto small neighbourhoods. (While the set $K_{M,x}$

of reference states at x in a curved spacetime M can simply be defined here as the image $(\Psi_x^*)^{-1}(K)$ of the corresponding set K on the flat space-time through the normal coordinates Exp_x in the C^∞-context, the intrinsic characterization of the notions in M related to the KMS condition will certainly require the analytic structure on M, as is the case in the discussion of analytic wavefront sets in Ref.[20])

—Interacting vs. free—

In view of our focus on small neighbourhoods of a spacetime point, the local normality allows us to treat the effects of *interactions* as a perturbation to the *free* dynamics, without complications related to the thermodynamic limit such as the Haag theorem, at least, at the abstract levels. Therefore, it will be very convenient if we can choose *free field* models for the reference system[12]; this will not only make the reference states more accessible to the practical computations, but also conceptually appealing in relation to the physical origin of temperature referring to the *ideal gas* in the equation of states as well as Bolzmann's kinetic definition of it. To implement this idea, however, we should solve the difficulty taking the familiar form of *ultraviolet divergences*. While the combined use of the regularity condition (6) imposed on the states of relevance and the normal products in OPE can remove divergences order by order systematically in the perturbation, the available methods are not powerful enough to remove the full-order divergences at once, in spite of all the up-dated attractive tools for regularizing and renormalizing these divergences (*e.g.*, the use of energy bounds, OPE,[9] local perturbation scheme[19] of Epstein–Glaser type and Connes–Kreimer method[21] of renormalization). If we succeed in finding a satisfactory reason for terminating the perturbative expansions at some finite orders, then the algebraically formulated *local* perturbation scheme[19] will become physically relevant. Aside from this long-standing problem, what is also important in the present context is to attain the effective separation between *order parameters* to describe non-trivial structures in phase diagrams due to interactions and small *fluctuations* within fixed phases which are expected to be described by particle-like modes (Ref.[22]). In any case, it seems still premature to expect a practical and satisfactory solution to our desire in this direction.

Extrapolating the above logical lines, we can formulate[11] a unified scheme for generalized sectors (discrete and/or continuous) based upon selection criteria, which provides new physical operational interpretations of superselection theory, extends it to spontaneously broken symmetries, and exhibits close relationship of basic notions in quantum measurement theory with those in control theory,[23] such as the notions of *measurement scheme*[24] and *state preparation* processes in connection with *realizability* and *reachability*, respectively, constituting the core of the latter. It would be important to note that *applicability domain* of a given theory is, in principle, encoded in this scheme in the *matching* relation between a chosen selection criterion for relevant states and the available observables.

References

1. D. Buchholz, I. Ojima and H. Roos, Ann. Phys. (N.Y.), **297**, 219 (2002).
2. I. Ojima, Non-Equilibrium Local States in Relativistic Quantum Field Theory, to appear in Proc. of Japan–Italy Joint Workshop on Fundamental Problems in Quantum Mechanics, September 2001, Waseda University.
3. R. Haag, N.M. Hugenholtz and M. Winnink, Comm. Math. Phys. **5**, 215 (1967).
4. O. Bratteli and D. W. Robinson, *Operator Algebras and Quantum Statistical Mechanics, Vol.2*, Springer, 1996.
5. J. Bros and D. Buchholz, Nucl. Phys. **B429**, 291 (1994).
6. I. Ojima, Lett. Math. Phys. **11**, 73 (1986).
7. K. Fredenhagen and J. Hertel, Comm. Math. Phys. **80**, 555(1981); W. Driessler and J. Fröhlich, Ann. Inst. H. Poincaré **27**, 221 (1997).
8. R. Haag and I. Ojima, Ann. Inst. H. Poincaré **64**, 385 (1996).
9. H. Bostelmann, Lokale Algebren und Operatorprodukte am Punkt, Ph.D. Thesis, Universität Göttingen, 2000.
10. M. Ohya and D. Petz, *Quantum Entropy and Its Use*, Springer-Verlag (1993).
11. I. Ojima, A Unified Scheme for Generalized Sectors based on Selection Criteria, math-ph/0303009.
12. Discussion with Prof. Kossakowski in October, 2002.
13. Discussion with Prof. Fredenhagen in October, 2002.
14. R. Haag, *Local Quantum Physics* (2nd. ed.), Springer-Verlag (1996).
15. R. Haag, H. Narnhofer and U. Stein, Comm. Math. Phys. **94**, 219 (1984).
16. See *e.g.*, S. Kobayashi, and K. Nomizu, *Foundations of Differential Geometry, Vol.1*, Interscience, 1963.
17. I. Ojima, Quantum-field theoretical approach to non-equilibrium dynamics in curved space-time, pp.91–96 in Proc. 2nd Int. Symp. Foundations of Quantum Mechanics, Tokyo, 1986 (Physical Society of Japan, 1987).
18. R. Brunetti, K. Fredenhagen and R. Verch, The Generally Covariant Locality Principle, to appear in Comm. Math. Phys. 2003.
19. R. Brunetti, K Fredenhagen and M. Köhler, Commun. Math. Phys. **180**, 633 (1996).
20. A. Strohmaier, R. Verch and M. Wollenberg, Microlocal analysis of quantum fields on curved spacetimes: Analytic wavefront sets and Reeh–Schlieder theorems, math-ph/0202003.
21. A. Connes and D. Kreimer, Commun.Math.Phys. **199**, 203 (1998).

22. J. Bros and D. Buchholz, Z. f. Physik C Particles and Fields, **55**, 509 (1992).
23. M.A. Arbib and E.G. Manes, *Arrows, Structures, and Functors*, Academic Press (1975).
24. M. Ozawa, J. Math. Phys. **25**, 79 (1984); Publ. RIMS, Kyoto Univ. **21**, 279 (1985); Ann. Phys. (N.Y.) **259**, 121 (1997).

Some Applications of Renormalization Group Analysis

Hiroshi Watanabe

Department of Mathematics, Nippon Medical School
2-297-2, Kosugi, Nakahara, Kawasaki 211-0063, Japan
E-mail: watmath@nms.ac.jp

A few examples of renormalization group analysis are given. After two pedagogical examples, a new approach to renormalization group analysis for hierarchical spin models is proposed. In this approach, renormalization group transformation is written in terms of characteristic functions and analyzed for the hierarchical Ising model in 4 dimensions. Finally, an extension to the hierarchical $O(N)$ model is briefly mentioned.

1. Introduction

The renormalization group (RG) analysis has been mainly developed for studies of infinite dimensional integrals in mathematical physics. In the RG analysis, infinite dimensional integrals are decomposed into inductive steps from the UV direction to the IR direction based on the picture of scaling decompositions.

It seems that the RG analysis has matured as long as it concerns with the weakly coupled systems. We however have several open problems which requires a RG analysis of *strongly coupled systems*. For example, it is widely believed that the Ising model in four dimensions will be Gaussian in the continuum limit (*triviality*). There have been accumulated a number of facts indicating the triviality,[1] but no mathematically rigorous proof has been obtained so far. In fact, in order to study the Ising model, we have to perform a RG analysis *in the strong coupling region*, since the Ising model is a strong coupling limit of the ϕ^4 model.

In this note, after describing intuitive picture of RG analysis, I introduce a new approach to the RG analysis in the strong coupling region using characteristic functions which is applicable to the so-called *hierarchical models*.[2]

2. Intuitive picture of RG analysis

In general, implementation of RG analysis with mathematical rigor is technically complicated, but the RG philosophy itself is quite clear. It would be desirable to get an intuition of RG philosophy through simple examples without mathematical difficulty.

2.1. *Harmonic series*

Let us begin with the asymptotic property of the harmonic series:

$$S_N = \frac{1}{1} + \frac{1}{2} + \frac{1}{3} + \dots + \frac{1}{N}, \quad N = 1, 2, 3, \cdots. \tag{1}$$

Using the equalities

$$\frac{1}{2k-1} + \frac{1}{2k} = \frac{1}{k} + \frac{1}{(2k-1)(2k)} \tag{2}$$

for $k = 1, 2, \cdots, N$, we obtain

$$S_{2N} = S_N + r_N, \tag{3}$$

where

$$r_N = \sum_{k=1}^{N} \frac{1}{(2k-1)(2k)} \tag{4}$$

with the bound

$$\alpha < r_N < \beta. \tag{5}$$

Here, α and β are positive constants. Then, (3) implies

$$1 + \alpha n \leq S_{2^n} \leq 1 + \beta n, \quad n = 1, 2, 3, \cdots. \tag{6}$$

This shows the logarithmic divergence of $\lim_{N\to\infty} S_N$.

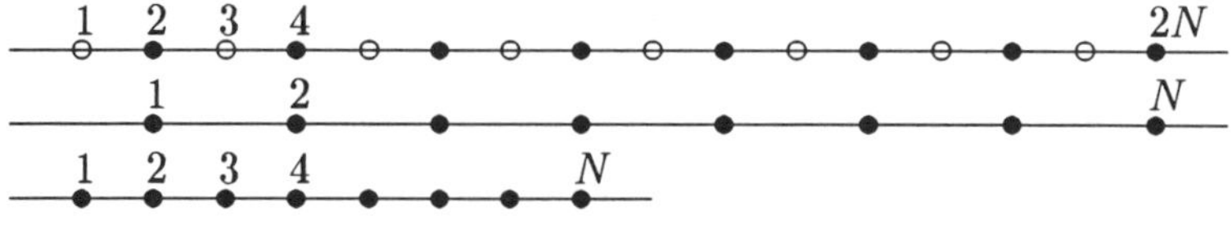

Fig. 1.   The scale transformation of N

The above calculation is related to the scaling transformation of $\mathbb{N} = \{1, 2, 3, \cdots\}$. Define the mapping $\mathcal{R} : \mathbb{N} \to \mathbb{N}$ by

$$\mathcal{R}(2k - 1) = \mathcal{R}(2k) = k , \quad k \in \mathbb{N} , \tag{7}$$

and the action of $\mathcal{R}$ on the set of all sequences $a = (a(n))_{n \in \mathbb{N}}$ by

$$(\mathcal{R}a)(k) = \sum_{j : \mathcal{R}(j) = k} a(j) = a(2k - 1) + a(2k). \tag{8}$$

Then, it holds that

$$S_{2^n} = \sum_{k=1}^{2^n} a(k) = (\mathcal{R}^n a)(1) . \tag{9}$$

This means that the summation of 2^n numbers is decomposed into successive processes reducing the set $\mathbb{N}$ by half as described in Fig. 1. A transformation, like $\mathcal{R}$, which reduces the geometric scale together with the degree of freedoms of the system is called a RG transformation. Note that $\mathcal{R}$ has an approximate fixed point $a(k) = \frac{1}{k}$ with the error r :

$$\mathcal{R}a = a + r . \tag{10}$$

from which we obtain

$$\mathcal{R}^n a = a + \sum_{k=0}^{n-1} \mathcal{R}^k r , \quad n = 1, 2, \cdots . \tag{11}$$

Since r has the bound

$$\alpha < (\mathcal{R}^n r)(1) < \beta , \tag{12}$$

we get the result $(\mathcal{R}^n a)(1) = \mathcal{O}(n)$.

2.2. *Gaussian chain*

The next example of RG analysis is the equality

$$\sum_{\substack{-\pi < q \leq \pi \\ \ell q \equiv p \pmod{2\pi}}} \frac{1}{1 - \cos q} = \frac{\ell^2}{1 - \cos p} , \tag{13}$$

where $\ell = 1, 2, 3, \cdots$. Let us show this equality from the RG point of view.

Define the matrix $K = (K_{x,y})_{x,y \in \mathbb{Z}}$ by

$$K_{x,y} = \begin{cases} J + m , & x = y , \\ -\frac{1}{2}J , & |x - y| = 1 , \\ 0 , & |x - y| > 1 , \end{cases} \tag{14}$$

and put

$$H(\phi) = \sum_{x,y \in \mathbb{Z}} \phi_x K_{x,y} \phi_y \ , \tag{15}$$

where $J, m > 0$ and $\phi = (\phi_x)_{x \in \mathbb{Z}}$. Then, there exists a Gaussian measure $d\mu(\phi)$ formally written as

$$d\mu(\phi) = \frac{1}{Z} \exp(-\frac{1}{2} H(\phi)) \prod_{x \in \mathbb{Z}} d\phi_x \ , \tag{16}$$

where Z is the normalization. Introduce the Fourier kernel $D(p)$ of $K_{x,y}$ by

$$K_{x,y} = \frac{1}{2\pi} \int_{-\pi}^{\pi} D(p) e^{-ip(x-y)} dp \ , \tag{17}$$

$$D(p) = J(1 - \cos p) + m \ . \tag{18}$$

Then, the covariance $C = K^{-1}$ of the Gaussian measure $d\mu(\phi)$ has the following expression :

$$C_{x,y} = \frac{1}{2\pi} \int_{-\pi}^{\pi} \frac{1}{D(p)} e^{-ip(x-y)} dp \ . \tag{19}$$

Let us derive (13) as an equality of RG transformation for $1/D(p)$ at $m = 0$ (the massless propagator).

Consider the subset $\ell\mathbb{Z} = \{\ell x \mid x \in \mathbb{Z}\}$ of $\mathbb{Z}$ for $\ell = 2, 3, \cdots$. If we integrate the both sides of (16) with respect to variables $\phi_x, x \notin \ell\mathbb{Z}$, we obtain a Gaussian measure on $\ell\mathbb{Z}$. Furthermore, if we reduce $\ell\mathbb{Z}$ by $1/\ell$ to reproduce $\mathbb{Z}$ similarly as Fig. 1, we obtain a Gaussian measure $d\tilde{\mu}$ on $\mathbb{Z}$. It is easily seen that the covariance of $d\tilde{\mu}$ is

$$\tilde{C}_{x,y} = C_{\ell x, \ell y} \ , \quad x, y \in \mathbb{Z}, \tag{20}$$

and that $d\tilde{\mu}$ is formally written again as (14), (15), (16) with suitably changed J and m. In case $\ell = 2$, they should be replaced by

$$\tilde{J} = \frac{J^2}{2(J + m)} \ , \tag{21}$$

$$\tilde{m} = \frac{2Jm + m^2}{(J + m)^2} \ , \tag{22}$$

respectively, and in general we have

$$\lim_{m \to 0} \tilde{J} = \frac{J}{\ell} \ , \tag{23}$$

$$\lim_{m \to 0} \tilde{m} = 0 \tag{24}$$

for $\ell = 2, 3, 4, \cdots$. Now, the Fourier kernel $D(p)$ of K is transformed into

$$\tilde{D}(p) = \tilde{J}(1 - \cos p) + \tilde{m} , \qquad (25)$$

and (19) implies that the Fourier kernel of $\tilde{C}$ is written as

$$\frac{1}{\tilde{D}(p)} = \sum_{\substack{-\pi < q \le \pi \\ \ell q \equiv p \pmod{2\pi}}} \frac{1}{D(p)} . \qquad (26)$$

Thus, we have (13) as the massless limit $m \to 0$ of (26).

3. Hierarchical model

Let us introduce the hierarchical Ising model and formulate the RG transformation to study the continuum limit.

As is well-known, one dimensional Ising model with nearest neighbor interactions does not undergo a phase transition. But if a model has long range interactions, a phase transition can occur. Dyson[3] showed this fact by introducing a special spin system nowadays called a hierarchical model. In fact, the hierarchical model has a 'simplified' Gaussian measure that is suitable for the RG analysis.

Hierarchical model is defined as follows. Let Λ be a positive integer, and consider the 2^Λ variables (spin variables) $\phi_\theta = \phi_{\theta_\Lambda,\ldots,\theta_1}$ labeled by

$$\theta = (\theta_\Lambda, \ldots, \theta_1) \in \{0, 1\}^\Lambda, \qquad (27)$$

Let us define the Hamiltonian H_Λ and the expectation values $\langle \cdot \rangle$, respectively, by

$$H_\Lambda(\phi) = -\frac{1}{2} \sum_{n=1}^{\Lambda} \left(\frac{c}{4}\right)^n \sum_{\theta_\Lambda,\ldots,\theta_{n+1}} \left(\sum_{\theta_n,\ldots,\theta_1} \phi_{\theta_\Lambda,\ldots,\theta_1} \right)^2 , \qquad (28)$$

$$\langle F \rangle_{\Lambda,h} = \frac{1}{Z_{\Lambda,h}} \int d\phi F(\phi) \exp(-\beta H_\Lambda(\phi)) \prod_\theta h(\phi_\theta), \qquad (29)$$

$$Z_{\Lambda,h} = \int d\phi \exp(-\beta H_\Lambda(\phi)) \prod_\theta h(\phi_\theta), \qquad (30)$$

where c is a constant satisfying $0 < c < 2$, and h is a single spin measure density normalized as

$$\int_{\mathbb{R}} h(x)dx = 1 . \qquad (31)$$

In the following, we shall fix the so far arbitrary normalization of the spin variables by

$$\beta = \frac{1}{c} - \frac{1}{2}. \tag{32}$$

3.1. *RG transformation*

Hierarchical models are so designed that the RG transformation has a simple form (see (38)). Define the block spins ϕ' by

$$\phi'_\tau = \frac{\sqrt{c}}{2} \sum_{\theta_1 = 0,1} \phi_{\tau \theta_1} \, , \quad \tau = (\tau_{\Lambda - 1}, ..., \tau_1). \tag{33}$$

Then, the equality

$$\sum_{\theta_n, ..., \theta_1} \phi_{\theta_N, ..., \theta_1} = \sum_{\theta_n, ..., \theta_2} \frac{\sqrt{c}}{2} \phi'_{\theta_N, ..., \theta_2} \tag{34}$$

implies

$$H_N(\phi) = H_{N-1}(\phi') - \frac{1}{2} \sum_\tau {\phi'_\tau}^2. \tag{35}$$

Suppose that a function $F(\phi)$ depends on ϕ through ϕ' only, namely, there is a function $F'(\phi')$ on the block spins such that

$$F(\phi) = F'(\phi'). \tag{36}$$

Then it holds that

$$\langle F \rangle_{\Lambda, h} = \langle F' \rangle_{\Lambda - 1, \mathcal{R}h} \, , \tag{37}$$

where $\mathcal{R}h$ is defined by

$$\mathcal{R}h(x) = \text{const.} \exp(\frac{\beta}{2} x^2) \int_{\mathbb{R}} h(\frac{x}{\sqrt{c}} + y) h(\frac{x}{\sqrt{c}} - y) \, dy, \ x \in \mathbb{R}. \tag{38}$$

Note that

$$h_G(x) = \text{const.} \exp(-\frac{1}{4} x^2) \tag{39}$$

is a fixed point of $\mathcal{R}$, which we shall refer to as the density function of *hierarchical massless Gaussian measure*. By looking into the asymptotics of *e.g.*, susceptibility for the hierarchical massless Gaussian measure and comparing it with that of the standard nearest neighbor massless Gaussian measure on d-dimensional regular lattice, we see that the dimensionality d of the system may be identified (at least for the Gaussian fixed point) as

$$c = 2^{1 - 2/d}. \tag{40}$$

We shall extend the correspondence to hierarchical models with non-Gaussian measures, and use the terminology *d-dimensional hierarchical models* whenever (40) holds.

3.2. *RG trajectory*

The 'continuum limit' of the hierarchical model is analyzed through the asymptotic property of the RG trajectory

$$h_N = \mathcal{R}^N h_0 \,, \quad N = 0, 1, 2, \cdots . \tag{41}$$

The RG trajectory (41) has been extensively investigated in a weak coupling region[4,5,6] *i.e.*, in a 'neighborhood' of h_G. In particular, it is known that, if $d \geq 4$, then there are no non-Gaussian fixed points in a 'neighborhood' of h_G, and that a 'continuum limit' constructed from a critical trajectory with an initial function in the 'neighborhood' of h_G is trivial (Gaussian).

3.3. *Hierarchical Ising model*

Our concern is the hierarchical Ising model, which is defined by the following single spin measure density

$$h_{\mathrm{I},s}(x) = \frac{1}{2}(\delta(x - s) + \delta(x + s)), \tag{42}$$

where $s > 0$. Hierarchical Ising model has an infinite volume limit $\Lambda \to \infty$, if $0 < c < 2 \ (d > 0)$, and has a phase transition, if $1 < c < 2 \ (d > 2)$.[3]

Note that the density (42) is regarded as a strong coupling limit $\lambda \to \infty$ of the ϕ^4 densities

$$h_{\mu,\lambda}(x) = \mathrm{const.} \exp(-\mu x^2 - \lambda x^4), \quad \mu = -2\lambda s^2, \tag{43}$$

and an investigation of the RG trajectory for hierarchical Ising model requires an analysis in the 'strong coupling region' far away from the Gaussian fixed point.

This problem is solved and the triviality of the four dimensional hierarchical Ising model is shown by using characteristic functions of single spin distributions.

4. Triviality of hierarchical Ising model in four dimensions

As mentioned in Introduction, there is the longstanding conjecture that

the continuum limit of four dimensional Ising model will be trivial (Gaussian).

The following theorem claims that the *hierarchical version* of the above conjecture is true.

Theorem 1: [2] *If $d \geq 4$ (i.e. $c \geq \sqrt{2}$), there exists a critical trajectory converging to the Gaussian fixed point starting at a hierarchical Ising model. Namely, there exists a positive real number s_c such that if h_m, $m = 0, 1, 2, \cdots$, are defined by (41) with $h_0 = h_{\mathrm{I},s_c}$, then the sequence of measures $h_m(x)\,dx$, $m = 0, 1, 2, \cdots$, converges weakly to the massless Gaussian measure $h_G(x)\,dx$.*

4.1. *Characteristic function*

Main idea of our proof is to use characteristic functions of single spin distributions:

$$\hat{h}_m(\xi) = \mathcal{F}h_m(\xi) = \int_{\mathbb{R}} e^{i\xi x} h_m(x)\,dx \,. \tag{44}$$

The RG transformation for $\hat{h}_m$ is

$$\hat{h}_{m+1} = \mathcal{F}\mathcal{R}\mathcal{F}^{-1}\hat{h}_m \,, \tag{45}$$

which has a decomposition

$$\mathcal{F}\mathcal{R}\mathcal{F}^{-1} = \mathcal{T}\mathcal{S}, \tag{46}$$

where

$$\mathcal{S}g(\xi) = g\left(\frac{\sqrt{c}}{2}\xi\right)^2, \tag{47}$$

$$\mathcal{T}g(\xi) = \text{const. } \exp\left(-\frac{\beta}{2}\triangle\right)g(\xi), \tag{48}$$

and the constant is so defined that

$$\mathcal{T}g(0) = 1\,. \tag{49}$$

4.2. *Newman's inequalities*

Let us introduce a 'potential' V_m for the characteristic function $\hat{h}_m$ and its Taylor coefficients $\mu_{n,m}$, respectively, by

$$\hat{h}_m(\xi) = e^{-V_m(\xi)}, \tag{50}$$

$$V_m(\xi) = \sum_{n=1}^{\infty} \mu_{n,m}\xi^n. \tag{51}$$

Note that

$$V_m(0) = 0 \, , \tag{52}$$

$$\mu_{n,m} = 0, \quad n = 1, 3, 5, \dots \tag{53}$$

hold, since $\hat{h}_m(0) = 1$ and $\hat{h}_m(\xi)$ is an even function. The coefficient $\mu_{n,m}$ is called *a truncated n point correlation.*

The function V_m has a remarkable properties due to Newman[7]

$$\mu_{2n,m} \geq 0, \ n \geq 1, \tag{54}$$

$$\mu_{2n,m} \leq \frac{1}{n}(2\mu_{4,m})^{n/2}, \ n \geq 3. \tag{55}$$

These bounds follow from the Lee-Yang property for ferromagnetic systems.

Newman's inequalities are extensively used in our proof. We here note the following facts.

1. The right hand side of (51) has non-zero radius of convergence.
2. It suffices to prove $\mu_{2,m} \to 1$ and $\mu_{4,m} \to 0$ as $m \to \infty$ in order to show that the trajectory converges to the massless Gaussian measure.

4.3. *Strategy*

The proof of Theorem 1 is decomposed into two parts:

- *In the weak coupling region,* we control the RG flow by means of a *finite* number of truncated correlations ($\mu_{2n,m}, n = 1, 2, 3, 4$ for $d = 4$ and $\mu_{2n,m}, n = 1, 2$ for $d > 4$), in terms of which we give a criterion, a set of sufficient conditions, for the measure to be in a domain of attraction of the Gaussian fixed point.
- *In the strong coupling region,* we prove that there is a trajectory whose initial point is an Ising measure and for which the criterion in the weak coupling region is satisfied after a small number (in fact 70 for $d = 4$) of iterations. This part of our proof is partially computer-aided and shows that the critical value of s for $d = 4$ lies in the interval $[1.792\,567\,117\,009\,262\,4, 1.792\,567\,117\,009\,262\,5]$.

The essential point is that we can control the RG flow through *finite* degrees of freedom. This is why we can obtain rigorous bounds by means of computer.

5. Hierarchical $O(N)$ model

The hierarchical model considered in the previous sections is extended to $N > 1$ component spins $\phi_\theta \in \mathbb{R}^N$. This model is called the hierarchical $O(N)$ model and is defined by the following formulae corresponding to (28), (29), (30), (42), respectively:

$$H_\Lambda(\phi) = -\frac{1}{2} \sum_{n=1}^{\Lambda} \left(\frac{c}{4}\right)^n \sum_{\theta_\Lambda,\ldots,\theta_{n+1}} \left| \sum_{\theta_n,\ldots,\theta_1} \phi_{\theta_\Lambda,\ldots,\theta_1} \right|^2, \tag{56}$$

$$\langle F \rangle_{\Lambda,h} = \frac{1}{Z_{\Lambda,h}} \int d\phi F(\phi) \exp(-\beta H_\Lambda(\phi)) \prod_\theta h(\phi_\theta), \tag{57}$$

$$Z_{\Lambda,h} = \int d\phi \exp(-\beta H_\Lambda(\phi)) \prod_\theta h(\phi_\theta), \tag{58}$$

$$h(x) = \text{const.}\delta(|x| - s) , \quad x \in \mathbb{R}^N , \tag{59}$$

where $s > 0$

The analysis in the preceding section may be extended to the hierarchical $O(N)$ model, if we perform the necessary numerical calculations for each N. Furthermore it seems plausible that the triviality may be proved without computer, if N is sufficiently large. In fact, as in shown below, the model is exactly solvable in the limit $N \to \infty$, which is called the (hierarchical) spherical model.

5.1. *RG transformations for $O(N)$ and spherical models*

The operators $\mathcal{S}$ and $\mathcal{T}$ for $N > 1$ have the same form as (47) and (48), respectively, where ξ runs through $\mathbb{R}^N$ and $\triangle$ is the N–dimensional Laplacian. Introduce the t–dependent potential $V_m(t, \xi)$ by

$$\exp(-V_m(t,\xi)) = \exp(-t\triangle) \exp(-2V_m(\frac{\sqrt{c}}{2}\xi)) . \tag{60}$$

Then, $V_m(t,\xi)$ obeys the differential equation

$$\frac{\partial V_m(t,\xi)}{\partial t} = -\triangle V_m(t,\xi) + |\text{grad}\, V_m(t,\xi)|^2 \tag{61}$$

and, the next effective potential V_{m+1} is given by

$$V_{m+1}(\xi) = V_m(\frac{\beta}{2},\xi) . \tag{62}$$

Note that the characteristic functions and potentials have spherical symmetry, hence we can regard them as functions of t and $|\xi|$. In particular,

(61) is written as

$$\dot{V}_m = {V'_m}^2 - (N-1)\frac{1}{|\xi|}V'_m - V''_m \ , \tag{63}$$

where $\dot{V}_m$ and V'_m stand for derivatives with respect to t and $|\xi|$, respectively.

On the other hand, the initial characteristic function $\hat{h}_0(\xi)$, the Fourier transform of (59), is explicitly written in terms of Bessel function as

$$\hat{h}_0(\xi) = \Gamma(N/2)\frac{J_{N/2-1}(s|\xi|)}{(s|\xi|/2)^{N/2-1}} \ . \tag{64}$$

From this expression, we see that, in order to take a sensible limit $N \to \infty$ of $V_0 = -\log \hat{h}_0$, it is necessary to put $s = \sqrt{N}\sigma$ and at the same time to scale the $V_0(\xi)$ as follows :

$$U_0(\eta) = \frac{1}{N}V_0(\sqrt{N}\eta) \ . \tag{65}$$

Then, the limit

$$u_0(\eta) = \lim_{N \to \infty} U_0(\eta) \tag{66}$$

satisfies

$$u'_0(z) = \frac{2\sigma^2 z}{1 + \sqrt{1 - 4\sigma^2 z^2}} \ , \tag{67}$$

where u_0 is regarded as a function of $z = |\eta|$.

Now, the scaled m-th effective potential

$$U_m(t,\eta) = \frac{1}{N}V_m(t,\sqrt{N}\eta) \tag{68}$$

obviously satisfies

$$\dot{U}_m(t,z) = U'_m(t,z)^2 - (1 - \frac{1}{N})\frac{1}{z}U'_m(t,z) - \frac{1}{N}U''_m(t,z) \tag{69}$$

as a function of t and $z = |\eta|$. Then, we *define* the recursion of the spherical model as follows:

$$u_m(0,z) = 2u_m(\frac{\sqrt{c}}{2}z) \ , \tag{70}$$

$$u_{m+1}(z) = u_m(\frac{\beta}{2},z) \ , \tag{71}$$

where $u_m(t,z)$ obeys

$$\dot{u}_m(t,z) = u'_m(t,z)^2 - \frac{1}{z}u'_m(t,z) \ . \tag{72}$$

Note that (72) is a formal limit $N \to \infty$ of (69).

5.2. *RG trajectory for the spherical model*

Since the differential equation (72) is solved by means of characteristic curves, the recursion for the spherical model is exactly solved. In fact, for $d = 4$ $(c = \sqrt{2})$, the function $w_m(x)$ defined by

$$w_m(z^2) = u_m(z) \tag{73}$$

is determined by

$$\frac{1}{1-q} = w'_m(x) \quad \Longleftrightarrow \quad q = \frac{1}{m/2+1}\Big(\frac{2x}{(1-q)^2} + \gamma_m - R_m(q)\Big) \tag{74}$$

where

$$\gamma_m = \frac{1}{\sqrt{2}} + \Big(\frac{1}{2-\sqrt{2}} - \sigma^2\Big)2^{m/2} \;, \tag{75}$$

$$R_m(q) = q^2\Big(\frac{1}{1-q} + \frac{1}{2}\sum_{j=1}^{m} \frac{1}{2^{j/2}-q}\Big) \;. \tag{76}$$

From this result, we can prove that the critical value of σ is

$$\sigma_c = \frac{1}{\sqrt{2-\sqrt{2}}} \tag{77}$$

and that, for $\sigma = \sigma_c$, w'_m converges to 1, which means the triviality $u_m(z) \to z^2$ of the spherical model.

In order to show the triviality of the hierarchical $O(N)$ model in four dimensions, it is critical to show that U_m converges to u_m as $N \to \infty$. This problem is under investigation.

References

1. R. Fernández, J. Fröhlich, A. D. Sokal, *Random Walks, Critical Phenomena, and Triviality in Quantum Field Theory,* Springer, 1992.

2. T. Hara, T. Hattori, H. Watanabe, Triviality of hierarchical Ising model in four dimensions, Commun. Math. Phys., **220**, 13–40 (2001).

3. F. J. Dyson, *Existence of a Phase–Transition in a One–Dimensional Ising Ferromagnet,* Commun. Math. Phys.**12** 91–107 (1969).

4. Ya. G. Sinai, *Theory of Phase Transition: Rigorous Results,* Pergamon Press, 1982.

5. P. Collet, J. -P. Eckmann, *A Renormalization Group Analysis of the Hierarchical Model in Statistical Mechanics,* Springer Lecture Note in Physics 74.

6. K. Gawedzki, A. Kupiainen, *Non-Gaussian Fixed Point of the Block Spin Transformation. Hierarchical Model Approximation,* Commun. Math. Phys.**89** (1983) 191–220.

7. C. M. Newman, Inequalities for Ising models and field theories which obey the Lee–Yang theorem, Commun. Math. Phys., **41**, 1–9 (1975).

Seven-Vertex Solutions of the Colored Yang–Baxter Equation

Shi-kun Wang

Institute of Applied Mathematics, AMSS, CAS
Beijing 100080, China
E-mail: wsk@amss.ac.cn

Ke Wu

Department of Mathematics, Capital Normal University
Beijing 100037
E-mail: wuke@itp.ac.cn

In this paper all seven-vertex type solutions of the colored Yang–Baxter equation dependent on spectral as well as colored parameters are given. It is proved that they are composed of six groups of basic solutions up to five solution transformations. Moreover, all solutions can be classified into two types called Baxter type and free-fermion type.

1. Preliminary

The Yang–Baxter equation, first appeared in Refs.[1,2,3] is an important subject related to many other branches, such as factorized S-matrix,[4] exactly solvable models of statistical physics,[5] complete integrable quantum and classical systems,[6] quantum groups,[7] conformal field theory, link invariants[8−11] and so on. Motivated by the important role of the equation, much attention has been attracted to the search for the solutions.

The colored Yang–Baxter equation (CYBE) dependent on spectral as well as colored parameters is a generalization of the usual Yang–Baxter equation. Because the colored Yang–Baxter equation concerns the free-fermion model in a magnetic field, multi-variable invariants of links and representations of quantum algebras and so on,[12−16] it has also attracted a lot of research to find the solutions of this type of Yang–Baxter equation.[17−22]

The main theme of this paper is to give and classify seven-vertex solutions of CYBE by a computer algebraic method. Moreover, it is proved from a theorem in Ref.[23] that all seven-vertex type solutions of CYBE have indeed been obtained in this paper. In section 2 we will introduce the symmetries or solution transformation for CYBE. Then by the symmetries and a computer algebraic method, we find the most simple system of equations, including differential equation, the solution set of which covers the one of CYBE. In section 3 and section 4 we will seperately give all solutions of the seven-vertex CYBE and classify them into two kinds called Baxter and Free-Fermion types.

2. The symmetries and initial conditions of CYBE

The colored Yang–Baxter equation means the following matrix equation:

$$\begin{aligned}
\check{R}_{12}(u,\xi,\eta)&\check{R}_{23}(u+v,\xi,\lambda)\check{R}_{12}(v,\eta,\lambda) \\
&= \check{R}_{23}(v,\eta,\lambda)\check{R}_{12}(u+v,\xi,\lambda)\check{R}_{23}(u,\xi,\eta), \\
\check{R}_{12}(u,\xi,\eta) &= \check{R}(u,\xi,\eta) \otimes E, \\
\check{R}_{23}(u,\xi,\eta) &= E \otimes \check{R}(u,\xi,\eta),
\end{aligned} \tag{1}$$

where $\check{R}(u,\xi,\eta)$ is a $N \times N$-matrix function of u, ξ and η. E is the unit matrix of order N and $\otimes$ means the tensor product of two matrices. u, v are spectral parameters and ξ, η are colored parameters. If we let $\xi = \eta = 0$, we can get the usual Yang–Baxter equation. Similarly, if we have $u = v = 0$, then (1) will be reduced to the pure colored Yang–Baxter equation.

In this paper, the main interest for the equation (1) is to discuss the solutions of seven-vertex type

$$\check{R}(u,\xi,\eta) = \begin{pmatrix} a_1(u,\xi,\eta) & 0 & 0 & a_7(u,\xi,\eta) \\ 0 & a_2(u,\xi,\eta) & a_6(u,\xi,\eta) & 0 \\ 0 & a_5(u,\xi,\eta) & a_3(u,\xi,\eta) & 0 \\ 0 & 0 & 0 & a_4(u,\xi,\eta) \end{pmatrix}. \tag{2}$$

The elements in this matrix are called weight functions in (2). In this paper, we only consider the case where weight functions $a_i(u,\xi,\eta)(i = 1,2,\ldots,7)$ are meromorphic functions of u,ξ,η and $a_i(u,\xi,\eta) \not\equiv 0$. Throughout this paper, we let

$$u_i = a_i(u,\xi,\eta), \quad v_i = a_i(v,\eta,\lambda), \quad w_i = a_i(u+v,\xi,\lambda), \quad i = 1,2,\ldots,7.$$

For the seven-vertex-type solutions, the matrix equation (1) is equivalent to the following 19 equations:

$$u_2 w_3 v_2 - u_3 w_2 v_3 = 0, \tag{3a}$$

$$\left.\begin{array}{l} u_1v_1w_2 - u_2v_2w_1 - u_5v_6w_2 = 0, \\ u_2v_1w_5 - u_2v_5w_1 - u_5v_3w_2 = 0, \\ u_1v_2w_6 - u_3v_6w_2 - u_6v_2w_1 = 0, \end{array}\right\} \quad (3b)$$

$$\left.\begin{array}{l} u_4v_4w_2 - u_2v_2w_4 - u_6v_5w_2 = 0, \\ u_2v_4w_6 - u_2v_6w_4 - u_6v_3w_2 = 0, \\ u_4v_2w_5 - u_3v_5w_2 - u_5v_2w_4 = 0, \end{array}\right\} \quad (3c)$$

$$\left.\begin{array}{l} u_1v_1w_3 - u_3v_3w_1 - u_5v_6w_3 = 0, \\ u_3v_1w_5 - u_3v_5w_1 - u_5v_2w_3 = 0, \\ u_1v_3w_6 - u_2v_6w_3 - u_6v_3w_1 = 0, \\ u_1v_2w_7 - u_4v_7w_2 - u_7v_1w_1 + u_7v_5w_5 = 0, \\ u_1v_6w_7 - u_2v_7w_6 - u_5v_1w_7 + u_7v_3w_5 = 0, \\ u_1v_7w_1 - u_3v_1w_7 - u_6v_7w_6 + u_7v_4w_3 = 0, \end{array}\right\} \quad (3d)$$

$$\left.\begin{array}{l} u_4v_4w_3 - u_3v_3w_4 - u_6v_5w_3 = 0, \\ u_3v_4w_6 - u_3v_6w_4 - u_6v_2w_3 = 0, \\ u_4v_3w_5 - u_2v_5w_3 - u_5v_3w_4 = 0, \\ u_4v_2w_7 - u_1v_7w_2 - u_7v_4w_4 + u_7v_6w_6 = 0, \\ u_4v_5w_7 - u_2v_7w_5 - u_6v_4w_7 + u_7v_3w_6 = 0, \\ u_4v_7w_4 - u_3v_4w_7 - u_5v_7w_5 + u_7v_1w_3 = 0. \end{array}\right\} \quad (3e)$$

Assume $\check{R}(u,\xi,\eta)$ is the solution of equation(1). We can find there are five symmetries in the system of equations (3):

(A) Symmetry of interchanging indices. The system of equations (3) is invariant if we interchange the two sub-indices 1 and 4 as well as the two sub-indices 5 and 6.

(B) The scaling symmetry. Multiplication of the solution $\check{R}(u,\xi,\eta)$ by an arbitrary function $\mathcal{F}(u,\xi,\eta)$ is still a solution of the equation (1).

(C) Symmetry of the weight functions. If the weight functions

$$a_2(u,\xi,\eta), \quad a_3(u,\xi,\eta), \quad a_7(u,\xi,\eta)$$

are replaced by the new weight functions:

$$\tilde{a}_2(u,\xi,\eta) = \frac{N(\xi)}{N(\eta)}a_2(u,\xi,\eta) \qquad \tilde{a}_3(u,\xi,\eta) = \frac{N(\eta)}{N(\xi)}a_3(u,\xi,\eta),$$

$$\tilde{a}_7(u,\xi,\eta) = \frac{1}{sN(\xi)N(\eta)}a_7(u,\xi,\eta)$$

respectively, or $a_7(u,\xi,\eta), a_5(u,\xi,\eta)$ and $a_6(u,\xi,\eta)$ are replaced by $-a_7(u,\xi,\eta), -a_5(u,\xi,\eta)$ and $-a_6(u,\xi,\eta)$, where $N(\xi)$ is an arbitrary function of ξ and s is a complex constant, the new matrix $\check{R}(u, \xi, \eta)$ is still a solution of (1).

(D) Symmetry of spectral parameters. If we take the new spectral parameter $\tilde{u} = \mu u$ where μ is a complex constant, the new matrix $\check{R}(\tilde{u}, \xi, \eta)$ is still a solution of (1).

(E) Symmetry of the colored parameters. If we take the new colored parameters $\zeta = f(\xi), \theta = f(\eta)$, where $f(\xi)$ is an arbitrary function, then the new matrix $\check{R}(u, \zeta, \theta)$ is also a solution of the equation (1).

The five symmetries (A)–(E) are called solution transformations **A–E** of seven-vertex-type solutions of colored Yang–Baxter equation (1), respectively.

Dividing both sides of equation (3a) by $u_2 w_2 v_2$, we get

$$f(u + v, \xi, \lambda) = f(u, \xi, \eta) f(v, \eta, \lambda), \tag{4}$$

where $f(u, \xi, \eta) = \frac{u_3}{u_2}$. Taking $u = v = \eta = 0$ in (4) we get

$$f(0, \xi, \lambda) = f(0, \xi, 0) f(0, 0, \lambda).$$

Taking $u = v = \xi = 0$ in (4) we have

$$f(0, 0, \lambda) = f(0, 0, \eta) f(0, \eta, \lambda).$$

Then we have

$$f(0, 0, \lambda) = f(0, 0, \eta) f(0, \eta, 0) f(0, 0, \lambda).$$

This means

$$f(0, 0, \eta) f(0, \eta, 0) = 1.$$

Otherwise, it is easy to show that $f(u, \xi, \eta) = 0$, *i.e.* $a_3(u, \xi, \eta) = 0$. Therefore, we have

$$f(0, \xi, \eta) = \frac{M(\xi)}{M(\eta)}, \tag{5}$$

where $M(\xi) = f(0, \xi, 0)$. On the other hand, if we differentiate both sides of (4) with respect to the spectral variable v and then set $v = 0$, $\lambda = \eta$, then

$$f'(u, \xi, \eta) = f(u, \xi, \eta) f'(0, \eta, \eta)$$

holds, where the prime means derivative with respect to u and the simple formula

$$\left. \frac{dH(u + v)}{dv} \right|_{v=0} = \frac{dH(u)}{du}$$

for any function $H(u)$ is used. Similarly, if we differentiate (4) with respect to u and then set $u = 0$ and $\eta = \xi$ we have

$$f'(v, \xi, \lambda) = f(v, \xi, \lambda) f'(0, \xi, \xi).$$

The above two formulas imply $f'(0, \xi, \xi)$ is a constant independent of colored parameter ξ. Hence

$$f(u, \xi, \eta) = \frac{M(\xi)}{M(\eta)} \exp(ku), \qquad (6)$$

where k is a complex constant.

So, up to the solution transformation $\mathbf{B}$ and $\mathbf{C}$, we can assume

$$a_2(u, \xi, \eta) = 1, \quad a_3(u, \xi, \eta) = \exp(ku)$$

without losing generality. Then the system of equations (3) can be simplified to the following 12 equations:

$$
\begin{aligned}
&u_1 w_5 - u_5 w_1 - v_5 \exp(ku) = 0, \\
&w_6 v_1 - w_1 v_6 - u_6 \exp(kv) = 0, \\
&u_1 v_1 - w_1 - u_6 v_5 = 0, \\
&u_1 w_1 v_7 + u_7 v_4 - u_5 w_5 v_7 - w_7 v_1 = 0, \\
&u_1 w_7 v_5 + u_7 w_6 - u_6 w_7 v_1 - w_5 v_7 \exp(ku) = 0, \\
&u_7 w_6 v_6 - u_7 w_1 v_1 + u_1 w_7 \exp(kv) - u_4 v_7 \exp(ku + kv) = 0
\end{aligned}
\qquad (7)
$$

plus six equations which are called the counterparts of (7) obtained by interchanging the sub-indices 1 and 4 as well as 5 and 6 in each of the equations in (7).

Now we solve the equations obtained by setting $u = 0$ and $\eta = \xi$ in (7). It is easy to get

$$
\begin{aligned}
a_1(0, \xi, \xi) = a_4(0, \xi, \xi) = 1, \\
a_5(0, \xi, \xi) = a_6(0, \xi, \xi) = a_7(0, \xi, \xi) = 0
\end{aligned}
\qquad (8)
$$

which are called the initial conditions of (1) in this paper. Substituting (8) into (7) after letting $v = -u, \lambda = \xi$ we obtain

$$
\begin{aligned}
a_5(u, \xi, \eta) &= -a_5(-u, \eta, \xi) \exp(ku), \\
a_6(u, \xi, \eta) &= -a_6(-u, \eta, \xi) \exp(ku), \\
a_4(u, \xi, \eta) a_4(-u, \eta, \xi) &= a_1(u, \xi, \eta) a_1(-u, \eta, \xi), \\
a_7(u, \xi, \eta) a_1(-u, \eta, \xi) &= -a_7(-u, \eta, \xi) a_4(u, \xi, \eta).
\end{aligned}
\qquad (9)
$$

Differentiating both sides of all equations in (7) and their counterparts with respect to the variable v and letting $v = 0, \lambda = \eta$, by virtue of initial conditions (8) we can get

$$u_1 u_5' - u_5 u_1' - m_5(\eta)\exp(ku) = 0,$$
$$u_6' + u_6 m_1(\eta) - u_1 m_6(\eta) - ku_6 = 0,$$
$$u_1 m_1(\eta) - u_1' - u_6 m_5(\eta) = 0,$$
$$u_1 u_7 m_5(\eta) + u_7 u_6' - u_5 m_7(\eta)\exp(ku) - u_6 u_7 m_1(\eta) - u_6 u_7' = 0,$$
$$ku_1 u_7 + u_1 u_7' + u_6 u_7 m_6(\eta) - u_1 u_7 m_1(\eta) - u_7 u_1' - u_4 m_7(\eta)\exp(ku) = 0,$$
$$(u_1^2 - u_5^2)m_7(\eta) + (m_4(\eta) - m_1(\eta))u_7 - u_7' = 0$$

$$(10a)$$

and their counterparts, where and throughout this paper unless otherwise stated we denote

$$a_i'(u,\xi,\eta) = \frac{\partial a_i(u,\xi,\eta)}{\partial u}, \quad m_i(\xi) = a_i'(u,\xi,\eta)\Big|_{(u=0,\eta=\xi)} \quad . \quad (i = 1,2,\ldots,7)$$

We call $m_i(\xi)$ Hamiltonian coefficients of weight functions with respect to the spectral parameter or simply coefficients. Sometimes we write m_i instead of $m_i(\xi)$ for brevity.

If we differentiate (7) with respect to u and let $u = 0, \eta = \xi$ and then replace the variables v and λ by u and η, we can get from the initial condition (8)

$$u_5 m_1(\xi) + u_5' - u_1 m_5(\xi) - ku_5 = 0,$$
$$u_1 u_6' - u_1' u_6 - m_6(\xi)\exp(ku) = 0,$$
$$u_1 m_1(\xi) - u_1' - u_5 m_6(\xi) = 0,$$
$$u_1 u_7 m_1(\xi) - u_1 u_7' + u_1' u_7 + u_4 m_7(\xi) - u_5 u_7 m_5(\xi) = 0, \qquad (10b)$$
$$u_5 u_7 m_1(\xi) + u_5 u_7' - u_5' u_7 + u_6 m_8(\xi) - ku_5 u_7 - u_1 u_7 m_6(\xi) = 0,$$
$$(m_1(\xi) - m_4(\xi))u_7 \exp(ku)$$
$$\qquad\qquad + (u_6^2 - u_1^2)m_7(\xi) + u_7'\exp(ku) - ku_7\exp(ku) = 0$$

and their counterparts. Similarly, if we differentiate (7) with respect to λ and then let $v = 0, \lambda = \eta$, we can get

$$u_1 u_5' - u_5 u_1' - \breve{m}_5(\eta)\exp(ku) = 0,$$
$$u_6' + u_6 \breve{m}_1(\eta) - u_1 \breve{m}_6(\eta) = 0,$$
$$u_1 \breve{m}_1(\eta) - u_1' - u_6 \breve{m}_5(\eta) = 0,$$
$$u_1 u_7 \breve{m}_5(\eta) + u_7 u_6' - u_5 \breve{m}_7(\eta)\exp(ku) - u_6 u_7 \breve{m}_1(\eta) - u_6 u_7' = 0, \qquad (10c)$$
$$u_1 u_7' + u_6 u_7 \breve{m}_6(\eta) - u_1 u_7 \breve{m}_1(\eta) - u_7 u_1' - u_4 \breve{m}_7(\eta)\exp(ku) = 0,$$
$$(u_1^2 - u_5^2)\breve{m}_7(\eta) + (\breve{m}_4(\eta) - \breve{m}_1(\eta))u_7 - u_7' = 0$$

and their counterparts. Here we denote

$$u_i' = \frac{\partial a_i(u,\xi,\eta)}{\partial \eta}, \quad \breve{m}_i(\eta) = \frac{\partial a_i(v,\eta,\lambda)}{\partial \lambda}\Big|_{(v=0,\lambda=\eta)}.$$

Similarly, if we differentiate (7) with respect to ξ and let $u = 0, \eta = \xi$ and then replace v, λ by u, η we have

$$
\begin{aligned}
&u_5 \hat{m}_1(\xi) + u_5' - u_1 \hat{m}_5(\xi) = 0, \\
&u_1 u_6' - u_1' u_6 - \hat{m}_6(\xi) \exp(ku) = 0, \\
&u_1 \hat{m}_1(\xi) - u_1' - u_5 \hat{m}_6(\xi) = 0, \\
&u_1 u_7 \hat{m}_1(\xi) - u_1 u_7' + u_1' u_7 + u_4 \hat{m}_7(\xi) - u_5 u_7 \hat{m}_5(\xi) = 0, \\
&u_5 u_7 \hat{m}_1(\xi) + u_5 u_7' - u_5' u_7 + u_6 \hat{m}_8(\xi) - u_1 u_7 \hat{m}_6(\xi) = 0, \\
&(\hat{m}_1(\xi) - \hat{m}_4(\xi)) u_7 \exp(ku) + (u_6^2 - u_1^2) \hat{m}_7(\xi) + u_7' \exp(ku) = 0
\end{aligned}
\tag{10d}
$$

and their counterparts. Here we denote

$$
u_i' = \frac{\partial a_i(u, \xi, \eta)}{\partial \xi}, \quad \hat{m}_i(\xi) = \left. \frac{\partial a_i(u, \xi, \eta)}{\partial \xi} \right|_{(u=0, \eta=\xi)}.
$$

From the last equation of (10a) and its counterpart we have

$$
2u_7' = m_7(\eta)(u_1^2 + u_4^2 - u_5^2 - u_6^2).
\tag{11a}
$$

We also get from the last equation of (10b) and its counterpart the following equation:

$$
2u_7' = m_7(\xi)(u_1^2 + u_4^2 - u_5^2 - u_6^2) \exp(-ku) + 2ku_7.
\tag{11b}
$$

Now there are two cases, $k = 0$ and $k \neq 0$. We will discuss them separately in the later sections.

3. The solutions of equation (1) in the case of $k \neq 0$

Now we first consider the case of $k \neq 0$. From equations (11a) and (11b), we can obtain

$$
u_7' \neq 0, \quad m_7(\xi) \neq 0, \quad m_7(\eta) \neq 0
\tag{12}
$$

and

$$
\frac{u_7' - ku_7}{u_7'} \exp(ku) = \frac{m_7(\xi)}{m_7(\eta)}.
$$

Solving the above equation we can get

$$
a_7(u, \xi, \eta) = F_7(\xi, \eta) \exp(ku) - \frac{F_7(\xi, \xi)}{F_7(\eta, \eta)}, \quad m_7(\eta) = kF_7(\eta, \eta).
\tag{13}
$$

The following work is to eliminate the five weight functions $\{w_1, w_4, w_5, w_6, w_7\}$ in the equation (7) and their counterparts. Then we get seven polynomial equations without weight functions $w_i (i = 1, 4, \ldots, 7)$. If we differentiate the obtained seven equations with respect to the spectral parameter v, and let $v = 0, \lambda = \eta$ and substitute the initial condition (8) into them, we obtain the following seven polynomial equations:

$$m_6(\eta)\exp(ku) + (m_1(\eta) + m_4(\eta))u_4u_6 - m_6(\eta)(u_1u_4 + u_5u_6) - ku_4u_6 = 0,$$
$$m_5(\eta)\exp(ku) + (m_4(\eta) + m_1(\eta))u_1u_5 - m_6(\eta)(u_1u_4 + u_5u_6) - ku_1u_6 = 0,$$
$$m_7(\eta)(u_1^2 - u_4^2 - u_5^2 + u_6^2) + 2u_7(m_4(\eta) - m_1(\eta)) = 0,$$
$$m_7(\eta)(u_1^2u_6 - u_5^2u_6 + u_5\exp(ku)) - (m_5(\eta) + m_6(\eta))u_1u_7$$
$$- (k - m_1(\eta) - m_4(\eta))u_6u_7 = 0,$$
$$m_7(\eta)(u_1u_5^3 - u_1^3u_5 - u_1u_6\exp(ku)) + m_5(\eta)u_7\exp(ku)$$
$$+ 2(m_1(\eta) - m_4(\eta)u_5u_7 + m_6(\eta)u_1u_4u_7 - m_5(\eta)u_5u_6u_7 = 0,$$
$$m_7(\eta)(u_1^3 - u_1u_5^2 - u_4\exp(ku)) + u_1u_7(k + m_4(\eta) - 3m_1(\eta)) + (m_5(\eta)$$
$$+ m_6(\eta))u_6u_7 = 0,$$
$$m_7(\eta)u_4(u_1^2 - u_5^2) + u_4u_7(k - m_1(\eta) - m_4(\eta)) + (m_5(\eta) + m_6(\eta))u_5u_7 = 0.$$
$$(14)$$

As the second step, we separate (14) into two systems of equations, which are equivalent to (14). The first is

$$m_5\exp(ku) + (m_1 + m_4)u_1u_5 - m_5(u_1u_4 + u_5u_6) - ku_1u_5 = 0,$$
$$m_7u_1u_5(u_4^2 - u_1^2 + u_5^2 - u_6^2) + 2m_5u_7(u_1u_4 + u_5u_6 - \exp(ku))$$
$$+ (2k - 4m_4)u_1u_5u_7 = 0,$$
$$m_7u_1(u_4^2u_5 - u_5^2u_6 + u_6\exp(ku)) - u_7(m_5\exp(ku) + m_6u_1u_4 - m_5u_5u_6) = 0,$$
$$(15)$$

which contains m_1, m_4 and m_5. The second is

$$(u_1u_4 + u_5u_6 - \exp(ku))(-m_7u_4u_6(u_4^2u_5 - u_5u_6^2 + u_6\exp(ku)),$$
$$+ m_6u_7(u_4^2u_6 - u_5^2u_6 + u_5\exp(ku))) = 0,$$
$$(u_1u_4 + u_5u_6 - \exp(ku))(m_7u_1u_5(u_6^2 - u_4^2) + m_7\exp(ku)(u_4u_5 - u_1u_6)$$
$$+ m_6u_7(u_1u_4 - u_5u_6)) = 0,$$
$$(u_1u_4 + u_5u_6 - \exp(ku))(m_7u_5u_6(u_4u_6 - u_1u_5) + m_7u_4u_5(u_5^2 - u_4^2)$$
$$+ m_7\exp(ku)(u_1u_5 - u_4u_6) + m_6u_7(u_4^2 - u_5^2)) = 0,$$
$$(u_1u_4 + u_5u_6 - \exp(ku))(m_7u_4u_5(u_1u_5 - u_4u_6)$$
$$+ m_7(u_5u_6 - \exp(ku))(u_6^2 - u_5^2) + m_6u_7(u_4u_6 - u_1u_5)) = 0,$$
$$(16)$$

which does not contain m_1, m_4 and m_5. So

$$u_1u_4 + u_5u_6 = \exp(ku) \tag{17}$$

or

$$-m_7 u_4 u_6 (u_4^2 u_5 - u_5 u_6^2 + u_6 \exp(ku))$$
$$+m_6 u_7 (u_4^2 u_6 - u_5^2 u_6 + u_5 \exp(ku)) = 0,$$
$$m_7 u_1 u_5 (u_6^2 - u_4^2) + m_7 \exp(ku)(u_4 u_5 - u_1 u_6) + m_6 u_7 (u_1 u_4 - u_5 u_6) = 0,$$
$$m_7 u_4 u_5 (u_1 u_5 - u_4 u_6) + m_7 (u_5 u_6 - \exp(ku))(u_6^2 - u_5^2)$$
$$+m_6 u_7 (u_4 u_6 - u_1 u_5) = 0,$$
$$m_7 u_5 u_6 (u_4 u_6 - u_1 u_5) + m_7 u_4 u_5 (u_5^2 - u_4^2) + m_7 \exp(ku)(u_1 u_5 - u_4 u_6)$$
$$+m_6 u_7 (u_4^2 - u_5^2) = 0$$

$$(18)$$

will hold. In the third step, applying the fourth equation in (18) as a main equation to kill m_6 in the three other equations in (18) and then performing factorization of the new polynomial equations, we can obtain

$$m_7 u_5 u_6 (u_4 u_6 - u_1 u_5) + m_7 u_4 u_5 (u_5^2 - u_4^2) + m_7 \exp(ku)(u_1 u_5 - u_4 u_6)$$
$$+m_6 u_7 (u_4^2 - u_5^2) = 0,$$
$$m_7 (u_5 u_6 - \exp(ku))(u_1^2 u_5 - 2u_1 u_4 u_6 + u_4^2 u_5 - u_5^3 + u_5 u_6^2)u_5 = 0,$$
$$m_7 (u_5 u_6 - \exp(ku))(-u_1 u_4^2 u_6 + u_1 u_5^2 u_6 + u_4^3 u_5 - u_4 u_5^3 - u_1 u_5 \exp(ku)$$
$$+u_4 u_6 \exp(ku))u_5 = 0,$$
$$m_7 (u_5 u_6 - \exp(ku))(-u_1^2 u_4 + 2u_1 u_5 u_6 + u_4^3 - u_4 u_5^2 - u_4 u_6^2)u_5 = 0$$

$$(19)$$

which are equivalent to (18). We have known that $m_7(\eta) \neq 0$ from (12) and $u_5 u_6 - \exp(ku) \neq 0$ from the initial conditions (8). Therefore, the following system of equations is equivalent to (19):

$$m_7 u_5 u_6 (u_4 u_6 - u_1 u_5) + m_7 u_4 u_5 (u_5^2 - u_4^2) + m_7 \exp(ku)(u_1 u_5 - u_4 u_6)$$
$$+m_6 u_7 (u_4^2 - u_5^2) = 0,$$
$$u_1^2 u_5 - 2u_1 u_4 u_6 + u_4^2 u_5 - u_5^3 + u_5 u_6^2 = 0,$$
$$-u_1 u_4^2 u_6 + u_1 u_5^2 u_6 + u_4^3 u_5 - u_4 u_5^3 - u_1 u_5 \exp(ku) + u_4 u_6 \exp(ku) = 0,$$
$$-u_1^2 u_4 + 2u_1 u_5 u_6 + u_4^3 - u_4 u_5^2 - u_4 u_6^2 = 0.$$

$$(20)$$

Remark 1. When we perform the operation of eliminating indeterminate in a system of equations, according to the theorem of zero structure of algebraic varieties,[16] the coefficient of the term with the highest degree of the indeterminate in the main polynomial equation (to be eliminated in the other polynomials) should not be identified with zero. In the event it is identified with zero, we should add the coefficient into the equations to produce a new system of equations. Otherwise, it is possible to lose some solutions. In our cases, we can discover that the coefficients of the terms which are eliminated are not identified with zero thanks to the initial conditions and the nondegenerate conditions we have set.

From the argument above, we have known the system of equations (14) is equivalent to two groups of equations. The first is (15) and (20) called Baxter case. The second is (15) and (17) called free-fermion case. We will discuss them separately.

3.1. *Baxter-type solutions*

We consider the first case in $k \neq 0$ case now *i.e.* (15) and (20). If we differentiate the second and the fourth equations in (20) and take $u = 0, \xi = \eta$ and substitute the initial conditions into the results, we can prove

$$m_5(\eta) = m_6(\eta), \quad m_1(\eta) = m_4(\eta). \tag{21}$$

By eliminating $m_4(\eta)$ we can factorize the last equation in (20) to be

$$u_5^2(u_6 - u_5)(u_6 + u_5)(u_6 - u_1)(u_6 + u_1)\exp(ku) = 0. \tag{22}$$

So

$$a_5(u, \xi, \eta) = a_6(u, \xi, \eta) \tag{23a}$$

or

$$u_5(u, \xi, \eta) = -u_6(u, \xi, \eta) \tag{23b}$$

will hold because of the initial conditions (8).

If $u_5 = -u_6$ holds, together with the second equation of (20) we have $u_1 = -u_4$ which is impossible for the initial conditions (8). Then, we have only $u_5 = u_6$. Substituting (23a) into the second equation of (20) we can obtain

$$a_1(u, \xi, \eta) = a_4(u, \xi, \eta). \tag{24}$$

Combining (23a), (24) and the first equation of (20), we can get

$$(m_7(\eta)u_1 u_5 - m_5(\eta)u_7)(u_5^2 - u_1^2) = 0. \tag{25}$$

Then the following equation is correct for the initial conditions (8)

$$m_7(\eta)u_1 u_5 = m_5(\eta)u_7. \tag{26}$$

So, we have

$$m_5(\eta) \neq 0 \tag{27}$$

because of (12). From (11a), (23a) and (24) we get

$$\frac{\partial a_7(u, \xi, \eta)}{\partial u} = m_7(\eta)(a_1^2(u, \xi, \eta) - a_5^2(u, \xi, \eta)).$$

Substituting (13) into it

$$F_7(\xi,\eta)\exp(ku) = F_7(\eta,\eta)(u_1^2 - u_5^2) \tag{28}$$

then differentiating the result above with respect to u and letting $u = 0$, $\xi = \eta$ it is easy to obtain

$$m_1(\eta) = \frac{k}{2}. \tag{29}$$

From (10a), (15), (21), (23a), (24) and (29) we also have

$$m_5(\eta)(u_5^2 + u_1^2 - \exp(ku)) = 0.$$

So there must be

$$u_5^2 + u_1^2 = \exp(ku) \tag{30}$$

for (27). But combining (28) and (30) by letting $\xi = \eta$ we have

$$a_1^2(u,\eta,\eta) - a_5^2(u,\eta,\eta) = a_1^2(u,\eta,\eta) + a_5^2(u,\eta,\eta) = \exp(ku).$$

It is to show that

$$a_5(u,\eta,\eta) = 0 \quad i.e. \quad m_5(\eta) = 0$$

which has discrepancy with (27). So, there is no solution in this case.

3.2. *Free-fermion-type solutions*

We consider the second case now *i.e.* (15) and (17). If we differentiate (17) with respect to u and then let $u = 0, \eta = \xi$ we can get

$$m_1(\eta) + m_4(\eta) = k. \tag{31}$$

Substituting (17) into (15) we obtain

$$\begin{aligned}
m_7(\eta)(u_1^2 + u_6^2 - u_4^2 - u_5^2) &= 2(m_1(\eta) - m_4(\eta))u_7, \\
m_7(\eta)(u_1 u_6 + u_4 u_5) &= (m_6(\eta) + m_5(\eta))u_7.
\end{aligned} \tag{32}$$

If we differentiate the system of equations after eliminating w_i ($i = 1, 4, \ldots, 7$) from (7) and then set $u = 0, \eta = \xi$ and replace v and λ by

u and η respectively, we can get

$$
\begin{aligned}
&m_6(\xi)(u_1u_4 + u_5u_6 - \exp(ku)) + (k - m_1(\xi) - m_4(\xi))u_1u_6 = 0, \\
&m_5(\xi)(u_1u_4 + u_5u_6 - \exp(ku)) + (k - m_1(\xi) - m_4(\xi))u_4u_6 = 0, \\
&m_7(\xi)(u_1^2 - u_4^2) - (m_5(\xi) + m_6(\xi))(u_1u_6 - u_4u_5)u_7 \\
&\qquad + 2(m_4(\xi) - m_1(\xi))u_1u_4u_7 = 0, \\
&m_7(\xi)(u_1u_6 + u_4u_5) - (m_5(\xi) + m_6(\xi))(u_1^2 + u_5^2)u_7 \\
&\qquad - 2(m_4(\xi) - m_1(\xi))u_1u_5u_7 = 0, \\
&m_7(\xi)(u_1u_5 + u_4u_6) - m_5(\xi) + m_6(\xi))u_7\exp(ku) = 0, \\
&m_7(\xi)u_1(u_6^2 - u_1^2) - (m_5(\xi) + m_6(\xi))u_5u_7\exp(ku) \\
&\qquad - 2(m_4(\xi) - m_1(\xi))u_1u_7\exp(ku) + m_7(\xi)u_4\exp(ku) = 0, \\
&m_7(\xi)u_1(u_5^2 - u_4^2) - (m_5(\xi) + m_6(\xi))u_5u_7\exp(ku) + m_7(\xi)u_4\exp(ku) = 0.
\end{aligned}
\tag{33}
$$

From the equations above, we can obtain

$$
\begin{aligned}
&m_7(\xi)(u_1^2 + u_5^2 - u_4^2 - u_6^2) = 2(m_1(\xi) - m_4(\xi))u_7\exp(ku), \\
&m_7(\xi)(u_1u_5 + u_4u_6) = (m_5(\xi) + m_6(\xi))u_7\exp(ku).
\end{aligned}
\tag{34}
$$

If we set $\eta = \xi$ in (32) and (34), then the two systems of equations can yield the following equation:

$$
(\hat{u}_1\hat{u}_5 + \hat{u}_4\hat{u}_6)(\hat{u}_1^2 + \hat{u}_6^2 - \hat{u}_4^2 - \hat{u}_5^2) = (\hat{u}_1\hat{u}_6 + \hat{u}_4\hat{u}_5)(\hat{u}_1^2 + \hat{u}_5^2 - \hat{u}_4^2 - \hat{u}_6^2).
$$

We can write it in the factorized form

$$
(\hat{u}_5 - \hat{u}_6)(\hat{u}_1 + \hat{u}_4)(\hat{u}_1 - \hat{u}_4 - \hat{u}_5 - \hat{u}_6)(\hat{u}_1 - \hat{u}_4 + \hat{u}_5 + \hat{u}_6) = 0,
$$

where we have used the notation

$$
\hat{u}_i = a_i(u, \xi, \xi).
$$

Then, by the initial conditions (8), we get

$$
a_5(u, \xi, \xi) = a_6(u, \xi, \xi),
\tag{35}
$$

$$
m_5(\xi) = m_6(\xi).
\tag{36}
$$

Setting $\eta = \xi$ in the second formulas of (32) and (34) and substituting (35) and (36) we have

$$
(m_5(\xi) + m_6(\xi))\hat{u}_7 = (m_5(\xi) + m_6(\xi))\hat{u}_7\exp(ku).
$$

Then we obtain

$$
m_5(\xi) = -m_6(\xi),
\tag{37}
$$

which means

$$
m_5(\xi) = m_6(\xi) = 0.
$$

So, from the second equations of (32) and (34) it is easy to show

$$m_7(\eta)(u_1 u_6 + u_4 u_5) = m_7(\xi)(u_1 u_5 + u_4 u_6) = 0,$$

i.e.

$$(u_1 - u_4)(u_5 - u_6) = 0, \quad u_1 u_6 + u_4 u_5 = 0.$$

If we take $u_5 = u_6$ then $u_1 + u_4 = 0$ which disagrees with the initial conditions (8). Therefore, we can only choose

$$u_1 = u_4, \quad u_5 = -u_6. \tag{38}$$

Combining (38) and (31) we arrive at the following conclusion:

$$m_1(\xi) = m_4(\xi) = \frac{k}{2}, \quad m_5(\xi) = m_6(\xi) = 0. \tag{39}$$

Substituting (38) and (39) into (10a) we can get

$$\frac{\partial}{\partial u} a_1(u, \xi, \eta) = \frac{k}{2} a_1(u, \xi, \eta),$$

$$\frac{\partial}{\partial u} a_5(u, \xi, \eta) = \frac{k}{2} a_5(u, \xi, \eta), \tag{40}$$

$$\frac{\partial}{\partial u} a_7(u, \xi, \eta) = m_7(\eta) \exp(ku).$$

Using the initial conditions (8) and the equation (13) to solve them we can write

$$\begin{aligned}
a_1(u, \xi, \eta) &= F_1(\xi, \eta) \exp(ku/2), \quad F_1(\xi, \xi) = 1, \\
a_5(u, \xi, \eta) &= F_5(\xi, \eta) \exp(ku/2), \quad F_5(\xi, \xi) = 1, \\
a_7(u, \xi, \eta) &= F_7(\eta) \exp(ku) - F_7(\xi).
\end{aligned} \tag{41}$$

On the other hand, from (17) and (38) we can obtain

$$\breve{m}_1(\eta) = \breve{m}_4(\eta) = 0. \tag{42}$$

Then substituting (38) and (42) into (10c) we have

$$\frac{\partial}{\partial \eta} a_1(u, \xi, \eta) = m_5(\eta) a_5(u, \xi, \eta),$$

$$\frac{\partial}{\partial \eta} a_5(u, \xi, \eta) = m_5(\eta) a_1(u, \xi, \eta). \tag{43}$$

We can continually get

$$u_1 + u_5 = H_1(u, \xi) \exp(F(\eta)), \quad u_1 - u_5 = H_2(u, \xi) \exp(F(\eta)).$$

By the initial conditions and (41), we find

$$H_1(u, \xi) = \exp(\frac{k}{2}u - F(\xi)),$$

$$H_2(u, \xi) = \exp(\frac{k}{2}u + F(\xi)).$$

So, we can write down the solution

$$
\begin{aligned}
&a_1(u, \xi, \eta) = a_4(u, \xi, \eta) = \cosh(F(\xi) - F(\eta)) \exp(ku/2), \\
&a_2(u, \xi, \eta) = 1, \\
&a_3(u, \xi, \eta) = \exp(ku), \\
&a_5(u, \xi, \eta) = -a_6(u, \xi, \eta) = \sinh(F(\xi) - F(\eta)) \exp(ku/2), \\
&a_7(u, \xi, \eta) = F_7(\eta) \exp(ku) - F_7(\xi),
\end{aligned}
\tag{44}
$$

where k is a non-zero complex constant and $F(\xi), F_7(\xi)$ are two arbitrary function of colored parameter.

4. The solutions of equation (1) in the case of $k = 0$

Now equation (11) changes to

$$2u_7' = m_7(\eta)(u_1^2 + u_4^2 - u_5^2 - u_6^2) = m_7(\xi)(u_1^2 + u_4^2 - u_5^2 - u_6^2). \tag{45}$$

It is obviously that

$$m_7(\xi) = m_7(\eta) = \alpha \tag{46}$$

due to the initial conditions (8). Here α is a complex constant independent of colored parameters.

Proposition 1: *In the case of $k = 0$, either m_7 or m_5 does not vanish identically, and the same is true for m_6 replacing m_5. Otherwise, the solution will be independent of spectral parameter.*

Proof. If $m_7(\xi) = 0$ we have $u_7' = 0$ because of (45). Now substituting them into the fourth and fifth equation of (10a) $(k = 0)$, we have

$$
\begin{aligned}
u_1' &= u_6 m_6(\eta) - u_1 m_1(\eta), \\
u_6' &= u_6 m_1(\eta) - u_1 m_5(\eta)
\end{aligned}
\tag{47}
$$

which means

$$m_1(\eta) = 0, \quad m_5(\eta) + m_6(\eta) = 0. \tag{48}$$

Differentiating both sides of (47) and letting $u = 0, \eta = \xi$. Substituting the results into the first three equations of (10a) we obtain

$$u_1' = -m_5(\eta)u_6, \quad u_4' = m_5(\eta)u_5,$$
$$u_5' = -m_5(\eta)u_4, \quad u_6' = -m_5(\eta)u_1.$$

So

$$u_i'' = m_5^2(\eta)u_i \ (i = 1, 4, 5, 6)$$

will hold. If we start from (10b) to do the same work as above, we can get

$$u_i'' = m_5^2(\xi)u_i \ (i = 1, 4, 5, 6).$$

Then we can know

$$m_5(\eta) = m_5(\xi) = \beta,$$

where β is a complex constant independent of colored parameter. Now we come to the conclusion of Proposition 1.

Remark 2. From (10) we can observe that if $k = 0$

- *(10a) is the same as (10c) if $m_i(\eta)$ is replaced by $\check{m}_i(\eta)$.*
- *(10b) is the same as (10d) if $m_i(\eta)$ is replaced by $\hat{m}_i(\eta)$.*
- *(10a) is the same as (10b) if we interchange sub-indices 5 and 6 and replace $m_i(\eta)$ by $m_i(\xi)$.*
- *(10c) is the same as (10d) if we interchange sub-indices 5 and 6 and replace $\check{m}_i(\eta)$ by $\hat{m}_i(\xi)$.*

In fact, those tricks are always sound in the following of this paper and we will often employ them. We call the kind of transformation from (10a) to (10b) as symmetric operation.

As the case of $k \neq 0$, we have

$$
\begin{aligned}
&m_5 + (m_1 + m_4)u_1u_5 - m_5(u_1u_4 + u_5u_6) = 0, \\
&m_7u_1u_5(u_4^2 - u_1^2 + u_5^2 - u_6^2) + 2m_5u_7(u_1u_4 + u_5u_6 - 1) \\
&\qquad +(-4m_4)u_1u_5u_7 = 0, \\
&m_7u_1(u_4^2u_5 - u_5^2u_6 + u_6 - u_7(m_5 + m_6u_1u_4 - m_5u_5u_6) = 0,
\end{aligned}
\tag{49}
$$

which contains m_1, m_4 and m_5, and

$$u_1u_4 + u_5u_6 = 1 \tag{50}$$

or

$$(m_7 u_5 u_6 (u_4 u_6 - u_1 u_5) + m_7 u_4 u_5 (u_5^2 - u_4^2) + m_7 (u_1 u_5 - u_4 u_6)$$
$$+ m_6 u_7 (u_4^2 - u_5^2) = 0,$$
$$u_1^2 u_5 - 2 u_1 u_4 u_6 + u_4^2 u_5 - u_5^3 + u_5 u_6^2 = 0, \tag{51}$$
$$-u_1 u_4^2 u_6 + u_1 u_5^2 u_6 + u_4^3 u_5 - u_4 u_5^3 - u_1 u_5 + u_4 u_6 = 0,$$
$$-u_1^2 u_4 + 2 u_1 u_5 u_6 + u_4^3 - u_4 u_5^2 - u_4 u_6^2 = 0,$$

where we have used proposition 1.

Now there are also two cases: the first is (49) as well as (51) (Baxter case), and the second is (49) as well as (50) (the free-fermion condition[12]).

4.1. *Baxter-type solutions*

From (51), it is easy to obtain

$$m_5(\eta) = m_6(\eta), \quad m_1(\eta) = m_4(\eta), \quad m_7 = \alpha \neq 0. \tag{52}$$

From (51), by eliminating $m_4(\eta)$, we also have

$$u_5^2 (u_6 - u_5)(u_6 + u_5)(u_6 - u_1)(u_6 + u_1) = 0. \tag{53}$$

So

$$a_5(u, \xi, \eta) = a_6(u, \xi, \eta) \tag{54a}$$

or

$$u_5(u, \xi, \eta) = -u_6(u, \xi, \eta) \tag{54b}$$

will hold because of the initial conditions (8).

If $u_5 = -u_6$ holds, together with the second equation of (51) we have $u_1 = -u_4$ which is impossible for the initial conditions (8). Then, we have only $u_5 = u_6$. Substituting (54a) into the second equation of (51) we can obtain

$$a_1(u, \xi, \eta) = a_4(u, \xi, \eta). \tag{55}$$

Combining (54a), (55) and the first equation of (51), we can get

$$(m_7(\eta) u_1 u_5 - m_5(\eta) u_7)(u_5^2 - u_1^2) = 0. \tag{56}$$

Then the following equation is correct for the initial conditions (8) and (46)

$$\alpha u_1 u_5 = m_5(\eta) u_7. \tag{57}$$

Using the symmetric operation, we also get

$$\alpha u_1 u_6 = \alpha u_1 u_5 = m_6(\xi) u_7 = m_5(\xi) u_7.$$

So, there holds

$$m_5(\xi) = \beta, \tag{58}$$

where β is a complex constant independent of colored parameter. There also is

$$u_7 = \frac{\alpha}{\beta} u_1 u_5. \tag{59}$$

Substituting (54a) and (55) into (10a) and (49), we can obtain the following conclusion through some calculations:

$$\begin{aligned}
(u_5')^2 &= \beta^2 - (\beta^2 - m_1(\eta)^2)u_5^2, \\
(u_1')^2 &= \beta^2 - (\beta^2 - m_1(\eta)^2)u_1^2, \\
\beta^2(1 - u_5^2 - u_1^2) &+ 2\beta m_1(\eta)u_1 u_5 = 0.
\end{aligned} \tag{60}$$

If we use the symmetric operation, we also get

$$\begin{aligned}
(u_5')^2 &= \beta^2 - (\beta^2 - m_1(\xi)^2)u_5^2, \\
(u_1')^2 &= \beta^2 - (\beta^2 - m_1(\xi)^2)u_1^2, \\
\beta^2(1 - u_5^2 - u_1^2) &+ 2\beta m_1(\xi)u_1 u_5 = 0.
\end{aligned} \tag{61}$$

So we can affirm

$$m_1(\xi) = \gamma. \tag{62}$$

Then, (60) can be rewritten

$$\begin{aligned}
(u_5')^2 &= \beta^2 - (\beta^2 - \gamma^2)u_5^2, \\
(u_1')^2 &= \beta^2 - (\beta^2 - \gamma^2)u_1^2, \\
\beta^2(1 - u_5^2 - u_1^2) &+ 2\beta\gamma u_1 u_5 = 0.
\end{aligned} \tag{63}$$

Substituting (54a) and (55) into (10c) and (49), we can obtain the following conclusion through some calculation:

$$\begin{aligned}
\left(\frac{\partial \check{u}_5}{\partial \eta}\right)^2 &= \check{m}_5(\eta)^2 - (\check{m}_5(\eta)^2 - \check{m}_1(\eta)^2)\check{u}_5^2, \\
\left(\frac{\partial \check{u}_1}{\partial \eta}\right)^2 &= \check{m}_5(\eta)^2 - (\check{m}_5(\eta)^2 - \check{m}_1(\eta)^2)\check{u}_1^2
\end{aligned} \tag{64}$$

and from the last equation of (60) we also have

$$\frac{(\check{m}_5(\eta))^2 - (\check{m}_1(\eta))^2}{(\check{m}_5(\eta))^2} = 1 - \frac{(u_5^2 + u_1^2 - 1)^2}{4u_1^2 u_5^2} = \frac{\beta^2 - \gamma^2}{\beta^2}. \tag{65}$$

So, (64) can be rewritten

$$\left(\frac{\partial \breve{u}_5}{\partial \eta}\right)^2 = \breve{m}_5(\eta)^2\left(1 - \frac{\beta^2 - \gamma^2}{\beta^2}\breve{u}_5^2\right),$$

$$\left(\frac{\partial \breve{u}_1}{\partial \eta}\right)^2 = \breve{m}_5(\eta)^2\left(1 - \frac{\beta^2 - \gamma^2}{\beta^2}\breve{u}_1^2\right). \tag{66}$$

Combining (60), (66), (59) and the initial conditions (8), we can immediately write down the solutions of (1) in this case.

4.1.1. *Subcase of $\beta^2 = \gamma^2$*

We can let $\beta = \gamma$ according to the solution transformation **C**.

$$
\begin{aligned}
a_1(u,\xi,\eta) &= a_4(u,\xi,\eta) = \beta u + F(\xi) - F(\eta) + 1, \\
a_2(u,\xi,\eta) &= a_3(u,\xi,\eta) = 1, \\
a_5(u,\xi,\eta) &= a_6(u,\xi,\eta) = \beta u + F(\xi) - F(\eta), \\
a_7(u,\xi,\eta) &= \frac{\alpha}{\beta}(\beta u + F(\xi) - F(\eta) + 1)(\beta u + F(\xi) - F(\eta).
\end{aligned}
\tag{67a}
$$

4.1.2. *Subcase of $\beta^2 \neq \gamma^2$*

$$
\begin{aligned}
a_1(u,\xi,\eta) &= a_4(u,\xi,\eta) = \frac{\cos(\sqrt{\beta^2 - \gamma^2}u + F(\xi) - F(\eta) - \theta)}{\cos\theta}, \\
a_2(u,\xi,\eta) &= a_3(u,\xi,\eta) = 1, \\
a_5(u,\xi,\eta) &= a_6(u,\xi,\eta) = \frac{\sin(\sqrt{\beta^2 - \gamma^2}u + F(\xi) - F(\eta))}{\cos\theta}, \\
a_7(u,\xi,\eta) &= \frac{\alpha\beta}{\beta^2 - \gamma^2}\cos(\sqrt{\beta^2 - \gamma^2}u + F(\xi) - F(\eta) - \theta) \\
&\quad \times \sin(\sqrt{\beta^2 - \gamma^2}u + F(\xi) - F(\eta)),
\end{aligned}
\tag{67b}
$$

where the definition of θ is

$$\sin\theta = \frac{\gamma}{\beta}, \quad \cos\theta = \frac{\sqrt{\beta^2 - \gamma^2}}{\beta}.$$

4.2. *Free-fermion-type solutions*

Now, let's consider the second case of $k = 0$ *i.e.* equation (49) and (50). From (50) we can affirm

$$m_1(\eta) + m_4(\eta) = 0 \tag{68}$$

by differentiating it and then setting $u = 0$, $\eta = \xi$.

Substituting (50) into (49) we get

$$u_1 u_4 + u_5 u_6 = 1, \qquad (69a)$$

$$\alpha(u_1^2 + u_6^2 - u_4^2 - u_5^2) = 2(m_1(\eta) - m_4(\eta))u_7, \qquad (69b)$$

$$\alpha(u_1 u_6 + u_4 u_5) = (m_6(\eta) + m_5(\eta))u_7. \qquad (69c)$$

Using the symmetric operation, there are

$$\begin{aligned}
u_1 u_4 + u_5 u_6 &= 1, \\
\alpha(u_1^2 + u_6^2 - u_4^2 - u_5^2) &= 2(m_1(\xi) - m_4(\xi))u_7, \\
\alpha m_7(u_1 u_6 + u_4 u_5) &= (m_6(\xi) + m_5(\xi))u_7.
\end{aligned} \qquad (70)$$

If we set $\eta = \xi$ in the second one of (69) and (70) then we can get

$$(a_6(u, \xi, \xi))^2 = (a_5(u, \xi, \xi))^2 \qquad (71a)$$

i.e.

$$m_5(\eta)^2 = m_6(\eta)^2. \qquad (71b)$$

From (69) and the last equation in (10a) as well as its counterpart, we can write down

$$(u_7')^2 = \alpha^2 - ((m_5(\eta) + m_6(\eta))^2 - 4m_1(\eta)^2)u_7^2. \qquad (72a)$$

Using the symmetric operation

$$(u_7')^2 = \alpha^2 - ((m_5(\xi) + m_6(\xi))^2 - 4m_1(\xi)^2)u_7^2. \qquad (72b)$$

So we can obtain

$$\delta^2 = (m_5(\xi) + m_6(\xi))^2 - 4m_1(\xi)^2, \qquad (73)$$

where δ is a complex constant in dependent of colored parameters. In the following, we will respectively discuss the two cases $m_5(\xi) = m_6(\xi)$ and $m_5(\xi) = -m_6(\xi)$

4.2.1. *Subcase: $m_5(\xi) = -m_6(\xi)$*

From (69) and (70) we have

$$u_1 = u_4, \quad u_5 = -u_6, \quad u_1^2 - u_5^2 = 1. \qquad (74)$$

Together with (68), there will be

$$m_1(\xi) = m_4(\xi) = 0. \qquad (75)$$

So, from (10a) we have

$$\begin{aligned}
u_7' &= \alpha, \\
(u_5')^2 &= (m_5(\eta))^2(1 + u_5^2), \\
(u_1')^2 &= (m_5(\eta))^2(u_1^2 - 1).
\end{aligned} \tag{76}$$

Using the symmetric operation, we also can affirm

$$m_5(\eta) = \beta. \tag{77}$$

So, (76) can be rewritten as

$$\begin{aligned}
u_7' &= \alpha, \\
(u_5')^2 &= \beta^2(1 + u_5^2), \\
(u_1')^2 &= \beta^2(u_1^2 - 1).
\end{aligned} \tag{78a}$$

Thanks to the remark 2 if we apply the same process to equation (10c), we can obtain

$$\begin{aligned}
\frac{\partial u_7}{\partial \eta} &= (\check{m}_7(\eta))^2, \\
\left(\frac{\partial u_1}{\partial \eta}\right)^2 &= (\check{m}_5(\eta))^2(u_1^2 - 1), \\
\left(\frac{\partial u_5}{\partial \eta}\right)^2 &= (\check{m}_5(\eta))^2(u_5^2 + 1).
\end{aligned} \tag{78b}$$

Now, from (66) and the initial conditions (8) the solution of (1) in this case is

$$\begin{aligned}
a_1(u, \xi, \eta) &= a_4(u, \xi, \eta) = \cosh(\beta u + F(\xi) - F(\eta)), \\
a_2(u, \xi, \eta) &= a_3(u, \xi, \eta) = 1, \\
a_5(u, \xi, \eta) &= -a_6(u, \xi, \eta) = \sinh(\beta u + F(\xi) - F(\eta)), \\
a_7(u, \xi, \eta) &= \alpha u + F(\xi) - F(\eta).
\end{aligned} \tag{79}$$

4.2.2. *Subcase:* $m_5(\xi) = m_6(\xi)$

Now, from equation (72) and (73) change to be

$$(u_7')^2 = \alpha^2 - \delta^2 u_7^2 \tag{80}$$

and

$$\delta^2 = 4m_5(\xi)^2 - 4m_1(\xi)^2. \tag{81}$$

And from (69c) we can know in this case

$$\alpha \neq 0. \tag{82}$$

Otherwise, there holds $m_5(\xi) = 0$ which has discrepancy with proposition 1. If we try to solve equation (80), we find there are two branches which is related to $\delta = 0$ and $\delta \neq 0$.

(A) $\delta = 0$ *subcase*

We can see that $\delta = 0$ means $m_1(\xi)^2 = m_5(\xi)^2$. Thanks to the solution transformation **C** and equation (68), we can set

$$m_1(\xi) = m_5(\xi) = m_6(\xi) = -m_4(\xi). \tag{83}$$

At this time, (80) changes to

$$u_7' = \alpha \tag{84}$$

thanks to the solution transformation **C**. Now, let's consider (69b) and (69c). Using (83) we can get

$$u_1^2 + u_6^2 - u_4^2 - u_5^2 - 2u_1 u_6 - 2u_4 u_5 = 0.$$

Considering the second and the third formulas of (70) and using (83) we can also get

$$u_1^2 + u_5^2 - u_4^2 - u_6^2 - 2u_1 u_5 - 2u_4 u_6 = 0.$$

From these two equations, it is easy to obtain

$$2(u_1 + u_4)(u_1 - u_4 - u_5 - u_6) = 0,$$

which means

$$u_1 = u_4 + u_5 + u_6. \tag{85}$$

Together with (69a) we immediately get

$$\begin{aligned}
\breve{m}_1(\eta) + \breve{m}_4(\eta) &= 0, \\
2\breve{m}_1(\eta) &= \breve{m}_5(\eta) + \breve{m}_6(\eta).
\end{aligned} \tag{86}$$

Substituting it into the fifth equation of (10c) and its counterpart, we get

$$2\breve{m}_1 u_7 (1 - u_6) = u_1 \frac{\partial u_7}{\partial \eta} - u_4 \breve{m}_7 = u_1 \breve{m}_7 - u_4 \frac{\partial u_7}{\partial \eta}.$$

Using the initial conditions (8) there is

$$\frac{\partial u_7}{\partial \eta} = \breve{m}_7(\eta). \tag{87}$$

So, combining (84), (87) and the initial conditions (8), we can write down

$$u_7 = \alpha u + F(\xi) - F(\eta). \tag{88}$$

Substituting all the results into the sixth equation of (10a) as well as its counterpart, we have

$$(u_1^2 - u_5^2)\alpha - 2m_1(\eta)u_7 - \alpha = 0,$$
$$(u_4^2 - u_6^2)\alpha + 2m_1(\eta)u_7 - \alpha = 0,$$
$$u_1 u_4 + u_5 u_6 = 1,$$
$$(u_1 u_6 + u_4 u_5)\alpha = 2m_1(\eta)u_7. \tag{89}$$

Using the symmetric operation, we also have

$$(u_1^2 - u_6^2)\alpha - 2m_1(\xi)u_7 - \alpha = 0,$$
$$(u_4^2 - u_5^2)\alpha + 2m_1(\xi)u_7 - \alpha = 0,$$
$$u_1 u_4 + u_5 u_6 = 1,$$
$$(u_1 u_5 + u_4 u_6)\alpha = 2m_1(\xi)u_7. \tag{90}$$

From (89) we have

$$u_4(\alpha + 2m_1(\eta)u_7) = \alpha u_1 - 2m_1(\eta)u_5 u_7,$$
$$u_5(2m_1(\eta)u_7 - \alpha) = \alpha u_6 - 2m_1(\eta)u_4 u_7. \tag{91a}$$

From (90) we also get

$$u_1(\alpha - 2m_1(\xi)u_7) = \alpha u_4 - 2m_1(\xi)u_5 u_7,$$
$$u_6(2m_1(\xi)u_7 - \alpha) = \alpha u_5 - 2m_1(\xi)u_4 u_7. \tag{91b}$$

Solving the system of equation (91a) and (91b), we get

$$\frac{u_4}{u_1} = \frac{\alpha(m_1(\xi) + m_1(\eta)) - 2m_1(\xi)m_1(\eta)u_7}{\alpha(m_1(\xi) + m_1(\eta)) + 2m_1(\xi)m_1(\eta)u_7},$$
$$\frac{u_6}{u_5} = \frac{-\alpha(m_1(\xi) - m_1(\eta)) + 2m_1(\xi)m_1(\eta)u_7}{\alpha(m_1(\xi) - m_1(\eta)) + 2m_1(\xi)m_1(\eta)u_7}. \tag{92}$$

We denote here

$$H_1 = \alpha(m_1(\xi) + m_1(\eta)) + 2m_1(\xi)m_1(\eta)u_7,$$
$$H_4 = \alpha(m_1(\xi) + m_1(\eta)) - 2m_1(\xi)m_1(\eta)u_7,$$
$$H_5 = \alpha(m_1(\xi) - m_1(\eta)) + 2m_1(\xi)m_1(\eta)u_7,$$
$$H_6 = -\alpha(m_1(\xi) - m_1(\eta)) + 2m_1(\xi)m_1(\eta)u_7.$$

Then we can assume

$$u_1 = H_1 X, \qquad u_4 = H_4 X, \qquad u_5 = H_5 Y, \qquad u_6 = H_6 Y. \tag{93}$$

On the other hand, from (89) and (90), we have

$$\alpha(u_1^2 - u_4^2) = 2u_7(m_1(\xi) + m_1(\eta)).$$

Similarly, from the second formulas of (89) and (90), we also have

$$\alpha(u_5^2 - u_6^2) = 2u_7(m_1(\xi) - m_1(\eta)).$$

From them, we can get

$$X = Y = \frac{1}{2\alpha\sqrt{m_1(\xi)m_1(\eta)}}.$$

So, we have obtained the solution of this case

$$a_1(u,\xi,\eta)$$
$$= \frac{1}{2\alpha\sqrt{G(\xi)G(\eta)}}(\alpha(G(\xi) + G(\eta)) + 2G(\xi)G(\eta)(\alpha u + F(\xi) - F(\eta))),$$
$$a_2(u,\xi,\eta) = a_3(u,\xi,\eta) = 1,$$
$$a_4(u,\xi,\eta)$$
$$= \frac{1}{2\alpha\sqrt{G(\xi)G(\eta)}}(\alpha(G(\xi) + G(\eta)) - 2G(\xi)G(\eta)(\alpha u + F(\xi) - F(\eta))),$$
$$a_5(u,\xi,\eta)$$
$$= \frac{1}{2\alpha\sqrt{G(\xi)G(\eta)}}(\alpha(G(\xi) - G(\eta)) + 2G(\xi)G(\eta)(\alpha u + F(\xi) - F(\eta))),$$
$$a_6(u,\xi,\eta)$$
$$= \frac{1}{2\alpha\sqrt{G(\xi)G(\eta)}}(-\alpha(G(\xi) - G(\eta)) + 2G(\xi)G(\eta)(\alpha u + F(\xi) - F(\eta))).$$

$$\tag{94}$$

(B) $\delta \neq 0$ *subcase*

At this time, (80) can changed to be

$$(u_7')^2 = \alpha^2 - \delta^2 u_7^2. \tag{95}$$

Thanks to remark 2, we also have

$$\left(\frac{\partial u_7}{\partial \eta}\right)^2 = \breve{m}_7(\eta)^2 - ((\breve{m}_5(\eta) + \breve{m}_6(\eta))^2 - 4\breve{m}_1(\eta)^2)u_7^2. \tag{96}$$

On the other hand, still because of remark 2 we have

$$(\breve{m}_5(\eta) + \breve{m}_6(\eta))^2 - 4\breve{m}_1(\eta)^2$$
$$= \frac{\breve{m}_7(\eta)^2}{u_7^2}((u_1 u_6 + u_4 u_5)^2 - \frac{1}{4}(u_1^2 + u_6^2 - u_4^2 - u_5^2)) = \frac{\delta^2}{\alpha^2}\breve{m}_7(\eta)^2.$$

So, (96) changes to be

$$\left(\frac{\partial u_7}{\partial \eta}\right)^2 = \breve{m}_7(\eta)^2 - \frac{\delta^2}{\alpha^2}\breve{m}_7(\eta)^2 u_7^2. \tag{97}$$

Together with (96) and the initial conditions (8), we have

$$a_7(u,\xi,\eta) = \frac{\alpha}{\delta}\sin(\delta u + F(\xi) - F(\eta)), \tag{98}$$

where $F(\xi)$ is an arbitrary function of colored parameter.

Substituting all the results into the sixth equation of (10a) as well as its counterpart, (69a) and (69c), we have

$$\delta \cos(\delta u + F(\xi) - F(\eta)) + 2m_1(\eta) \sin(\delta u + F(\xi) - F(\eta)) - \delta(u_1^2 - u_5^2) = 0,$$
$$\delta \cos(\delta u + F(\xi) - F(\eta)) + 2m_1(\eta) \sin(\delta u + F(\xi) - F(\eta)) - \delta(u_4^2 - u_6^2) = 0,$$
$$u_1 u_4 + u_5 u_6 = 1,$$
$$\delta(u_1 u_6 + u_4 u_5) - 2m_5(\eta) \sin(\delta u + F(\xi) - F(\eta)) = 0. \tag{99}$$

Using the symmetric operation, we can also have

$$\delta \cos(\delta u + F(\xi) - F(\eta)) + 2m_1(\xi) \sin(\delta u + F(\xi) - F(\eta)) - \delta(u_1^2 - u_6^2) = 0,$$
$$\delta \cos(\delta u + F(\xi) - F(\eta)) + 2m_1(\xi) \sin(\delta u + F(\xi) - F(\eta)) - \delta(u_4^2 - u_5^2) = 0,$$
$$u_1 u_4 + u_5 u_6 = 1,$$
$$\delta(u_1 u_4 + u_4 u_6) - 2m_5(\xi) \sin(\delta u + F(\xi) - F(\eta)) = 0. \tag{100}$$

Like in (i) ($\delta = 0$ subcase), when we do the same procedures, at last we can get

$$\frac{u_4}{u_1} = \frac{\delta m_5(\xi) + \delta m_5(\eta) \cos \theta - 2m_1(\xi) m_5(\eta) \sin \theta}{\delta m_5(\eta) + \delta m_5(\xi) \cos \theta + 2m_5(\xi) m_1(\eta) \sin \theta},$$

$$\frac{u_6}{u_5} = \frac{-\delta m_5(\eta) + \delta m_5(\xi) \cos \theta - 2m_5(\xi) m_1(\eta) \sin \theta}{-\delta m_5(\xi) + \delta m_5(\eta) \cos \theta - 2m_1(\xi) m_5(\xi) \sin \theta}, \tag{101}$$

where $\theta = \delta u + F(\xi) - F(\eta)$. Similarly we set here

$$\begin{aligned}
H_1 &= \delta m_5(\eta) + \delta m_5(\xi) \cos \theta + 2m_5(\xi) m_1(\eta) \sin \theta, \\
H_4 &= \delta m_5(\xi) + \delta m_5(\eta) \cos \theta - 2m_1(\xi) m_5(\eta) \sin \theta, \\
H_5 &= -\delta m_5(\xi) + \delta m_5(\eta) \cos \theta - 2m_1(\xi) m_5(\eta) \sin \theta, \\
H_6 &= -\delta m_5(\eta) + \delta m_5(\xi) \cos \theta - 2m_5(\xi) m_1(\eta) \sin \theta.
\end{aligned} \tag{102}$$

We also assume

$$u_1 = H_1 X, \quad u_4 = H_4 X, \quad u_5 = H_5 Y, \quad u_6 = H_6 Y. \tag{103}$$

Then from (99) and (100) we have

$$X = \sqrt{\frac{2(m_1(\xi) + m_1(\eta)) \sin(\delta u + F(\xi) - F(\eta))}{\delta(H_1^2 - H_4^2)}},$$

$$Y = \sqrt{\frac{2(m_1(\xi) - m_1(\eta)) \sin(\delta u + F(\xi) - F(\eta))}{\delta(H_5^2 - H_6^2)}}. \tag{104}$$

So, we at last write down the solution in this case

$$
\begin{aligned}
a_1(u,\xi,\eta) &= X(\delta G(\eta) + \delta G(\xi)\cos\theta + 2G(\xi)H(\eta)\sin\theta),\\
a_2(u,\xi,\eta) &= a_3(u,\xi,\eta) = 1,\\
a_4(u,\xi,\eta) &= X(\delta G(\xi) + \delta G(\eta)\cos\theta - 2G(\eta)H(\eta)\sin\theta),\\
a_5(u,\xi,\eta) &= -Y(\delta G(\xi) + \delta G(\eta)\cos\theta - 2G(\eta)H(\xi)\sin\theta),\\
a_6(u,\xi,\eta) &= -\delta G(\eta) + \delta G(\xi)\cos\theta - 2G(\xi)H(\eta)\sin\theta,\\
a_7(u,\xi,\eta) &= \frac{\alpha}{\delta}\sin\theta,
\end{aligned}
\tag{105}
$$

where $F(\xi), G(\xi)$ and $H(\xi)$ are arbitrary functions of colored parameters and X, Y are defined by equation (104).

5. General solutions

In this paper, we have given six basic solutions of equation (1) and also classify them into two types. These six solutions *i.e.* (44), (67a), (67b), (79), (94) and (104) together with the five solution transformations **A-E** will give all seven-vertex-type solutions of colored Yang–Baxter equation (1) and the general solutions can also be classified into two types. The first are Baxter-type solutions which can be got from the basic Baxter-type solution by some solution transformations. The second are free-fermion-type which can be obtained *via* the basic free-fermion-type solutions by some solution transformations.

According to the standard model given by Baxter, for a given R matrix the spin-chain Hamiltonian is generally of the following form,

$$
H = \sum_{j=1}^{N}(J_x\sigma_j^x\sigma_{j+1}^x + J_y\sigma_j^y\sigma_{j+1}^y + J_z\sigma_j^z\sigma_{j+1}^z + \frac{1}{2}(\sigma_j^z + \sigma j + 1^z)),
$$

where σ^x, σ^y and σ^z are Pauli matrices and the coupling constant are

$$
J_x = \frac{1}{4}(m_5 + m_6 + m_7), \quad J_y = \frac{1}{4}(m_5 + m_6 - m_7),
$$

$$
J_z = \frac{1}{4}(m_1 - m_3 + m_4 - m_2), \quad h = \frac{1}{4}(m_1 - m_3 - m_4 + m_2).
$$

In this paper, we have proved that in the six basic solutions the Hamiltonian coefficients obey

$$
m_1^2 = m_4^2, \quad m_5^2 = m_6^2.
$$

It follows from the solution transformations **B** and **D** that

$$
(m_1 - m_3)^2 = (m_4 - m_2)^2, \quad m_5^2 = m_6^2
$$

for general solutions. This clearly describes the relation between classification of seven-vertex-type solutions and spin-chain Hamiltonian.

References

1. C.N.Yang 1967 *Phys. Rev. Lett.* **19** 1312–1314.
2. C.N.Yang 1968 *Phys. Rev. Lett.* **168** 1920–1923.
3. R.J.Baxter 1972 *Ann. Phys. Lpz.* **70** 193–288.
4. A.B.Zamolodchikov 1979 *Ann. Phys. Lpz.* **120** 253–291.
5. R.J.Baxter 1982 *Exactly Solved Models in statistical Mechanics* (London: Academic).
6. M.Jimbo 1989 *Yang–Baxter equation in integrable systems* (Singapore: World Scientific).
7. V.G.Drinfel'd 1987 *Quantum groups Proc. Int. Cong. of Mathematicians* (Berkeley, CA, 1987) 789–820.
8. Alvarez-Gaumé L, Cómez C and Sierra G 1989 *Nucl. Phys.* **319** 155; 1990 *Nucl. Phys.* B **330** 347; 1989 *Phys. Lett.***220** 142.
9. B.Frenkel and Reshetikhin N Yu 1992 *Commu. Math. Phys.* B **146** 1.
10. V.G.Turaev 1988 *Inven, Math.* **92** 527.
11. Y.Akutsu and M.Wadati 1987 *J. Phys. Soc. Japan* **56** 839–42.
12. C.Fan and F.Y.Wu 1996 *Phys. Rev.* B **2** 723.
13. J.Murakami 1990 Astate model for the multi-variable Alexander polynomial *preprint* Osaka Univ.
14. R.Cuerno, C.Gómez, E.López and G. sierra 1993 *Phys. Lett.* **307** B 56–60.
15. J.Murakami 1992 *Int. J. Mod. Phys.* A **7** 765.
16. M.Ruiz-Altaba 1992 *Phys. Lett.* **277** 326.
17. S.K.Wang 1996 Classification of eight-vertex solutions of colored Yang–Baxter equation *J.Phys. A: Math. Gen.* **29** 2259–2277.
18. V.V.Bazhanov and Y.G.Stroganov 1985 *Theor. Math. Fiz* **62** 253–260.
19. M.L.Ge and K.Xue 1993 *J. Phys. A:Math.Gen.* **26** 281.
20. G.W.Delius, M.D.Gould and Yao-zhong Zhang 1994 *Nucl Phys.* B **432** 377.
21. A.J.Brachen, M.D.Gould, Yao-zhong Zhang and G.W.Delius 1994 *J. Math. Phys* A **27** 6551.
22. X.D.Sun, S.K.Wang and K.Wu 1996 Classification of six-vertex type solutions of the colored Yang–Baxter equation, *J. Math. Phys.* **36** 6043–6063.
23. Wu Wen-tsun 1978 *Sci. Sinica* **21** 157–179.

Brownian Motion as a Model for B-Cell Movement

F.W. Wiegel

Institute of Theoretical Physics, University of Amsterdam
Valckenierstraat 65, 1018 XE Amsterdam, The Netherlands
and
Department of Applied Physics, University of Twente
P.O. Box 217, 7500 AE Enschede, The Netherlands

Brownian motion is used to describe the movement of the B-cells, which are the monads of the immune system. The "signal" measured by a specific B-cell is a stochastic quantity, the probability distribution of which is shown to be the solution of a simple partial differential equation. For the case of a uniformly infected region in the body this equation is solved explicitly. In the final section of the essay it is shown how the immune system can use the statistical properties of the signals carried by the B-cells, to acquire relevant information about the infected region: How large is this region? How high is the density of disease-causing agents in it? Where in the body is this region located?

1. Introduction: a new monadology

Brownian motion has been a paradigm for a variety of physical phenomena for most of the last century. During the last two decades Brownian motion is also finding applications to other parts of natural science, especially to medicine and biology. In the present essay I shall outline some of these applications to the physics of chemoreception and to theoretical immunology. In this way one hopes to identify some new questions about Brownian motion; questions that are meaningless in the context of physics but meaningful for a biological system. These new questions might be the motivation to find exactly solvable models, of the type which have been explored recently by Hiroshi Ezawa.[1,2]

In order to guide the reader this introduction will comment on the concept of chemoreception and on the B-cells of immunology, and why this author feels that the study of the Brownian motion of a B-cell amounts to the rise of a new monadology. In later sections the treatment will be somewhat more quantitative, but in general we aim at the formulation of the proper questions rather than at answering them in a quantitative way.

Chemoreception is the ability of most biological cells to "sense" those properties of their environment that are essential to their survival. The physical principles which are involved in the processes of chemoreception are fairly well understood by the community of biophysicists; the literature on this topic is reviewed in Ref.[3] For convenience we shall make all order-of-magnitude estimates for the "average biological cell". It is characterized in the following way:

Shape :   spherical
Radius : $4 \times 10^{-6}\,\mathrm{m}$
Volume : $2.68 \times 10^{-16}\,\mathrm{m}^3$
Density : $1.03 \times 10^3\,\mathrm{kg\,m}^{-3}$
Mass : $2.76 \times 10^{-13}\,\mathrm{kg}$

These numbers imply that the "average human" with a body mass of 75 kg, will consist of approximately 2.72×10^{14} cells. Most of these cells are continuously involved in some form of chemoreception.

The immune system consists of several types of cells; its function is to maintain the health of the body by guarding its integrity. The number of cells in the immune system is of the order of 5% of all cells, for easy estimates we shall use the number 2×10^{13}. One of the most important groups of cells in the immune system are the B-cells, which have a role in the maintenance of health which is similar to the information-gathering department of a national government. For a summary of immunology for physicists the reader might want to consult a review paper by Perelson and Weisbuch.[4]

From the preceding estimates the number of B-cells in an adult human can be taken to be of order 5×10^{12}. A particular B-cell will perform a Brownian motion during which it is constantly colliding with a variety of biopolymers. The cell carries receptor molecules in its outer membrane, which can bind selectively to a particular biopolymer. That biopolymer is called the antigen for which this particular B-cell is sensitive. A particular B-cell will "count" the number of these binding events during a long time interval. In order to complete this very rough picture of B-cell function, it should be pointed out that the total population of $\approx 5 \times 10^{12}$ B-cells consists of $\approx 5 \times 10^7$ different clones; each clone consists of $\approx 10^5$ genetically identical cells. All the B-cells in one particular clone will bind to one particular antigen; the B-cells that belong to another clone will specifically bind to another antigen, and so on. Obviously, new questions about the mathematics of Brownian motion shall present themselves in this context.

Before these new questions are outlined in the next three sections it should be pointed out why I call this approach the rise of a new monadology. This phrase is inspired by the fact that the immune system is nowadays viewed as our second cognitive system (the neural system is our first cognitive system). This new paradigm for the immune system has been described, in general terms, by one of its originators I. R. Cohen, in Ref.[5] In this cognitive system the most elementary information-gathering entity is the individual B-cell. The elementary unit of information which a single B-cell contributes to the cognitive ego of the immune system, is the statement: "I have performed Brownian motion for a long time, in such-and-such a part of the body, and I have encounted so-and-so many copies of the antigen which I can bind specifically at my outer membrane: here they are!" Obviously, the B-cells are the monads of the immune ego, and the theoretician is faced with the question how the immune ego can build up a more sophisticated image of the body, using these elementary units of information.

2. Mathematical intermezzo: the signal-distribution equation

It will be shown in Sec. 4 that the "signal" which a B-cell registers when it performs Brownian motion during a time interval T, can be written in the form of an integral

$$X \equiv \int_0^T V\big(\vec{r}^{\,1}(t)\big)\, dt,$$

where V is a function of the position $\vec{r}^{\,1}$ of the Brownian walker at time t. The explicit form of V depends on the specific application one has in mind; in this section we shall not yet specialize but study the mathematics for its own sake. If an ensemble of random walks is considered, all starting at position $\vec{r}$ at time 0, the resulting values of X will have a probability density which we shall denote by $P(X, \vec{r}, T)$. For one-dimensional systems some remarkable properties of P were proved by Ezawa in Refs.[1,2], and for two-dimensional systems similar properties of P were proved by Ezawa, Nakamura, Watanabe and the author in Ref.[6] In this essay I consider the three-dimensional case, which is most relevant to the biophysical applications.

Firstly, one can show that the signal-distribution function $P(X, \vec{r}, T)$ is a solution of the partial differential equation

$$\frac{\partial P}{\partial T} = D \, \Delta P - V(\vec{r}) \frac{\partial P}{\partial X}, \tag{1}$$

where Δ denotes the Laplace operator with respect to the coordinates of $\vec{r}$, and where D is the diffusion coefficient of the random walk.

There are several ways to derive this equation. The simplest derivation would represent the random walk by a discrete random flight, in which the walker takes a step of length l after each time interval ε; at the end of the calculations the limit $l \to 0$, $\varepsilon \to 0$ is taken in such a way that the diffusion coefficient

$$D = \frac{l^2}{6\varepsilon} \tag{2}$$

stays constant. This random flight model immediately gives the integral equation

$$P(X, \vec{r}, T + \varepsilon) = \frac{1}{4\pi l^2} \int \delta\big(|\vec{r}^{\,1} - \vec{r}| - l\big) \, P\big(X - \varepsilon V(\vec{r}), \vec{r}^{\,1}, T\big) \, d^3 \vec{r}^{\,1}. \tag{3}$$

If now one expands the left-hand side to first order in ε, and the right-hand side to first order in ε and to second order in l, then the limit mentioned above leads to (1) after some straightforward algebra.

The signal-distribution equation (1) has several simple properties which shall be listed here in a more or less arbitrary order:

(a) The initial condition is always

$$\lim_{T \downarrow 0} P(X, \vec{r}, T) = \delta(X), \quad (\text{all } \vec{r}). \tag{4}$$

Usually the function $V(\vec{r})$ will be $\neq 0$ only in a finite part of space. In that case the random walk will eventually always drift into the signal-free part of space, and hence one has the boundary conditions

$$\lim_{X \to \pm\infty} P(X, \vec{r}, T) = 0, \quad (\text{all } \vec{r} \text{ and } T), \tag{5}$$

$$\lim_{r \to \infty} P(X, \vec{r}, T) = \delta(X), \quad (\text{all } X \text{ and } T). \tag{6}$$

(b) The normalization

$$F(\vec{r}, T) \equiv \int_{-\infty}^{\infty} P(X, \vec{r}, T) dX \tag{7}$$

obeys the partial differential equation

$$\frac{\partial F}{\partial T} = D\,\Delta F.$$

Because of (4) the initial value of F is:

$$F(\vec{r},0) = 1, \quad (\text{all } \vec{r}),$$

and hence at all times you have

$$F(\vec{r},T) = 1, \quad (\text{all } T). \tag{8}$$

(c) For $D \to 0$ the signal-distribution equation becomes

$$\frac{\partial P}{\partial T} = -V(\vec{r})\frac{\partial P}{\partial X}. \tag{9}$$

The general solution is $P(X,\vec{r},T) = f\big(X - V(\vec{r})T\big)$ for any function f. Because of the initial value (4) f equals a δ-function, and you find for this limiting case

$$P(X,\vec{r},T) = \delta\big(X - V(\vec{r})T\big); \tag{10}$$

a result that could have been written down without any calculations at all, and which serves only as a check against calculational errors.

(d) For any value of D there is a simple equation for the average value of X at time T. Denote this average by

$$H(\vec{r},T) \equiv \int_{-\infty}^{\infty} XP(X,\vec{r},T)\,dX. \tag{11}$$

Using (1) one easily finds

$$\frac{\partial H}{\partial T} = D\,\Delta H + V(\vec{r}). \tag{12}$$

One can write down the explicit solution of this equation, for any (sufficiently well-behaved!) choice of $V(\vec{r})$, both for the time-independent case and for the time-dependent case. I omit these results here because this essay should not be burdened by too much calculational detail.

(e) Finally one should note the most beautiful property of the signal-distribution equation: for a sufficiently well-behaved function $V(\vec{r})$, the solution will have a limit for $T \to \infty$. Calling this the "equilibrium" solution

$$P^{\text{eq}}(X,\vec{r}) \equiv \lim_{T \to \infty} P(X,\vec{r},T), \tag{13}$$

one has a simple, partial differential equation from which it can be solved:

$$D\,\Delta P^{\mathrm{eq}} = V(\vec{r})\frac{\partial P^{\mathrm{eq}}}{\partial X}.\tag{14}$$

The boundary conditions on P^{eq} follow from (5) and (6):

$$\lim_{X\to\pm\infty} P^{\mathrm{eq}}(X,\vec{r}) = 0,\tag{15}$$

$$\lim_{r\to\infty} P^{\mathrm{eq}}(X,\vec{r}) = \delta(X).\tag{16}$$

In the next section we shall demonstrate the method of solution of (14) for a specific case which is the most useful one in biophysical applications. In the last section of the essay we return to the immune system, and to B-cell Brownian motion as a new monadology.

3. The case of an uniform sphere

The most important example, which shall be used in Sec. 4 to model B-cells diffusing in a part of the body, is the case where $V(\vec{r})$ is a constant inside a sphere of radius a:

$$V(\vec{r}) = V_0, \quad (0 < r < a),\tag{17}$$

$$V(\vec{r}) = 0, \quad (r > a),\tag{18}$$

This case was recently solved by Nakamura, Ezawa and Watanabe[7]; we shall reconstruct their solution for $P^{\mathrm{eq}}(X,\vec{r})$ with the method of Sec. 2 (e).

First, one should realize that, in the case where $r > a$, the Brownian walker will miss the uniform sphere with a non-zero probability $\Pi_{\mathrm{esc}}(r)$, and it will hit this sphere with probability $\Pi_{\mathrm{cap}}(r) = 1 - \Pi_{\mathrm{esc}}(r)$. These probabilities for escape c.q. capture can be calculated by simple means; for diffusion in a space of dimensionality $n' = 3, 4, 5, \ldots$ one finds

$$\Pi_{\mathrm{cap}}(r) = \left(\frac{a}{r}\right)^{n'-2},\tag{19}$$

$$\Pi_{\mathrm{esc}}(r) = 1 - \left(\frac{a}{r}\right)^{n'-2},\tag{20}$$

For the three-dimensional case one has, for $r > a$,

$$\Pi_{\mathrm{cap}}(r) = \frac{a}{r},\tag{21}$$

$$\Pi_{\mathrm{esc}}(r) = 1 - \frac{a}{r}.\tag{22}$$

But the walkers that miss the sphere altogether will lead to a probability distribution $\delta(X)$, whereas those that hit the sphere at some time will all give a probability distribution $h(X)$ that is independent of the initial coordinate $\vec{r}$ because of the Markov property of Brownian motion. Hence the function $P^{\mathrm{eq}}(X, \vec{r})$ must have the following form (where we take $V_0 > 0$). For $0 < r < a$: P^{eq} is some smooth spherically symmetric function of X and r. The physically relevant values of X are $0 \le X < \infty$.

For $a < r$ one has

$$P^{\mathrm{eq}}(X, \vec{r}) = \left(1 - \frac{a}{r}\right)\delta(X) + \frac{a}{r}h(X), \tag{23}$$

where $h(X)$ is a smooth function of X.

In order to find the smooth part $h(X)$ one tries to find special solutions of (14) of the form $f_n(r)\exp(-\lambda_n X)$. Substitution of (17) and (18) into (14) gives

$$\left(\frac{d^2}{dr^2} + \frac{2}{r}\frac{d}{dr}\right)f_n(r) = \begin{cases} -\lambda_n \dfrac{V_0}{D} f_n(r) & (0 < r < a), \\[2mm] 0 & (r > a). \end{cases} \tag{24}$$

Note that one needs only find the spherically symmetric solutions because of the a-priori form (23) of the distribution.

The (unnormalized) solution of (24) is found to be given by:

$$f_n(r) = \begin{cases} \dfrac{C}{r}\sin\left[\pi\left(n + \dfrac{1}{2}\right)\dfrac{r}{a}\right] & (0 < r < a), \\[4mm] \dfrac{C}{r}(-1)^n & (r > a), \end{cases} \tag{25}$$

$$\lambda_n = \left(n + \frac{1}{2}\right)^2 \frac{\pi^2 D}{a^2 V_0} \quad (n = 1, 2, \ldots). \tag{26}$$

Here C denotes an arbitrary constant. The value of λ_n followed from the requirement that $f_n(r)$ and $f_n'(r)$ should be continuous at $r = a$.

At this stage one can write down the series expansion:

$$P^{\mathrm{eq}}(X, \vec{r}) = \sum_{n=0}^{\infty} \frac{C_n}{r}\sin\left[\pi\left(n + \frac{1}{2}\right)\frac{r}{a}\right]\exp\left[-\left(n + \frac{1}{2}\right)^2 \frac{\pi^2 D}{a^2 V_0}X\right] \tag{27}$$

for $0 < r < a$, and

$$P^{\mathrm{eq}}(X, \vec{r}) = \left(1 - \frac{a}{r}\right)\delta(X) + \sum_{n=0}^{\infty}\frac{C_n}{r}(-1)^n \exp\left[-\left(n + \frac{1}{2}\right)^2 \frac{\pi^2 D}{a^2 V_0}X\right] \tag{28}$$

for $r > a$. This is the solution of our problem provided the expansion coefficients C_n can be chosen in such a way that the normalization condition (7) and (8)

$$\int_0^\infty P^{\mathrm{eq}}(X,\vec{r})\,dX = 1 \tag{29}$$

holds true for any value of $\vec{r}$. Using the series (27) one gets the condition

$$\sum_{n=0}^\infty \frac{C_n}{r}\frac{a^2 V_0}{\pi^2 D(n+\frac{1}{2})^2}\sin\left[\pi\left(n+\frac{1}{2}\right)\frac{r}{a}\right] = 1, \tag{30}$$

which should hold for $0 < r < a$. Multiply this equation with a function $\frac{1}{r}\sin\left[\pi\left(m+\frac{1}{2}\right)\frac{r}{a}\right]$, multiply again with a volume-weight factor r^2, and integrate r from 0 to a. The integral

$$\int_0^a \sin\left[\pi\left(n+\frac{1}{2}\right)\frac{r}{a}\right]\sin\left[\pi\left(m+\frac{1}{2}\right)\frac{r}{a}\right]dr = \begin{cases} 0 & (m \neq n), \\ \dfrac{1}{2}a & (m = n), \end{cases} \tag{31}$$

then gives the intermediate result

$$C_m \frac{a}{2}\frac{a^2 V_0}{\pi^2 D(m+\frac{1}{2})^2} = \int_0^a r\sin\left[\pi\left(m+\frac{1}{2}\right)\frac{r}{a}\right]dr. \tag{32}$$

The integral on the right hand side equals

$$\frac{a^2}{\pi^2}\frac{(-1)^m}{(m+\frac{1}{2})^2},$$

so one has found the value of the expansion coefficient

$$C_m = \frac{2D}{aV_0}(-1)^m. \tag{33}$$

In other words, the solution for the case of a uniform sphere can be written in the form of a series:

$$P^{\mathrm{eq}}(X,\vec{r}) = \frac{2D}{aV_0}\frac{1}{r}\sum_{n=0}^\infty (-1)^n \sin\left[\pi\left(n+\frac{1}{2}\right)\frac{r}{a}\right]\exp\left[-\left(n+\frac{1}{2}\right)^2\frac{\pi^2 D}{a^2 V_0}X\right] \tag{34}$$

for $0 < r < a$, and

$$P^{\mathrm{eq}}(X,\vec{r}) = \left(1 - \frac{a}{r}\right)\delta(X) + \frac{2D}{aV_0}\frac{1}{r}\sum_{n=0}^\infty \exp\left[-\left(n+\frac{1}{2}\right)^2\frac{\pi^2 D}{a^2 V_0}X\right] \tag{35}$$

for $r > a$. As a check on the absence of calculational errors one easily verifies the normalization condition (29) to hold true also in the case $r > a$.

4. B-cell movement: its information-gathering function

Now return to the B-cell as an information-gathering monad in the immune system, and consider a particular B-cell. This cell is released (from the blood stream) at some arbitrary location in the body, diffuses through the tissues for a relatively long time, and is absorbed into the lymphatic system, at some arbitrary position, through the walls of some lymphatic capillary. If the cell has encountered the antigen for which it is specific, it will carry a certain number of these antigens in its outer membrane; if the cell did not encounter its antigen then its membrane will be free of bound antigens. This process is repeated for each one of the 10^5 B-cells which form the particular clone that is "sensitive" to the antigen. One now asks for the type of information which the immune system can extract from these monads after they have all returned to the lymph nodes by way of the lymphatic capillaries.

In order to keep the analysis as simple as possible I shall consider the case in which the antigens are due to a diseased part of the body, which will be represented by a sphere of radius a. Let the antigens be scattered uniformly throughout this sphere, with number density n_0. For the same reason we shall represent the body by a second, concentric sphere of radius R. For a 75 kg human $R \simeq 2.62 \times 10^{-1}$ m. What follows is a list of characteristics about the disease which the immune system can extract from the statistics of the B-cells only. Note that a will be very small as compared to R, so one can always assume that the B-cell starts its random walk <u>outside</u> the infected region.

(A) The immune system, especially in the lymphnode(s) in which the B-cells are collected after their travels through the body, can determine the fraction π_A of B-cells with antigen. For B-cells that were released at a distance r from the diseased region, this fraction was given as $\Pi_{\text{cap}}(r) = a/r$ in Eq. (21). As the initial point of B-cell release can be anywhere inside the body, one has to average (21) over a sphere of radius R. This gives

$$\pi_A = \int_0^R \frac{3r^2}{R^3} \frac{a}{r} \, dr = \frac{3a}{2R}, \tag{36}$$

where the contribution from $0 < r < a$ is negligible anyhow. Similarly, the fraction (π_0) of B-cells that carry no antigens in their outer membrane follows from (22)

$$\pi_0 = \int_0^R \frac{3r^2}{R^3} \left(1 - \frac{a}{r}\right) dr = 1 - \frac{3a}{2R}. \tag{37}$$

Because of the fact that R is determined by human physiology, a simple statistical treatment of the incoming B-cells would enable the immune system to determine a, the linear size of the infected region.

It should be noted how sensitive this method of detection is. The immune system probably needs the presence of only a few antigen-binding B-cells to trigger an immune response. Estimating this number to be unity, π_A can be as small as 10^{-5} and hence, with Eq. (36) you find that a can be as small as 10^{-6} m. In other words: the infected region can be smaller than a typical cell, yet the immune system will find out about its existence.

It should also be pointed out that the probability π_A consists of contributions from all values of r. Actually, B-cells released near to the infected area contribute least to the integral (36) and those released far from the infected area contribute most.

(B) Now suppose that the immune system observes the distribution of the numbers of bound antigen for those B-cells that show up at the lymphnodes with bound antigen in their cell membrane. Equation (35) tells you that the variable X has a probability distribution of the form

$$\pi_A(X) = \frac{2D}{a^2 V_0} \sum_{n=0}^{\infty} \exp\left[-\left(n + \frac{1}{2}\right)^2 \frac{\pi^2 D}{a^2 V_0} X\right]. \tag{38}$$

This equation shows that the immune system can also determine the value of the constant $D/(a^2 V_0)$.

It is useful to make the analysis a bit more concrete, and to go back to the discrete random flight model of Sec. 2. As one wants $V_0\, dt$ to represent the number of antigens encountered by a B-cell during a short time interval dt, one calculates this number in the following way, which is a bit qualitative but which should be a fair approximation of what a more detailed analysis could give. A B-cell performs a random walk by crawling around between the other cells; each antigen that touches its surface is bound to it almost irreversibly. During a time interval dt the B-cell will take dt/ε steps; as each step has length l the center of the B-cell moves a distance $l\, dt/\varepsilon$. If we call ρ the radius of the B-cell then the volume of space swept through by this cell equals $\pi \rho^2 l\, dt/\varepsilon$. As the antigens have a number density n_0 the number of antigens "caught" by the B-cell in time dt equals $\pi \rho^2 l n_0\, dt/\varepsilon$. Identification of this number with $V_0 dt$ gives

$$V_0 = \frac{\pi \rho^2 l n_0}{\varepsilon}. \tag{39}$$

Combining this with Eq. (2) one has found the fact that the immune system can determine the value of the parameter

$$\frac{D}{a^2 V_0} = \frac{l}{6\pi a^2 \rho^2} \frac{1}{n_0}.$$ (40)

Note that this parameter is independent of the time interval ε between jumps. Also note that ρ is the known radius of a B-cell, and l is another physiological parameter (l is of the order 2ρ). As a is known from the estimate described in (A) this shows that the immune system can determine the value of n_0.

This list of characteristics of the disease, which the immune system can extract from the statistics of the B-cells, could be expanded considerably by adjusting the model in various ways. I shall not pursue these applications in the present essay, but hope to have demonstrated how the simple paradigm of the Brownian motion of a B-cell is sufficient to explain some of the information-gathering activities of the immune system, our second cognitive system.

References

1. H. Ezawa, "Ornstein–Uhlenbeck path integral and its applications", *Acta Aplicandae Mathematicae* **63**, 119–135 (2000).
2. H. Ezawa, "The long time average of field values measured by a Brownian wanderer", *J. Phys. Soc. Jpn.* **71**, 35–42 (2002).
3. F.W. Wiegel, *Physical Principles in Chemoreception*, Springer (Berlin and New York, 1991).
4. A.S. Perelson and G. Weisbuch, "Immunology for physicists", *Rev. Mod. Phys.* **69**, 1219–1267 (1997).
5. I.R. Cohen, *Tending Adam's Garden*, Academic Press (London, 2000).
6. H. Ezawa, T. Nakamura, K. Watanabe and F. W. Wiegel, "Long-time average of the field measured by a two-dimensional Brownian wanderer", preprint (2002).
7. T. Nakamura, H. Ezawa and K. Watanabe, "Long-time average of the field measured by a Brownian wanderer; the case of three dimensions", preprint (2002).

The Lorentz Force and the Casimir Force at Finite Temperature and Casimir Entropy

M. Revzen

Department of Physics, Technion–Israel Institute of Technology,
Haifa 32000, Israel

K. Nakamura

Division of Natural Science, Meiji University,
Izumi Campus, Eifuku, Suginami-ku, Tokyo 168, Japan

A. Mann

Department of Physics, Technion–Israel Institute of Technology,
Haifa 32000, Israel

Following the work of Ezawa *et al.*[3], we calculate directly the Casimir force at any temperature, and point out how one may obtain the Casimir free energy and the Casimir entropy from the expression for the force.

1. Introduction

The "Casimir force" was first calculated by Casimir,[1] who considered the changes in the zero-point energy of the electromagnetic field caused by changing the boundary conditions on that field. The Casimir effect is considered one of the direct manifestations of the zero-point energy. It has attracted a great deal of attention over the years, including many reviews and books [Reference[2] is the most recent book, which includes many references to previous work].

Also over the years, various sophisticated methods (which may be found in Ref.[2] and references therein) have been developed and applied to the analysis of the Casimir effect . However, it seems that a rather simple, intuitive and direct approach to the calculation of the Casimir force was missing, until the rather recent work of Ezawa *et al.*[3] In this work, the

Casimir force at zero-temperature was derived directly and simply as the *Lorentz* force acting on the plates, without recourse to (or need of) the zero point energy. In the present work we (a) generalize that derivation of the Casimir force to finite temperature, and (b) apply the same method to the calculation of the force between a conducting plate and a permeable plate. The results we obtain are not new, since they were obtained earlier by different methods. However, we believe the *new* derivation is of interest, affording a different point of view of the origin of the Casimir force. For example, the "problem" of apparently a smaller number of degrees of freedom within the confined region (as compared with the number in the outside region) which seemed to imply (wrongly) that Casimir forces are always attractive, does not arise at all. Also, perhaps considering the Casimir force as being simply the Lorentz force may help our intuition regarding the sign of the force ("understanding the sign is a sign of understanding"). Another possibly useful feature is related to the issue of the fluctuations of Casimir forces, discussed recently in Ref.[4] Recognizing the Casimir force as the expectation value of the Lorentz force (in the appropriate quantum state), fluctuations may be calculated simply as higher-order correlation functions of the Lorentz force (in the proper quantum state).

2. Casimir force as the Lorentz Force at finite temperature for two conducting plates

We shall use the notations and conventions of Ezawa *et al.*[3] As a matter of fact, the only change we need to make is that instead of calculating the expectation value of the Lorentz force in the vacuum state, we have to calculate it in the thermal equilibrium state. In practical terms, this means replacing terms of the general form [Note that in the following equation we have suppressed any explicit reference to the various indices and exponential factors that accompany each operator, since they cancel when the expectation value in the (free) thermal state is taken]

$$(b^\dagger + b)(b^\dagger + b) = b^{\dagger 2} + b^2 + 2b^\dagger b + 1$$

by their expectation value in the thermal state of the free electromagnetic field. Hence we see immediately that the *vacuum* expectation value, which is represented by the term "1" in the above equation, should be replaced by

$$1 + \frac{2}{e^{\beta\epsilon} - 1}$$

where the added term $2/(e^{\beta\epsilon} - 1)$ arises from the expectation value in the thermal state of the number operator $b^\dagger b$.

Taking into account this simple change, the Lorentz–Casimir force at finite temperature on the plate at $z = -a$ becomes (from Eq.(34) of Ref.[3], correcting for some typos there) in the limit $L \to \infty$

$$\langle F_z \rangle_T = \frac{\hbar c}{2} \left(\frac{L}{\pi}\right)^2 \int_0^\infty \frac{\pi}{2} k_\parallel dk_\parallel$$

$$\times \left[\frac{1}{\pi} \int_{-\infty}^\infty dk_z \frac{k_z^2}{\sqrt{k_\parallel^2 + k_z^2}} \left(1 + \frac{2}{e^{\beta\hbar c\sqrt{k_\parallel^2 + k_z^2}} - 1}\right) \right.$$

$$\left. - \frac{1}{2a} \sum_{n=-\infty}^\infty \frac{\left(\frac{n\pi}{2a}\right)^2}{\sqrt{k_\parallel^2 + \left(\frac{n\pi}{2a}\right)^2}} \left(1 + \frac{2}{e^{\beta\hbar c\sqrt{k_\parallel^2 + \left(\frac{n\pi}{2a}\right)^2}} - 1}\right) \right]. \quad (1)$$

Note that in the limit $T \to 0$ $(\beta \to \infty)$ this reduces to Eq.(34) of Ref.[3] — after correcting for a superfluous factor of 2 in the denominator of the second term there.

We therefore see that at finite temperature the Casimir force per unit area may be written as a sum of two terms, as follows

$$\frac{\langle F_z \rangle_T}{L^2} = \frac{\pi^2 \hbar c}{2(2a)^4}(I_0 + I_T). \quad (2)$$

Here I_0 pertains to the zero-temperature contribution, and was calculated in Ref.[3] (where it was denoted by I) to be $1/120$. The term I_T is given by

$$I_T = \int_{-\infty}^\infty U_T(\nu)d\nu - \sum_{n=-\infty}^\infty U_T(n),$$

where, with $\gamma = \pi\hbar c\beta/2a$,

$$U_T(\nu) = \nu^2 \int_0^\infty \frac{k dk}{\sqrt{\nu^2 + k^2}} \cdot \frac{1}{e^{\gamma\sqrt{\nu^2 + k^2}} - 1}$$

$$= \nu^2 \int_{|\nu|}^\infty \frac{dx}{e^{\gamma x} - 1} = -\frac{\nu^2}{\gamma} \ln(1 - e^{-\gamma|\nu|})$$

$$= \frac{\nu^2}{\gamma} \sum_{m=1}^\infty \frac{e^{-\gamma|\nu|m}}{m}.$$

Hence we obtain the following form for the temperature-dependent contribution to the force:

$$I_T = \sum_{m=1}^{\infty} \frac{1}{\gamma m} \left[\int_0^{\infty} \nu^2 e^{-\gamma \nu m} d\nu - \sum_{n=0}^{\infty} n^2 e^{-\gamma n m} \right]. \tag{3}$$

We note that these sums and integrals are convergent without the need for any regularization. To proceed further, and in order to compare with the results of previous work, we use the Poisson summation formula

$$\sum_{n=-\infty}^{\infty} F(n) = \sqrt{2\pi} \sum_{n=-\infty}^{\infty} f(2\pi n),$$

where

$$f(y) = \frac{1}{\sqrt{2\pi}} \int_{-\infty}^{\infty} F(x) e^{-ixy} dx.$$

For $F(x) = x^2 e^{-\gamma |x| m}$, we obtain after some calculations

$$f(y) = -\sqrt{\frac{2}{\pi}} \frac{\partial^2}{\partial y^2} \left(\frac{\gamma m}{y^2 + (\gamma m)^2} \right), \tag{4}$$

Using the relations (valid in our case)

$$F(0) = 0, \quad F(x) = F(-x), \quad f(y) = f(-y),$$

$$F(0) + 2 \sum_{n=1}^{\infty} F(n) = \sqrt{2\pi} f(0) + 2\sqrt{2\pi} \sum_{m=1}^{\infty} f(2\pi m),$$

and

$$\sqrt{2\pi} f(0) = \int_{-\infty}^{\infty} F(x) dx = 2 \int_0^{\infty} F(x) dx.$$

We obtain

$$\frac{2}{\gamma m} \left[\int_0^{\infty} \nu^2 e^{-\gamma \nu m} d\nu - \sum_{n=1}^{\infty} n^2 e^{-\gamma n m} \right] = -\frac{2}{\gamma m} \sqrt{2\pi} \sum_{n=1}^{\infty} f(2\pi n).$$

Using Eq.(4), and the formula

$$\sum_{n=1}^{\infty} \frac{1}{n^2 + a^2} = -\frac{1}{2a^2} + \frac{\pi}{2a} \coth(\pi a),$$

we finally obtain

$$I_T = \frac{-12}{(2\pi)^4} \sum_{n=1}^{\infty} \frac{1}{n^4} + \frac{1}{\gamma^2} \sum_{n=1}^{\infty} \frac{1}{n^2} \mathrm{cosech}^2\Big(\frac{\pi}{\gamma} 2\pi n\Big)$$

$$+ \frac{2}{\gamma^2} \sum_{n=1}^{\infty} \Big(\frac{\pi}{\gamma}\Big)^2 \frac{1}{n} \mathrm{cosech}^2\Big(\frac{\pi}{\gamma} 2\pi n\Big) \coth\Big(\frac{\pi}{\gamma} 2\pi n\Big)$$

$$+ \frac{1}{\gamma^2} \sum_{n=1}^{\infty} \frac{\gamma}{2\pi^2} \frac{1}{n^3} \coth\Big(\frac{\pi}{\gamma} 2\pi n\Big).$$

The sum in the first term is well-known to equal $\pi^4/90$, and therefore the first term simply cancels I_0 in the sum $I = I_0 + I_T$. To compare our result with that of Ref.[5], we introduce the notation

$$t = \frac{\pi}{\gamma} = \frac{2a}{\hbar c \beta}, \quad \lambda = 2\pi n.$$

We then have for the force per unit area at finite temperature

$$\frac{\langle F_z \rangle_T}{L^2} = \frac{-2\pi^2}{\beta(\hbar c \beta)^3} \sum^{\infty} \Big[\frac{1}{(t\lambda)^3} \coth(t\lambda)$$

$$+ \frac{1}{(\tau\lambda)^2} \mathrm{cosech}^2(t\lambda) + \frac{1}{\tau\lambda} \mathrm{cosech}^2(t\lambda) \cdot \coth(t\lambda) \Big] . \qquad (5)$$

In Ref.[5] we calculated an expression for the Casimir free energy. As is well-known, to obtain the force one should differentiate the free energy with respect to the distance between the plates. Carrying out this calculation for the Casimir free energy of Ref.[5], we indeed obtain Eq.(5) (after correcting for various factors of 2 missing in Ref.[5], in particular in Eq.(15) there).

3. On the Lorentz–Casimir force for a conducting and a permeable plate at finite temperature

This case was first treated (at zero-temperature) by Boyer[6] in 1974. Recently it was treated at finite temperature, including the classical limit, by da Silva *et al.*[7] Really, the only (but crucial!) difference between this case and the previous one of two conducting plates follows very simply from the different boundary condition at the permeable plate. Indeed, the boundary condition at that plate entails the substitution

$$n \to n + \frac{1}{2}$$

in Eq.(1). The calculation of the Lorentz–Casimir force, after carrying out this change, follows essentially the same pattern as the calculation in Ref.[3]

and in the previous section. Not surprisingly, the resulting Lorentz–Casimir force per unit area agrees with the Casimir force (per unit area) obtained explicitly in Ref.[7] by differentiating the free energy. It turns out to be repulsive.

4. Casimir free energy and Casimir entropy from the Lorentz–Casimir force

The usual derivation of the Casimir force at any temperature starts with the calculation of the Casimir free energy. The Casimir force is then obtained by differentiation with respect to the relevant distance. Our present approach follows the inverse route. The measurable quantity, which is the force, is calculated directly as the expectation value of the Lorentz force in the thermal vacuum state. The resulting expression for the Casimir force (which, as mentioned, is the derivative of the Casimir free energy) may then be integrated to yield the Casimir free-energy. The Casimir energy, as well as the Casimir entropy, may then be obtained following the procedures of Ref.[5] (We note that there is no room in the Casimir free-energy for a distance-independent but temperature-dependent constant of integration.) In Refs.[5,7,8], the classical limit of the Casimir force was discussed by following the usual method of first calculating the Casimir free-energy. It was pointed out there that in the classical limit the force should be proportional to the temperature. This of course may be verified directly once the Lorentz–Casimir force is calculated following the method of the present work.

5. Conclusions

Following the lead of Ezawa *et al.*[3], we have presented a new approach to the Casimir effect, whereby the Casimir force at any temperature is obtained directly as the expectation value of the Lorentz force in the appropriate state. Both the zero-temperature and finite temperature direct calculations agree with the results of other methods. While the results are not new, they were derived in what we believe is a new and intuitively satisfying way. We hope that this approach, pioneered by Ezawa *et al.*, will also yield an insight into the *sign* of the force.

Dedication

We would like to dedicate this work, inspired by Professor Ezawa, to Professor Hiroshi Ezawa upon his seventieth birthday and on the occasion of his going into *formal* retirement. We hope and believe that his creativity will lead and guide us for many years to come.

References

1. H.B.G. Casimir, Proc. K. Ned. Akad. Wet. **51**, 793 (1948).
2. K.A. Milton, *The Casimir Effect* (World Scientific, New Jersey, 2001).
3. H. Ezawa, K. Nakamura, and K. Watanabe, The Casimir Force from Lorentz's, in *Frontiers in Quantum Physics*, ed. S.C. Lim, R. Abd-Shukor, K.H. Kwek (Springer, Singapore, 1998).
4. D. Bartolo, A. Ajdare, J.-B. Fournier, and R. Golestanian, Phys. Rev. Lett. **89**, 230601 (2002).
5. M. Revzen, R. Opher, M. Opher and A. Mann, J. Phys. A. **30**, 7783 (1997).
6. T.H. Boyer, Phys. Rev. **A9**, 2078 (1974).
7. J.C. da Silva, A. Matos Neto, H. Q. Placido, M. Revzen and A.E. Santana, Physica **A 292**, 411 (2001).
8. J. Feinberg, A. Mann and M. Revzen, Ann. Phys. (NY) **288**, 103 (2001).

Information Dynamics and its Application to Recognition Process

Masanori Ohya

Department of Information Sciences,
Science University of Tokyo
278 Noda City, Chiba, Japan

1. Introduction

H. Ezawa has worked on vast area related with fundamental physics. His interests and pioneering works are not limited to "elementary-particle-oriented" works[4] but also cover more *complex* system, such as finite temparature quantum field theory and nonequilibrium theory.[5] It is, therefore, great honor of mine to contribute to this volume by presenting an attempt, information dynamics, to treat various complex systems mathematically and its one of the latest application, a description of recognition process[8] based on the works.[10,11] In this section we give a review on what is the complex system. The discussion leads the introduction of Information Dynamics in natural way.

The complex system has been considered in Santafe research center as follows:

(1) A system is composed of several elements called agents. The size of the system (the number of the elements) is medium.

(2) The agent has intellegence.

(3) Each agent has interaction due to local information. The decision of each agent is determined by not all information but the limited information of the system.

Under a small modification, I define the complex system as follows:

(1) A system is composed of several elements. The scale of the system is often large but not always, in some cases one.

(2) Some elements of the system have special (self) interactions (rela-

tions), which produce a dynamics of the system.

(3) The system shows a particular character (not sum of the characters of all elements) due to (2).

Definition 1: *A system having the above three properties is called "complex system". The "complexity"of such a complex system is a quantity measuring that complexity, and its change describes the appearance of the particular character of the system.*

There exist such measures describing the complexity for a system, for instance, variance, correlation, level - statistics, fluctuation, randomness, multiplicity, entropy, fuzzy, fractal dimension, ergodicity (mixing, flow), bifurcation, localization, computational complexity (Kolmogorov's or Chaitin's), catastrophy, dynamical entropy, Lyapunov exponent, etc. These quantities are used case by case and they are often difficult to compute. Moreover, the relations among these are lacking (not clear enough). Therefore it is important to find common property or expression of these quantities. In this paper, we introduce such a common degree to describe the chaotic aspect of quantum dynamical systems. Further we describe the function of barin in the framework of information dynamics[19] (ID for short) and we discuss the value of information attached to the brain in terms of the complexity in ID and the chaos degree.[22,23]

2. Information dynamics

There are two aspects for the complexity, that is, the complexity of a state describing the system itself and that of a dynamics causing the change of the system (state). The former complexity is simply called the "complexity" of the state, and the later is called the "chaos degree" of the dynamics in this paper. Therefore the examples of the complexity are entropy, fractal dominion, and those of the chaos degree are Lyapunov exponent, dynamical entropy, computational complexity. Let us discuss a common quantity measuring the complexity of a system so that we can easily handle. The complexity of a general quantum state was introduced in the frame of ID[19,12] and the quantum chaos degree was defined in Ref.[13], which we will review in this section.

Information Dynamics is a synthesis of dynamics of state change and complexity of state. More precisely, let $(\mathcal{A}, \mathfrak{S}, \alpha(G))$ be an input (or initial) system and $(\overline{\mathcal{A}}, \overline{\mathfrak{S}}, \overline{\alpha}(\overline{G}))$ be an output (or final) system. Here $\mathcal{A}$ is the set of all objects to be observed and $\mathfrak{S}$ is the set of all means for measurement of $\mathcal{A}$, $\alpha(G)$ is a certain evolution of system. Once an input and an output systems are set, the situation of the input system is described by a state, an element of $\mathfrak{S}$, and the change of the state is expressed by a mapping from $\mathfrak{S}$ to $\overline{\mathfrak{S}}$, called a channel, $\Lambda^* : \mathfrak{S} \to \overline{\mathfrak{S}}$. Often we have $\mathcal{A} = \overline{\mathcal{A}}$, $\mathfrak{S} = \overline{\mathfrak{S}}$, $\alpha = \overline{\alpha}$, which is assumed in the sequel. Thus we claim

$$[\text{Giving a mathematical structure to input and output triples} \\ \equiv \text{Having a theory}]$$

For instance, when $\mathcal{A}$ is the set $M(\Omega)$ of all measurable functions on a measurable space $(\Omega, \mathcal{F})$ and $\mathfrak{S}(\mathcal{A})$ is the set $P(\Omega)$ of all probability measures on Ω , we have usual probability theory, by which the classical dynamical system is described. When $\mathcal{A} = B(\mathcal{H})$, the set of all bounded linear operators on a Hilbert space $\mathcal{H}$, and $\mathfrak{S}(\mathcal{A}) = \mathfrak{S}(\mathcal{H})$, the set of density operators on $\mathcal{H}$, we have a usual quantum dynamical system. In this paper, we assume that both the input and output triple $(\mathcal{A}, \mathfrak{S}, \alpha(G))$ is a C*-dynamical system or the usual quantum system as above, and a channel, $\Lambda^* : \mathfrak{S} \to \mathfrak{S}$ is a completely positive map.

There exist two complexities in ID, which are axiomatically given as follows:

Let $(\mathcal{A}_t, \mathfrak{S}_t, \alpha^t(G^t))$ be the total system of both input and output systems; $\mathcal{A}_t \equiv \mathcal{A} \otimes \mathcal{A}, \mathfrak{S}_t \equiv \mathfrak{S} \otimes \mathfrak{S}, \alpha^t \equiv \alpha \otimes \alpha$ with suitable tensor products $\otimes$. Further, let $C(\varphi)$ be the complexity of a state $\varphi \in \mathfrak{S}$ and $T(\varphi; \Lambda^*)$ be the transmitted complexity associated with the state change $\varphi \to \Lambda^* \varphi$. These complexities C and T are the quantities satisfying the following conditions:

(i) For any $\varphi \in \mathfrak{S}$,

$$C(\varphi) \geq 0, \ T(\varphi; \Lambda^*) \geq 0.$$

(ii) For any orthogonal bijection $j : ex\mathfrak{S} \to ex\mathfrak{S}$ (the set of all extreme points in $\mathfrak{S}$),

$$C(j(\varphi)) = C(\varphi),$$

$$T(j(\varphi); \Lambda^*) = T(\varphi; \Lambda^*).$$

(iii) For $\Phi \equiv \varphi \otimes \psi \in \mathfrak{S}_t$,

$$C(\Phi) = C(\varphi) + C(\psi).$$

(iv) For any state φ and a channel Λ^*,

$$T(\varphi; \Lambda^*) \leq C(\varphi).$$

(v) For the identity map "id" from $\mathfrak{S}$ to $\mathfrak{S}$.

$$T(\varphi; id) = C(\varphi).$$

Definition 2: *Quantum Information Dynamics (QID) is defined by*

$$(\mathcal{A}, \mathfrak{S}, \alpha(G); \; \Lambda^*; \; C(\varphi), T(\varphi; \Lambda^*)) \tag{1}$$

and some relations R among them.

There are several examples of the above complexities C and T such as quantum entropy and quantum mutual entropy.[17,21] Information Dynamics can be applied to the study of chaos in the following sense:

Definition 3: [22,23,12] *ψ is more chaotic than φ as seen from the reference system $\mathcal{S}$ if $C(\psi) \geq C(\varphi)$.*

When φ changes to $\Lambda^*\varphi$, the degree of chaos associated to this state change(dynamics) Λ^* is given by

$$D(\varphi; \Lambda^*) = \inf\left\{ \int_{\mathfrak{S}} C(\Lambda^*\omega)d\mu; \mu \in M(\varphi) \right\},$$

where $\varphi = \int_{\mathfrak{S}} \omega d\mu$ is a maximal extremal decomposition of φ and $M(\varphi)$ is the set of such measures. In some cases such that Λ^* is linear, this chaos degree $D(\varphi; \Lambda^*)$ can be written as $C(\Lambda^*\varphi) - T(\varphi; \Lambda^*)$.

Since ID has hierarchy (hierarchical structure), it can be applied several open systems. Later we apply ID to Brain Dynamics.

3. Entropic chaos degree (ECD)

In the context of information dynamics, a chaos degree associated with a dynamics in classical systems was introduced in Ref.[22] It has been applied to several dynamical maps such logistic map, Baker's transformation and Tinkerbel map with succesful explainations of their chaotic characters.[15] This chaos degree has several merits compared with usual measures such as Lyapunov exponent.

Here we discuss the quantum version of the classical chaos degree, which is defined by quantum entropies in Section 2, and we call the quantum chaos degree the entropic quantum chaos degree. In order to contain both classical and quantum cases, we define the entropic chaos degree (ECD) in C*-algebraic terninology. This setting will not be used in the sequel application, but for mathematical completeness we first discuss the C*-algebraic setting.

Let $(\mathcal{A}, \mathfrak{S})$ be an input C* system and $(\overline{\mathcal{A}}, \overline{\mathfrak{S}})$ be an output C* system; namely, $\mathcal{A}$ is a C* algebra with unit I and $\mathfrak{S}$ is the set of all states on $\mathcal{A}$. We assume $\overline{\mathcal{A}} = \mathcal{A}$ for simlicity. For a weak* compact convex subset $\mathcal{S}$ (called the reference space) of $\mathfrak{S}$, take a state φ from the set $\mathcal{S}$ and let

$$\varphi = \int_{\mathcal{S}} \omega d\mu_\varphi$$

be an extremal orthogonal decomposition of φ in $\mathcal{S}$, which describes the degree of mixture of φ in the reference space. $\mathcal{S}$[18,25] The measure μ_φ is not uniquely determined unless $\mathcal{S}$ is the Schoque simplex, so that the set of all such measures is denoted by $M_\varphi(\mathcal{S})$. The entropic chaos degree with respect to $\varphi \in \mathcal{S}$ and a channel Λ^* is defined by

$$D^{\mathcal{S}}(\varphi; \Lambda^*) \equiv \inf\left\{\int_{\mathcal{S}} S^{\mathcal{S}}(\Lambda^*\varphi)\, d\mu_\varphi; \mu_\varphi \in M_\varphi(\mathcal{S})\right\} \tag{2}$$

where $S^{\mathcal{S}}(\Lambda^*\varphi)$ is the mixing entropy of a state φ in the reference space $\mathcal{S}$[20,12]. When $\mathcal{S} = \mathfrak{S}$, $D^{\mathcal{S}}(\varphi; \Lambda^*)$ is simply written as $D(\varphi; \Lambda^*)$. This $D^{\mathcal{S}}(\varphi; \Lambda^*)$ contains both the classical chaos degree and the quantum one.

In usual quantum system including classical discrete system, $\mathcal{A}$ is the set $\mathbf{B}(\mathcal{H})$ of all bounded operators on a Hilbert space $\mathcal{H}$ and $\mathfrak{S}$ is the set $\mathfrak{S}(\mathcal{H})$ of all density operators on $\mathcal{H}$, in which an extreme decomposition of $\rho \in \mathfrak{S}(\mathcal{H})$ is a Schatten decomposition $\rho = \sum_k p_k E_k$ (*i.e.*, $\{E_k\}$ are one dimensional orthogonal projections with $\sum E_k = I$), so that the entropic chaos degree is written as

$$D(\rho; \Lambda^*) \equiv \inf\left\{\sum_k p_k S(\Lambda^* E_k); \{E_k\}\right\}, \tag{3}$$

where the infimum is taken over all possible Schatten decompositions and S is von Neumann entropy. Note that in classical discrete case, the Schatten decomposition is unique $\rho = \sum_k p_k \delta_k$ with the delta measure $\delta_k(j) \equiv \begin{cases} 1 \ (k = j) \\ 0 \ (k \neq j) \end{cases}$, and the entropic chaos degree is written by

$$D(\varphi; \Lambda^*) = \sum_k p_k S(\Lambda^* \delta_k), \tag{4}$$

where ρ is the probability distribution of the orbit obtained from a dynamics of a system and the channel Λ^* is generated from the dynamics.

We can judge whether the dynamics Λ^* causes a chaos or not by the value of D as

$$D > 0 \text{ and not constant} \Longleftrightarrow \text{chaotic,}$$
$$D = \text{constant} \Longleftrightarrow \text{weak stable,}$$
$$D = 0 \Longleftrightarrow \text{stable.}$$

The classical version of this degree was applied to study the chaotic behaviors of several nonlinear dynamics.[15,22] The quantum entropic chaos degree is applied to the analysis of quantum spin system[13] and quantum Baker's type transformation[16], and we could measure the chaos of these systems. The information theoretical meaning of this degree was explained in Ref.[23]

The ECD can resolve some inconvenient properties of the Lyapunov exponent, another degree of chaos[15,14]:

(1) Lyapunov exponent takes negative value and sometimes $-\infty$, but the ECD is always positive for any $a \geq 0$.

(2) It is difficult to compute the Lyapunov exponent for some maps like Tinkerbell map f because it is difficult to compute f^n for large n. On the other hand, the ECD of f is easily computed.

(3) Generally, the algorithm for the ECD is much easier than that for the Lyapunov exponent.

4. Quantum Information dynamic description of brain

The Information Dynamics can be employed to describe not only several classical and quantum physical physics but also life sciences. We will construct a model describimg the function of brain in the context of Quantum Inforamtion Dynamics (QID).

We study a possible function of brain, in particular, we try to describe several aspects of the process of recognition. In order to understand the fundamental parts of the recognition process, the quantum teleportation scheme[3,2,10,11] seems to be useful. We consider a channel expression of the teleportation process that serves for a simplified description of the recognition process in brain.

It is the processing speed that we take as a particular character of the brain, so that the high speed of processing in the brain is here supposed to come from the coherent effects of substances in the brain like quantum computer, as was pointed out by Penrose. Having this in our mind, we propose a model of brain describing its function as follows:

The brain system $BS = \mathfrak{X}$ is supposed to be described by a triple ($B(\mathcal{H})$, $\mathcal{S}(\mathcal{H})$, $\Lambda^*(G)$) on a certain Hilbert space $\mathcal{H}$ where $B(\mathcal{H})$ is the set of all bounded operators on $\mathcal{H}$, $\mathcal{S}(\mathcal{H})$ is the set of all density operators and $\Lambda^*(G)$ is a channel giving a state change with a group G.

Further we assume the following:

(1) *BS* is described by a quantum state and the brain itself is divided into several parts, each of which corresponds to a Hilbert space so that $\mathcal{H} = \oplus_k \mathcal{H}_k$ and $\varphi = \oplus_k \varphi_k$, $\varphi_k \in \mathfrak{S}(\mathcal{H}_k)$. However, in this paper we simply assume that the brain is in one Hilbert space $\mathcal{H}$ because we only consider the basic mechanism of recognition.

(2) The function (action) of the brain is described by a channel $\Lambda^* = \oplus_k \Lambda_k^*$. Here as in (1) we take only one channel Λ^*.

(3) *BS* is composed of two parts; information processing part "P" and others "O" (consciousness, memory, recognition) so that $\mathfrak{X} = \mathfrak{X}_P \otimes \mathfrak{X}_O$, $\mathcal{H} = \mathcal{H}_P \otimes \mathcal{H}_O$.

Thus in our model the whole brain may be considered as a parallel quantum computer,[24] but we here explain the function of the brain as a quantum computer, more precisely, a quantum communication process with entanglements like in a quantum teleportation process. We will explain the mathematical structure of our model.

Let $s = \{s^1, s^2, \cdots, s^n\}$ be a given (input) signal (perception) and $\bar{s} = \{\bar{s}^1, \bar{s}^2, \cdots, \bar{s}^n\}$ the output signal. After the signal s enters the brain, each element s^j of s is coded into a proper quantum state $\rho^j \in \mathcal{S}(\mathcal{H}_P)$, so that the state corresponding to the signal s is $\rho = \otimes_j \rho^j$. This state may be regarded as a state processed by the brain and it is coupled to a state ρ_O stored as a memory (pre-conciousness) in brain. The processing in the brain is expressed by a properly chosen quantum channel Λ^* (or $\Lambda_P^* \otimes \Lambda_O^*$). The channel is determined by the form of the network of neurons and some other biochemical actions, and its function is like a (quantum) gate in quantum computer.[17,26] The outcome state $\bar{\rho}$ contacts with an operator F describing the work as noema of consciousness (Husserl's noema), after the contact a certain reduction of state is occured, which may correspond to the noesis (Husserl's) of consciousness. A part of the reduced state is stored in brain as a memory. The scheme of our model is represented in the following figure.

5. Value of information in brain

The complex system responses to the information and has a particular role to choose the information (value of information). Brain selects some infor-

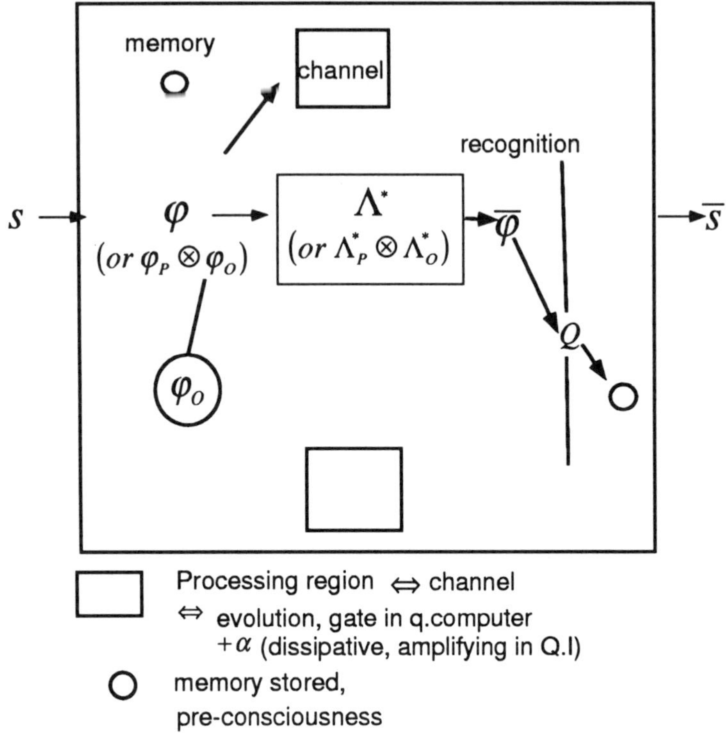

mation (inputs) from huge flow of information (inputs). It will be important to find a rule or rules of such selection mechanisum. In the model of Sec.4, an output signal s (information) is somehow coded into a quantum state φ, then it runs in brain with a certain processing effect Λ^* and a memory stored, and it changes its own figure. Thus we have two standpoints to catch the value of information in brain. Suppose that we have a fixed purpose (intention) described by an operator Q, then one view of the value of information is whether the signal s is important for the purpose Q and the processing Λ^* and another is whether the processing Λ^* chosen in brain is effective for s and Q. From these considerations, that value should be estimated by a function of the state $\varphi \otimes \varphi_O$, a channel Λ^* and an operator Q, so that one possibility to define a measure $V(\varphi \otimes \varphi_O, \Lambda^*, Q)$ estimating the effect of a signal and a function of brain is as follows:

Define

$$V\left(\varphi \otimes \varphi_O, \Lambda^*, Q\right) = tr\Lambda^* \varphi \otimes \varphi_O Q$$

Definition 4: *Value of Information:*

(1) $s = \left\{s^1, s^2, \cdots, s^n\right\}$ *is more valuable than* $s^{'} = \left\{s^{'1}, s^{'2}, \cdots, s^{'n}\right\}$ *for* Λ^* *and* Q *iff*

$$V(\varphi \otimes \varphi_O, \Lambda^*, Q) \geqq V(\varphi^{'} \otimes \varphi_O, \Lambda^*, Q).$$

(2) Λ^* *is more valuable than* $\Lambda^{'*}$ *for given* $s = \left\{s^1, s^2, \cdots, s^n\right\}$ *and* Q *iff*

$$V(\varphi \otimes \varphi_O, \Lambda^*, Q) \geqq V(\varphi \otimes \varphi_O, \Lambda^{'*}, Q).$$

The details of this estimator is discussed in Ref.[9], where there exist some relations between the information of value and the complexity or the chaos degree under properly chosen complexity C and transmitted complexity T. For instance, with entropy type complexities and a certain Q, we conjecture (partially proved so far)

$$D\left(\varphi \otimes \varphi_O, \Lambda^*; Q\right) \leq D\left(\varphi \otimes \varphi_O, \Lambda^{'*}; Q\right)$$

$$\Longleftrightarrow V\left(\varphi \otimes \varphi_O, \Lambda^*, Q\right) \geq V\left(\varphi \otimes \varphi_O, \Lambda^{'*}, Q\right)$$

This result is quite natural because the more chaos a processing produces, the less value it has.

6. A speculation of brain function

The set of neurons in brain is divided into several parts and each part corresponds to a configuration domain G, each point in which has two states, excited or not. Thus $G \equiv \cup_k G_k$ with $G_k \cap G_j = \emptyset$ for any $k \neq j$. Let assume that the Hilbert space $\mathcal{H}$ describing the barin is Fock space on the square integrable random variables $L_2(G, \mu)$ with the counting measure μ, so that the whole Hilbert space $\mathcal{H}$ is decomposed as

$$\mathcal{H} \equiv \Gamma\left(L_2(G, \mu)\right) = \otimes_k \Gamma\left(L_2(G_k, \mu)\right).$$

Let $\{x_1, x_2, \cdots, x_n\}$ describes the (positions of) excited neurons in as certain domain G_k, so that the vector in $L^2(G_k, \mu)$ corresponding this configuration is denoted by $\sum_{j=1}^{n} \delta_{x_j}$ by the delta measure δ_x corresponding to x.

When we consider only one domain, for simplicity, denoted by the same G and it is decomposed ito the processing part G^P and other part G^O including the effect of conciousness as in the previous section, our Hilbert space of the brain is $\mathcal{H} \equiv \Gamma\left(L_2(G,\mu)\right) = \Gamma\left(L_2(G^P,\mu)\right) \otimes \Gamma\left(L_2(G^O,\mu)\right)$. Along the above settings we may explain some functions of brain in the terminologies of Fock space and quantum teleportation,[10,11] on which we are working now.[9]

In the sequel, we will explain the first trial explaining the brain function, in particular the memory change due to recognition, based on the quantum teleportation scheme done in Ref.[8]

Let us assume the Hilbert space $\mathcal{H}_O$ is composed of two parts, before and after recognition. For notational simplicity, we denote the Hilbert spaces by $\mathcal{H}_1, \mathcal{H}_2, \mathcal{H}_3$ where $\mathcal{H}_1$ represents the processing part, $\mathcal{H}_2$ the memory before recognition and $\mathcal{H}_3 = \mathcal{H}_2$ the memory after recognition. Throughout this paper we will have in mind this interpretation of the Hilbert spaces $\mathcal{H}_j\ (j = 1, 2, 3)$. However, this is just an illustration of what we are going to do, and the teleportation scheme may be applied to very different situations.

We are mainly interested in the changes of the memory after the process of recognition. For that reason we consider channels from the set of states on $\mathcal{H}_1 \otimes \mathcal{H}_2$ into $\mathcal{H}_3$. Main object to be measured causing the recognition is here assumed to be a self-adjoint operator

$$F = \sum_{k,l=1}^{n} z_{k,l} F_{k,l}$$

on $\mathcal{H}_1 \otimes \mathcal{H}_2$ where the operators $F_{k,l}$ are orthogonal projections (alternatively, we may take $F_{k,l}$ as an operator valued measure). The channel $\Lambda_{k,l}$ describes the state of the memory after the process of recognition if the outcome of the measurement according to F was $z_{k,l}$ and is given by

$$\Lambda_{k,l}(\rho \otimes \gamma) := \frac{\mathrm{Tr}_{1,2}(F_{k,l} \otimes 1)(\rho \otimes J\gamma J^*))(F_{k,l} \otimes 1)}{\mathrm{Tr}_{1,2,3}(F_{k,l} \otimes 1)(\rho \otimes J\gamma J^*))(F_{k,l} \otimes 1)}$$

where ρ and γ (denoted ρ_O above) are the state of the processing part and of the memory before recognition and J an isometry extending from $\mathcal{H}_2$ to $\mathcal{H}_2 \otimes \mathcal{H}_3$ and 1 denotes the identical operator. The value $\mathrm{Tr}_{1,2,3}(F_{k,l} \otimes 1)(\rho \otimes J\gamma J^*)(F_{k,l} \otimes 1)$ represents the probability to measure the value $z_{k,l}$. So, obviously, we have to assume that this probability is greater than 0. The state $\Lambda_{k,l}(\rho \otimes \gamma)$ gives the state of the memory after the process of recognition. The elements of a basis $(b_k)_{k=1}^{n}$ of $\mathcal{H}_j$ are interpreted as elementary signals.

In this first attempt to our model described above, there appear still a lot of effects being non-realistic for the process of recognition. Some examples (cf. the last subsection) show that with this model one can describe extreme cases such as storing the full information or total loss of memory, but - as mentioned above - that is still far from being a realistic description.

In the paper,[8] we restrict ourselves to finite dimensional Hilbert spaces. Moreover, we assume equal dimension of the Hilbert spaces $\mathcal{H}_j$ $(j = 1, 2, 3)$. It seems that infinite dimensional schemes will lead to more realistic models. However, this is just a first attempt to describe the brain function. Moreover, for finite dimensional Hilbert spaces the mathematical model becomes more transparent and one can obtain easily a general idea of the model. To indicate obvious generalizations to more general situations and especially to infinite dimensional Hilbert spaces we sometimes use notions and notations from the general functional analysis.[6,7]

6.1. *Basic notions*

Let $\mathcal{H}_1, \mathcal{H}_2, \mathcal{H}_3$ be Hilbert spaces with equal finite dimension:

$$\dim \mathcal{H}_j = n, \quad (j \in \{1, 2, 3\}).$$

First we will represent these Hilbert spaces in a way that it seems to be convenient for our considerations. Each of the spaces $\mathcal{H}_1, \mathcal{H}_2, \mathcal{H}_3$ can be identified with the space $\mathbb{C}^n$ of n-dimensional complex vectors. The space $\mathbb{C}^n$ again may be identified with the space $\{f : G \longrightarrow \mathcal{C}\}$ of all complex-valued function on $G := \{1, \ldots, n\}$. The scalar product then is given by

$$\langle f, g \rangle := \sum_{k=1}^{n} \overline{f(k)} g(k) = \int \overline{f(k)} g(k) \mu(dk)$$

where μ is the counting measure on G, i.e. $\mu = \sum_{k=1}^{n} \delta_k$ with δ_k denoting the DIRAC measure in k. So, each of the spaces $\mathcal{H}_j$ can be written formally as an L_2-space:

$$\mathcal{H}_j = L_2(G, \mu) := L_2(G) \quad (j \in \{1, 2, 3\}).$$

For the tensor product one obtains

$$f \otimes g(k, l) = f(k) g(l) \quad (f, g \in L_2(G), k, \in G),$$

and we have

$$\mathcal{H}_1 \otimes \mathcal{H}_2 = L_2(G \times G, \mu \times \mu) = \mathcal{H}_2 \otimes \mathcal{H}_3.$$

We will abbreviate this tensor product by $L_2(G^2, \mu^2)$ or just by $L_2(G^2)$.

By $\mathcal{B}(\mathcal{H})$ we denote the space of all bounded linear operators on a Hilbert space $\mathcal{H}$. In $\mathcal{B}(L_2(G))$ the operator of multiplication by a function $g \in L_2(G)$ is given by

$$(\mathcal{O}_g f)(k) = g(k)f(k) \quad (f \in L_2(G), k \in G).$$

Observe that for all $f, g \in L_2(G)$ one has

$$\mathcal{O}_f g = \mathcal{O}_g f, \quad \mathcal{O}_f^* = \mathcal{O}_{\overline{f}}$$

and for $f \in L_2(G)$ with $f(k) \neq 0$ for all $k \in G$ it holds $\mathcal{O}_f^{-1} = \mathcal{O}_{1/f}$.

The function $\mathbf{1}$, $\mathbf{1}(k) = 1$ for all $k \in G$, obviously belongs to $L_2(G)$ and $\mathbf{1} = \mathcal{O}_{\mathbf{1}}$ is the identity in $\mathcal{B}(L_2(G))$.

Consequently, an operator of multiplication $\mathcal{O}_f$ is unitary if and only if $|f(k)| = 1$ for all $k \in G$.

Further, we will use the mapping J from $L_2(G)$ into $L_2(G^2)$ given by

$$(J f)(k, l) = f(k)\delta_{k,l} \quad (f \in L_2(G), k, l \in G) \tag{5}$$

where $\delta_{k,l}$ denotes the KRONECKER symbol. It is immediate to see that J is an isometry. For the adjoint $J^* : L_2(G^2) \longrightarrow L_2(G)$ we obtain

$$(J^*\Phi)(k) = \Phi(k, k) \quad (\Phi \in L_2(G^2),\ k \in G). \tag{6}$$

Observe that G equiped with the operation $\oplus : G \times G \longrightarrow G$, $k \oplus l := (k + l)\mathrm{mod}\ n$ is a group. The operation inverse to $\oplus$ we denote by $\ominus$. Let us remark that $k \ominus l = k - l$ in the case $k > l$ and $k \ominus l = k - l + n$ if $k \leq l$. We conclude that for all $k \in G$ the operator $U_k \in \mathcal{B}(L_2(G))$ given by

$$(U_k f)(m) := f(k \oplus m) \quad (f \in L_2(G)) \tag{7}$$

is unitary.

Now, let $(b_k)_{k=1}^n$ be an orthonormal basis in $L_2(G)$, and denote by $(B_k)_{k=1}^n$ the sequence of multiplication operators corresponding to the elements of this basis, i.e. $B_k := \mathcal{O}_{b_k}, k \in G$. Then for $k, l \in G$ we put

$$\xi_{k,l} := (B_k \otimes U_l) J \mathbf{1}. \tag{8}$$

One can show that the sequence $(\xi_{k,l})_{k,l \in G}$ is an orthonormal basis in $L_2(G^2)$. And we denote by $F_{i,j} \in \mathcal{B}(L_2(G^2))$ the projection onto $\xi_{i,j}$, i.e.

$$F_{i,j} := |\xi_{i,j}\rangle\langle\xi_{i,j}| = \langle\xi_{i,j}, \cdot\rangle\xi_{i,j}. \tag{9}$$

6.2. *Channels*

Definition 5: *Let γ be a state on $\mathcal{H}_2 = L_2(G)$ (i.e. γ is a positive trace-class operator with $\mathrm{Tr}(\gamma) = 1$). The state $\mathbf{e}(\gamma)$ on $L_2(G^2) = \mathcal{H}_2 \otimes \mathcal{H}_3$ given by*

$$\mathbf{e}(\gamma) = J\gamma J^* \tag{10}$$

where J is the isometry given by (5) we call the entangled state corresponding to γ.

Now, let ρ and γ be states on $\mathcal{H}_1$ resp. $\mathcal{H}_2$, the state $\mathbf{e}(\gamma)$ (usually denoted by σ^{10}) will be a state on $\mathcal{H}_2 \otimes \mathcal{H}_3$. . Remember that we assumed $\mathcal{H}_1 = \mathcal{H}_2 = \mathcal{H}_3 = L_2(G)$. The numbering only indicates the meaning of the states (we recall that $\mathcal{H}_1$ represents the processing part, $\mathcal{H}_2$ the memory before and $\mathcal{H}_3$ the memory after the recognition process.) Then $\rho \otimes \mathbf{e}(\gamma)$ is a state on $\mathcal{H}_1 \otimes \mathcal{H}_2 \otimes \mathcal{H}_3$ and we observe immediately

$$\rho \otimes \mathbf{e}(\gamma) = (\mathbf{1} \otimes J)(\rho \otimes \gamma)(\mathbf{1} \otimes J^*). \tag{11}$$

In subsection 6.3 we calculate explicitly the trace of

$$(F_{i,j} \otimes \mathbf{1})(\rho \otimes \mathbf{e}(\gamma))(F_{i,j} \otimes \mathbf{1}) = (F_{i,j} \otimes \mathbf{1})(\mathbf{1} \otimes J)(\rho \otimes \gamma)(\mathbf{1} \otimes J^*)(F_{i,j} \otimes \mathbf{1}). \tag{12}$$

The following proposition will be very useful for this.

Proposition 1: [8] *Let $(g_k)_{k=1}^n$ and $(h_k)_{k=1}^n$ be orthonormal systems in $L_2(G)$ and ρ and γ states on $L_2(G)$ having the following representations:*

$$\rho = \sum_{k=1}^n \alpha_k |g_k> < g_k|, \qquad \gamma = \sum_{k=1}^n \beta_k |h_k> < h_k|,$$

$$\alpha_k \geq 0, \beta_k \geq 0, \sum_{k=1}^n \alpha_k = \sum_{k=1}^n \beta_k = 1.$$

Then for all $i, j \in G$

$$(F_{i,j} \otimes \mathbf{1})(\rho \otimes \mathbf{e}(\gamma))(F_{i,j} \otimes \mathbf{1}) = F_{i,j} \otimes \sum_{k,l=1}^n \alpha_k \beta_l |G_{i,j} g_k \otimes h_l> < G_{i,j} g_k \otimes h_l|. \tag{13}$$

where $G_{i,j}$ is given by, for $i, j \in G$

$$G_{i,j} := J^*(U_j \otimes \mathbf{1})(B_i^* \otimes \mathbf{1}) = J^*(U_j B_i^* \otimes \mathbf{1}) \tag{14}$$

where $B_i^* = \mathcal{O}_{b_i}^* = \mathcal{O}_{\bar{b}_i}$.

Denote by $\mathcal{T}$ the set of all positive trace-class operators on $L_2(G)$ including the null operator $\mathbf{0}$,

$$\mathbf{0}(f) = 0 \qquad (f \in L_2(G)).$$

We fix an operator $\tau \in \mathcal{T}$ having the representation

$$\tau = \sum_{k=1}^{n} \gamma_k |h_k ><h_k| \tag{15}$$

with $(\gamma_k)_{k \in G} \subseteq [0, \infty)$ and $(h_k)_{k \in G}$ being an orthonormal basis in $L_2(G)$.

The linear mapping $K_\tau : \mathcal{T} \longrightarrow \mathcal{T}$ given by

$$K_\tau(\rho) := \sum_{k=1}^{n} \gamma_k \mathcal{O}_{h_k} \rho \mathcal{O}_{h_k}^* \qquad (\rho \in \mathcal{T}) \tag{16}$$

depends only on the operator τ but not on its special representation.

Definition 6: *Denote by $\mathcal{S}$ the set of all states on $L_2(G)$ and for $\tau \in \mathcal{T}$ by $\mathcal{S}_\tau$ the set of all states ρ from $\mathcal{S}$ with the property that $\mathrm{Tr}K_\tau(\rho)$ is positive:*

$$\mathcal{S}_\tau := \{\rho \in \mathcal{S} : \mathrm{Tr}K_\tau(\rho) > 0\}. \tag{17}$$

For $\tau \in \mathcal{T}$ the mapping $\hat{K}_\tau : \mathcal{S}_\tau \longrightarrow \mathcal{S}$ given by

$$\hat{K}_\tau(\rho) := \frac{1}{\mathrm{Tr}K_\tau(\rho)} K_\tau(\rho) \qquad (\rho \in \mathcal{S}_\tau) \tag{18}$$

is called the CHANNEL *corresponding to τ. The channel corresponding to τ is called* UNITARY *if there exists an unitary operator U on $L_2(G)$ such that $\hat{K}_\tau(\rho) = U\rho U^*$*

Observe that the channel $\hat{K}_\tau$ is in general nonlinear.

Let us make some remarks on the physical meaning of the channels K_τ and $\hat{K}_\tau$. The channels K_τ are mixtures of linear channels of the type

$$K^h(\rho) := \mathcal{O}_h \rho \mathcal{O}_h^* \qquad (\rho \in) \tag{19}$$

with $h \in L_2(G)$, $\|h\| = 1$. Let us consider the more general case

$$\|h\| > 0, \ |h(k)| \leq 1 \qquad (k \in G). \tag{20}$$

We define an operator $t_h : L_2(G) \longrightarrow L_2(\{1, 2\} \times G)$ by setting for all $f \in L_2(G)$ and $k \in G$

$$(t_h f)(l, k) = \begin{cases} h(k)f(k) & \text{for } l = 1 \\ \sqrt{1 - |h(k)|^2}f(k) & \text{for } l = 2. \end{cases}$$

The operator t_h is an isometry from $L_2(G)$ to $L_2(\{1,2\} \times G) \cong L_2(\{1,2\}) \otimes L_2(G)$. Indeed,

$$\|t_h\, f\|^2 = \sum_{l=1}^{2} \sum_{k=1}^{n} |t_h\, f(l,k)|^2$$
$$= \sum_{k=1}^{n} \left(|h(k)|^2 + 1 - |h(k)|^2 \right) |f(k)|^2 = \|f\|^2. \qquad (21)$$

Consequently, the mapping $E_h : \mathcal{B}(L_2(\{1,2\} \times G)) \longrightarrow \mathcal{B}(L_2(G))$ given by

$$E_h(B) := t_h^* B t_h$$

is completely positive and identity preserving. The channel $E_h^*(\rho) = t_h \rho t_h^*$ is the corresponding linear channel from the set of states on $L_2(G)$ into the set of states on $L_2(\{1,2\} \times G)$. The space $L_2(\{1,2\} \times G)$ has an orthogonal decomposition into $L_2(\{1\} \times G)$ and $L_2(\{2\} \times G)$ both being trivially isomorphic to $L_2(G)$. Performing a measurement according to the projection onto $L_2(\{1\} \times G) \cong L_2(G)$ given the state $E_h^*(\rho)$ one obtains the state $\hat{K}^h(\rho)$. A measurement according to the projection onto $L_2(\{2\} \times G) \cong L_2(G)$ leads to the state $\hat{K}^{\sqrt{1-|h|^2}}(\rho)$.

6.3. *The state of the memory after recognition*

Let us recall that for states ρ, γ on $L_2(G)$ and $i, j \in G$

$$(F_{i,j} \otimes 1)(\rho \otimes \mathbf{e}(\gamma))(F_{i,j} \otimes 1)$$

is a linear operator from $L_2(G^3)$ into $L_2(G^2)$, and that (cf. (12)) it is equal

$$(F_{i,j} \otimes 1)(1 \otimes J)(\rho \otimes \gamma)(1 \otimes J^*)(F_{i,j} \otimes 1).$$

In the following we consider the family of channels $(\Lambda_{i,j})_{i,j \in G}$ from the set of product states $\rho \otimes \gamma$ on $\mathcal{H}_1 \otimes \mathcal{H}_2$ into the states on $\mathcal{H}_3$ given by

$$\Lambda_{i,j}(\rho \otimes \gamma) := \frac{\mathrm{Tr}_{1,2}(F_{i,j} \otimes 1)(\rho \otimes \mathbf{e}(\gamma))(F_{i,j} \otimes 1)}{\mathrm{Tr}_{1,2,3}(F_{i,j} \otimes 1)(\rho \otimes \mathbf{e}(\gamma))(F_{i,j} \otimes 1)} \qquad (22)$$

where $\mathrm{Tr}_{1,2}$ resp. $\mathrm{Tr}_{1,2,3}$ denotes the partial trace with respect to the first two components resp. the full trace with respect to all three spaces. In the sequel we always will assume that

$$\mathrm{Tr}_{1,2,3}(F_{i,j} \otimes 1)(\rho \otimes \mathbf{e}(\gamma))(F_{i,j} \otimes 1) > 0. \qquad (23)$$

Let ρ and γ are given as in Proposition 1. Since $(\xi_{i,j})_{i,j\in G}$ is an orthonormal basis in $L_2(G^2)$ we get from Proposition 1

$$\mathrm{Tr}_{1,2}(F_{i,j}\otimes 1)(\rho\otimes e(\gamma))(F_{i,j}\otimes 1) = \sum_{k,l=1}^{n} \alpha_k\beta_l\langle G_{i,j}g_k\otimes h_l,\cdot\rangle\, G_{i,j}g_k\otimes h_l \quad (24)$$

Summarizing, we get the following representation of $\Lambda_{i,j}$:

Proposition 2: [8] *Let ρ and γ be given as in Proposition 1. Further, assume (23). Then*

$$\Lambda_{i,j}(\rho\otimes\gamma) = \frac{\sum_{k,l=1}^{n}\alpha_k\beta_l\langle G_{i,j}g_k\otimes h_l,\cdot\rangle\, G_{i,j}g_k\otimes h_l}{\sum_{k,l=1}^{n}\alpha_k\beta_l\|G_{i,j}g_k\otimes h_l\|^2} \quad (25)$$

where for $\Phi\in\mathcal{L}_2(G^2)$

$$\|G_{i,j}\Phi\|^2 = \sum_{m=1}^{n}|b_i|^2(m\oplus j)|\Phi(m\oplus j,m)|^2. \quad (26)$$

Fortunately, we can find expressions for the state $\Lambda_{i,j}(\rho\otimes\gamma)$ of the memory after the recognition process being in many cases simpler. We can express the teleportation channel $\Lambda_{i,j}$ with the help of the channels K_τ we introduced in the previous subsection.

Proposition 3: [8] *Let $i,j\in G$ and let ρ be a state from $\mathcal{S}_{|\overline{b_i}\rangle<\overline{b_i}|}$ (cf. (16) and Definition 6). Further, let γ be a state from $\mathcal{S}$ such that*

$$U_j K_{|\overline{b_i}\rangle<\overline{b_i}|}(\rho)U_j^* \in \mathcal{S}_\gamma. \quad (27)$$

Then

$$\Lambda_{i,j}(\rho\otimes\gamma) = \hat{K}_\gamma \circ K^j \circ \hat{K}_{|\overline{b_i}\rangle<\overline{b_i}|}(\rho) \quad (28)$$

where K^j denotes the unitary channel given by $K^j(\rho) = U_j\rho U_j^*$.

Remark 13: All proofs of this paper can be seen in Ref.[8]

Concluding remarks: We touched the problem of finding simplified models for the recognition process. We were interested in how the input signal arriving at the brain is entangled (connected) to the memory already stored and the consciousness that existed in the brain, and how a part of the signal will be finally stored as a memory. It is clear that this simple model is just for illustration and can not serve for describing realistic aspects of recognition. Choosing a more complex basis one obtains expressions depending heavily on the states ρ and γ. Though the above presented model is only a first attempt it shows that there are possibilities to model the process of recognition. To get closer to realistic models we will try to refine the above models by

- passing over to infinite Hilbert spaces,
- replacing pure states by coherent states on the Fock space,
- making more complex measurements than simple one-dimensional projections $F_{i,j}$,
- replacing the trivial entanglement J by a more complex one based on beam splitting procedures, and finally
- examing whether some symmetry breaking as in Ref.[4] will occur in the process of recognition and storing memory.

References

1. L. Accardi and M. Ohya, Compound channels, transition expectations, and liftings, *Appl. Math. Optim.* **39**, 33–59 (1999).
2. L. Accardi and M. Ohya, Teleportation of general quantum states, *quant-ph/9912087* (1999).
3. C.H. Bennett, G. Brassard, C. Crepeau, R. Jozsa, A. Peres and W.K. Wootters, Teleporting an unknown quantum state *via* Dual Classical and Einstein-Podolsky-Rosen channels, *Phys. Rev. Lett.* **70**, 1895–1899 (1993).
4. H. Ezawa and J.A. Swieca, Spontaneous breakdown of symmetries and zero-mass states. *Commun. Math. Phys.* **5**, 330–336 (1967).
5. H. Ezawa, Thermo field dynamics of heat conduction. Progress in quantum field theory, North-Holland, Amsterdam, 305–324 (1986).
6. K-H. Fichtner and W. Freudenberg, Characterization of states of infinite Boson systems I.- On the construction of states, *Commun. Math. Phys.* **137**, 315–357 (1991).
7. K-H. Fichtner, W. Freudenberg and V. Liebscher, Time evolution and invariance of Boson systems given by beam splittings, *Infinite Dimensional Analysis, Quantum Probability and Related Topics* **1** no. 4, 511–533 (1998).
8. K-H. Fichtner, W. Freudenberg and M. Ohya, Recognition and Teleportation, to be published.
9. K-H. Fichtner, W. Freudenberg and M. Ohya, On Functions of Brain, in preparation.
10. K-H. Fichtner and M. Ohya, Quantum teleportation with entangled states given by beam splittings, *Commun. Math. Phys.* **222**, 229–247 (2001).
11. K-H. Fichtner and M. Ohya, Quantum Teleportation and Beam Splitting, *Commun. Math. Phys.* **225**, 67–89 (2002).
12. R.S. Ingarden, A. Kossakowski and M. Ohya, Information Dynamics and Open Systems, Kluwer Academic Publishers, 1997.
13. K. Inoue, A. Kossakowski and M. Ohya, A treatment of quantum chaos, to be submitted.

14. A. Kossakowski and M. Ohya, Y. Togawa, How can we observe and measure chaos?, to be published.

15. K. Inoue, M. Ohya and K. Sato, Application od chaos degree to some dynamical systems, Chaos, Solitons & Fractals. 11, 1377–1385, 2000.

16. K. Inoue, M. Ohya and I.V. Volovich, Semiclassical properties and chaos degree for the quantum baker's map, *J. Math. Phys.*, 43–2, 734–755, 2002.

17. M. Ohya, On compound state and mutual information in quantum information theory, IEEE Transaction of Information Theory, 29, 770–777, 1983.

18. M. Ohya, Some aspects of quantum information theory and their applications to irreversible processes, Reports on Mathematical Physics, 27, 19–47, 1989.

19. M. Ohya, Information dynamics and its application to optical communication processes, Springer Lecture Notes in Physics, 378, 81–92, 1991.

20. M. Ohya, Entropy transmission in C*-dynamical systems, *J. Math. Anal. Appl.*, 100, 222–235, 1984.

21. M. Ohya, Complexity and fractal dimensions for quantum states, Open Systems and Information Dynamics, 4, 141–157, 1997.

22. M. Ohya, Complexities and their applications to characterization of chaos, International Journal of Theoretical Physics, 37, No.1, 495–505 (1998).

23. M. Ohya, Complexity in dynamics and computation, *Acta App. Math.* **63**, 293–306 (2000).

24. M. Ohya, Complexity in quantum system and its application to brain function, *Quantum Information II* (ed. T. Hida and K. Saito, published by World Scientific), 149–160 (2000).

25. M. Ohya and D. Petz, Quantum Entropy and Its Use, Springer-Verlag, 1993.

26. M. Ohya, I.V. Volovich,*Quantum Computers, Teleportations and Cryptography*, Springer-Verlag, to be published.

Amenability for Weighted Hopf C^*-Algebras

Yoshiomi Nakagami

Department of Mathematical and Physical Science, Faculty of Science,
Japan Women's University,
8-1, Mejirodai 2 chome, Bunkyo-ku, Tokyo, 112-8681 Japan.
E-mail: nakagami@fc.jwu.ac.jp

The concept of amenability for locally compact groups is generalized to weighted Hopf C^*-algebras which are interpreted to be locally compact quantum groups.

1. Introduction

Let A be a C^*-algebra endowed with a coassociative coproduct δ. The pair (A, δ) is said to be a compact quantum group[8] if

(1) A is unital and
(2) $\delta(A)(\mathbb{C}1 \otimes A)$ and $(A \otimes \mathbb{C}1) \otimes A$ are linearly dense in $A \otimes_{\min} A$.

It is known by Refs.[7,8] that a compact quantum group has a Haar state, *i.e.* a state h on A with

$$(h \otimes \varphi)(\delta(a)) = h(a)\varphi(1) = (\varphi \otimes h)(\delta(a))$$

for $a \in A$ and $\varphi \in A^*$. A locally compact version of this category is not yet successfully defined in full generality. The first trial for this direction was performed by Ref.[4] assuming the existence of a Haar weight, which is named a Woronowicz algebra. This category is interpreted as a modification of Kac algebra by introducing a scaling automorphism groups coming from a deformation parameter q of the concrete quantum groups $SU_q(2)$ and *etc.* As a second stage we generalized this category to C^*-algebraic framework and named it a wighted Hopf C^*-algebra. An analogous approach has been done by Ref.[3], which was named a locally compact quantum group. These

"

two objects assume the existence of Haar weights and define essentially the same category.

2. Terminology and definition

Throughout the paper we use the minimal tensor product for C^*-algebras.

Definition 1: *A pair (A, δ) where A is a C^*-algebra and $\delta \in \mathrm{Mor}(A, A \otimes A)$ satisfying the coassociativity condition $(\delta \otimes \mathrm{id}_A) \circ \delta = (\mathrm{id}_A \otimes \delta) \circ \delta$ is called a C^*-bialgebra.*

The mapping δ will be called a coproduct. In what follows we denote by $1 = 1_A$ the unit of the multiplier algebra $M(A)$. The relation $\delta \in \mathrm{Mor}(A, A \otimes A)$ means that $\delta(a) \in M(A \otimes A)$ and the linear span of $\delta(A)(A \otimes A)$ is dense in $A \otimes A$.

Definition 2: *A C^*-bialgebra (A, δ) is said to be* proper *if $\delta(a)(1 \otimes b)$ and $(b \otimes 1)\delta(a)$ belong to $A \otimes A$ for any $a, b \in A$.*

Definition 3: *Let (A, δ) be a proper C^*-bialgebra. We say that (A, δ) has the* cancellation property *if the linear spans of the sets $\{\delta(a)(1 \otimes b) : a, b \in A\}$ and $\{(b \otimes 1)\delta(a) : a, b \in A\}$ are dense in $A \otimes A$.*

Let (A, δ) be a proper C^*-bialgebra. Then for any continuous linear functionals $\varphi, \psi \in A^*$ and any $a \in A$, the convolution products

$$\varphi * a := (\mathrm{id} \otimes \varphi)\delta(a), \quad a * \psi := (\psi \otimes \mathrm{id})\delta(a)$$

belong to A. We shall also use the convolution product of functionals:

$$\varphi * \psi := (\varphi \otimes \psi) \circ \delta \in A^*.$$

Throughout the paper, the concept of a right invariant weight on a C^*-bialgebra plays an essential role. In brief, h is right invariant if $(h \otimes \mathrm{id})\delta(a) = h(a)\,1$.

Definition 4: *Let (A, δ) be a C^*-bialgebra and h be a locally finite, lower semicontinuous weight on A. We say that h is right invariant if*

$$h(\varphi * a) = \varphi(1)h(a)$$

for any $a \in A_+$ with $h(a) < \infty$ and for any $\varphi \in A^*_+$.

The object which we study in this paper is defined as follows.

Definition 5: *Let (A, δ) with A separable be a proper C^*-bialgebra with the cancellation property. We say that (A, δ) is a* weighted Hopf C^*-algebra *if there exist a closed densely defined linear map κ acting on A (called* antipode*) and a locally finite strictly faithful lower semicontinuous right invariant weight h on A (called* Haar weight*) such that*

(i) *The operator κ admits the following polar decomposition:*

$$\kappa = R \circ \tau_{i/2}$$

where $\tau_{i/2}$ is the analytic generator of a one parameter group $\tau = \{\tau_t\}_{t \in \mathbb{R}}$ of automorphisms of the C^*-algebra A (called scaling group*) and R is an involutive antiautomorphism of A (called* unitary antipode*) commuting with automorphisms τ_t for all $t \in \mathbb{R}$.*

(ii) *The weight h is relatively invariant under $\{\tau_t\}_{t \in \mathbb{R}}$: there exists a positive scalar λ such that $h \circ \tau_t = \lambda^t h$ for all $t \in \mathbb{R}$.*

(iii) *(Strong right invariance) For any $\varphi \in A^*$ with $\varphi \circ \kappa \in A^*$ we have*

$$h((\varphi * a^*)b) = h(a^*((\varphi \circ \kappa) * b)))$$

for all a, $b \in \mathfrak{n}_h$.

A compact quantum group is a weighted Hopf C^*-algebra. In this case the Haar weight turns out to be a Haar state.

Recall that R is an involutive antiautomorphism of A commuting with τ_t we conclude that

$$\kappa(ab) = \kappa(b)\kappa(a), \quad \kappa(\kappa(a)^*)^* = a$$

for any $a, b \in \mathcal{D}(\kappa)$. In other words the mapping $a \mapsto \kappa(a)^*$ is a conjugate linear multiplicative involution acting on $\mathcal{D}(\kappa)$.

Let $\{\pi_h, \mathcal{H}_h, \eta_h\}$ be the GNS construction of the Haar weight h. The mapping $\eta = \eta_h$ is called a GNS-mapping of A to $\mathcal{H} = \mathcal{H}_h$. The left ideal $\{a \in A : h(a^*a) < \infty\}$ is denoted by $\mathcal{D}(\eta)$. In what follows we identify A with $\pi_h(A)$. The unitary W corresponding to the right regular representation defined by

$$W(\eta(a) \otimes \eta(b)) = (\eta \otimes \eta)(\delta(a)(1 \otimes b))$$

is called the Kac–Takesaki operator. It satisfies the pentagonal equation. We often use the operators

$$W(\xi, \xi') = (\omega_{\xi,\xi'} \otimes \mathrm{id})(W), \quad W_\varphi = (\mathrm{id} \otimes \varphi)(W)$$

for $\xi, \xi' \in \mathcal{H}$ and $\varphi \in A^*$. Then the set $\{W(\xi, \xi') : \xi, \xi' \in \mathcal{H}\}$ of matrix elements of W is linearly dense in A and the dual C^*-algebra $\widehat{A}$ is defined to be the closure of the set $\{W_\varphi : \varphi \in A_*\}$.

Let $\widehat{W} = \sigma(W)$, where σ is the flip. Then $\widehat{W}$ satisfies the pentagonal equation. The dual weighted Hopf C^*-algebra $(\widehat{A}, \widehat{\delta})$ is defined by

$$\widehat{\delta}(b) = \widehat{W}(b \otimes 1)\widehat{W}^*, \quad \widehat{R}(b) = Jb^*J, \quad \widehat{\tau}_t(b) = Q^{2it}bQ^{-2it}$$

for $b \in \widehat{A}$, where J is the modular conjugation operator for h and Q is the implementing positive self-adjoint operator of the scaling automorphism group : $\tau_t(a) = Q^{2it}aQ^{-2it}$ for $a \in A$. For the definition of the dual Haar weight we refer to Ref.[5] We often use a formula

$$(\widehat{\eta}(W_\varphi^*)|\eta(a)) = \varphi(a), \quad a \in \mathcal{D}(\eta)$$

for any L^2-bounded $\varphi \in A^*$, where $\widehat{\eta}$ is the GNS-mapping for the dual Haar weight.

3. Amenability

The coproduct δ, the unitary antipode R and scaling group $\{\tau_t\}_{t \in \mathbb{R}}$ on a weighted Hopf C^*-algebra (A, δ) are uniquely extended to the multiplier algebra $M(A)$ (or the von Neumann algebra A''). We use the same symbols:

$$\delta(a) = W(a \otimes 1)W^*, \quad R(a) = \widehat{J}a^*\widehat{J}, \quad \tau_t(a) = Q^{2it}aQ^{-2it}$$

for $a \in M(A)$ (or $\in A''$) and $t \in \mathbb{R}$, where $\widehat{J}$ is the modular conjugation for the dual Haar weight.

For each C^*-subalgebra B of $\mathcal{L}(\mathcal{H})$, we denote the set $\{\omega|_B : \omega \in \mathcal{L}(\mathcal{H})_*\}$ by B_*. A linear functional φ of A is said to be L^2-bounded if there exists an element $z_\varphi \in \mathcal{H}$ with $\varphi(a) = (z_\varphi|\eta(a))$ for any $a \in \mathcal{D}(\eta)$. It is known that the set $\{\varphi \in A^* : L^2\text{-bounded}\}$ is norm dense in A_*.

For any element φ in $M(A)_*$ the extension of $\varphi|_A$ to $M(A)$ coincides with the functional φ to start with. Each linear functional in A_* has a unique extension to $M(A)$. We sometimes use the same symbol for the extension.

Definition 6: *A weighted Hopf C^*-algebra (A, δ) is said to be amenable if there exists a right (resp. left) invariant mean, i.e. there exists a state m on $M(A)$ such that*

$$a * m = m(a)1 \quad (resp. \ m * a = m(a)1), \quad a \in M(A).$$

Hence $(m * \varphi)(a) = \varphi(1)m(a)$ (or $(\varphi * m)(a) = \varphi(1)m(a)$) for $\varphi \in A_*$ or $\varphi \in M(A)^*$.

Let m and m' be a right and a left invariant means on (A, δ). Then

$$m(a) = \|m'\|m(a) = m * m'(a) = \|m\|m'(a) = m'(a).$$

for $a \in M(A)$.

For a weighted Hopf C^*-algebra (A, δ), it is known that $\sigma \circ (R \otimes R) \circ \delta = \delta \circ R$. Hence, if $m \in M(A)^*$ is a right invarinat state, then $m \circ R$ is a left invarinat state and vice versa. Thus they satisfy $m = m \circ R$.

4. Compact quantum groups

In this section we will discuss several equivalent conditions for weighted Hopf C^*-algebras to be compact.

Theorem 1: *Let (A, δ) be a weighted Hopf C^*-bialgebra. Then the following six conditions are equivalent:*

(i) *The C^*-algebra A is unital, i.e. (A, δ) is a compact quantum group.*
(ii) *There exists a right (or left) invarinat mean m on (A, δ) with $m \in M(A)_*$ and $m|_A$ L^2-bounded.*
(iii) *There exists a unit vector $\xi \in \mathcal{H}$ such that*

$$W_\varphi \xi = \varphi(1)\xi, \quad \varphi \in A_*.$$

(i.e. the regular representation : $\varphi \mapsto W_\varphi$ of the involutive convolution algebra $A_0^* = \{\varphi \in A_* : \varphi^* \circ \kappa \in A_*\}$ has a one dimensional trivial subrepresentation : $\varphi \in A_0^* \mapsto \varphi(1) \in \mathbb{C}$, where the involution is given by $\varphi \to \varphi^* \circ \kappa$)
(iv) *There exists a unit vector $\xi \in \mathcal{H}$ such that $W(\xi, \zeta) = (\xi|\zeta)1$ for all $\zeta \in \mathcal{H}$.*
(v) *There exists a unit vector $\xi \in \mathcal{H}$ such that $W(\xi, \xi)$ is the identity of A.*
(vi) *There exists a bounded counit on the dual $(\widehat{A}, \widehat{\delta})$ belonging to $\widehat{A}_*$.*

Proof. (i)$\Rightarrow$(ii) If A is unital, then $M(A) = A$. By a Theorem of Van Daele or Woronowicz, there exists a right (or left) invariant state m on A :

$$a * m = m(a)1 \quad (\text{or } m * a = m(a)1), \quad a \in A.$$

Since m is the Haar state on (A, δ), for any $a \in A$ we have

$$|m(a)|^2 \leq m(1)m(a^*a) = \|m\|\|\eta(a)\|^2.$$

Hence m is L^2-bounded and $m \in A_*$.

(ii)$\Rightarrow$(iii) Let m be a right invariant state on $M(A)$ with $m \in M(A)_*$ and $m|_A$ L^2-bounded. The right invariance implies that $m * \varphi = \|\varphi\| m$ for $\varphi \in A_*^+$. Since $m|_A$ is L^2-bounded on A, m is considered to be an L^2-bounded element in A_*. Hence m is the Haar state h. Put $\xi = \widehat{\eta}(W_m^*)$. Then for any $\varphi \in A_+^*$ the element $m * \varphi$ is L^2-bounded. Indeed, we see that

$$|(m * \varphi)(a)| = |m(\varphi * a)| = |(\widehat{\eta}(W_m^*)|\eta(\varphi * a))|$$
$$= |(\widehat{\eta}(W_m^*)|W_\varphi \eta(a))| \leq \|\varphi\|\|\widehat{\eta}(W_m^*)\|\|\eta(a)\|.$$

Thus $m * \varphi$ is L^2-bounded. Hence we have

$$W_\varphi^* \xi = \widehat{\eta}(W_{m*\varphi}^*) = \|\varphi\|\widehat{\eta}(W_m^*) = \|\varphi\|\xi = \overline{\varphi(1)}\xi.$$

Since any element $\varphi \in A_*$ is a linear combination of four elements in A_*^+, the above equality holds for any $\varphi \in A_*$. Hence we have $W_\varphi \xi = \varphi(1)\xi$ for $\varphi \in A_*$.

(iii)$\Rightarrow$(iv) For any $\zeta \in \mathcal{H}$ and $\varphi \in A_*^+$ we have

$$\|\varphi\|(\xi|\zeta) = (W_\varphi^* \xi|\zeta) = (\xi|W_\varphi \zeta) = \varphi(W(\xi,\zeta)).$$

Hence

$$\varphi((\xi|\zeta)1) = \varphi(W(\xi,\zeta))$$

for all $\varphi \in A_*$. Thus we get assertion (iv).

(iv)$\Rightarrow$(v)$\Rightarrow$(i) It is clear.

(v)$\Leftrightarrow$(vi) By the pentagonal equation we have $\widehat{W}_{13}^* \widehat{W}_{12}^* = (\mathrm{id} \otimes \widehat{\delta})(\widehat{W}^*)$. Let $\varphi \in A^*$. Applying $\varphi \otimes \mathrm{id} \otimes \omega_\xi$ or $\varphi \otimes \omega_\xi \otimes \mathrm{id}$ to the above equality, we have formulae

$$(\mathrm{id} \otimes \varphi)((1 \otimes W(\xi,\xi))W) = (\mathrm{id} \otimes \omega_\xi)(\widehat{\delta}(W_\varphi)) \tag{1}$$

$$(\mathrm{id} \otimes \varphi)(W(1 \otimes W(\xi,\xi))) = (\omega_\xi \otimes \mathrm{id})(\widehat{\delta}(W_\varphi)). \tag{2}$$

Since the mappings $(\mathrm{id}\otimes\omega_\xi)\circ\widehat{\delta}$ and $(\omega_\xi\otimes\mathrm{id})\circ\widehat{\delta}$ are continuous, if $W(\xi,\xi) = 1$, then ω_ξ is a counit of $(\widehat{A},\widehat{\delta})$ belonging to $\widehat{A}_*$. Conversely, since $W_\varphi = (\mathrm{id}\otimes\varphi)(W)$, if ω_ξ is a counit of $(\widehat{A},\widehat{\delta})$, then $W(1\otimes W(\xi,\xi)) = W$ and hence $W(\xi,\xi) = 1$. ∎

Remark:

(1) In the above proof (ii) $\Rightarrow$ (iii), we can show that $\|\widehat{\eta}(W_m^*)\| = 1$. Indeed, in this case m is the Haar state h,

$$(\widehat{\eta}(W_m^*)|\eta(a)) = m(a) = h(a) = (\eta(1)|a\eta(1)) = (\eta(1)|\eta(a))$$

for any $a \in A$, and hence $\widehat{\eta}(W_m^*) = \eta(1)$. Thus $\|\widehat{\eta}(W_m^*)\| = \|\eta(1)\| = 1$.

(2) The counit ω_ξ on the dual $(\widehat{A}, \widehat{\delta})$ in assertion (vi) satisfies $\omega_\xi(W_\varphi) = \varphi(1)$ for $\varphi \in A_*$.

(3) For a noncompact (A, δ), we see that $m = 0$ on A. Indeed, $m = 0$ on $\mathcal{D}(\eta)^* \mathcal{D}(\eta)$ by the right invariance of the Haar weight h; hence $m(a) = 0$ and $m * a = 0$ by the local finiteness of h.

Corollary 1: *Let (A, δ) be a compact weighted Hopf C^*-algebra, h the Haar state and $\widehat{\varepsilon}$ the bounded counit on the dual $(\widehat{A}, \widehat{\delta})$. If $\xi = \widehat{\eta}(W_h^*)$, then $\xi = \eta(1)$ and*

$$h = \omega_\xi|_A, \quad \text{and} \quad \widehat{\varepsilon} = \omega_\xi|_{\widehat{A}}$$

5. Scaling invariance

Definition 7: *Let (A, δ) be a weighted Hopf C^*-algebra. A state on the multiplier algebra $M(A)$ is said to satisfy the* strong right invariance *if*

$$m((\varphi * a)b) = m(a(\varphi \circ \kappa)(b)), \quad a, \, b \in M(A)$$

for any $\varphi \in A_*$ with $\varphi \circ \kappa \in A_*$.

Lemma 1: *Let (A, δ) be a weighted Hopf C^*-algebra.*

(i) *If (A, δ) has an invariant mean, i.e. is amenable, then there exists an invariant mean invariant under the scaling group $\{\tau_t\}_{t \in \mathbb{R}}$.*

(ii) *If (A, δ) has a strong right invariant state h, then there exists a strong right invariant state invariant under the scaling group $\{\tau_t\}_{t \in \mathbb{R}}$.*

Proof. (i) Let m be an invariant mean of (A, δ). For each $t \in \mathbb{R}$ we denote $m \circ \tau_t$ by m_t. Put $f_a(t) = m_t(a)$ for $a \in M(A)$. Then $f_a \in C_b(\mathbb{R})$. Using an invariant mean M on $\mathbb{R}$, we define a state $\overline{m}$ on A by

$$\overline{m}(a) = M(f_a).$$

Then $\overline{m}$ is $\{\tau_t\}_{t \in \mathbb{R}}$ invariant and

$$m_t(\varphi * a) = (m \circ \tau_t \otimes \varphi)(\delta(a)) = (m \otimes \varphi \circ \tau_{-t})(\delta(\tau_t(a))$$
$$= (m * (\varphi \circ \tau_{-t}))(\tau_t(a)) = m(\tau_t(a)) = m_t(a)$$

and hence $f_{\varphi * a} = f_a$. Then $\overline{m}$ satisfies the right invariance :

$$(\overline{m} * \varphi)(a) = \overline{m}(\varphi * a) = M(f_{\varphi * a}) = M(f_a) = \overline{m}(a).$$

(ii) Since

$$\tau_t(\varphi * a) = \tau_t((\mathrm{id} \otimes \varphi)(\delta(a))) = (\mathrm{id} \otimes \varphi \circ \tau_{-t})(\delta(\tau_t(a)))$$
$$= (\varphi \circ \tau_{-t}) * (\tau_t(a))$$

and κ commutes with the scaling group, it follows that

$$m_t((\varphi * a)b) = m(\tau_t(\varphi * a)\tau_t(b)) = m(((\varphi \circ \tau_{-t}) * \tau_t(a))\tau_t(b))$$
$$= m(\tau_t(a)((\varphi \circ \tau_{-t}) * \kappa)(\tau_t(b)))$$
$$= m(\tau_t(a)\tau_t((\varphi \circ \kappa) * b)) = m_t(a((\varphi \circ \kappa) * b)).$$

Using the invariant mean M on $\mathbb{R}$, we find the desired result. ∎

6. Strong amenability

Definition 8: *A weighted Hopf C^*-algebra (A, δ) is said to be* strongly amenable *if there exists a bounded counit on the dual weighted Hopf C^*-algebra $(\widehat{A}, \widehat{\delta})$.*

Theorem 2: *Let (A, δ) be a weighted Hopf C^*-algebra. Then the following seven conditions are equivalent:*

(i) *There exists a net $\{\xi_i\}_{i \in I}$ of unit vectors in $\mathcal{H}$ such that*

$$\|W_\varphi \xi_i - \varphi(1)\xi_i\| \to 0$$

  for all $\varphi \in A_*$.

(ii) *There exists a net $\{\xi_i\}_{i \in I}$ of unit vectors in $\mathcal{H}$ such that for each $a \in A$ the both nets $\{aW(\xi_i, \xi_i)\}_{i \in I}$ and $\{W(\xi_i, \xi_i)a\}_{i \in I}$ in A converge to a in the $\sigma(A, A_*)$-topology.*

(iii) *There exists a net $\{\xi_i\}_{i \in I}$ of unit vectors in $\mathcal{H}$ such that the both nets $\{(\mathrm{id} \otimes \omega_{\xi_i}) \circ \widehat{\delta}\}_{i \in I}$ and $\{(\omega_{\xi_i} \otimes \mathrm{id}) \circ \widehat{\delta}\}_{i \in I}$ of completely positive mappings converge to the identity on $\widehat{A}$ in the pointwise norm topology, i.e. $\{\omega_{\xi_i}\}_{i \in I}$ is an approximate counit on $\widehat{A}$ with respect to the pointwise norm topology.*

(iv) *There exists a net $\{\xi_i\}_{i \in I}$ of unit vectors in $\mathcal{H}$ such that the net $\{W(\xi_i, \xi_i)\}_{i \in I}$ of matrix elements is an approximate identity of A.*

(v) *The regular representation: $\varphi \mapsto W_\varphi$ of the involutive convolution algebra $A_0^* = \{\varphi \in A_* : \varphi^* \circ \kappa \in A_*\}$ contains weakly the trivial representation: $\varphi \in A_0^* \mapsto \varphi(1) \in \mathbb{C}$.*

(vi) *(A, δ) is strongly amenable.*

(vii) *There exists a net $\{\xi_i\}_{i \in I}$ of unit vectors in $\mathcal{H}$ such that for any normal representation $\{\pi, \mathcal{H}_\pi\}$ of the C^*-algebra A*

$$\|(\mathrm{id} \otimes \pi)(W)(\xi_i \otimes \zeta) - \xi_i \otimes \zeta\| \to 0$$

  for all $\zeta \in \mathcal{H}_\pi$.

Proof. (i)$\Rightarrow$(ii) Let $\{\xi_i\}_{i\in I}$ be a net of unit vectors such that $\{W_\varphi \xi_i - \|\varphi\|\xi_i\}_{i\in I}$ converges in norm to 0 for $\varphi \in A_*$. Assume that φ is a state on A. For each $b \in A$

$$
\begin{aligned}
|\varphi(b^* W(\xi_i, \xi_i) b - b^* b)| &= |(b\varphi b^*)(W(\xi_i, \xi_i)) - \|b\varphi b^*\|(\xi_i|\xi_i)| \\
&= |(\xi_i | W_{b\varphi b^*} \xi_i - \|b\varphi b^*\|\xi_i)| \\
&\le \|W_{b\varphi b^*}\xi_i - \|b\varphi b^*\|\xi_i\| \to 0.
\end{aligned}
$$

Since a bounded linear functional is a linear combination of four states, the net $\{aW(\xi_i, \xi_i)b - ab\}_{i\in I}$ for $a,\ b \in A$ converges to 0 in $\sigma(A, A_*)$-topology by the polarization argument. Hence the Doran–Wichmann's factorization theorem tells us that the nets $\{W(\xi_i, \xi_i)a\}_{i\in I}$ and $\{aW(\xi_i, \xi_i)\}_{i\in I}$ converge to a in the $\sigma(A, A_*)$-topology.

(ii)$\Rightarrow$(iii) Let $\{\xi_j'\}_{j\in J}$ be a net of unit vectors such that $\{aW(\xi_j', \xi_j')\}_{j\in J}$ and $\{W(\xi_j', \xi_j')a\}_{j\in J}$ converges to a in the $\sigma(A, A_*)$-topology. By formulae (1) and (2) in the proof of Theorem 1, we see that

$$
(\mathrm{id} \otimes \varphi)((1 \otimes W(\xi_j', \xi_j'))W) = (\mathrm{id} \otimes \omega_{\xi_j'})(\widehat{\delta}(W_\varphi))
$$

$$
(\mathrm{id} \otimes \varphi)(W(1 \otimes W(\xi_j', \xi_j'))) = (\omega_{\xi_j'} \otimes \mathrm{id})(\widehat{\delta}(W_\varphi))
$$

for any $\varphi \in A^*$. For any $\omega \in \widehat{A}^*$ we see that

$$
\begin{aligned}
\omega &\left((\mathrm{id} \otimes \omega_{\xi_j'})(\widehat{\delta}(W_\varphi)) - W_\varphi \right) \\
&= (\omega \otimes \varphi)((1 \otimes W(\xi_j', \xi_j'))W - W) \\
&= \varphi(W(\xi_j', \xi_j')(\omega \otimes \mathrm{id})(W) - (\omega \otimes \mathrm{id})(W)). \qquad (3)
\end{aligned}
$$

Thus for each $\varphi \in A_*$ the right hand side converges to 0 by assertion (ii). Since the set $\{W_\varphi : \varphi \in A_*\}$ is norm dense in $\widehat{A}$ and the linear functional $b \in \widehat{A} \mapsto \omega((\mathrm{id} \otimes \omega_{\xi_j'})(\widehat{\delta}(b)) - b)$ is bounded, the net $\{(\mathrm{id} \otimes \omega_{\xi_j'})(\widehat{\delta}(b))\}_{j\in J}$ in $\widehat{A}$ for $b \in \widehat{A}$ converges to b in the $\sigma(\widehat{A}, \widehat{A}^*)$-topology. In the same way we can prove that the net $\{(\omega_{\xi_j'} \otimes \mathrm{id})(\widehat{\delta}(b))\}_{j\in J}$ for $b \in \widehat{A}$ converges weakly to b.

Let E be the product space of copies of the Banach space $(\widehat{A}, \|\ \|)$:

$$
E = \prod_{b\in \widehat{A}_1} (\widehat{A}, \|\ \|),
$$

where $\widehat{A}_1$ is the unit ball of $\widehat{A}$. Let ρ be a linear mapping of $\widehat{A}^*$ to E defined by $\rho(\omega) = (\omega * b)_{b\in\widehat{A}_1}$. Since the $\sigma(E, E^*)$-closure of the convex cone of $\{\rho(\omega_{\xi_j'}) : j \in J\}$ contains the element $(b)_{b\in\widehat{A}_1}$ in E, there exists a net $\{\omega_k'\}_{k\in K}$ in $S(\widehat{A}) \cap \widehat{A}_*$ of finite convex combinations of elements $\omega_{\xi_j'}$, $j \in J$ such that for each $b \in \widehat{A}_1$

$$\|(\mathrm{id} \otimes \omega_k') \circ \widehat{\delta}(b) - b\| \to 0.$$

In the same way we can choose a subnet $\{\omega_i\}_{i\in I}$ of $\{\omega_k'\}_{k\in K}$ satisfying

$$\|(\omega_i \otimes \mathrm{id}) \circ \widehat{\delta}(b) - b\| \to 0.$$

Hence there exists a net $\{\omega_i\}_{i\in I}$ in $S(\widehat{A}) \cap \widehat{A}_*$ such that the nets $\{(\omega_i \otimes \mathrm{id}) \circ \widehat{\delta}(b)\}_{i\in I}$ and $\{(\mathrm{id} \otimes \omega_i) \circ \widehat{\delta}(b)\}_{i\in I}$ converge in norm to b. Since each state ω_i on $\widehat{A}$ is normal, there exists a unit vector $\xi_i \in \mathcal{H}$ such that $\omega_i = \omega_{\xi_i}$. Thus we obtain assertion (iii).

(iii)$\Rightarrow$(iv) Let $\{\xi_j'\}_{j\in J}$ be a net of unit vectors such that the net $\{(\omega_{\xi_j'} \otimes \mathrm{id})(\widehat{\delta}(b))\}_{j\in J}$ and $\{(\mathrm{id} \otimes \omega_{\xi_j'})(\widehat{\delta}(b))\}_{j\in J}$ for $b \in \widehat{A}$ converge in norm to b. By formula (3) we have for any $\varphi \in A^*$ and $\omega \in \widehat{A}^*$

$$\varphi(W(\xi_j',\xi_j')(\omega \otimes \mathrm{id})(W) - (\omega \otimes \mathrm{id})(W))$$
$$= \omega\left((\mathrm{id} \otimes \omega_{\xi_j'}) \circ \widehat{\delta}(W_\varphi) - W_\varphi\right)$$

and

$$\varphi((\omega \otimes \mathrm{id})(W)W(\xi_j',\xi_j') - (\omega \otimes \mathrm{id})(W))$$
$$= \omega\left((\omega_{\xi_j'} \otimes \mathrm{id}) \circ \widehat{\delta}(W_\varphi) - W_\varphi\right).$$

The right hand sides converge to 0 by assertion (iii). Since the set $\{(\omega \otimes \mathrm{id})(W) : \omega \in \widehat{A}^*\}$ is linearly dense in A in norm, we can replace $(\omega \otimes \mathrm{id})(W)$ by any element a in A. Hence the nets $\{W(\xi_j',\xi_j')a\}_{j\in J}$ and $\{aW(\xi_j',\xi_j')\}_{j\in J}$ converge to a in the $\sigma(A, A^*)$-topology.

Let E be the product space of copies of the Banach space $(A, \| \ \|)$:

$$E = \prod_{a \in A_1} (A, \| \ \|),$$

where A_1 is the unit ball of A. Let ρ be a linear mapping of $\widehat{A}^*$ to E defined by $\rho(\omega) = ((\omega \otimes \mathrm{id})(W)a)_{a\in A_1}$. Since the $\sigma(E, E^*)$-closure of the convex cone of $\{\rho(\omega_{\xi_j'}) : j \in J\}$ contains the element $(a)_{a\in A_1}$ in E, there exists a net $\{\omega_k\}_{k\in K}$ in $S(\widehat{A}) \cap \widehat{A}_*$ of finite convex combinations of elements $\omega_{\xi_j'}$, $j \in J$ such that for each $a \in A_1$

$$\|(\omega_k \otimes \mathrm{id})(W)a - a\| \to 0.$$

In the same way we can choose a subnet $\{\omega_i\}_{i\in I}$ of $\{\omega_k\}_{k\in K}$ satisfying

$$\|a(\omega_i \otimes \mathrm{id})(W) - a\| \to 0.$$

Hence there exists a net $\{\omega_i\}_{i\in I}$ in $S(\widehat{A}) \cap \widehat{A}_*$ such that the net $\{(\omega_i \otimes \mathrm{id})(W)\}_{i\in I}$ converges strictly to the identity of $M(A)$. Since each state ω_i on $\widehat{A}$ is normal, it is of the form ω_{ξ_i} for some unit vector $\xi_i \in \mathcal{H}$. Thus $\{W(\xi_i,\xi_i)\}_{i\in I}$ is an approximate identity of A.

(iv)$\Rightarrow$(v) Let $\{\xi_i\}_{i\in I}$ be a net of unit vectors in $\mathcal{H}$ such that $\{W(\xi_i,\xi_i)\}_{i\in I}$ is an approximate identity of A. Then for any $\varphi \in A_*$ we find that

$$\omega_{\xi_i}(W_\varphi) = \varphi(W(\xi_i,\xi_i)) \to \varphi(1)$$

and hence that

$$|\varphi(1)| \leq \overline{\lim}_{i\in I}|\omega_{\xi_i}(W_\varphi)| \leq \|W_\varphi\|.$$

Therefore there exists a bounded linear mapping π of $\widehat{A}$ to $\mathbb{C}$ with

$$\pi(W_\varphi) = \lim_{i\in I}\omega_{\xi_i}(W_\varphi) = \varphi(1), \quad \varphi \in A_*. \tag{4}$$

It is easy to see that π is a homomorphism of $\widehat{A}$ to $\mathbb{C}$. Hence the regular representation : $\varphi \in A_0^* \mapsto W_\varphi \in \widehat{A}$ contains weakly the trivial representation $\varphi \in A_0^* \mapsto \varphi(1) \in \mathbb{C}$.

(iv)$\Rightarrow$(vi) In the above proof we can show that the net $\{\omega_{\xi_i}\}$ converges weakly* to π. Indeed, the set $\{W_\varphi : \varphi \in A_*\}$ is norm dense in $\widehat{A}$, we can replace W_φ by arbitrary element b in $\widehat{A}$ in the convergence (4). Since the net $\{W(\xi_i,\xi_i)\}$ is approximate identity of A by assertion (iv), we see that

$$(\pi \otimes \mathrm{id})(\widehat{\delta}(W_\varphi)) = \lim_i(\omega_{\xi_i} \otimes \mathrm{id})(\widehat{\delta}(W_\varphi)) = \lim_i(\mathrm{id} \otimes \varphi)(W(1 \otimes W(\xi_i,\xi_i)))$$

$$= \lim_i(\mathrm{id} \otimes W(\xi_i,\xi_i)\varphi)(W) = W_\varphi$$

by formula (2) and that

$$(\mathrm{id} \otimes \pi)(\widehat{\delta}(W_\varphi)) = W_\varphi.$$

Thus π is the counit of $(\widehat{A},\widehat{\delta})$.

((v) or (vi))$\Rightarrow$(vii) There exists a homomorphism π of $\widehat{A}$ to $\mathbb{C}$ with $\pi(W_\varphi) = \varphi(1)$ for $\varphi \in A_0^*$. Since π is interpreted to be a state on $\widehat{A}$, there exists a net $\{\omega_i\}_{i\in I}$ of states in $\widehat{A}_*$ converging weakly* to π. Since each ω_i is normal, it is of the form ω_{ξ_i} for some unit vector ξ_i. Thus for each $\varphi \in A_0^*$

$$\varphi(W(\xi_i,\xi_i)) = \omega_{\xi_i}(W_\varphi) \to \pi(W_\varphi) = \varphi(1).$$

Since A_0^* is norm dense in A_* and the net $\{W(\xi_i,\xi_i)\}_{i\in I}$ is bounded, φ can be replaced by any elements in A_*. Hence for any state $\varphi \in A_*$ we denote the GNS construction by $\{\pi_\varphi, \mathcal{H}_\varphi, \eta_\varphi\}$. Put $\zeta = \eta_\varphi(a)$ for $a \in \mathcal{D}(\eta_\varphi)$. Then we see that

$$\|(\mathrm{id} \otimes \pi_\varphi)(W)(\xi_i \otimes \zeta) - (\xi_i \otimes \zeta)\|^2$$
$$\leq 2|(\xi_i \otimes \zeta|(\mathrm{id} \otimes \pi_\varphi)(W)(\xi_i \otimes \zeta) - (\xi_i \otimes \zeta))|$$
$$= (a^*\varphi a)(W(\xi_i, \xi_i)) - (a^*\varphi a)(1)$$

converges to 0. Since $(\mathrm{id} \otimes \pi_\varphi)(W)$ is unitary, we can replace ζ by any vector in $\mathcal{H}_\varphi$. Thus we get assertion (vii).

(vii)$\Rightarrow$(i) Let $\varphi \in A_*^+$ and $\{\pi_\varphi, \mathcal{H}_\varphi\}$ the GNS representation. Then φ is of the form $\omega_\zeta \circ \pi_\varphi$ for some $\zeta \in \mathcal{H}_\varphi$. Since

$$(\xi|W_\varphi\xi_i - \varphi(1)\xi_i) = (\xi \otimes \zeta|(\mathrm{id} \otimes \pi_\varphi)(W)(\xi_i \otimes \zeta) - (\xi_i \otimes \zeta))$$

for any $\xi \in \mathcal{H}$, it follows that

$$\|W_\varphi\xi_i - \varphi(1)\xi_i\| \leq \|\zeta\|\|(\mathrm{id} \otimes \pi_\varphi)(W)(\xi_i \otimes \zeta) - (\xi_i \otimes \zeta)\|.$$

Hence $\|W_\varphi\xi_i - \varphi(1)\xi_i\| \to 0$. By polarization argument we see that this convergence holds for any $\varphi \in A_*$. ∎

Theorem 3: *Let (A, δ) be a weighted Hopf C^*-algebra. If it is strongly amenable, then assertion* (i) *holds. If assertion* (i) *holds, then assertion* (ii) *holds. If assertion* (ii) *holds, then assertion* (iii) *holds :*

(i) *There exists a state on $M(A)$ which satisfies the strong right invariance.*
(ii) *(A, δ) is amenable.*
(iii) *There exists a net $\{\psi_i\}_{i \in I}$ of L^2-bounded states on A such that*

$$\|\psi_i * \varphi - \psi_i\| \to 0$$

for all $\varphi \in S(A)$.

Proof. (Strong amenability)$\Rightarrow$(i) Let $\{\xi_i\}_{i \in I}$ be a net of unit vectors satisfying assertion (vii) in the previous theorem. For any state φ in A_* we denote the GNS-representation of $(M(A), \varphi)$ by $\{\pi_\varphi, \mathcal{H}_\varphi, \xi_\varphi\}$. For any $a, b \in M(A)$ we see that

$$\omega_{\xi_i}((\varphi * a)b) - \omega_{\xi_i}(aW_{\varphi^*}^* b)$$
$$= \omega_{\xi_i, b\xi_i}((\mathrm{id} \otimes \varphi)(\delta(a))) - \omega_{\xi_i}(a(\mathrm{id} \otimes \varphi)(W^*)b)$$
$$= (\xi_i \otimes \xi_\varphi|(\mathrm{id} \otimes \pi_\varphi)(W)(a \otimes 1)(\mathrm{id} \otimes \pi_\varphi)(W)^*(b\xi_i \otimes \xi_\varphi))$$
$$\quad - (a^*\xi_i \otimes \xi_\varphi|(\mathrm{id} \otimes \pi_\varphi)(W^*)(b\xi_i \otimes \xi_\varphi))$$
$$= ((\mathrm{id} \otimes \pi_\varphi)(W)^*(\xi_i \otimes \xi_\varphi) - (\xi_i \otimes \xi_\varphi)|(a \otimes 1)(\mathrm{id} \otimes \pi_\varphi)(W)^*(b\xi_i \otimes \xi_\varphi)).$$

Hence we have

$$|\omega_{\xi_i}((\varphi * a)b) - \omega_{\xi_i}(aW_{\varphi^*}^* b)|$$
$$\leq \|(\mathrm{id} \otimes \pi_\varphi)(W)^*(\xi_i \otimes \xi_\varphi) - \xi_i \otimes \xi_\varphi\| \|a\| \|b\|$$

and the net $\{\omega_{\xi_i}((\varphi * a)b) - \omega_{\xi_i}(aW_{\varphi^*}^* b)\}_{i \in I}$ converges to 0. Repeating the similar argument for a state ψ in A_*, we obtain

$$\omega_{\xi_i}(a(\psi * b)) - \omega_{\xi_i}(aW_\psi b)$$
$$= \omega_{a^* \xi_i, \xi_i}((\mathrm{id} \otimes \psi)(\delta(b))) - \omega_{a^* \xi_i, b\xi_i}(W_\psi)$$
$$= (a^* \xi_i \otimes \xi_\psi | (\mathrm{id} \otimes \pi_\psi)(W)(b \otimes 1)(\mathrm{id} \otimes \pi_\psi)(W)^*(\xi_i \otimes \xi_\psi))$$
$$- (a^* \xi_i \otimes \xi_\psi | (\mathrm{id} \otimes \pi_\psi)(W)(b \otimes 1)(\xi_i \otimes \xi_\psi)).$$

Hence we have

$$|\omega_{\xi_i}(a(\psi * b)) - \omega_{\xi_i}(aW_\psi b)|$$
$$\leq \|a\| \|b\| \|(\mathrm{id} \otimes \pi_\psi)(W)^*(\xi_i \otimes \xi_\psi) - \xi_i \otimes \xi_\psi\|.$$

Since a bounded linear functional on $M(A)$ is a linear combination of four states, the net $\{\omega_{\xi_i}(a(\psi * b)) - \omega_{\xi_i}(aW_\psi b)\}_{i \in I}$ converges to 0 for any functional $\psi \in A_*$. Since $W_\phi^* = W_{\phi^* \circ \kappa}$ for any stete ϕ with $\phi^* \circ \kappa \in A^*$, we find that the net $\{\omega_{\xi_i}((\varphi * a)b) - \omega_{\xi_i}(a((\varphi \circ \kappa) * b))\}_{i \in I}$ converges to 0 for $\varphi \in A_*$ with $\varphi \circ \kappa \in A_*$. Since the set $\{\omega_{\xi_i} : i \in I\}$ of states on $M(A)$ is precompact with respect to the weak* topology, there exists a subnet converging to a state m on $M(A)$. From the above estimate we find that $m((\varphi * a)b) = m(a((\varphi \circ \kappa) * b))$ for all state $\varphi \in A_*$ with $\varphi \circ \kappa \in A_*$ and $a, b \in M(A)$. Since the both sides of this equality is linear, we can replace φ by any functoinal in A_* with $\varphi \circ \kappa \in A_*$.

(i)$\Rightarrow$(ii) Let $b = 1$. Then $(\varphi \circ \kappa)(b) = \varphi(1)$ for any $\varphi \in A_*$ with $\varphi \circ \kappa \in A_*$. Hence we obtain the right invarance of the state m on $M(A)$.

(ii)$\Rightarrow$(iii) Let m be an invariant mean of (A, δ). Let ψ be the restriction of $\omega_{\xi, \eta'(b)}$ to A. Then ψ is an L^2-bounded element in A^* and $\widehat{\eta}(W_\psi^*) = b^* \xi$. Hence the set $\{\psi \in A^* : L^2\text{-bounded}\}$ is weakly* dense in A^*. Then there exists a net $\{\psi_i\}_{i \in I}$ of L^2-bounded elements in $S(A)$ such that $\{\psi_i\}_{i \in I}$ extended to $M(A)$ converges weakly* to m. For any L^2-bounded $\psi \in A^*$ we see that $\psi * \varphi$ is L^2-bounded by the same argument as in the proof (ii)$\Rightarrow$(iii) of Theorem 1. Then for any state φ in A^* the net

$$\psi_i * \varphi - \psi_i = (\psi_i - m) * \varphi + (m - \psi_i)$$

converges weakly* to 0 on $M(A)$. Hence for each $i \in I$ the functional $\psi_i * \varphi - \psi_i$ is L^2-bounded for $\varphi \in S(A)$ and hence belongs to A_*. The net $\{\psi_i * \varphi - \psi_i\}_{i \in I}$ converges weakly* to 0 as well.

Let E be the product space of copies of the Banach space $(A_*, \| \ \|)$:

$$E = \prod_{\varphi \in S(A)} (A_*, \| \ \|).$$

Then E is a locally convex space. The (topological) dual space E^* is identified with

$$\coprod_{\varphi \in S(A)} (A_*, \| \ \|)^* = \coprod_{\varphi \in S(A)} (A'', \sigma\text{-weak topology})$$

and the pairing is given by

$$\langle (\omega_\varphi)_{\varphi \in S(A)}, (a_\varphi)_{\varphi \in S(A)} \rangle = \sum_{\varphi \in S(A)} \omega_\varphi(a_\varphi)$$

for $(\omega_\varphi)_{\varphi \in S(A)} \in E$ and $(a_\varphi)_{\varphi \in S(A)} \in E^*$. Notice that the right hand side is finite sum. Hence the $\sigma(E, E^*)$-topology on E is identified with the product topology

$$\prod_{\varphi \in S(A)} (A_*, \text{weak topology}).$$

Let ρ be the linear mapping from $\{\psi \in A^* : L^2\text{-bounded}\}$ to E defined by

$$\rho(\psi) = (\psi * \varphi - \psi)_{\varphi \in S(A)}.$$

Then the $\sigma(E, E^*)$-closure of the range of $\{\psi \in S(A) : L^2\text{-bounded}\}$ by ρ contains the zero element $(0)_{\varphi \in S(A)} \in E$. Since the range is convex in E, the Hahn-Banach's separation theorem tells us that the closure of the range in E with respect to the product topology in E contains the zero element. Hence there exists a net $\{\phi_j\}_{j \in J}$ of L^2-bounded states on A such that

$$\|\phi_j * \varphi - \phi_j\| \to 0$$

for each $\varphi \in S(A)$. ∎

Remark: In Theorem 3 it is not known whether assertion (ii) implies assertion (i).

7. Discrete weighted Hopf C^*-algebras

Definition 9: *A weighted Hopf C^*-algebra (A, δ) is said to be* discrete *if the dual weighted Hopf C^*-algebra $(\widehat{A}, \widehat{\delta})$ is compact.*

For any discrete locally compact group the following assertion holds. It is open problem whether it holds for any weighted Hopf C^*-algebras.

Problem: Let (A, δ) be a weighted Hopf C^*-algebra. If (A, δ) is discrete, then the following three conditions are equivalent :

(1) (A, δ) is amenable.

(2) (A, δ) is strongly amenable.

(3) The C^*-algebra $\widehat{A}$ is nuclear.

By private communication,[6] Tomatsu proved that the amenability and the strong amenability is equivalent for a discrete weighted Hopf C^*-algebra. Further he proved that the above condition holds if and only if $\widehat{A}$ is nuclear and it has a character as well.

Finally the author would like to thank Tomatsu for his critical reading of the first draft of the manuscript and for indicating him some references.

References

1. Enock, M. and J-M. Schwartz : *Algebres de Kac moyennables*, Pacific J. Math., **125** (1986), 363–379.
2. Greenleaf, F. P. : Invarinat means on topological groups and their applications, Van Nostrand, 1969.
3. Kustermans, J. and S. Vaes : *Locally compact quantum groups*, Ann. Scient. Ec. Norm. Sup., (2000).
4. Masuda, T. and Y. Nakagami, *A von Neumann algebra framework for the duality of the quantum groups,*, Publ. RIMS Kyoto Univ., **30**(1994), 799–850.
5. Masuda, T., Y. Nakagami and S. L. Woronowicz : *A C^*-algebraic framework for quantum groups*, to appear.
6. Tomatsu, R. Private communication, January 29, 2003.
7. Van Daele, A. : *The Haar measure on a compact quantum group,* Proc. Amer. Math. Soc., **123**(1995), 3125–3128.
8. Woronowicz, S. L. : *Compact quantum groups*, Quantum Symmetries, (1998), 845-884.
9. Voiclescu, Dan. : *Amenability and Katz algebras*, "Algebres d'Operatours et leurs Applications en Physique Mathematique", (1978), 451–457.

The Bessel Equation and Dissipation

Eleonora Alfinito

INFM, Sezione di Lecce, 73100 Lecce, Italy
E-mail: eleonora.alfinito@unile.it

Giuseppe Vitiello

Dipartimento di Fisica "E.R.Caianiello", Università di Salerno,
84080 Salerno, Italy
INFN, Gruppo Collegato di Salerno and INFM, Sezione di Salerno
E-mail: vitiello@sa.infn.it

The Bessel equation can be cast, by means of suitable transformations, into a system of two damped/amplified parametric oscillator equations. The role of group contraction and the breakdown of loop–antiloop symmetry is discussed. The relation between the Virasoro algebra and the Euclidean algebras $e(2)$ and $e(3)$ is also presented.

1. Introduction

Professor Hiroshi Ezawa belongs to that restricted family of mathematical physicists who are not satisfied by the mathematical "rigor" if this is not justified and accompanied by a *clear* physical vision. In our opinion this is a very important teaching. Too many times we have seen respectable colleagues who entered the fascinating forest of Mathematics and... got lost in it... On the contrary, we have been always pleasantly guided by Professor Ezawa through unexplored paths of that forest with the certainty of the physical discovery. We would thank him for giving us such a pleasure and we are glad to contribute with this paper to the book in his honor. We hope he will not dislike much our rudimentary way of analyzing here some of the properties of the Bessel functions in connection with group contraction and damped/amplified oscillators.

We wish to show that the Bessel equation can be cast, by means of suit-

able transformations, into a system of two parametric oscillator equations, one for a damped oscillator, the other one for an amplified one. We will see that the group contraction mechanism is involved in such a relation of the Bessel equation with the dissipation/amplification system.

Since Bateman,[1] the interest in the damped/amplified oscillator system has been growing[2] and it has been used to describe thermal field theories,[3] Chern-Simons gauge theory,[4] Bloch electrons in metals,[4] *etc.* It also presents features common to the formalism of two-dimension gravity models.[5] In particular, the equivalence, under suitable parametrization, between the spherical Bessel equation and the damped/amplified oscillators was firstly recognized in the study of expanding geometry models[6] and of ordered domain formation in brain models.[7,8,9]

The plan of the paper is the following. In Section 2 we consider the Bessel equation and derive from it the set of two damped/amplified parametric oscillators. In Section 3 we show that the mechanism of group contraction,[10] by which one gets the Euclidean groups $E(2)$ and $E(3)$, whose representations are given in terms of planar and spherical Bessel functions, respectively, introduces the breakdown of the loop–antiloop symmetry around a preferred axis. In turn, this can be read off, in a given re-parametrization, as the breakdown of the time-reversal symmetry. In Section 4, in view of the relevance of infinite dimensional algebraic structures in many physical problems, some results on the relation between the Virasoro-like algebra and the $e(2)$ and $e(3)$ algebras are presented. Section 5 is devoted to conclusions.

2. The Bessel equation and the breaking of time-reversal symmetry

The spherical Bessel equation of order n (also said of fractional order[11]), is

$$\eta^2\, J_{n;\eta\eta} + 2\eta\, J_{n;\eta} + [\eta^2 - n(n+1)]\, J_n = 0 \,, \tag{1}$$

where $n = 0, \pm 1, \pm 2, \ldots$ and "$;\eta$" and "$;\eta\eta$" denote first and second order derivatives, respectively. The solutions of Eq. (1), the so-called spherical Bessel functions, can be expressed in terms of the first and second kind Bessel functions and their linear combinations (the Hankel functions), and constitute a complete set of (parametric) decaying functions.[11]

Equation (1) is invariant under the transformation $n \to -(n+1)$ and J_n and $J_{-(n+1)}$ are both solutions of the same equation: J_n and $J_{-(n+1)}$ can be thought to be degenerate solutions corresponding to the same eigenvalue $n(n+1)$ of the operator $\eta^2\, \frac{d^2}{d\eta^2} + 2\eta\, \frac{d}{d\eta} + \eta^2$.

We now perform in Eq. (1) the change of variables[12]: $\eta \to \eta \equiv \epsilon x$ with $x \equiv e^{-t/\alpha}$, ϵ and α arbitrary parameters. The new variable t may be thought to denote, e.g., the time variable. By using $w_{n,l} \equiv J_n \cdot (x)^{-l}$, Eq. (1) then goes into the following one:

$$\ddot{w}_{n,l} \; - \; \frac{2l+1}{\alpha} \; \dot{w}_{n,l} + \left[\frac{l(l+1) - n(n+1)}{\alpha^2} \; + \; \left(\frac{\epsilon}{\alpha}\right)^2 e^{-2t/\alpha} \right] w_{n,l} \; = \; 0,$$

$$(2)$$

where $\dot{w}$ denotes derivative of w with respect to time t. By choosing $l(l+1) = n(n+1)$ the degeneracy between the solutions J_n and $J_{-(n+1)}$ is removed: two *different* sets of equations are obtained, one for $w_{n,l}$ and the other one for $w_{-(n+1),l}$, respectively, each set being composed by two *different* equations, one for $l = -(n+1)$ and the other one for $l = n$. The set for $w_{n,l}$ is

$$\ddot{w}_{n,-(n+1)} \; + \; \frac{2n+1}{\alpha} \; \dot{w}_{n,-(n+1)} + [(\tfrac{\epsilon}{\alpha})^2 \, e^{-2t/\alpha}] \; w_{n,-(n+1)} \; = \; 0,$$

$$\ddot{w}_{n,n} \; - \; \frac{2n+1}{\alpha} \; \dot{w}_{n,n} + [(\tfrac{\epsilon}{\alpha})^2 \, e^{-2t/\alpha}] w_{n,n} \; = \; 0 \; , \qquad (3)$$

for $l = -(n+1)$ and for $l = n$, respectively. Similarly, two equations for $w_{-(n+1),l}$ are obtained:

$$\ddot{w}_{-(n+1),n} \; - \; \frac{2n+1}{\alpha} \; \dot{w}_{-(n+1),n} + [(\tfrac{\epsilon}{\alpha})^2 e^{-2t/\alpha}] w_{-(n+1),n} = 0,$$

$$\ddot{w}_{-(n+1),-(n+1)} \; + \; \frac{2n+1}{\alpha} \; \dot{w}_{-(n+1),-(n+1)} + [(\tfrac{\epsilon}{\alpha})^2 e^{-2t/\alpha}] w_{-(n+1),-(n+1)} = 0,$$

$$(4)$$

for $l = n$ and $l = -(n+1)$, respectively. Inspection of Eqs. (3) and Eqs. (4) shows that the symmetry under the transformation $n \to -(n+1)$ has been broken.

Let us now choose the arbitrary parameters α and ϵ to be n-dependent: $\alpha \to \alpha_n$ and $\epsilon \to \epsilon_n$ (such a choice means that we use $\eta \to \eta_n \equiv \epsilon_n x_n$ with $x_n \equiv e^{-t/\alpha_n}$). We also perform our choice in such a way that $\frac{2n+1}{\alpha_n} \equiv L$ and $\frac{\epsilon_n}{\alpha_n} \equiv \omega_0$ do not depend on n (and on time). By setting $u_n \equiv w_{n,-(n+1)}$, and $v_n \equiv w_{n,n}$, we then see that Eqs. (3) are nothing else than the couple of equations for the damped/amplified parametric oscillators (sometimes called Hill-type equations[13]):

$$\ddot{u}_n \; + L \, \dot{u}_n \; + \omega_n{}^2(t) u_n \; = \; 0,$$

$$\ddot{v}_n \; - L \, \dot{v}_n \; + \omega_n{}^2(t) v_n \; = \; 0, \qquad (5)$$

with

$$\omega_n(t) \; = \; \omega_0 \, e^{-\frac{Lt}{2n+1}}. \qquad (6)$$

We see that $\omega_n(t)$ approaches to the time-independent value ω_0 for $n \to \infty$: the frequency time-dependence is thus "graded" by the order n of the original Bessel equation. L and ω_0, which may be arbitrarily chosen, are characteristic parameters of the oscillator system.

Our choice of keeping L independent of n implies that $\alpha_{-(n+1)} = -\alpha_n$. Then the transformation $n \to -(n+1)$ leads to solutions (corresponding to $J_{-(n+1)}$) which have frequencies exponentially increasing in time (cf. Eq. (6)). These solutions can be respectively obtained from the ones of Eqs. (5) by time-reversal $t \to -t$ and exchanging u with v (which we refer to as "charge conjugation"). In the large n limit $(\omega_n \to \omega_0)$ u_n and v_n are each the time-reversed of the other one and in that limit the two sectors $\{J_i\}$, $i = n, -(n+1)$ are mapped one into the other one by time-reversal.

The core of the relation between the spherical Bessel equation and the system of damped/amplified oscillators is thus revealed: the breakdown of the $n \to -(n+1)$ symmetry of the original spherical Bessel equation corresponds in our representation to the breakdown of time-reversal symmetry in the manifold of the spherical Bessel functions $\{J_i\}$, $i = n, -(n+1)$.

In conclusion, the spherical Bessel equation represents, under suitable transformations, a two-fold hierarchy of couples of parametric oscillators with damping/amplification constant given by L and with time dependent frequency graded by n and by $-(n+1)$. Transition from one tower to the other one $(n \to -(n+1))$ is induced by time-reversal $t \to -t$ combined with "charge conjugation" $u \to v$.

Remarkably, the first of Eqs. (5), with $n = 1$ is commonly used in expanding geometry (inflationary) models of the Universe.[6] In that case L denotes the Hubble constant.

We can apply the procedure followed above for the spherical Bessel equation, also to the planar Bessel equation[12]

$$\eta^2 J_{n;\eta\eta} + \eta \, J_{n;\eta} + [\eta^2 - n^2] \, J_n = 0, \tag{7}$$

By setting $w_{n,l}(\eta) = (\eta)^{-l} J_n(\eta)$, $\eta = \epsilon e^{-t/\alpha}$, and by introducing the mirror parameter $l = \pm n$, Eq. (7) can be cast into the couple of damped/amplified harmonic oscillators:

$$l = -n : \ddot{w}_{n,-n} + \tfrac{2n}{\alpha} \, \dot{w}_{n,-n} + \left[(\tfrac{\epsilon}{\alpha})^2 \, e^{-2t/\alpha} \right] w_{n,-n} = 0,$$
$$l = n : \quad \ddot{w}_{n,n} - \tfrac{2n}{\alpha} \, \dot{w}_{n,n} + \left[(\tfrac{\epsilon}{\alpha})^2 \, e^{-2t/\alpha} \right] w_{n,n} = 0. \tag{8}$$

3. Group contraction and breaking of loop–antiloop symmetry

It is well known that the representations of the Euclidean groups $E(2)$ and $E(3)$ can be constructed in terms of the planar and spherical Bessel functions, respectively.[11,14] Let us shortly summarize how this goes.

We start with $E(2)$, which is the group of the $T(\vec{v})R(\theta)$ transformations, where $T(\vec{v})$ is the translation in the plane by the vector $\vec{v}$ ($\vec{v} \equiv (a,b)$) and $R(\theta)$ is the rotation of the plane around the origin by the angle θ. The associated Lie algebra is given in terms of the two translation generators P_a, P_b and of the rotation generator M:

$$[P_a, P_b] = 0, \quad [P_a, M] = -P_b, \quad [P_b, M] = P_a. \tag{9}$$

The invariant operator is $P^2 = P_a^2 + P_b^2 = P_+ \, P_- = P_- \, P_+$, with $P_\pm \equiv P_a \pm iP_b$, which has non-positive eigenvalue, $-p^2$. The group transformations on square integrable functions f are represented by the action of the operator $D(\vec{v}, \theta)$:

$$\begin{aligned} D(\vec{v}, 0)f(\phi) &= e^{i\vec{p}\cdot\vec{v}} f(\phi), \\ D(0, \theta)f(\phi) &= f(\phi - \theta) \, . \end{aligned} \tag{10}$$

The representation of $E(2)$ in terms of the planar Bessel functions J_m[14] is easily obtained by assuming as basis functions the complete set of normalized eigenfunctions of the rotation subgroup $\{f_n(\phi)\}$, $f_n \equiv (2\pi)^{-\frac{1}{2}} \, i^{-n} \, e^{in\phi}$: $D(0, \theta)f_n(\phi) = e^{in\theta} f_n(\phi)$. The complete representation of $E(2)$ can be expressed in the form:

$$\begin{aligned} D(\vec{v}, \theta)f_n &= \sum_m \Delta(\vec{v}, \theta)_{mn} f_m, \\ \Delta(\vec{v}, \theta)_{mn} &= (-1)^{m-n} e^{-im\beta} \, J_{m-n}(pr) e^{in(\beta-\theta)}, \end{aligned} \tag{11}$$

where (r, β) are the polar coordinates of $\vec{v}$.

We consider now the Laplace equation in the isotropic and homogeneous 3D-space in order to study the P^2 eigenvalue equation:

$$\nabla^2\psi \equiv \left(\frac{\partial^2}{\partial x_1^2} + \frac{\partial^2}{\partial x_2^2} + \frac{\partial^2}{\partial x_3^2} \right) \psi = 0, \tag{12}$$

By choosing, instead of the spherical or rectangular coordinates, the cylindrical ones, we have:

$$\left(\frac{\partial^2}{\partial r^2} + \frac{\partial}{r\partial r} + \frac{\partial^2}{r^2\partial\theta^2} + \frac{\partial^2}{\partial x_3^2} \right) \psi = 0, \tag{13}$$

and we search for solutions of the type:

$$\psi(r, \theta, x_3) = \varphi(r, \theta) \cdot \sigma(x_3), \qquad \frac{\partial^2}{\partial x_3^2}\sigma(x_3) \equiv p^2\sigma(x_3). \qquad (14)$$

Square integrable eigenfunctions (for p positive) are obtained by selecting, for positive x_3, the solution: $\sigma = e^{-x_3 p}$ and for negative x_3 the solution: $\sigma = e^{x_3 p}$. We remark that one can identify x_3 with the "time" t coordinate. Then p has the dimensions of an energy over an action $\epsilon/\hbar$.

The choice of the cylindrical, instead of the spherical coordinates, is not a trivial one: it breaks the symmetry of the 3D-spatial rotation group $SO(3)$. Indeed, when cylindrical coordinates are chosen, x_3 becomes a privileged axis for rotations and thus it is differently treated with respect to the two remaining coordinates. The resulting symmetry group is $E(2)$, the group contraction of $SO(3)$. This is manifest in the fact that the Laplace equation (13) reduces to the eigenvalue equation for P^2 (realized in polar coordinates), *i.e.* the 2D-Helmholtz equation:

$$P_+ P_- \varphi = \left(\frac{\partial^2}{\partial r^2} + \frac{\partial}{r\partial r} + \frac{\partial^2}{r^2\partial\theta^2} \right) \varphi = -p^2\varphi. \qquad (15)$$

Searching for solutions of the type $\varphi(r, \theta) = f(r) \cdot e^{in\theta}$, we obtain $f(r) = J_n(pr)$, being $J_n(pr)$ the solution of the planar Bessel equation (7) with $\eta = pr$.

Positive/negative values of n in Eq. (7) correspond to positive/negative rotations (loop/antiloop) around the x_3 axis, *i.e.*, they correspond to different orientations of the x_3 axis (or of the time axis when x_3 is identified with time t) . The related solutions are, therefore, different.[11] Notice that, in spite of the fact that the solutions are not symmetric under the reversal of the x_3 axis (under time-reversal), Eq. (7) is invariant under the $n \to -n$ exchange. The $n \to -n$ invariance of Eq. (7) actually reflects the existence of the two sets of the $SO(3)$ independent representations: the D^n and the D^{-n-1}.[15]

Summarizing, the breakdown of the rotational symmetry of $SO(3)$ which leads to its group contraction $E(2)$ introduces a crucial difference (loss of loop–antiloop symmetry, indeed) in the double choice of the x_3 axis orientation (or *time-arrow* orientation). This, in turn, results in the difference between the planar Bessel functions or order $+n$ and the ones of order $-n$, in terms of which the $E(2)$ representations can be built.

We can express this in different terms: we may introduce $J_x \equiv \rho \cdot P_a$, $J_y \equiv \rho \cdot P_b$, $J_z \equiv M$ with the Js denoting the generators of the algebra

$so(3)$ and ρ the radius of the sphere S_2 (corresponding to $SO(3)$). By recalling that the contraction of $SO(3)$ to $E(2)$ can be geometrically depicted as the projection of the sphere on the plane (*e.g.* the plane tangent to one of the poles of the sphere), we see that ρ acts as a "scale": the $E(2)$ translations in the tangent plane are "good" approximations to rotations around x_1 and x_2 in the limit $\rho \to \infty$, namely for distances much smaller than ρ. This is the limit of "local" observations. Thus, in physical terms, the $SO(3)$ contraction to $E(2)$ manifests itself in local observations. However, in the local observation process the x_3 (or time) axis orientation gets "locked" (rotations around x_1 and x_2 are frozen and approximated by translations in the plane) . Which amounts to the loss of symmetry of the solutions under the $n \to -n$ exchange: breakdown of the loop–antiloop symmetry. Specifying the direction of the x_3 axis (or of the *arrow of time*), *i.e.* choosing one of the two possible forms for σ (cf. Eq. (14)), selects topologically inequivalent sets of solutions.

Similar considerations can be made for the case of $E(3)$ and of the spherical Bessel equation. $E(3)$, the Euclidean group in the space, is the group contraction of $SO(4)$. The algebra $e(3)$ has six generators P_i and M_i, $i = 1, 2, 3$, corresponding to the translation and to the rotation generators, respectively. The commutation relations are:

$$[P_i, P_j] = 0, \quad [M_i, M_j] = \epsilon_{ijk} M_k, \quad [P_i, M_j] = \epsilon_{ijk} P_k; \qquad (16)$$

The algebra $e(3)$ has two invariants, $P^2 = \Sigma P_i^2$ and $\Sigma P_i \cdot M_i$.

In the 4D-space the Laplace equation for the function $\psi = \psi(x_1, x_2, x_3, x_4)$ may be solved by assuming: $\psi = \varphi(r, \theta, \phi) \cdot \sigma(x_4)$, $(r, \theta, \phi$ spherical coordinates, $\sigma = e^{\pm x_4 p}$). The resulting equation is solved by the function $\varphi = Y_{n,m}(\theta, \phi) \cdot J_n(pr)$ where $Y_{n,m}$ is the spherical harmonics and J_n, which depends on the continuous eigenvalue p^2 of P^2, is the solution of the spherical Bessel equation (1). Also in the present case, we could identify the x_4 coordinate with the "time" coordinate.

The order n is of course related with the discrete eigenvalue $n(n + 1)$ of the rotation operator $\mathbf{M}^2$ and classifies the representations D^n of the compact subgroup $SO(3)$ of $E(3)$. Actually, as for the planar case, the existence of two sets of $SO(3)$ independent representations (the D^n and the D^{-n-1})[15] is reflected in Eq.(1) through its invariance under the transformation $n \to -(n+1)$. Again, the breakdown of the symmetry under the transformation $n \to -(n+1)$ is built in the geometrical structure of the $E(3)$ group: the breakdown of the $n \to -(n+1)$ symmetry is nothing but the breakdown of the x_4 axis (or time-) reversal symmetry (breakdown of the loop–antiloop symmetry). Also in the present case, the $SO(4)$ contraction to $E(3)$ manifests itself in local observations and the x_4 axis orientation (the *arrow of time* orientation) then gets "locked". Which amounts to the loss of symmetry of the solutions under the $n \to -(n+1)$ exchange.

In order to better clarify the topological properties of Bessel functions related with loop operators and loop–algebras let us recall that Bessel functions may be associated to different values of the Pontryagin number in the punctured plane $R^2/(0)$, where the elements of the homotopy group, Π_n, may be represented by differential operators acting on analytic functions:

$$\Pi_n \equiv \frac{\partial^n}{\partial z^n} \, , \quad n \in \mathsf{N} \, , \tag{17}$$

with $\Pi_n \cdot \Pi_m = \Pi_{n+m}$ and n is the loop number around the hole. There are two different kinds of behavior, corresponding to two different functions:

$$\frac{\partial^n}{z \partial z^n} \, \varphi_m(z) = (-)^m \, \varphi_{m+n}(z) \, , \quad \varphi_m(z) = \frac{J_m(z)}{z^m} \tag{18}$$

and

$$\frac{\partial^n}{z \partial z^n} \, \psi_m(z) = \psi_{m-n}(z) \, , \quad \psi_m(z) = z^m J_m(z) \, , \tag{19}$$

so that on φ Π_n acts in counter-clockwise way while on ψ it acts in clockwise way. $J_m(z)$ is the planar Bessel function (Bessel function of integer order).

Eqs.(18) and (19) are the well known differential formulae for the planar Bessel functions[11]; analogous formulae are true for the functions

$$\varphi_m(z) = j_m(z) z^{-m} \, , \qquad \psi_m(z) = z^{(m+1)} j_m(z) \, , \tag{20}$$

where the j_m are the spherical Bessel functions[16] and we notice that while for the planar Bessel functions, the "raising" and "lowering" functions coincide for $m = 0$ (no loops), this is not true with the spherical Bessel functions.

These topological properties, in the planar and in the spherical case, of the φ and ψ functions clearly remind us of the x_3- and x_4-reversal symmetry breakdown discussed above. And this brings us to consider the loop algebras as extensions of the Euclidean algebras.

4. Loop algebras and Euclidean algebras

Let us focus our attention in particular on the Virasoro algebra which plays a central role in the conformal field theories. By following a general construction based on the so-called graded contraction method,[17,18] it is possible to follow a contraction procedure which maps the Virasoro algebra into a sort of generalization of the Euclidean algebra $e(3)$. The Virasoro algebra $\mathcal{L}$ of central charge c (c commuting with all the T's) is

$$[T_n, T_m] = (n - m)T_{n+m} + \frac{c}{12}(n^3 - n)\delta_{n+m,0} , \quad m, n \in \mathsf{Z} . \tag{21}$$

The Z_2-grading of the algebra consists in dividing the set of the T_n generators into an even set $L_0 \equiv \{A_n, c\}$ and an odd set $L_1 \equiv \{B_n\}$, with

$$A_n = \frac{1}{2}\left(T_{2n} + \frac{c}{8}\delta_{n,0}\right) , \quad B_n = \frac{1}{2}T_{2n+1} , \tag{22}$$

so that $\mathcal{L} = L_0 \bigoplus L_1$ and

$$[L_0, L_0] \subseteq L_0 , \quad [L_0, L_1] \subseteq L_1 , \quad [L_1, L_1] \subseteq L_0.$$

The commutation relations of the graded generators are explicitly given by

$$[A_n, A_m] = (n - m)A_{n+m} + \frac{2c}{12}(n^3 - n)\delta_{n+m,0} , \tag{23}$$

$$[B_n, B_m] = (n - m)A_{n+m+1} + \frac{2c}{12}\left(n - \frac{1}{2}\right)\left(n + \frac{1}{2}\right)\left(n + \frac{3}{2}\right)\delta_{n+m+1,0} , \tag{24}$$

$$[A_n, B_m] = \left(n - m - \frac{1}{2}\right)B_{n+m} . \tag{25}$$

Equation (23) shows that $\{A_n, c\}$ is again a Virasoro algebra but with central charge $2c$.

The Z_2-graded contraction of the algebra (23)–(25) is obtained[19] by putting

$$[B_n, B_m] = 0 \tag{26}$$

and leaving all the other commutators the same as in (23) and (25). Then, by setting:

$$M_+ \equiv A_1 , \quad M_- \equiv A_{-1} , \quad M_3 \equiv iA_0 ,$$
$$P_+ \equiv B_{\frac{1}{2}}, \quad P_- \equiv B_{-\frac{3}{2}} , \quad P_3 \equiv iB_{-\frac{1}{2}} , \tag{27}$$

we see that the Ms and Ps generators satisfy the commutation relations (16) in the centerless case ($c = 0$): thus, for ($c = 0$) the A_0 and $A_{\pm 1}$ generators in (27) close the algebra isomorphic to $so(3) \sim su(2)$ and the set of these three generators and the operators $B_{-\frac{1}{2}}$, $B_{\frac{1}{2}}$ and $B_{-\frac{3}{2}}$ close the $e(3)$ isomorphic algebra.

This result has a general extension, *i.e.* the algebra $\mathcal{E}_n \equiv \{A_0, A_{\pm n}\} \bigoplus \{B_{-\frac{1}{2}}, B_{\pm n - \frac{1}{2}}\}$ reproduces the $e(3)$ algebra for each integer non-zero value of n, provided the following positions are assumed:

$$M_+ \equiv \tfrac{1}{n} A_n \ , \quad M_- \equiv \tfrac{1}{n} A_{-n} \ , \quad M_3 \equiv \tfrac{i}{n} A_0 \ ,$$
$$P_+ \equiv B_{n-\frac{1}{2}} \ , \quad P_- \equiv B_{-n-\frac{1}{2}} \ , \quad P_3 \equiv iB_{-\frac{1}{2}} \ . \tag{28}$$

Finally, we notice that the $e(2)$-algebra can be obtained as a subalgebra of (28) by choosing $A_{\pm n} = 0$, for non-zero values of n.

In conclusion, the extension of the Virasoro algebra by means of its Z_2-grading with the subsequent step of the Z_2-graded contraction appears as a n-graded hierarchy of Euclidean algebras.

5. Conclusions

We have shown that, after a suitable re-parametrization, the spherical as well as the planar Bessel equation can be cast in a couple of equations for the damped/amplified oscillators. The breakdown of the $n \to -(n+1)$ $(n \to -n)$ symmetry of the original spherical (planar) Bessel equation corresponds to the breakdown of time-reversal symmetry in the manifold of the solution $\{J_i\}$, $i = n, -(n+1)$ $(i = n, -n)$. This process is controlled by the mechanism of group contraction and the original Bessel equation, conveniently reparametrized, "splits" into a couple of damped/amplified oscillators.

The quantization of the damped harmonic oscillator, a prototype of dissipative system, has been extensively studied in a number of papers.[2−4,7] Here we have not considered this quantization problem. We only recall that in order to perform the canonical quantization of the simple damped harmonic oscillator, one "doubles" the system by introducing the companion amplified oscillator.[2,3] Canonical quantization is only possible for Hamiltonian systems, and the doubling of the dissipative (non-hamiltonian, open) system amounts to "closing" it by inclusion of the environment (represented by the doubled, amplified oscillator, indeed). At a classical level, however, it is possible, in principle, to treat the dissipative system by ignoring the environment (*i.e.* there is no necessity of introducing the amplified oscillator as instead required in the canonical quantization procedure). Remarkably, from our discussion in the present paper it emerges that Bessel equation may represent both the damped and the amplified oscillator as an inseparable "double-face" unity also at the classical level.

We have also considered some topological properties of the Bessel functions in connection with Virasoro-like loop–algebras and we have presented some results on the relation between such algebras and the algebras $e(2)$ and $e(3)$.

We close with a remark on some features of our result which introduce analogies with the spontaneous breakdown of symmetry in Quantum Field Theory (QFT). The phenomenon of the spontaneous breakdown of symmetry in QFT occurs when the *continuous* symmetry of the dynamical equations is not the symmetry of the physical vacuum. We have seen that in the Bessel equation case the *discrete* time-reversal symmetry may get broken and the degeneracy between solutions of n- and $(n+1)$-index (or n- and $-n$-index) may be removed. However, time-reversal symmetry breakdown manifests itself in the appearing of dissipation/amplification phenomena, thus continuous time translational symmetry is also broken (energy nonconservation for the damped (or amplified) system). One may thus think of a resemblance of the symmetry breakdown in the Bessel equation case with the QFT case. The analogy also appears if one considers that in QFT the effects of spontaneous breakdown of symmetry are solely observable in the contraction limit,[20] namely at the observation scale ("locality") which is always "small" with respect to the system volume (the infinite volume limit). Also in the Bessel case time-reversal symmetry is broken at local, observational level and there dissipation/amplification phenomena become observable phenomena.

Acknowledgments

We acknowledge partial financial support by the ESF Program COSLAB.

References

1. H. Bateman, *Phys. Rev.* **38** (1931) 815.
2. H. Feshbach and Y. Tikochinski, *Trans. NY Acad. Sci.* **38** (1977) 44.
3. E. Celeghini, M. Rasetti and G. Vitiello, *Ann. Phys.* **215** (1992) 156.
4. M. Blasone, E. Graziano, O.K. Pashaev and G. Vitiello, *Ann. Phys. (N.Y.)* **252** (1996) 115.
5. D. Cangemi, R. Jackiw and B. Zwiebach, *Ann. Phys.* **245** (1996) 408.
6. E. Alfinito, R. Manka and G. Vitiello, *Class. Quant. Grav.* **17** (2000) 93.
7. E. Alfinito and G. Vitiello, *Int. J. Mod. Phys.* B **14** (2000) 853 [Erratum-ibid. B **14** (2000) 1613].
8. G. Vitiello, *Int. J. Mod. Phys.* B **9** (1995) 973.
9. G. Vitiello, *My Double Unveiled*, (John Benjamins, Amsterdam 2001).
10. E. Inönü and E.P. Wigner, *Proc. Nat. Acad. Sci. U.S.A.* **39** (1953) 510.
11. M. Abramowitz, I. Stegun (Eds.), *Handbook of Mathematical Functions*, (Dover Inc., New York 1970).

12. E. Alfinito and G. Vitiello, arXiv:hep-th/0210129.
13. G. Profilo and G. Soliani, *Phys. Rev.* **A44** (1991) 2057.
14. J.D. Talman, *Special Functions. A group theoretical approach*, (Benjamin, New York, 1968).
15. B.G. Wybourne, *Classical groups for physicists*, (J.Wiley and Sons, 1974).
16. M. Mekhfi, *Int.J. Theor. Phys.* **39**, No.4 (2000) 1163.
17. M. de Montigny and J. Patera, *J. Phys.* A **24** (1991) 525.
18. R. V. Moody and J. Patera, *J. Phys.* A **24** (1991) 2227.
19. I.V. Kostyakov, N.A. Gromov and V.V. Kuratov, *Nucl.Phys.* **B102** (Proc. Suppl.) (2001) 316.
20. C. De Concini, G. Vitiello, *Nucl. Phys.* **B116** (1976) 146.

Simon Stevin and the Cultural Revolution in the 16th Century

Yoshitaka Yamamoto

Department of Physics, Sundai Preparatory School,
Kanda–Surugadai, Chiyoda-ku, Tokyo 101–0062, Japan

Dedicated to Professor Hiroshi Ezawa on the occasion of his seventieth birthday

In early modern times, especially in the latter half of the 16th century, the style of scientific research changed from interpretations and readings of old doctrines to observations of and experiments on nature itself. In a word, the object to which people attached importance began to shift from documents to experiences. This change was supported by the participation of engineers, artisans, artists, sailors and soldiers in the research activities. The main factor which brought such a situation about was the use of the vernaculars, namely, native languages instead of Latin as the language of scholarly communication. Indeed, the use of native languages in Europe destroyed the monopoly on knowledge of scholars and the clergy educated at the universities and monasteries, and thus brought "diastrophism" to the world of intellectuals, which may be called a cultural revolution. Simon Stevin should be remembered as a conscious proponent of this change.

1. Simon Stevin and the new mechanics

Simon Stevin was born in 1548 at the dawn of modern science. Indeed, five years earlier, the works of Archimedes were published along with Copernicus's *Revolution*. Giordano Bruno, who advocated the heliocentric system and the infinity of the universe was the same age as he. Tycho Brahe and Thomas Digges were born 2 years before him. Thus, Stevin was of the same generation of pioneers of the scientific revolution.

What kind of education Stevin got in his younger days is not known, but it is inferred that he was trained in the practice of business. In 1582 he published the pamphlets *Tafelen van Interest* (*Table of Interest*) and *Problemata Geometrica* (*Geometrical Problems*). After the Netherlands gained their independence, Stevin served in the State Army as an engineer. Later he was appointed Quartermaster of the Army, and also acted as a tutor in mathematics and natural sciences to Prince Maurice, the Commander-in-Chief of the Dutch Army, and he wrote many books on mathematics, mechanics and engineering. Thus, the stage of his activities was entirely outside of the academy and the university.

Among his works, the most important in the context of the history of physics is a trilogy published in 1586, namely *De Behinselen der Weeghconst* (*The Elements of the Art of Weighing*), *De Weeghdaet* (*The Practice of Weighing*) and *De Behinselen des Waterwichts* (*The Elements of Hydrostatics*). Here, the balance of a body on a slope is discussed, and the figure sketched on the cover is famous and has been reproduced in books on the history of physics frequently. By using a thought experiment and an argument based on the impossibility of perpetual motion, Stevin concluded that the effective gravity acting on the body on an inclined plane is inversely proportional to the length of the slope (proportional to the sine of the inclination of the slope). However, we will not examine this theme here, because it has been discussed by many authors.[1]

Another important contribution of Stevin to mechanics is the fact that he denied Aristotle's theory of falling bodies, which asserted that a heavy body falls more swiftly than a light body, or to put it more precisely, the falling velocity is proportional to the weight. In the Appendix of his *De Behinselen der Weeghconst*, he describes his own experiment as follows:

> The experience against Aristotle is the following: Let us take (as the very learned Mr. Jan Cornets de Groot, most industrious investigator of the secrets of Nature, and myself have done) two spheres of lead, the one ten times larger and heavier than the other, and drop them together from a height of 30 feet on to a board or something on which they give a perceptible sound. Then it will be found that the lighter will not be ten times longer on its way than the heavier, but that they fall together on to the board so simultaneously that their two sounds seem to be one and the same rap. The same is found also to happen in practice with two equally large bodies whose gravities are in the ratio of one to ten: therefore Aristotle's aforesaid proportion is incorrect.[2]

Usually this type of experiment is said to have been done by Galilei at the Leaning Tower of Pisa, but no evidence has been given to support this assertion. Whether it is true or not, Galilei stayed at Pisa from 1589 to 1592, therefore, even if Galilei had actually done this, the experiment of Stevin and de Groot was performed at least 3 years earlier, and as a criticism of Aristotle's theory of motion it is the first experiment which is recorded explicitly.

In this way, Stevin's mechanics is on the one hand based on purposive experiments and on the other hand is constructed by rigorous mathematical reasoning, so he is undoubtedly a pioneer of modern mechanics, and is said to be the precursor of Galilei, Descartes and Huygens.

2. Simon Stevin and the use of the Dutch vernacular

Stevin's contribution to the new science is not confined to particular theses or individual experiments on mechanics, however. Indeed, he foresaw clearly how and by whom the new science should be performed and conducted.

The interesting and remarkable argument for the present purpose is Stevin's peculiar assertion about the use of Dutch as a technical language for natural sciences and engineering.

In the Preface of *De Behinselen der Weeghconst*, "the Worth of the Dutch Language" is described. There he argues ardently that Dutch is one of the oldest languages in the world and has been a powerful language from ancient times, and also that it is an excellent and suitable language for the sciences because of its brevity and clarity. The reasoning of his assertion is that in Dutch there are more monosyllabic words than in Greek or in Latin, and therefore Dutch has an advantage in that it is easy to construct a clear compound, but such reasoning is rather complacent and has no meaning except for having inspired the patriotism of the newly born Netherlands Republic.

In any case, Stevin's works are all written in Dutch except for the earliest two books. And in writing his books in Dutch, he coined many Dutch words for sciences and technologies, and some of those technical terms such as 'driehoek (triangle)', 'aftrekken (subtract)', 'wortel (root)' and so on are being used even today. In such a way, he founded the rise and the rapid development of natural sciences in the Netherlands after the 17th century.

But the reason for attaching importance to the use of Dutch by Stevin was not restricted to such a nationalistic one as mentioned above. The more important fact is that he recognized that the participation of the wide range of artisans and engineers who were outside of academism and could read only vernaculars was decisively important to the construction and propulsion of new sciences. Indeed, as his *De Thiende* (*The Decimal System*) begins with the address "to Astronomers, Land-meters, Measurers of Tapestry, Gaugers, Stereometers in general, Money-masters and to all Merchants", his works, even his mathematical books, were all written not for scholars in universities but for artisans, artists, soldiers, sailors and merchants. Moreover, the reason for the use of Dutch is not only the negative one that they were written for people who could not read Latin, but also the positive one that the new sciences had to be based on many practical experiences and observations, and so, had to be performed by a large number of people.

And this argument is, of course, not restricted to Dutch only, but is applicable to vernaculars in general, and so is more important. Indeed, Stevin clearly says:

> What we lack is a large body of data obtained by practical experience, on which science can be firmly founded. In order to arrive at such a body of data, a great many people would have to apply themselves jointly to this task. To arrive at so great a number of men as is needed for this, the aforesaid experiences and pursuit of science would have to be practiced by a nation in its own native language.[3]

In short, "If ordinary people were to study a science, they would have to understand the language in which it was written, which should be their own language".[4]

These paragraphs are cited from "The Age of the Sages" contained in his *Wisconstighe Gedachatenissen* (*Mathematical Memories*). This point of issue is argued more concretely with regard to the example of astronomy as follows:

> In the first place, one man cannot continually, by night and by day, year in and year out, observe the positions of the planets and all that is necessary; but when a large number of people do this, what is lacking in the observations of one man will be found in those of another. Secondly, the data obtained by one man, even if they were exact in themselves, yet do not serve others as a sure basis on which to proceed in framing their theories, because they have not been checked. But when the data obtained by a great many different people, having been compared with each other, are found to agree as closely as the matter requires, one can

rely thereon .... Thirdly, the sky will often be overcast in some places, so that for some weeks no heavenly bodies are seen; in such a case one may then rely on the data obtained by others in the countries where the sky was clear. Fourthly, ambition and rivalry would be apt to arise among observers, each wishing to prove his own work best, owing to which the sciences usually make no inconsiderable progress, whilst on the other hand, if the branch of science is practiced by few people, each of them will keep his findings to himself and conceal them.[5]

The facts pointed out here may seem to be trivial to us now, but they were far ahead of Stevin's time.

The most accurate, and almost only, accumulated data of astronomical observations of the time of Stevin was that obtained by Tycho Brahe's life-long work. It is known that Johannes Kepler derived his famous three laws about the motions of planets from these data. Indeed, Tycho's observations were far superior to previous ones not only in their precision, but also in their continuation, so the reliability of his data was outstanding in that time.

However, Tycho had been able to devote himself to the observation of the sky day after day, and thus to observe for long years without a break, only because as a feudal lord he had immense wealth and power and could employ many people to assist him and set them to work at will. Such a state of affairs is, of course, impossible for a scholar as a citizen of modern times, not to mention for an artisan. Stevin, recognizing the value of Tycho's data, understood that for citizens to obtain data which ranks with Tycho's, the cooperation of many people is absolutely indispensable. Moreover, when Stevin said that data obtained by one man must be subject to confirmation by others, and if not it does not serve others as a sure basis for their theory, he became aware of the fundamental defect of Tycho's data, that is, its impossibility of verification.

Moreover, Stevin's indication that data obtained by few men is apt to be hoarded up surely holds true for Tycho's data. Indeed, in 1600, Kepler, just after his first meeting with Tycho, complained to a friend, "Tycho did not give me the chance to share his practical knowledge except in conversation during meals, today something about the Apogee, tomorrow something about the knots of other planets", and, in the next year, after he was employed by Tycho as an assistant, wrote "Tycho is very stingy as to communicating his ovservations".[6] To Tycho, his own observational data was his important private property and to open it to the public was out of the question. And, that such an attitude must first of all be denied, Stevin became firmly aware.

Stevin was only two years younger than Tycho, yet Stevin, a citizen of the newly born republic, and Tycho, a feudal noble of the old monarchy, belonged to different times and different societies. To Stevin, observations and experiments should be performed by a large number of artisans, engineers and scholars jointly, and also the results of such studies should be opened to the public as common properties. Stevin envisioned the investigation of nature as such cooperative and open work, and just for that reason, the language of science should be not Latin but the vernaculars. And, as a result, Stevin has changed and renovated the sciences fundamentally, because, as Ernest Cassirer said, "the liberation from medieval Latin, the gradual construction and development of the 'volgare' as an independent scientific form of expression was a necessary prerequisite for the free development of scientific thought and its methodological ideals".[7]

3. Simon Stevin and Francis Bacon

It is usually said that the man who first groped for and envisioned the new style to carry out science was Francis Bacon. The new science which he pictured, especially in his *Novum Organum* written in 1620, had the character of being able to be renovated step by step by many people's hands. Namely, what Bacon demanded were theories which would be improved along with the development of technology and the widening of experience of mankind. The evolution of such new science is, following Bacon, like that of mechanical art, which is "at first crude, then adequate, later refined and always progressing".[8] To acquire such a character, the research must be accomplished by the cooperation of a great number of people. Indeed, if research is performed by such cooperative work, the outcomes are accumulated and the theory based on them will be given adjustments constantly. And, the renewal of science cannot be finished in the course of one lifetime, but is taken up by successors.

Such a style of performing science necessarily leads to the negation of old philosophies. The philosophy advocated by an ancient great thinker was a closed system which claimed that it was constructed from the first principle and every problem could be explained and derived by a rigorous reasoning without a break from this absolutely correct principle. It is a theory to which, after it is constructed by its founder, only to interpret is left and permitted to the followers and, as a result, the founder becomes, following Bacon's expression, a "dictator". So, according to Bacon, "Philosophy and the intellectual science are like statues admired and venerated but not improved".[9] Therefore, "They stand still in their own footsteps and remain in practically the same state: they have made no notable progress".[10]

In Bacon's *Advancement of Learning*, publihed in 1605, we find the following passage:

> In arts mechanical the first devisor comes shortest, and time addeth and perfecteth; but in sciences the first author goeth furthest, and time leeseth and corrupteth. So we see, artillery, sailing, printing, and the like, were grossly managed at the first, and by time accommodated and refined: but contrariwise, the philosophies and sciences of Aristotle, Plato, Democritus, Hippocrates, Euclides, Archimedes, of most vigour at the first and by time degenerate and imbased; whereof the reason is no other, but that in the former many wits and industries have contributed in one; and in the latter many wits and industries have been spent about the wit of some one.[11]

Bacon groped for the science which consisted of plastic and expandable theories. But this ideal of Bacon's science was also practiced by Stevin. Stevin's *Wiscontighe Ghedachatenisse*, written in the same period as that of Bacon's *Advancement of Learning*, namely in the early 17th century, contains "On the Theory of Ebb and Flow", and at the opening of it he explained his opinion about natural science as follows:

> Since experience is the surest ground from which to draw general rules in order to gain knowledge of things, and since thanks to the great navigations of these countries we have better means than before for obtaining many sure experiences of the properties of ebb and flow, it seems suitable to me, in order to promote this, to describe a theory of this subject matter, based in part on experiences now available and in part on suppositions which seem to be in accordance with natural reason, whose description may serve as a starting point, in order to deal with this in the manner of a textbook and properly strive to gain fuller knowledge by means of ampler experience that may be obtained later. If anyone should be of the opinion that it would have been more fitting if, before publishing this treatise, I had first examined these things with certainty or caused them to be so examined, I say to this that since that is not the work of one man or small number of men, this seemed to me the best method for getting much information and certainly in a short time, for

when many people have been admonished to make the above-mentioned observations, it may happen that in different places more people will proceed to do so than would be possible by my private admonition to some private persons.[12]

The viewpoint of Stevin about science expressed here can be summed up as follows. At first, science must be founded upon experiences, but experiences have increased and changed on a large scale so far and will go on increasing hereafter. So, each scientific theory must be constructed upon the experiences obtained until that time, and therefore it should be readjusted when new experience is gained, and also should be opened to most people and be verified by them. In a word, every scientific theory is a working hypothesis to be verified which should be adjusted and renovated in the future. And the contribution of a single person is to secure the foothold for further progress which shall be made by later generations. This is nothing but the present comprehension about scientific theories and what Bacon envisaged.

But Bacon was only a propagandist of new science and was not himself a scientist. Moreover, he wrote many of his books in Latin, and did not comprehend both the importance of the use of vernaculars and the power of mathematics in natural sciences. On the other hand, Stevin was in the first place himself a scientist and recognized the importance of mathematics, and indeed established a mathematical science. Moreover, he wrote in a vernacular, namely Dutch, intentionally. Therefore Stevin not only anticipated, but also practiced and surpassed Bacon's idea of new science in the true sense.

4. Cultural revolution in the 16th century

In the last quarter of the 16th century, engineers and artisans took part in scientific research outside universities or academic society and published their results in the vernaculars in all parts of Europe.

In England, Robert Norman, who became an instrument maker in London after 20 years of the seaman's life, discovered the dip (inclination) of magnetic needles and published *The newe Attractive* in English. In the preface of this book he said the following:

> Albeeit it maie bee saied by the learned in the mathematicalles, as hath
> been alreadie written by some, that this no question or matter for a me-
> chanician or mariner to meddle with, no more than is finding of the lon-

gitude, for that it must bee handled exquisitely by geometricall demonstration, and arithmeticall calculation, in whiche artes they would have all mechanicians and seamen to bee ignorant, or at least insufficiently furnished to performe suche a matter, alledgyng against them the Latine proverbe of Apelles, Ne sutor ultra crepidam .... Yet there are in this lande divers mechanicians, that in their severall faculties and professions, have the use of those artes at their fingers endes, and can applie them to their severall purposes, as effectually and more readily, than those that would moste condemne them. For albeit thei have not the use of the Greeke and Latine tongues, to searche the varietie of authors in those artes, yet have they in Englishe for Geometrie Euclides *Elementes* with absolute demonstrations: and for Arithmeticke Recordes woorkes, bothe his first and seconde part: and divers others, bothe in Englishe and in other vulgar languages, that have also written of them, whiche bookes are sufficient to the industrious mechanicians, to make hym perfecte and ready in thoses sciences, but especially to applei the same to the artes of facultie whiche he cheefly professeth.[13]

In 1580, a year before the publication of *The newe Attractive* by Norman, a French potter, Bernard Palissy (1510c-89) who had risen from being an apprentice glass maker declared, "I wish to say that one can fully understand and argue about natural effects, even if he can't read Latin, the language of philosophers. In reality, I have shown by experiments that theories of many philosophers and also doctrines of the ancients of world wide reputation contained not a few mistakes".[14]

In France, Ambroise Paré (1510c-90) and several other surgeons published many medical books written in French from the 1540s to the 1580s.[15] Those surgeons were called "barber-surgeons", and were disdained by physicians educated at universities as artisans who meddled with manual work. The medical science taught at the universities at that time was that of Galenos or of Avicenna, and they were taught mainly through the lectures and readings of those classics, while clinical instructions were almost neglected. On the other hand, Paré, a self-taught surgeon, had much practical experience as a military surgeon. Especially, the beginning of the use of firearms in battle inflicted upon soldiers complex wounds, and in this unprecedented situation, old medicine was entirely useless and helpless. Although Paré knew no Latin, he did not stick to the old authorities like Galenos or Avicenna, and based only on his own experiences and observations in the battlefields and on his practice at the field hospitals, he completely reformed surgery. His books were written for the benefit of young surgeons outside universities.

Of course, such writings by sailors, artisans or surgeons became possible because of the appearance of printed books and publishing. And in the 16th century, Germany and Italy took the lead in printing and publishing businesses in Europe.

In Germany, the translation of the *New Testament* into German by Martin Luther greatly promoted the establishment of German as a written language. At the same time, the language for sciences was created by a painter and printmaker named Albrecht Dürer (1471-1528). In 1525, Dürer wrote a painter's manual, *Unterweysung der Messung mit der Zirkel und Richitscheit* (*A Manual of Measurement by means of Compass and Rulers*) in German. Of course, he studied geometry by himself. In the forward of this book it is written that geometry is the true basis of all painting, and he made an effort to teach the elements and principles of it to all young people who intended to become artisans. And in 1574, a skilled metallurgist named Lazarus Ercker (1530-94) wrote a treatise on ores and assaying, *Beschreibung der allervornehmsten mineralischen Erze und Bergwerksarten* (Account of important Ores and Minings), in German.

In Italy, an unschooled engineer named Vannoccio Biringuccio (1480-1539) wrote *De la Pirotechnia* (Of the Pyrotechnics) in Italian, which was published in 1540, a year after his death. As an English translator of this book wrote, "Biringuccio's source of information is entirely his own observation and experience in the shops where metals were smelted, worked and cast",[16] this book recorded for the first time all the knowledge and techniques about smelting and casting which had been handed down orally within the closed circle of guild members and to which academicians had not paid attention.

On the other hand, among the peoples who were educated at the universities and were proficient in Latin appeared a few men who began to write in the vernacular for artisans and engineers.

In England, as was referred to in Robert Norman's previous quotation, the English translation of Euclid's *Elements* was published in 1570, to which was added the "Mathematical Preface" written by John Dee (1527-1608). Dee, an Elizabethan polymath, himself had graduated from the Cambridge University, but declared in the "Preface" that he described it for "unlatined people and not Universitie Scholars".[17] And also, in 1542, a mathematician named Robert Recorde (1510-1558), who also had graduated from the Oxford University and is known to be the first proponent of the Copernican theory in England and have created the equal sign (=), wrote *The Ground of Artes* in English, and this was enlarged and reprinted in 1561 by Dee and was read by many artisans, seamen and engineers. And in 1584, Giordano Bruno, while staying in London, published his main works, *De l'infinito universo et Mondi* (Of the infinite Universe and the World) and *De la causa, principio, et Uno* (Of the Cause, Principle and One), both written in Italian.

In Germany, Paracelsus (1493-1541), who engaged in medical practices as a wanderer, wrote many books in German. He said the following:

> The physician does not learn everything he must know and master at high college alone; from time to time he must consult old women, gypsies, magicians, wayfarers, and all manner of peasant folk and random people, and learn from them; for these have more knowledge about such things than all the high colleges.

Therefore, according to him, "medicine should be taught so cleanly and clearly in the language of the homeland".[18]

In such a way, in the 16th century, a large class of people began to take part in scientific researches in many branches of different fields, and scientific activity was no longer confined to academic scholars, and new science was thus built outside the academic culture. This change was brought about mainly by the use of the vernaculars, which destroyed the monopoly on knowledge of the scholars and the clergy, and may be called a sort of cultural revolution. The 17th century is usually called a century of geniuses, such as Kepler, Descartes, Galilei, Newton and Leibniz, and others, and they conducted the Scientific Revolution. But, the appearance of those big names was prepared by this cultural revolution of the previous century. And Simon Stevin should be called the true precursor, because he became aware of the significance of the use of the vernacular in the sciences and really paved the way for the cultural revolution.

Acknowledgement

The author wishes to thank Mr. T. Takahasi and Mr. J. Bosna for correcting English.

References

1. See *e.g.* P. Duhem, *Les Origines de la Statique*, Tome 1 (Paris, 1905), Ch.12, pp.263–289, and also, E. Mach, *Die Mechanik in ihrer Entwickelung*, 4te Aufl. (Leipzig, 1901), Kap.1-2, pp.26–37.

2. *The Principal Works of Simon Stevin*, Vol.1, ed. by E.J. Dijksterhuis (Amsterdam, 1955), p.510f.

3. *The Principal Works of Simon Stevin*, Vol.3, ed. by A. Pannekoek and E. Crone (Amsterdam, 1961), p.608f.

4. *Ibid.*, p.612f.

5. *Ibid.*, p.610f.

6. Kepler to Herwart, Jul.12 (1600) and Kepler to Mästlin, Feb.8 (1601); from C. Baumgardt, *Johanns Kepler: Life and Letters*, (New York, 1951), pp.61, 64.

7. E. Cassirer, *The Individual and the Cosmos in Renaissance Philosophy*, tr. by M. Domandi (Philadelphia, 1963), p.56f.

8. F. Bacon, *The New Organon*, ed. by L. Jardine and M. Silverthorne (Cambridge University Press, 200), Bk.1, LXXIV, p.61.

9. *Ibid.*, Preface, p.7.

10. *Ibid.*, Bk.1, LXXIV, p.61.

11. F. Bacon, *The Advancement of Learning*, (Modern Library Paperback ed., 2001), Bk.1, IV-12, p.32.

12. *The Principal Works of Simon Stevin*, Vol.3, p.331.

13. R. Norman, *The newe Attractive* (London, 1581, reprinted Amsterdam, 1974), p.Bi.

14. *Les Oeuvres de Bernard Palissy*, par A. France (Paris, 1880), p.166.

15. H. Stone, 'The French Language in Renaissance Medicine', *Bibliotheque d'Humanisme et Renaissance*, Vol.15 (1953), pp.315–343.

16. C.S. Smith, 'Introduction', *The Pirotechnia of Vannoccio Biringuccio* (M.I.T.Press, 1942, reprinted 1966), p.xiv.

17. J. Dee, 'Mathematical Preface to the Elements of Geometrie of Euclid of Megara (1570)' with an Introduction by A. Debus (Science History Pub., 1975), p.Aiii–v.

18. Paracelsus, *Selected Writings*, ed. with an Introduction by J. Jacobi, tr. by N. Guterman (Princeton University Press, 1951), pp.57, 61.